Andre Yandle

D1616782

Matrices and MATLAB™:
A Tutorial

Matrices and MATLAB™: A Tutorial

Marvin Marcus
Microcomputer Laboratory
University of California

PRENTICE HALL
ENGLEWOOD CLIFFS, NEW JERSEY 07632

The manuscript for this book was prepared on a Macintosh® IIci using the word processing program WriteNow™ v.2.2, and the equation processing program Expressionist® v.2.07. Line drawings were prepared with MacDraw II® v.1.1. Camera ready copy was produced on a LaserWriter® II modified with a 600 d.p.i. Accel-a-Writer™ controller.

© 1993 by Prentice-Hall, Inc.
A Simon & Schuster Company
Englewood Cliffs, New Jersey 07632

All rights reserved. No part of this book may be
reproduced, in any form or by any means,
without permission in writing from the publisher.

Printed in the United Stats of America

10 9 8 7 6 5 4 3 2 1

ISBN 0-13-562901-2

Prentice-Hall International (UK) Limited, *London*
Prentice-Hall of Australia Pty. Limited, *Sydney*
Prentice-Hall Canada Inc., *Toronto*
Prentice-Hall Hispanoamericana, S.A., *Mexico*
Prentice-Hall of India Private Limited, *New Delhi*
Prentice-Hall of Japan, Inc., *Tokyo*
Simon & Schuster Asia Pte. Ltd., *Singapore*
Editora Prentice-Hall do Brasil, Ltda., *Rio de Janeiro*

Accel-a-Writer™ *is a registered trademark of the XANTE Corporation.*
Expressionist® *is a registered trademark of Prescience Corporation.*
LaserWriter® *II is a registered trademark of Apple Computer, Inc.*
MacDraw II® *is a registered trademark of Claris Corporation.*
Macintosh® *is a registered trademark of Apple Computer, Inc.*
MatLab™ *is a registered trademark of The Math Works, Inc.*
WriteNow™ *is a registered trademark of T/Maker Company.*

This book is dedicated with love to my wife, Rebecca Elizabeth Marcus.

ACKNOWLEDGMENTS

The author wishes to acknowledge the invaluable assistance of Susan L. Franklin, who helped type this manuscript and assisted with many of the MatLab™ programs. He also wishes to express his thanks to Kathleen A. Redmond for her help in preparing the manuscript for publication. The author is very grateful for the technical advice and assistance that he received from the staff of the Microcomputer Laboratory at the University of California, Santa Barbara, in particular, William G. Koseluk, Gregory A. Corgain, and Scott J. Schlieman.

DISCLAIMER

The author shall have no liability or responsibility to the purchaser or any other person or entity with respect to any liability, loss, or damage caused or alleged to be caused directly or indirectly by the use of this book and computer programs described herein, including but not limited to any interruption in service, loss of business and anticipatory profits, or consequential damages.

Contents

Preface and Guide xv

Chapter 1. Matrices 1
    1.1   Matrix Notation 1
    1.1   Exercises 5
    1.1   MatLab 6
    1.1   Glossary 22
    1.2   Basic Matrix Operations 23
    1.2   Exercises 31
    1.2   MatLab 37
    1.2   Glossary 49

Chapter 2. Complex Numbers 51
    2.1   Algebra of Complex Numbers 51
    2.1   Exercises 60
    2.1   MatLab 64
    2.1   Glossary 76
    2.2   Elementary Complex Functions 77
    2.2   Exercises 100
    2.2   MatLab 105
    2.2   Glossary 117

Chapter 3. Partitioned Matrices **119**
    3.1   Sequences and Submatrices 119
    3.1   Exercises 123
    3.1   MatLab 127
    3.1   Glossary 135
    3.2   Conformal Partitioning 136
    3.2   Exercises 139
    3.2   MatLab 139
    3.2   Glossary 147
    3.3   Kronecker Products 148
    3.3   Exercises 158
    3.3   MatLab 165
    3.3   Glossary 169

Chapter 4. Elementary Matrices and Rank 171
- 4.1 Inverses 171
- 4.1 Exercises 173
- 4.1 MatLab 175
- 4.1 Glossary 192
- 4.2 Elementary Matrices 193
- 4.2 Exercises 195
- 4.2 MatLab 197
- 4.2 Glossary 200
- 4.3 Canonical Forms and Generalized Inverses 201
- 4.3 Exercises 213
- 4.3 MatLab 222
- 4.3 Glossary 227

Chapter 5. Permutations 229
- 5.1 Introduction to Permutations 229
- 5.1 Exercises 237
- 5.1 MatLab 240
- 5.1 Glossary 247
- 5.2 The Cauchy Index and Conjugacy 247
- 5.2 Exercises 256
- 5.2 MatLab 260
- 5.2 Glossary 267
- 5.3 Some Special Results 267
- 5.3 Exercises 273
- 5.3 MatLab 274
- 5.3 Glossary 276

Chapter 6. Determinants 277
- 6.1 Generalized Matrix Functions 277
- 6.1 Exercises 285
- 6.1 MatLab 291
- 6.1 Glossary 299
- 6.2 Two Classical Determinant Theorems 299
- 6.2 Exercises 313
- 6.2 MatLab 324
- 6.2 Glossary 334

6.3	Compound Matrices	335
6.3	Exercises	347
6.3	MatLab	351
6.3	Glossary	356

Chapter 7. Eigenvalues — 357

7.1	The Characteristic Polynomial	357
7.1	Exercises	364
7.1	MatLab	369
7.1	Glossary	376
7.2	The Gram-Schmidt Process	377
7.2	Exercises	384
7.2	MatLab	402
7.2	Glossary	408

Chapter 8. Triangularization — 409

8.1	The Triangular Form	409
8.1	Exercises	422
8.1	MatLab	426
8.1	Glossary	434
8.2	Normal Matrices	434
8.2	Exercises	440
8.2	MatLab	444
8.2	Glossary	451
8.3	Singular Values	451
8.3	Exercises	474
8.3	MatLab	484
8.3	Glossary	494

Chapter 9. Congruence — 495

9.1	Forms	495
9.1	Exercises	511
9.1	MatLab	517
9.1	Glossary	525
9.2	Geometry of Forms	526
9.2	Exercises	548
9.2	MatLab	563
9.2	Glossary	569

9.3	The Toeplitz-Hausdorff Theorem	570
9.3	Exercises	588
9.3	MatLab	595
9.3	Glossary	604

Chapter 10. Matrix Polynomials and Similarity **605**

10.1	Equivalence	605
10.1	Exercises	610
10.1	MatLab	617
10.1	Glossary	628
10.2	Similarity	629
10.2	Exercises	657
10.2	MatLab	677
10.2	Glossary	690

References 693

Symbol Index 695

Index 699

Preface and Guide

More than 30 years ago Richard Bellman [*Introduction to Matrix Analysis*, McGraw-Hill Book Company, New York, 1960] called Matrix Theory the "arithmetic of higher mathematics." As Bellman observed, matrices are effective representations of linear functions, and such functions are at the center of modern mathematics. Matrix theory is now taught in virtually every postsecondary school in the United States. Apart from the essential role matrices play in pure mathematics, courses in the subject are important because matrices are basic in all scientific computation. Since the end of World War II the development of high-speed digital computers has had a profound effect on both pure and applied matrix theory. The solution of large-scale systems of linear equations, matrix factorization techniques, and eigenvalue and eigenvector methods are all subjects of study and computation in every scientific computing facility.

This book is designed to provide the diligent reader with a thorough understanding of the foundations of modern matrix theory. There are a number of good books on this subject, and an even larger number of mediocre ones. So, what justifies the publication of this book? There are several answers to this question. First, the selection of topics is wider and more pertinent to the current mathematical applications of the subject than is found in most elementary books. Second, a thorough introduction to complex numbers and elementary functions appears near the beginning of the book. It is important for the reader to understand precisely why and how any respectable scientific software package produces such output as

$$\log(i) = i\frac{\pi}{2} .$$

Beyond their essential role in matrix theory, complex numbers allow us to examine such interesting and contemporary applications as Julia sets and fractals. But perhaps the most unique aspect of this book is the complete integration of the software package MatLab™ into the study of elementary matrix theory. MatLab™ is a product of *The MathWorks, Inc.* and was originally written by Dr. Cleve Moler. The name MatLab is a compression of "matrix laboratory." MatLab™ is an interactive software package specially designed for scientific and engineering numerical computations. The program is currently used in nearly every engineering school in the United States. The superb built-in **Help** facility in MatLab makes it simple, and nearly instantaneous, for the reader to look up any available command or structure. This means that neither the author nor the reader needs to be preoccupied with the deadly task of organizing and remembering every detail of yet another arcane programming language. In fact, programming in MatLab™ should be relatively transparent to anyone who has done a month of BASIC programming in high school. Of course, MatLab™ is a sophisticated numerical package with dozens of special built-in commands and functions. But in this book the meaning and use of many of these commands are explained in the natural context of learning the relevant matrix ideas. Thus MatLab™ provides the student with a powerful interactive tool to examine significant examples, to strengthen intuitive insight, and to formulate and study plausible conjectures. These are "mathematical" rather than "computing" activities, but they cannot be carried out in any but the most trivial situations without an effective scientific computing package.

Each chapter of the book is organized into sections. Each section has four subsections. For example, Chapter 2 is entitled, **Complex Numbers**. The second section of Chapter 2, Section 2.2, consists of four subsections: 2.2 Elementary Complex Functions; 2.2 Exercises; 2.2 MatLab; 2.2 Glossary. The first subsection is a narration about elementary complex functions. The Exercises are of the ordinary "paper and pencil" variety. The MatLab section consists of hands-on exercises that cover and extend the narrative material. The Glossary is simply an index for the important ideas in the section. In working on the MatLab sections the student is repeatedly instructed to use the on-line Help facility, frequently more than once for the same item. Most exercises are accompanied by "Hints," which for nonroutine problems, are usually complete solutions.

It is absolutely essential for the reader to work on both the Exercises and the MatLab problems if he is to gain any mastery of this subject. As has been observed many times before, mathematics is not a spectator sport. Although most solutions are available immediately as Hints, this should not impede the learning process. The proximity of the solution to the problem reduces the reader's anxiety. Moreover, anyone using this book *must* be serious, so what would be the point of mindlessly copying a solution without first trying to work the problem? A further point: the reader should regard a Hint as a challenge. Can he improve on it? (Probably!). Can he extend the result stated in the problem? (Frequently). Can he devise MatLab experiments to test his own ideas about the problem? (Almost always). The author makes no claim that the MatLab functions and scripts that appear herein are examples of commercial-quality "bulletproof" programming. For example, not every contingency in input is necessarily accommodated, nor are the algorithms necessarily the most economical or sophisticated.

Matrices and MatLab: A Tutorial covers all the classical topics that should appear in a matrix theory course, including, in Chapter 10, the elementary divisor theory and resulting canonical forms under similarity. There are also several specialized items of more recent interest, e.g., the numerical range, the matrix sign function, Hadamard products, and compound matrices. Throughout the book, the MatLab exercises are used to reinforce the narrative material and to provide the student with an opportunity to make educated conjectures. But it is important to state that this is *not* a book on matrix computations. There are plenty of these, and some are first rate (see References.). Rather, an understanding of the material herein will enable the reader to move on painlessly to the current literature on linear numerical analysis. Another disclaimer: this book does not cover the endless applications of matrix theory: cryptography; pin-jointed frameworks; graphics; tomography, etc. Once again, there is no shortage of reasonable books covering these topics.

This book is designed for a serious one-semester or two-quarter course suitable for undergraduate or graduate students who have successfully completed a standard one-year calculus sequence and who are reasonably adept at high school algebra. It is assumed that the reader has access to MatLab™ operating on a Macintosh computer. However, except for superficial differences in user interface, the Macintosh platform is not important and this book can be used with any of the second-generation MatLab™ versions that run on MS-DOS compatible computers, Sun Workstations and VAX computers. Students can readily move from a course based on this book to more specialized texts. Some of the best of these appear in the References.

                                            Marvin Marcus
                                            University of California Santa Barbara
                                            1992

Matrices and MATLAB™: A Tutorial

Chapter 1

Matrices

Topics
- *basic matrix notation*
- *MatLab ™ fundamentals*
- *matrix algebra*
- *Kronecker product*
- *direct sum*
- *Hadamard product*

1.1 Matrix Notation

A *matrix* is a rectangular array of items. Thus

$$A = \begin{bmatrix} 1 & 2 \\ 3 & 4 \end{bmatrix}, \quad B = \begin{bmatrix} 0 & 1 & -1 \\ 2 & 3 & 5 \end{bmatrix}, \quad C = \begin{bmatrix} 1 \\ 0 \\ 1 \end{bmatrix},$$

$$D = \begin{bmatrix} \sqrt{2} & \pi & -1 & 4 \\ 3 & 7 & 0 & 9 \\ 8 & -1 & 6 & 2 \end{bmatrix}, \quad E = [\,1\ 0\ 2\ 3\ -1\,] \qquad (1)$$

are all matrices: A is 2×2 (read "two by two"), B is 2×3, C is 3×1, D is 3×4, E is 1×5. The second *row* of the matrix D is the 1×4 matrix

$$[\,3\ 7\ 0\ 9\,];$$

the first *column* of the matrix B is the 2 × 1 matrix

$$\begin{bmatrix} 0 \\ 2 \end{bmatrix}.$$

The items appearing in a matrix are called *entries* or *elements* or *components* of the matrix. If A is a matrix, then the entry lying at the intersection of row i and column j is denoted in any of the following ways:

$$a_{ij}, \ a(i,j), \ a_i^j, \ A_{ij}, \ A[i,j], \ A(i,j). \tag{2}$$

The last of the notations in (2) is the one used in MatLab™, and the first is most frequently used in mathematical writing. The size or *dimension* of a matrix is the pair of numbers giving the counts of rows and columns. Thus the dimensions of A, B, C, D, E in (1) are, respectively, 2 × 2, 2 × 3, 3 × 1, 3 × 4, 1 × 5. The letters i and j in (2) are called *subscripts* and sometimes, if the size of a matrix A is understood in a discussion, the matrix A is written simply as

$$A = [a_{ij}].$$

For example, the 3 × 3 matrix

$$A = \left[\frac{1}{i+j} \right]$$

is

$$A = \begin{bmatrix} \dfrac{1}{1+1} & \dfrac{1}{1+2} & \dfrac{1}{1+3} \\ \dfrac{1}{2+1} & \dfrac{1}{2+2} & \dfrac{1}{2+3} \\ \dfrac{1}{3+1} & \dfrac{1}{3+2} & \dfrac{1}{3+3} \end{bmatrix}.$$

Two matrices A and B are *equal*, written

$$A = B,$$

1.1 Matrix Notation

if they are the same size and corresponding entries are equal. Thus

$$\begin{bmatrix} 2 & 1 \\ 3 & 2 \end{bmatrix} = \begin{bmatrix} 5-3 & 2-1 \\ 2+1 & 1+1 \end{bmatrix},$$

whereas

$$\begin{bmatrix} 2 & 1 \\ 3 & 2 \end{bmatrix}, \quad \begin{bmatrix} 2 & 1 & 0 \\ 3 & 2 & 0 \end{bmatrix}$$

are not equal.

A matrix such as C or E in (1) is called a *vector*: C is a *column vector*, E is a *row vector*. It may or may not be necessary to distinguish a vector as being either a row or a column and MatLab allows for subscripting a vector with a single subscript. Thus for the vector C in (1) above we can write

$$C(1) = 1, \; C(2) = 0, \; C(3) = 1,$$

and for E,

$$E(1) = 1, \; E(2) = 0, \; E(3) = 2, \; E(4) = 3, \; E(5) = -1.$$

Row or column vectors are sometimes referred to as n-*tuples*, where n is the number of entries in the vector. Thus C is a 3-tuple and E is a 5-tuple.

If A is an m × n matrix then the pth *row* of A is the 1 × n matrix

$$A_{(p)} = \begin{bmatrix} a_{p1} & a_{p2} & a_{p3} & \cdots & a_{pn} \end{bmatrix}$$

and the qth column is the m × 1 matrix

$$A^{(q)} = \begin{bmatrix} a_{1q} \\ a_{2q} \\ a_{3q} \\ \vdots \\ a_{mq} \end{bmatrix}.$$

A *line* in a matrix is either a row or column.

The *transpose* of an m × n matrix A is the n × m matrix B whose i, j entry is

$$b_{ij} = a_{ji}, \ i = 1, ..., n, \ j = 1, ..., m.$$

That is, the first column of B is the first row of A, the second column of B is the second row of A, etc. For example, the transpose of

$$A = \begin{bmatrix} 1 & 2 & 3 \\ 4 & 5 & 6 \end{bmatrix}$$

is

$$B = \begin{bmatrix} 1 & 4 \\ 2 & 5 \\ 3 & 6 \end{bmatrix}.$$

The transpose of a matrix A is denoted by A^T, read "A transpose". Note the following formulas:

$$A^{TT} = A, \tag{3}$$

$$A^T{}_{(p)} = A^{(p)T}, \tag{4}$$

$$A^{T(q)} = A_{(q)}{}^T. \tag{5}$$

An m × n matrix A is *square* if m = n; we frequently write "A is n-square." If $A = [a_{ij}]$ is a square matrix and

$$A = A^T \tag{6}$$

then A is said to be *symmetric*. If $A^T = [-a_{ij}]$, written

$$A^T = -A, \tag{7}$$

then A is said to be *skew-symmetric*.

The totality of m × n matrices A whose entries belong to a set R is denoted by $M_{m,n}(R)$. If m = n then $M_{m,n}(R)$ is abbreviated to $M_n(R)$. If $A \in M_{m,n}(R)$ and $a_{ij} = 0$, i = 1, ..., m, j = 1, ..., n, then A is called the *zero matrix* and is denoted by $0_{m,n}$ or simply by 0 if m and n are understood; if m = n, $0_{m,n}$ is abbreviated to 0_n. If $A \in M_n(R)$ and $a_{ij} = 0$ for all i ≠ j then A is called a *diagonal matrix*, written

$$\text{diag}(a_{11}, a_{22}, ..., a_{nn}). \tag{8}$$

The *main diagonal* of A is the sequence of entries $a_{11}, a_{22}, ..., a_{nn}$.

The n-square matrix

$$I_n = \text{diag}(1, 1, ..., 1) \tag{9}$$

is called the n-square *identity matrix*.

1.1 Exercises

1. For the matrices in (1) evaluate: a_{12}; b_{23}; C_{11}; D^T_{41} (i.e., the (4, 1) entry of D^T); E^T_{31}.

2. For the matrices in (1) evaluate: $A_{(1)}$; $A^T_{(1)}$; (i.e., row 1 of A^T); $B^{(2)}$; $B^{(3)}$; $B^{T(2)}$; $D^{(4)}$; $D^{T(3)}$.

3. Prove that $A^{TT} = A$.

4. How many m × n matrices are there whose entries are in the set of three numbers {-1, 0, 1}?

Identify the matrices in #5 - 15 as symmetric, skew-symmetric, both, or neither.

5. $[0]$

6. $\begin{bmatrix} 0 & 0 \\ 0 & 0 \end{bmatrix}$

7. $\begin{bmatrix} 1 & 1 \\ -1 & 1 \end{bmatrix}$

8. $\begin{bmatrix} 1 & 2 & 3 \\ 4 & 5 & 6 \\ 7 & 8 & 9 \end{bmatrix}$

9. $\begin{bmatrix} 0 & 1 \\ -1 & 0 \end{bmatrix}$

10. $A = \begin{bmatrix} \dfrac{1}{p+q} \end{bmatrix}$, i.e., $a_{pq} = \dfrac{1}{p+q}$ and A is n-square.

11. $A = \begin{bmatrix} \dfrac{1}{p^2+q} \end{bmatrix}$; i.e., $a_{pq} = \dfrac{1}{p^2+q}$ and A is n-square.

12. $\begin{bmatrix} 0 & 0 \\ 1 & 0 \end{bmatrix}$

13. $\begin{bmatrix} 1 & 0 & 1 \\ 0 & 2 & 0 \\ -1 & 0 & 3 \end{bmatrix}$

14. $\begin{bmatrix} 1 & 1 & 1 \\ 1 & 1 & 1 \end{bmatrix}$

15. $\begin{bmatrix} 1 & 1 \\ 1 & 1 \\ 1 & 1 \end{bmatrix}$

1.1 MatLab

When you open the MatLab folder, it will look something like the following:

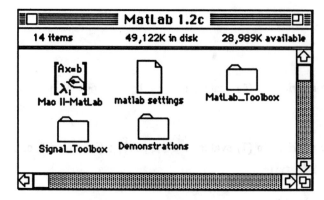

Fig. 1

1.1 MatLab

Double click on the Mac II - MatLab icon and the Command window will open with the Graph window behind it, as shown in Fig. 2.

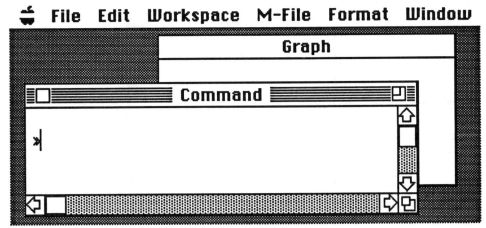

Fig. 2

The Command window prompt is ». If, as in Fig. 3, you type in

$$A = [1\ 2\ 3;\ 4\ 5\ 6;\ 7\ 8\ 9]$$

at the prompt and then press the return key, MatLab responds by exhibiting the matrix A (without matrix brackets). You can then change A(1, 3) to -5, press the return key, and MatLab will respond by exhibiting the modified matrix, all as indicated in Fig. 3.

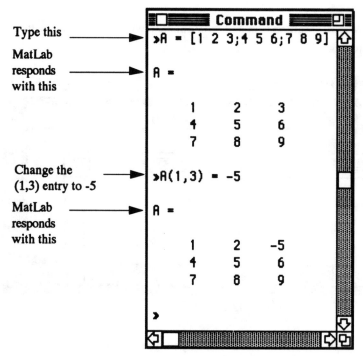

Fig. 3

MatLab has an excellent on-line **Help** system. To see how this works, select **About MacII-MatLab...** from the top of the Apple menu . A window similar to the one in Fig. 4 will appear:

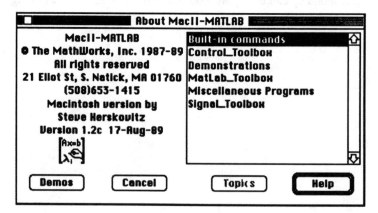

Fig. 4

1.1 MatLab

Click on the Help button and the window changes to Fig. 5.

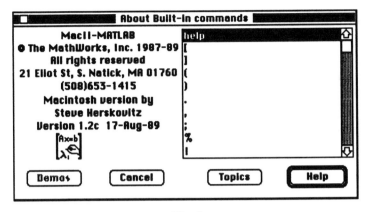

Fig. 5

Scroll down to the semicolon (;) and highlight it. Then click the Help button again and an explanation of the semicolon is given as in Fig. 6.

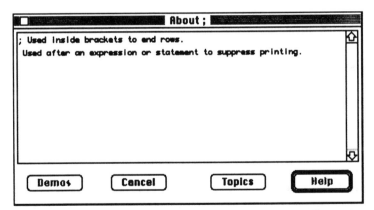

Fig. 6

Finally, if you click the Topics button, the window in Fig. 4 will return. You can close the window by clicking the close box.

To enter a matrix from the keyboard in MatLab: enclose the matrix in square brackets; type the entries in rowwise with elements separated by blanks or commas; separate the rows with a semicolon; after entering the closing bracket, press carriage return (<CR>) to exhibit the matrix. To assign a matrix to a variable name, e.g., A, type A = and then enter the matrix. If you choose not to assign the result of a computation to a variable, MatLab assigns it to the variable **ans**. Use Help to find out about ans.

In #1 - 4, enter the indicated matrices.

1. $A = \begin{bmatrix} 0 & 1 \\ -1 & 0 \end{bmatrix}$

2. $A = \begin{bmatrix} 1 & 0 \\ 0 & 1 \end{bmatrix}$

3. $A = \begin{bmatrix} 1 & 2 & 3 \\ 4 & 5 & 6 \\ 7 & 8 & 9 \end{bmatrix}$

4. $A = \begin{bmatrix} \frac{1}{2} & 0 \\ \frac{2}{3} & \frac{1}{4} \end{bmatrix}$

5. Select **Long** from the **Format** menu and type A at the prompt to exhibit the matrix in #4 with entries rounded to 14 digits to the right of the decimal point.

6. Select **Short** from the Format menu and enter

$$A = \begin{bmatrix} \frac{1}{3} & \sqrt{2} \\ \frac{1}{7} & 3^5 \end{bmatrix}.$$

Use **sqrt** for $\sqrt{}$ and ^ for exponentiation, i.e., 3^5 is 3 ^ 5. Note that Short format rounds to 4 digits to the right of the decimal point.

Addition, subtraction, multiplication, and division are denoted (as usual) by +, -, *, and /, respectively. Use Help to read about these operations in the Built-in commands.

7. Enter the matrix: $A = \begin{bmatrix} \sqrt{1 + \sqrt{2}} & \frac{1}{3} \\ \frac{1}{\sqrt{5}} & 5^3 \end{bmatrix}.$

1.1 MatLab

8. To continue a MatLab instruction to a second line, use two adjacent periods, e.g., enter

    A = [1 2 3; ..
            4 5 6].

9. Enter the matrix in #3 over 3 lines, one for each row.

10. Enter the matrix

$$A = \begin{bmatrix} \sqrt{2} & 1 & 2 & 4 \\ 3 & \frac{2}{3} & 5 & \frac{1}{2} \\ 2^5 & 3^{1/2} & 4 & 8 \end{bmatrix}$$

    over 3 lines.

11. Type **format long** at the prompt and exhibit A in #10.

12. Type **format short E** at the prompt and exhibit A in #10.

13. Type **format long E** at the prompt and exhibit A in #10.

14. Return to **format short** (how?). At successive prompts enter x = sqrt(2) and y = 1. Then enter the matrix

$$A = \begin{bmatrix} x & 2x & 3 \\ x+y & x-y & x+\pi y \\ \pi^2 & 2\pi x & x^2 + y^2 \end{bmatrix}.$$

    (pi = π is a built-in constant)

15. Copy and Paste the entry line for A in #14 opposite a new prompt - do not press return immediately. Change the entry $2\pi x$ to $2\pi x + y$ and then press <CR>.

16. Type A(2, 3) = 3 * y at the prompt and exhibit A.

17. Type A(1, 1) = 0 at the prompt and exhibit A.

18. Type A = 0 at the prompt and exhibit A.

19. Type A at the prompt to exhibit A.

20. Type x, y at the prompt to exhibit x and y.

21. Devise an experiment to determine if MatLab is case sensitive, e.g., enter A = 1 and then type a at the prompt. Then read the entry for **casesen** in the Built-in commands in the Help window.

22. Type **clc** at the prompt and then read the entry for clc in the Help window.

23. Get a new command window by: clicking the close box in the current window; selecting **New** from the **File** menu; clicking the Command button in the dialog box; clicking OK.

24. At 5 successive prompts enter casesen, a = 1, A = 2, a, A. Toggle casesen back on.

25. Enter the matrix

$$A = \begin{bmatrix} 1 & 2 \\ 3 & 4 \end{bmatrix}.$$

Then at successive prompts type: A(1, 3) = 6, A. Note that the contents of A have changed.

26. Next, type A(3, 3) = 1 and exhibit A. The contents of A change dynamically in response to various commands.

27. Use clc to clear the screen, and enter the matrix

$$A = \begin{bmatrix} 1 & 2 & 3 & 4 \\ 5 & 6 & 7 & 8 \end{bmatrix}.$$

1.1 MatLab

Then type A(1, :) and A(:, 1) at the next two prompts. Repeat with A(2, :) and A(:, 2). Then write commands to exhibit $A^{(3)}$ and $A^{(4)}$. Try exhibiting the nonexistent $A^{(5)}$.

28. Enter the matrix

$$A = \begin{bmatrix} 1 & 2 & 3 \\ 4 & 5 & 6 \\ 7 & 8 & 9 \end{bmatrix}.$$

Type A' at the next prompt. Use Help to find out about '.

29. Clear the screen and at successive Command window prompts (>>) enter:

   (a)　　u = 1 : 5
   (b)　　v = -5 : 2 : 8
   (c)　　w = (1 : 5)'　　　　　　　(the parentheses are necessary)
   (d)　　x = 5 : -1 : -5
   (e)　　y = -2 * pi : .1 : 2 * pi　　(pi = π is a built-in constant)
   (f)　　u(2)
   (g)　　v(3)
   (h)　　w(4)
   (i)　　y(1)
   (j)　　y(126)

   The "colon" notation is useful for generating vectors. Use Help to find out about the colon (:).

30. At successive prompts enter:

   (a)　　u .^ 2
   (b)　　u .^ 3
   (c)　　u .^ (1/ 2)
   (d)　　u .^ (-1)

   Use Help to find out about the period (.). MatLab requires that there be no space between . and ^.

31. Enter

$$A = \begin{bmatrix} 1 & 2 \\ 5 & 3 \end{bmatrix}$$

and at successive prompts enter:

(a) A.^2
(b) A.^3
(c) A.^(1/2)
(d) A.^(-1)
(e) A.^6
(f) (A.^2).^3
(g) sqrt(A)
(h) exp(A)
(i) log(A)
(j) log10(A)

The "period" notations result in entrywise operations. Use Help about Built-in commands to find out about **sqrt, exp,** and **log**. Use Help about MatLab_Toolbox to find out about **log10**.

32. Generate the vectors x = [1 2 3 4 5] and y = [-5 - 4 -3 -2 -1]. At the next prompt enter x .* y. Note that the multiplication is performed entrywise. MatLab requires that there be no spaces between . and *.

33. From the File menu choose New to obtain a window titled Edit1:Untitled.

1.1 MatLab

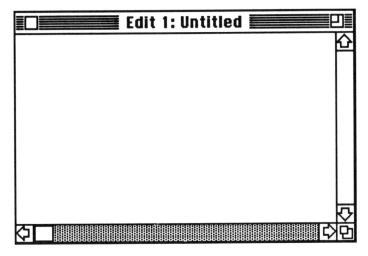

Fig. 7

At the flashing cursor enter the following lines of code:

```
x = 1:5
y = -5:-1
x .* y
```

Then from the **File** menu choose **Save And Go** and in response to the dialog box decide where the file is to be saved and give it the name **First**. Note that the result appears in the Command window.

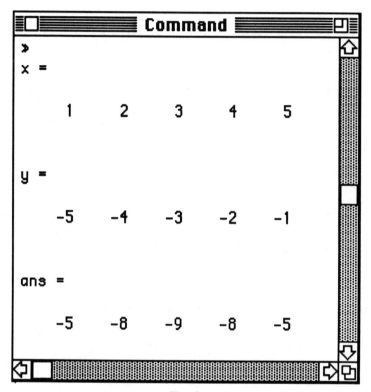

Fig. 8

The advantage of working in an Edit window is that the usual MacIntosh editing techniques are available. Note that the name of the file in the name box is Edit1:First. Close Edit1:First, open the File menu and select **Open...**. Find First in the resulting dialog box and open it. Insert a semicolon after each of the first two lines of code and then run the program by choosing Save and Go again from the File menu. The effect of the semicolon at the end of a line is to suppress printing the result of the execution of the line. Close First, and, this time from the **M-File** menu at the top of the screen, choose **Run Script**. In the resulting dialog box select First and click on the **Run** button. Again open the M-File menu, click on **Echo Scripts** and again choose Run Script to execute First. Notice that the output now contains a listing of the program (= script).

1.1 MatLab

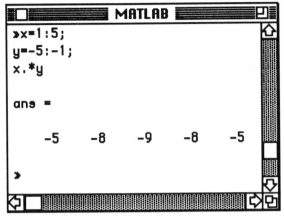

Fig. 9

From the M-File menu toggle back to Echo Scripts. Also from the File menu choose Open.. and open First in the resulting dialog box.

34. Get rid of the current code in the Edit1:First window by highlighting it and deleting, and then type in the following code:

   t = -2:.1:2;
   f = t .^ 2;
   plot(t, f)

Then choose Save and Go from the File menu. The Graph window opens with a plot of the parabola $f(t) = t^2$ on the interval $[-2, 2]$.

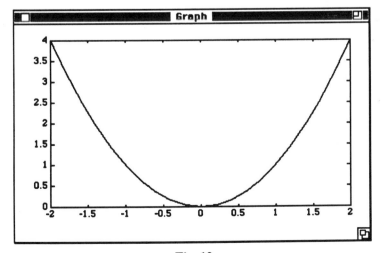

Fig. 10

Read the entry for **plot** in the Built-in commands. At this time the only pertinent information in the About plot window is that plot(t, f) plots vector t versus vector f. To explain this, note that $t = [t_1, t_2, t_3, \ldots]$ consists of the numbers -2, -1.9, -1.8, ..., 0, .1, ..., 2 while $f = [f_1, f_2, f_3, \ldots]$ consists of the numbers $(-2)^2, (-1.9)^2, (-1.8)^2, \ldots, 0, (.1)^2, \ldots, 2^2$. The command plot(t,f) plots the points (t_1, f_1), (t_2, f_2), (t_3, f_3), ..., joined by line segments. The visual effect is a relatively smooth curve.

35. Modify the script in #34 so that the last line becomes plot(t, f, t, 2*f) and run it.

36. On the basis of what you saw in #35, graph the unit circle in the Graph window. If you use the script

    t = -1: .1 :1;
    f = sqrt(1 - t .^ 2);
    plot(t, f, t, -f)

    the results are disappointing (see Fig. 11).

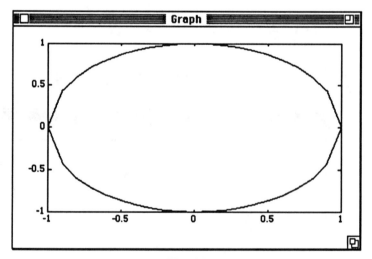

Fig. 11

Insert the line **axis('square')** immediately before the last line in the above script.

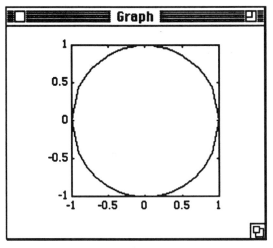

Fig. 12

Next type First at the prompt in the Command window to run the script again (See Fig. 12). Return the graph window to normal scaling by typing **axis('normal')** at the prompt in the Command window.

37. Write scripts in an Edit window or command sequences in the Command window to plot the following lines:

    (a) joining (1, -1) to (-3, 2). (hint: plot([1 -3], [-1 2])
    (b) slope = 2, y-intercept = -3.

38. Write and execute a script to plot $f(t) = \sin(t)$, $g(t) = \sin(2t)$, $h(t) = \sin(3t)$ over the interval $[0, 2\pi]$. Use an increment of .05 in generating the t vector. The Graph window can be adjusted in size with the size box.

39. Write and execute a script to plot

$$f(t) = \frac{1}{1 + t^2}$$

on the interval [-10, 10]. Use an increment of .05 in generating the t vector.

Hint: **ones(t)** generates a row vector of 1's having the same length as t.

40. With the Graph window open showing the graph of $f(t) = \dfrac{1}{1+t^2}$, select **Labels** from the Graph menu and enter the following:
    Title: graph of f(t) = 1/(1 +t^2)
    X-Label: t axis
    Y-Label: f(t) axis

    Then run the script for #39 and note the output.

41. It is required to graphically estimate the root(s) of the following equation:

    $$(x+1)^2 e^{x^2-2} - 1 = 0.$$

    (a) Rearrange the equation in the form $f_1(x) = f_2(x)$. Hint: $(x+1)^2 = e^{2-x^2}$.
    (b) Plot $y = f_1(x)$ and $y = f_2(x)$ on the same graph over the interval [-2, 2]. Hint: read the entry for **ones** in the Built-in commands. Use an increment of .05 in creating the x vector.
    (c) From (b) it is clear that [-2, -1.5] and [0.5, 1] are bracketing intervals for the roots. Insert the line axis ([-2, -1.5, -2, 2]) before the plot command. Read the entry for **axis** in the Built-in commands. What is the value of the root in [-2, -1.5], rounded to 2 places?
    (d) Graphically find the root of the equation that lies in [0.5, 1] rounded to 1 place.

42. Let $f(x) = e^x - x - 2$.

    (a) Use MatLab to graphically show that the equation $f(x) = 0$ has a root in $[-2, 0]$ and a root in $[0, 2]$.
    (b) Use the axis command (get Help on it) to bracket the roots in intervals sufficiently small that estimates may be made visually that are correct to 2 digits.

43. Suppose that it is required to solve an equation of the form $f(x) = 0$. Rewrite the equation in the form $x = g(x)$, select x_0 as an initial guess for a root, and successively compute $x_1 = g(x_0)$, $x_2 = g(x_1)$, ..., $x_{n+1} = g(x_n)$. If the sequence $x_0, x_1, x_2, \ldots$ converges to a limit r and g is a continuous function, then $r = \lim_{n \to \infty} x_n = \lim_{n \to \infty} g(x_{n-1}) = g(\lim_{n \to \infty} x_{n-1}) = g(r)$, i.e., $r = g(r)$ so that $f(r) = 0$. This idea is called *iteration*. A useful control structure for implementing iteration in

1.1 MatLab

MatLab is the **while..end** statement. Read the entry for the while..end statement in the Built-in commands. Suppose that $f(x) = x^2 - 2x - 8 = 0$ and write this equation in the form $x = g(x) = \frac{2x + 8}{x}$. By graphing $y = f(x)$, make a reasonable guess for x_0, an initial approximation to a root, say $x_0 = 3$. Then enter the following script in an Edit window and run it.

```
x = 3;
g = (2*x+8)/x;
r = [ x ];
n = 1;
c = [ n ];
while abs(g - x) > 1e-3
    x = g;
    g = (2*x+8)/x;
    n = n + 1;
    c = [ c n ];
    r = [ r x ];
end
disp (' count   iterate');
disp ('  ____   _____');
disp ( [ c' r' ] )
```

Read the entry for **disp** in the Built-in commands. The command is used to enhance the appearance of output. The effect of the statement $c = [\ c\ n\ ]$ is to create a row vector by successively appending the current value of n to the current value of c.

44. Rewrite the equation $f(x) = 0$ in #43 as $x = \sqrt{2x + 8}$. Modify the script in #43 and obtain a table of iterates. What is the least integer n for which x_{n+1} and x_n differ by less than 10^{-5}.

45. Rewrite the equation $f(x) = 0$ in #43 as $x = \frac{x^2 - 8}{2}$ and attempt to solve the problem in #44. Clearly, not all ways of converting a root-finding problem to an iteration will necessarily work.

46. Solve each of the following equations by iteration. Find the least n for which $|x_{n+1} - x_n| < 0.5 \cdot 10^{-2}$ in each case. The mathematical function ln (i.e., natural logarithm) is denoted by **log** in MatLab.

(a) $x^2 + \ln x = 0$; $x_0 = 1$
(b) $x - \cos x = 0$; $x_0 = 1$
(c) $x^5 + 5x + 1 = 0$; $x_0 = -0.5$
(d) $xe^x - 1 = 0$; $x_0 = 1$

1.1 Glossary

^	10
+, -, *, /	10
.	13
A'	13
$A_{(p)}$	3
$A^{(q)}$	4
A^T	4
axis('normal')	19
axis('square')	18
casesen	12
clc	12
colon notation	13
column vector	3
component	2
continuation	11
diag	5
diagonal matrix	5
dimension	2
disp	21
Echo Scripts	16
Edit window	15
element	2
entry	2
equality	2
exp(A)	14
Format	10
format long E	11

format short E	11
Help	8
identity matrix	5
I_n	5
iteration	20
Labels	20
log(A)	14
log10(A)	14
Long	9
M-File	16
main diagonal	5
matrix	1
$M_{m,n}(R)$	5
$M_n(R)$	5
n-tuple	3
$O_{m,n}$	5
O_n	5
pi	11
plot	18
row vector	3
Run Script	16
Short	10
skew-symmetric	5
sqrt	10
square	4
symmetric	5
transpose	4
vector	3
while..end	21
zero matrix	5

1.2 Basic Matrix Operations

There is a useful notation, based on the Greek capital sigma (Σ), for adding suitably labelled items. This is best explained by examples. For instance, suppose that $a = [\ a_1, ..., a_n\ ]$ is a vector and we want to indicate notationally the sum of its components: $s = a_1 + a_2 + a_3 + \cdots + a_n$. The sum s can be abbreviated to

$$s = \sum_{i=1}^{n} a_i . \tag{1}$$

This is read as "the sum from i equal 1 to n of a_i". Another example: suppose

$$A = \begin{bmatrix} 1 & 2 & 3 \\ 4 & 5 & 6 \end{bmatrix}, \quad B = \begin{bmatrix} -1 & 1 \\ 2 & 5 \\ 4 & -2 \end{bmatrix} . \tag{2}$$

If we write $A = [\, a_{ij} \,]$, $B = [\, b_{ij} \,]$, then the notation

$$\sum_{k=1}^{3} a_{ik} b_{kj} \tag{3}$$

is an abbreviation for

$$a_{i1}b_{1j} + a_{i2}b_{2j} + a_{i3}b_{3j} .$$

If $i = 2$ and $j = 1$ then (3) becomes

$$\sum_{k=1}^{3} a_{2k} b_{k1} = a_{21} b_{11} + a_{22} b_{21} + a_{23} b_{31}$$
$$= 4 \cdot (-1) + 5 \cdot 2 + 6 \cdot 4$$
$$= 30 .$$

If $A = [\, a_{ij} \,]$ is any n-square matrix then

$$\sum_{i=1}^{n} a_{ii} \tag{4}$$

is the sum of the main diagonal entries of A. For example, if $A = I_n$, the n-square identity matrix, then

$$\sum_{i=1}^{n} a_{ii} = \sum_{i=1}^{n} 1$$
$$= n .$$

The Σ notation is very flexible. For instance, the sum of all the entries of the matrix A in (2) can be written as

1.2 Basic Matrix Operations

$$\sum_{i=1}^{2}\sum_{j=1}^{3} a_{ij}. \qquad (5)$$

The *double sum* in (5) means that the sum of the entries in row i is computed for i = 1 and i = 2 and then these sums are added. The *columnwise* sum is

$$\sum_{j=1}^{3}\sum_{i=1}^{2} a_{ij}. \qquad (6)$$

Clearly (5) and (6) have the same value, namely the sum of all entries in A. There are all kinds of variations in the use of the Σ notation:

$$\sum_{i=1, i\neq 3}^{n} a_i$$

is the sum $a_1 + a_2 + a_4 + \cdots + a_n$, i.e., a_3 is omitted;

$$\sum_{i=1}^{3} \frac{1}{i^2}$$

is the sum

$$\frac{1}{1^2} + \frac{1}{2^2} + \frac{1}{3^2} = 1 + \frac{1}{4} + \frac{1}{9};$$

$$\sum_{i=-2}^{2} (-1)^i i^2$$

is the sum

$$(-1)^{-2} \cdot (-2)^2 + (-1)^{-1} \cdot (-1)^2 + (-1)^0 \cdot 0^2 + (-1)^1 \cdot 1^2 + (-1)^2 \cdot 2^2 = 6.$$

In analogy to the Σ notation for sums, there is a Π notation for products. For example, for the matrix B in (2), the product of the entries in the second column can be written as

$$\prod_{i=1}^{3} b_{i2}.$$

Another example: if the a_i are positive numbers, then

$$\log \prod_{i=1}^{n} a_i = \sum_{i=1}^{n} \log a_i .$$

Variations on these examples will be found in §1.2 Exercises.

We next consider several basic operations on matrices. The first of these is *addition*: if A and B are m × n matrices then their *sum* is the m × n matrix C defined by

$$c_{ij} = a_{ij} + b_{ij}, \quad i = 1, \ldots, m, \quad j = 1, \ldots, n;$$

C is written C = A + B. For example,

$$\begin{bmatrix} 0 & 1 \\ 2 & 3 \end{bmatrix} + \begin{bmatrix} 2 & -1 \\ 5 & 7 \end{bmatrix} = \begin{bmatrix} 2 & 0 \\ 7 & 10 \end{bmatrix},$$

$$\begin{bmatrix} 1 & 2 \\ 3 & 4 \end{bmatrix} + \begin{bmatrix} -1 & -2 \\ -3 & -4 \end{bmatrix} = \begin{bmatrix} 0 & 0 \\ 0 & 0 \end{bmatrix}. \tag{7}$$

In view of (7), we denote by -A the m × n matrix $[-a_{ij}]$. Thus, for example,

$$-\begin{bmatrix} 1 & 2 \\ 3 & 4 \end{bmatrix} = \begin{bmatrix} -1 & -2 \\ -3 & -4 \end{bmatrix}$$

and, in general, if $A \in M_{m,n}(R)$,

$$A + -A = -A + A = 0_{m,n} . \tag{8}$$

If α is a number (a number is frequently called a *scalar*) and A is an m × n matrix, then αA, *the scalar product of* A *by* α, is the matrix whose (i, j) entry is αa_{ij}. If the order in which α and the entries of A are multiplied does not matter, then we also write αA as $A\alpha$. Thus

$$2\begin{bmatrix} 3 & 4 \\ 5 & 6 \end{bmatrix} = \begin{bmatrix} 6 & 8 \\ 10 & 12 \end{bmatrix} = \begin{bmatrix} 3 & 4 \\ 5 & 6 \end{bmatrix} 2.$$

If $A = [a_{ij}]$ is m × n and $B = [b_{ij}]$ is n × p, then the m × p matrix C whose i, j entry

1.2 Basic Matrix Operations

is given by

$$c_{ij} = \sum_{k=1}^{n} a_{ik} b_{kj}, \quad i = 1, \cdots, m, \quad j = 1, \cdots, p \tag{9}$$

is called the *product of A and B* and is denoted by C = AB. For example, if
A = [2 1 3], B = [1 0 5]^T then

$$AB = [\, 2 \cdot 1 + 1 \cdot 0 + 3 \cdot 5 \,] = [\, 17 \,],$$

whereas

$$BA = \begin{bmatrix} 2 & 1 & 3 \\ 0 & 0 & 0 \\ 10 & 5 & 15 \end{bmatrix}. \tag{10}$$

Thus from (10) we see that it is not generally true that AB and BA are equal, i.e., matrix multiplication is not generally *commutative*. Even if A and B are n-square matrices, it is simple to construct examples for which AB ≠ BA, e.g.,

$$A = \begin{bmatrix} 0 & 1 \\ 0 & 0 \end{bmatrix}, \quad B = \begin{bmatrix} 0 & 0 \\ 1 & 0 \end{bmatrix}.$$

Two useful formulas for computing matrix products are

$$\begin{aligned}(AB)_{(i)} &= A_{(i)} B \\ &= \sum_{k=1}^{n} a_{ik} B_{(k)} \end{aligned} \tag{11}$$

and

$$\begin{aligned}(AB)^{(j)} &= A B^{(j)} \\ &= \sum_{k=1}^{n} A^{(k)} b_{kj}. \end{aligned} \tag{12}$$

The matrix A is m × n and the matrix B is n × p in (11) and (12), so that the product AB is defined. To verify (11) note that by (9) we have

$$(AB)_{(i)} = \left[\sum_{k=1}^{n} a_{ik}b_{k1} \quad \sum_{k=1}^{n} a_{ik}b_{k2} \quad \cdots \quad \sum_{k=1}^{n} a_{ik}b_{kp} \right]$$

$$= \sum_{k=1}^{n} a_{ik} \begin{bmatrix} b_{k1} & b_{k2} & \cdots & b_{kp} \end{bmatrix}$$

$$= \sum_{k=1}^{n} a_{ik} B_{(k)} \ .$$

The formula (12) is confirmed by a similar computation. As an example, if

$$A = \begin{bmatrix} 1 & 2 & 3 \\ 3 & 1 & -2 \\ 5 & -1 & 0 \end{bmatrix}, \quad B = \begin{bmatrix} 1 & 3 \\ -1 & 5 \\ 2 & -2 \end{bmatrix}$$

then the second row of AB can be computed using (11):

$$\begin{aligned}
(AB)_{(2)} &= A_{(2)}B \\
&= 3B_{(1)} + 1B_{(2)} + -2B_{(3)} \\
&= 3\begin{bmatrix} 1 & 3 \end{bmatrix} + \begin{bmatrix} -1 & 5 \end{bmatrix} + -2\begin{bmatrix} 2 & -2 \end{bmatrix} \\
&= \begin{bmatrix} 3 & 9 \end{bmatrix} + \begin{bmatrix} -1 & 5 \end{bmatrix} + \begin{bmatrix} -4 & 4 \end{bmatrix} \\
&= \begin{bmatrix} -2 & 18 \end{bmatrix} \ .
\end{aligned}$$

If A is m × n and B is p × q then the mp × nq matrix

$$\begin{bmatrix} a_{11}B & a_{12}B & \cdots & a_{1n}B \\ a_{21}B & a_{22}B & \cdots & a_{2n}B \\ \vdots & \vdots & & \vdots \\ a_{m1}B & a_{m2}B & \cdots & a_{mn}B \end{bmatrix} \tag{13}$$

is called the *Kronecker product* or *tensor product* or *direct product* of A and B and is denoted by A ⊗ B. For example, if

1.2 Basic Matrix Operations

$$A = \begin{bmatrix} a_{11} & a_{12} & a_{13} \\ a_{21} & a_{22} & a_{23} \end{bmatrix}, \quad B = \begin{bmatrix} b_{11} & b_{12} \\ b_{21} & b_{22} \end{bmatrix}$$

then

$$A \otimes B = \begin{bmatrix} a_{11}B & a_{12}B & a_{13}B \\ a_{21}B & a_{22}B & a_{23}B \end{bmatrix}$$

$$= \begin{bmatrix} a_{11}\begin{bmatrix} b_{11} & b_{12} \\ b_{21} & b_{22} \end{bmatrix} & a_{12}\begin{bmatrix} b_{11} & b_{12} \\ b_{21} & b_{22} \end{bmatrix} & a_{13}\begin{bmatrix} b_{11} & b_{12} \\ b_{21} & b_{22} \end{bmatrix} \\ a_{21}\begin{bmatrix} b_{11} & b_{12} \\ b_{21} & b_{22} \end{bmatrix} & a_{22}\begin{bmatrix} b_{11} & b_{12} \\ b_{21} & b_{22} \end{bmatrix} & a_{23}\begin{bmatrix} b_{11} & b_{12} \\ b_{21} & b_{22} \end{bmatrix} \end{bmatrix}$$

$$= \begin{bmatrix} a_{11}b_{11} & a_{11}b_{12} & a_{12}b_{11} & a_{12}b_{12} & a_{13}b_{11} & a_{13}b_{12} \\ a_{11}b_{21} & a_{11}b_{22} & a_{12}b_{21} & a_{12}b_{22} & a_{13}b_{21} & a_{13}b_{22} \\ a_{21}b_{11} & a_{21}b_{12} & a_{22}b_{11} & a_{22}b_{12} & a_{23}b_{11} & a_{23}b_{12} \\ a_{21}b_{21} & a_{21}b_{22} & a_{22}b_{21} & a_{22}b_{22} & a_{23}b_{21} & a_{23}b_{22} \end{bmatrix}.$$

If $A_1, \ldots, A_m$ are square matrices of dimensions $n_1, \ldots, n_m$ respectively then the $(n_1 + \cdots + n_m)$-square matrix

$$\begin{bmatrix} A_1 & & & \\ & A_2 & & 0 \\ & & \ddots & \\ & 0 & & A_m \end{bmatrix}$$

is called the *direct sum* of $A_1, \ldots, A_m$ and is denoted by

$$A_1 \oplus \cdots \oplus A_m$$

or sometimes as

$$\sum_{j=1}^{m} \oplus A_j \, .$$

Another frequently used notation for the direct sum is

$$A_1 \dotplus \cdots \dotplus A_m$$

or

$$\sum_{i=1}^{m} {}^{\bullet} A_i .$$

For example, if

$$A_1 = \begin{bmatrix} 1 & 2 \\ 3 & 4 \end{bmatrix}, \quad A_2 = \begin{bmatrix} 1 & 2 & 3 \\ 4 & 5 & 6 \\ 7 & 8 & 9 \end{bmatrix}$$

then

$$A_1 \oplus A_2 = \begin{bmatrix} 1 & 2 & 0 & 0 & 0 \\ 3 & 4 & 0 & 0 & 0 \\ 0 & 0 & 1 & 2 & 3 \\ 0 & 0 & 4 & 5 & 6 \\ 0 & 0 & 7 & 8 & 9 \end{bmatrix}.$$

If A and B are m × n matrices then the *Schur product* or *Hadamard product* of A and B is the m × n matrix C whose (i, j) entry is

$$c_{ij} = a_{ij} b_{ij}, \quad i = 1, \ldots, m, \quad j = 1, \ldots, n . \qquad (14)$$

The matrix C is denoted by

$$A \cdot B .$$

For example, if

$$A = \begin{bmatrix} 1 & 2 \\ 3 & 4 \end{bmatrix}, \quad B = \begin{bmatrix} 5 & 6 \\ 7 & 8 \end{bmatrix},$$

then

$$A \cdot B = \begin{bmatrix} 1 \cdot 5 & 2 \cdot 6 \\ 3 \cdot 7 & 4 \cdot 8 \end{bmatrix} = \begin{bmatrix} 5 & 12 \\ 21 & 32 \end{bmatrix}.$$

1.2 Exercises

1. Evaluate each of the following:

   (a) $\displaystyle\sum_{i=1}^{3} \frac{i}{i+1}$

   (b) $\displaystyle\prod_{i=1}^{3} \frac{i}{i^2+1}$

   (c) $\displaystyle\sum_{j=1}^{3} \left(\prod_{i=1}^{j} i\right)^j$

   (d) $\displaystyle\prod_{j=1}^{3} \left(\sum_{i=1}^{j} i\right)^j$

   (e) $\displaystyle\sum_{i=1}^{3} i \left(\prod_{j=1}^{i} j^2\right)$

   (f) $\displaystyle\prod_{i=1}^{2} \left(\prod_{j=1}^{i} \left(\sum_{k=1}^{j} k\right)\right)$

   (g) $\displaystyle\sum_{i=1}^{5} (-1)^i$

   (h) $\displaystyle\prod_{i=1}^{5} (-1)^i$

   (i) $\displaystyle\prod_{i=1}^{3} \left(\prod_{j=1}^{i} i\right)$

(j) $\prod_{i=1}^{4} (x)^i$

(k) $\prod_{i=1}^{4} (x_i)^i$

(l) $\prod_{i=1}^{4} x$

(m) $\prod_{i=1}^{4} (x)^{-i}$

(n) $\prod_{i=1}^{4} (x_i)^{-i}$

(o) $\prod_{i=1}^{4} (x_j)^i$

(p) $\prod_{i=1}^{4} (x_i)^j$

(q) $\prod_{i=1}^{4} (x)^{ij}$

(r) $\sum_{i=1}^{4} (x)^i$

(s) $\sum_{i=1}^{4} (x_i)^i$

(t) $\sum_{i=1}^{4} x$

(u) $\sum_{i=1}^{4} (x)^{-i}$

(v) $\sum_{i=1}^{4} (x_j)^i$

1.2 Exercises

(w) $\sum_{i=1}^{4} (x_i)^j$

In (x), (y), and (z) evaluate each of the sums. The subscripts $k_1, k_2, ..., k_n$ range over all nonnegative integers satisfying the equalities indicated below the Σ symbols:

(x) $\sum_{k_1 + k_2 = 3} 2^{k_1} 3^{k_2}$

(y) $\sum_{k_1 + k_2 + k_3 = 2} 2^{k_1} 3^{k_2} 4^{k_3}$

(z) $\sum_{k_1 + k_2 + k_3 = 1} (x)^{k_1}$

In exercises #2 - 20 evaluate the indicated matrix expressions.

2. $\begin{bmatrix} 1 & 2 & 3 \\ 4 & 5 & 6 \end{bmatrix} \begin{bmatrix} 2 & -1 \\ 3 & 0 \\ -1 & 2 \end{bmatrix}$

3. $\begin{bmatrix} 1 & 2 \\ 3 & 4 \\ -1 & -2 \end{bmatrix} \otimes \begin{bmatrix} 1 \\ 2 \end{bmatrix}$

4. $\begin{bmatrix} 1 & 2 \\ 3 & 4 \\ 5 & 6 \end{bmatrix} \cdot \begin{bmatrix} -1 & -2 \\ 0 & 3 \\ -1 & 1 \end{bmatrix}$

5. $\left(\begin{bmatrix} 1 & 2 \\ 3 & 4 \end{bmatrix} \begin{bmatrix} 1 & 2 \\ -3 & 5 \end{bmatrix} \right) \begin{bmatrix} 1 & 2 & 3 \\ 4 & 5 & 6 \end{bmatrix}$

6. $\begin{bmatrix} 1 & 2 \\ 3 & 4 \end{bmatrix} \left(\begin{bmatrix} 1 & 2 \\ -3 & 5 \end{bmatrix} \begin{bmatrix} 1 & 2 & 3 \\ 4 & 5 & 6 \end{bmatrix} \right)$

What do #5 and #6 suggest?

7. $\begin{bmatrix} 1 & 2 \\ 3 & 4 \end{bmatrix} \oplus \begin{bmatrix} 5 & 6 & 7 \\ 8 & 9 & 19 \\ -1 & -2 & -3 \end{bmatrix} \oplus \begin{bmatrix} 0 \end{bmatrix}$

8. $\text{diag}(1\ 2\ 3\ 4) \begin{bmatrix} 1 & 2 & 3 & 4 & 5 \\ 5 & 4 & 3 & 2 & 1 \\ -1 & -2 & -3 & 4 & 2 \\ 0 & 0 & 0 & 0 & 2 \end{bmatrix}$

9. $\text{diag}(1\ 2\ 3\ 4)\ \text{diag}(5\ 6\ 7\ 8)$

10. $\left(\begin{bmatrix} 1 & -1 \\ 1 & -1 \end{bmatrix} \otimes \begin{bmatrix} 2 & -2 \\ 0 & 2 \end{bmatrix} \right) \otimes \begin{bmatrix} 3 & 1 \\ 0 & 0 \end{bmatrix}$

11. $\begin{bmatrix} 1 & -1 \\ 1 & -1 \end{bmatrix} \otimes \left(\begin{bmatrix} 2 & -2 \\ 0 & 2 \end{bmatrix} \otimes \begin{bmatrix} 3 & 1 \\ 0 & 0 \end{bmatrix} \right)$

What do #10 and #11 suggest?

12. $(AB)_{(3)}$, where $A = \begin{bmatrix} 1 & 2 \\ 3 & 4 \\ 5 & 6 \end{bmatrix}$, $B = \begin{bmatrix} 1 & 0 & 2 & 4 \\ 3 & 5 & 6 & 8 \end{bmatrix}$.

13. $(AB)^{(4)}$ for the matrices in #12.

14. $AA^T \oplus BB^T$ for the matrices in #12.

15. $(A \otimes B)^T$ for the matrices in #12.

16. $A^T \otimes B^T$ for the matrices in #12.

What do #15 and #16 suggest?

17. $(A \otimes B)_{5,7}$ for the matrices in #12.

18. $(2A) \otimes (3B)$ for the matrices in #12. What general formula is suggested by this exercise?

1.2 Exercises

19. $(A \cdot A) \otimes (B \cdot B)$ for the matrices in #12.

20. $(A \otimes (A \otimes A))_{3,5}$ for the matrix in #12.

21. Prove the following general formulas for scalar multiplication and addition of m × n matrices (A and B are matrices, α, β scalars): $\alpha(A + B) = \alpha A + \alpha B$; $(\alpha + \beta)A = \alpha A + \beta A$; $\alpha(\beta A) = (\alpha\beta)A$; $1A = A$; $0A = 0_{m,n}$.

22. Find two 3 × 3 matrices A and B such that $A \cdot B = 0$ but neither A nor B is 0.

23. If $A = [a_{ij}]$ is 4 × 5 and $B = [b_{ij}]$ is 6 × 7, what is the (18, 23) entry of $A \otimes B$?

24. Prove that if $D = \text{diag}(d_1, d_2, \ldots, d_m)$ and A is m × n then $(DA)_{ij} = d_i a_{ij}$, $i = 1, \ldots, m$, $j = 1, \ldots, n$.

25. In each of the following formulas, the dimensions of the matrices permit the various products and sums to be defined. Prove that the following formulas are valid.

$A(B + C) = AB + AC$;
$(A + B)C = AC + BC$;
$I_m A I_n = A$ (A is m × n);
$A + -A = 0$;
$A(B - C) = AB - AC$ (where $B - C = B + (-C)$);
$(A - B)C = AC - BC$;
$A + (B + C) = (A + B) + C$ (this formula is called the *associative law for addition*);
$A(BC) = (AB)C$ (this formula is called the *associative law for multiplication*).

26. If x_1, x_2, x_3, x_4 are real numbers let

$$A(x_1, x_2, x_3, x_4) = \begin{bmatrix} x_1 & x_2 & -x_3 & -x_4 \\ -x_2 & x_1 & x_4 & -x_3 \\ x_3 & -x_4 & x_1 & -x_2 \\ x_4 & x_3 & x_2 & x_1 \end{bmatrix}$$

and set $e = A(1, 0, 0, 0)$, $i = A(0, 0, 1, 0)$, $j = A(0, 0, 0, 1)$, $k = A(0, 1, 0, 0)$. Verify that the multiplication table for these matrices (using matrix multiplication) is

	e	i	j	k
e	e	i	j	k
i	i	-e	k	-j
j	j	-k	-e	i
k	k	j	-i	-e

The matrices $A(x_1, x_2, x_3, x_4)$ are called *real quaternions*.

27. Show that for any two m × n matrices A and B, $A + B = B + A$, i.e., matrix addition is *commutative*.

28. Prove that for any matrix A, the matrices A^TA and AA^T are both defined. Exhibit an example of a square matrix A for which $AA^T \ne A^TA$.

29. Prove that if D and E are n-square diagonal matrices then $DE = ED$.

30. Let $A = [a_{ij}]$ be m × n, and let $D = \text{diag}(d_1, \ldots, d_m)$, $E = \text{diag}(e_1, \ldots, e_n)$. Show that $(DAE)_{ij} = d_i a_{ij} e_j$, $i = 1, \ldots, m$, $j = 1, \ldots, n$.

31. Construct two 3 × 3 matrices A and B such that $AB = 0$ but $A \ne 0$ and $B \ne 0$.

32. Construct a nonzero 2 × 2 matrix A such that $AA = 0$.

33. If $A = \begin{bmatrix} 2 & 0 & 0 \\ 1 & 0 & 0 \end{bmatrix}$ and $B = \begin{bmatrix} 2 & 0 & 0 & 0 \\ 1 & 0 & 0 & 0 \end{bmatrix}$ explain why $A \ne B$.

34. Show that if A is n × n then $A + A^T$ is symmetric and $A - A^T$ is skew-symmetric.

35. Show that if A is any n × n matrix then
$$A = \left(\frac{A + A^T}{2}\right) + \left(\frac{A - A^T}{2}\right).$$

1.2 MatLab

1. Read the entries for ·, +, -, *, ^, **zeros**, **eye**, **ones**, and **diag** in the Built-in commands in the Help window. Then use MatLab to write very brief sequences of commands that result in the display of each of the following matrices. Recall that a sequence of commands in MatLab is called a *script*.

   (a) $0_{2,3}$

   (b) I_4

   (c) $\begin{bmatrix} 1 & 1 \\ 1 & 1 \\ 1 & 1 \end{bmatrix}$

   (d) $\begin{bmatrix} 1 & 0 & 0 & 0 & 0 \\ 0 & 1 & 0 & 0 & 0 \\ 0 & 0 & 1 & 0 & 0 \end{bmatrix}$ Hint: eye(3,5);

   (e) $\begin{bmatrix} 0 & 1 & 1 & 1 & 1 \\ 1 & 0 & 1 & 1 & 1 \\ 1 & 1 & 0 & 1 & 1 \end{bmatrix}$ Hint: ones(3,5) - eye(3,5);

   (f) $\begin{bmatrix} 0 & 0 & 0 & 0 \\ 1 & 0 & 0 & 0 \\ 0 & 2 & 0 & 0 \\ 0 & 0 & 3 & 0 \end{bmatrix}$ Hint: v = [1 2 3];
   A = diag(v, -1)

   (g) $\begin{bmatrix} 0 & 0 & 0 & 0 \\ 1 & 0 & 0 & 0 \\ 0 & 1 & 0 & 0 \end{bmatrix}$

(h) $\begin{bmatrix} 0 & 1 & 0 \\ 0 & 0 & 2 \\ 0 & 0 & 0 \end{bmatrix}$ Hint: v = [1 2];
A = diag(v, 1)

(i) $\begin{bmatrix} 3 & 3 & 3 \\ 3 & 3 & 3 \end{bmatrix}$

(j) $\begin{bmatrix} 4 & 5 & 6 \\ 8 & 10 & 12 \\ 12 & 15 & 18 \end{bmatrix}$ Hint: u = [1; 2; 3] ;
v = [4, 5, 6];
A = u * v

(k) $\begin{bmatrix} 1 & \frac{1}{2} & \frac{1}{3} \\ \frac{1}{2} & \frac{1}{4} & \frac{1}{6} \\ \frac{1}{3} & \frac{1}{6} & \frac{1}{9} \end{bmatrix}$

(l)   The 5-square matrix all of whose rows are [1 2 3 4 5]

(m)   The 5-square matrix all of whose columns are [1 2 3 4 5]^T

(n)   The 3-square matrix whose (i, j) entry is i^j

Hint:
B = [1 2 3]' * ones (1, 3);
A = B .^ (B')

(o)   The 4-square matrix whose (i, j) entry is $\frac{1}{i+j}$

Hint:
B = [1: 4]' * ones (1, 4);
C = B + B';
A = C .^ (-1)

1.2 MatLab

(p) $\begin{bmatrix} 1 & 2 \\ 3 & 4 \end{bmatrix} \cdot \begin{bmatrix} 4 & -1 \\ 5 & 2 \end{bmatrix}$

(q) The matrix (AA)A where $A = \begin{bmatrix} 1 & 2 \\ 3 & 4 \end{bmatrix}$.

2. The colon (:) notation is useful for selecting parts of a matrix. Enter the matrix

$$A = \begin{bmatrix} 1 & 2 & 3 & 4 & 5 \\ 6 & 7 & 8 & 9 & 10 \\ 11 & 12 & 13 & 14 & 15 \\ 16 & 17 & 18 & 19 & 20 \end{bmatrix}.$$

Then enter each of the following expressions and observe the outputs:

(a) A(1, :)
(b) A(2, :)
(c) A(3, :)
(d) A(5, :)
(e) A(:, 1)
(f) A(:, 2)
(g) A(:, 5)
(h) A(1, 1: 2 :5)
(i) A([2 4], :)
(j) A([4 2], :)
(k) A(4: -1: 1, 5: -1:1)
(l) A([1 1], [2 2])
(m) A([1 1], [4 5])
(n) A([2 2], [5 4])
(o) A([3 2], [5 4])

3. Read the entries for **sum, cumsum, prod, cumprod, diff, while, size,** and **sqrt** in the Built-in commands in the Help window.

(a) Use cumsum and sum to compute the value of

$$\sum_{k=1}^{4}\sum_{i=1}^{k} a_i$$

where a = [1 2 3 4].

(b) What is the value of cumsum(sum(a)) for the matrix in (a)?

(c) Use sum and prod to compute the value of

$$\prod_{j=1}^{3}\sum_{i=1}^{3} a_{ij}$$

where $A = \begin{bmatrix} 1 & 4 & -2 \\ 3 & 5 & 8 \\ -1 & 7 & 6 \end{bmatrix}$.

(d) Use sum and prod to compute the value of

$$\sum_{j=1}^{3}\prod_{i=1}^{3} a_{ij}$$

for the matrix A in (c).

(e) Use sum and cumprod to compute

$$\sum_{j=1}^{3}\prod_{i=1}^{3} a_{ij}$$

for the matrix A in (c).

(f) Use while to write a MatLab script to compute $\sum_{k=1}^{n}\frac{1}{k(k+1)}$ for n = 10.
Evaluate the sum for n = 20, n = 100. Prove that $\lim_{n\to\infty}\sum_{k=1}^{n}\frac{1}{k(k+1)} = 1$.

Hint: Choose **New...** from the File menu and enter the following script in the resulting Edit window. Then choose Save and Go from the File menu to name the script and run it.

```
n = 10;
s = 0;
k = 1;
while k <= n
  s = s+1/(k*(k+1));
  k = k + 1;
end
s
```

Note: Without using while, an equivalent (and more efficient) script is:

```
n = 10;
s = sum(((1:n).*((1:n) + ones(1, n))).^(-1))
```

The semicolons in the script suppress printing. The while construction in MatLab always has the form

```
while (something is true)
  commands
end
```

The end defines the sequence of commands controlled by the while.

(g) In (f) we were required to run the script several times for various values of n. MatLab has an "input" statement that simplifies this. Replace the command n = 10; by

```
n = input ('Enter n ');
```

and run the program again. In the Command window, respond to the prompt "Enter n" by typing 10 followed by <CR>. At the resulting prompt (>>) in the Command window, type the name you gave the script in (f) and press return. Evaluate the sum in (f) for n = 50, 200, 1000.

(h) Modify the script in (f) as follows:

```
n = 1;
s = 0;
k = 1;
list = [ ];
```

```
        while n <= 20
          while k <= n
            s = s+1/(k*(k+1));
            k = k + 1;
          end
          n = n + 1;
          list = [ list s ];
        end
        list'
```

The statement list = [ ] initializes a matrix to be "empty", i.e., use two adjacent brackets. The outer while loop controls the inner while loop as it generates $\sum_{k=1}^{n} \frac{1}{k(k+1)}$ for each of the values n = 1, 2, 3, ..., 20. As each sum s is generated, it is appended to the matrix named list by the statement list = [list s]. Finally, after all values are in list, it is exhibited as a column of 20 numbers by the statement list'.

(i) The formatting of the output can be controlled to some extent in MatLab. Modify the script in (h) so that it has the following appearance in the Edit window.

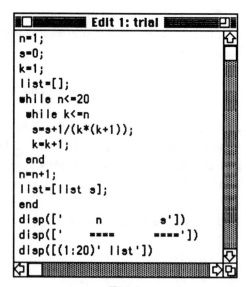

Fig. 1

1.2 MatLab

The output in the Command window will have the following appearance:

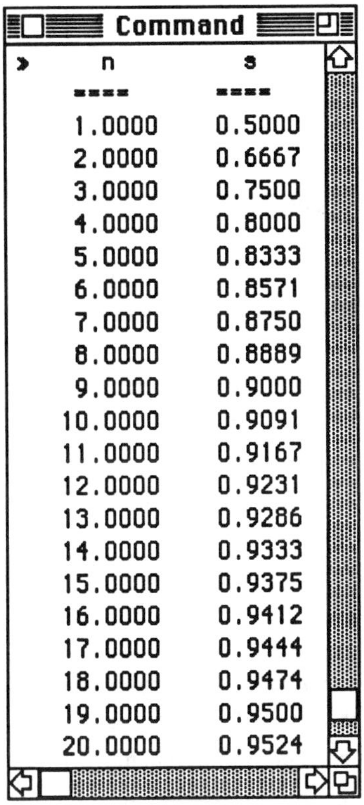

```
======= Command =======
>    n            s
    ----         ----
   1.0000       0.5000
   2.0000       0.6667
   3.0000       0.7500
   4.0000       0.8000
   5.0000       0.8333
   6.0000       0.8571
   7.0000       0.8750
   8.0000       0.8889
   9.0000       0.9000
  10.0000       0.9091
  11.0000       0.9167
  12.0000       0.9231
  13.0000       0.9286
  14.0000       0.9333
  15.0000       0.9375
  16.0000       0.9412
  17.0000       0.9444
  18.0000       0.9474
  19.0000       0.9500
  20.0000       0.9524
```

Fig. 2

In the Command window type "help disp" at the prompt (>>) to read the entry for disp.

(j) If A is the matrix in #2, what matrix is output by the following script?

```
n = 1;
T = A(:, 1)';
while n <= 4
  T = [ T;  A(:, n+1)' ];
  n = n + 1;
end
T
```

4. The notations for matrix addition and multiplication are the usual + and *. If A is the matrix in #2 and B is the matrix

$$B = \begin{bmatrix} 1 & -3 & 2 & 5 \\ -4 & 2 & 4 & 1 \\ 3 & 7 & -5 & 2 \end{bmatrix},$$

write a script that outputs $(BA)_{(2)}$. It is only necessary to input $B_{(2)}$.

5. Type help kron at the Command prompt (>>) and read the entry for **kron**. Use it to compute the Kronecker products in §1.2 Exercises #3, #10, #11.

6. Type help size at the Command prompt and read the entry for **size**. Use **size** to compute the dimensions of eye(p) ⊗ ones(m,n) for p = 3, m = 4, n = 5.

7. Type help input at the Command prompt (>>) and read the entry for **input**. Write and run a MatLab script that:

- calls for the input of a matrix A
- exhibits A
- calls for the input of a row index p
- calls for the input of a second row index q
- exhibits the result of interchanging rows p and q of A
- calls for the input of 1 to interchange another pair of rows of A, 0 to end
- repeats if 1 is entered, ends otherwise

Hint: Type help = at the Command prompt to read an explanation for ==, the equality relation (as contrasted to the = assignment operator). Then consider the following script:

```
A = input('enter a matrix A ')
resp = 1;
while resp == 1
    p = input('enter a row index: ');
    q = input('enter a second row index: ');
    A([p, q], :) = A([q, p], :)
    resp = input('enter 1 to interchange another pair of rows of A, 0 to end:');
end
```

8. Read the entries for **rand** and **floor**. Then write and run a MatLab script that calls for the input of two integers a and b, a < b, and then generates 10 random integers in the interval [a, b].

First note that

$$0 \leq \text{rand} < 1$$
$$0 \leq (b - a) * \text{rand} < b - a$$
$$a \leq a + (b - a) * \text{rand} < b.$$

Thus

$$a \leq \text{floor}(a + (b - a) * \text{rand}) < b.$$

Since the middle term in the above inequality is an integer it lies in the set $\{a, ..., b-1\}$.

Hint:
```
a = input('enter a: ');
b = input('enter b: ');
b = b + 1;
r = floor(a + (b - a)*rand(1, 10))
```

9. Modify the script for #8 so that it generates and exhibits ten 2-square random matrices with integer entries in the interval [a, b].

Hint:
```
a = input('enter a: ');
b = input('enter b: ');
r = [ ];
b = b + 1;
n = 1;
while n <= 10
        r = [r  floor(a + (b - a) * rand(2))];
        n = n + 1;
end
r
```

10. Read the Help entries for **tril** and **triu**.

   (a) Write and run a script that calls for the input of two real numbers a < b and then exhibits 10 random 2-square real matrices whose entries lie in the interval [a, b).

   Hint:
   ```
   a = input('enter a: ');
   b = input('enter b: ');
   r = [ ];
   n = 1;
   while n <= 10
           r = [ r   a + (b - a) * rand(2) ];
           n = n + 1;
   end
   r
   ```

   (b) Modify and run the script in (a) so that it exhibits 10 random 2-square real lower triangular matrices with entries in the interval [a, b).

   Hint:
   ```
   a = input ('enter a: ');
   b = input ('enter b: ');
   r = [ ];
   n = 1;
   while n <= 10
           r = [ r   tril(a + (b - a) * rand (2)) ];
           n = n + 1;
   end
   r
   ```

   (c) Modify and run the script in (a) so that it exhibits 10 random 2-square real symmetric matrices with entries in the interval [a, b).

   Hint:
   ```
   a = input ('enter a: ');
   b = input ('enter b: ');
   ```

```
r = [ ];
n = 1;
while n <= 10
        A = a + (b - a) * rand (2);
        L = tril(A, -1);
        r = [ r   L + L' + diag(diag(A)) ];
        n = n + 1;
end
r
```

11. Read the Help entries for **eps, pi, inf,** and **nan.** Then on successive prompts in the Command window, enter the following and observe the outputs.

   (a)   eps
   (b)   pi
   (c)   inf
   (d)   1 / 0
   (e)   0 / 0
   (f)   2 + nan
   (g)   nan / nan
   (h)   inf + inf
   (i)   1 / inf
   (j)   inf / 0
   (k)   2 * inf
   (l)   inf * 0
   (m) nan / inf
   (n)   inf / inf
   (o)   inf * inf

12. Read the Help entries for **clock** and **etime.** Write and run a script to determine the time it takes to generate (not exhibit!) 5000 random 2-square real matrices. Repeat the experiment with 5-square matrices.

   Hint:
```
n = 1;
t0 = clock;
while n <= 5000
        rand (2);
        n = n + 1;
end
t1 = clock;
etime (t1, t0)
```

13. As we learned, matrix multiplication of conformal matrices is denoted in MatLab by A * B. The pth power of a square matrix is denoted by A^p. The Hadamard product of two matrices of the same size is denoted A .* B. The matrix $\left[a_{ij}^r\right]$ is denoted by A .^ r in MatLab. Use MatLab to compute the answers to the following problems in §1.2 Exercises: #2, #4, #5, #6, #8, and #9.

14. Let
$$A = \begin{bmatrix} 0 & 1 \\ 1 & 0 \end{bmatrix}.$$

    Use MatLab to investigate the positive integral powers of A.

15. Compute A .^ r for A = 2 * ones(3), r = -1.

16. The empty matrix, [ ], is frequently useful for assignment purposes. For example, the sequence in { 1, 2, 3, 4, 5, 6 } complementary to { 2, 3, 6 } is easily computed by

    x = 1:6;
    x ([2 3 6]) = [ ].

    Write a MatLab script to generate a random 3-square matrix and then exhibit the three 2-square matrices obtained by deleting row and column i from A, i = 1, 2, 3.

    Hint:
    A = rand (3)
    i = 1;
    while i <= 3
        e = 1:3;
        e(i) = [ ];
        A(e,e)
        i = i + 1;
    end

17. Write a MatLab script that calls for the input of a real number r and an integer k and then prints out r rounded to k decimal places. Test the script with r = 3.14159, k = 2; r = 3.156, k = 2; r = 1552, k = -2, (i.e., round to the closest multiple of 100); r = 1549.3 , k = -2. Before running the program, type format long at the command prompt.

    Hint:
    r = input ('enter a real number: ');
    k = input ('enter an integer: ');
    floor ((10 ^ k) * r + 0.5) / 10^k

1.2 Glossary

$\dotplus$	30
., +, -, *, ^	37
==	44
Π	25
$\oplus$	29
Σ	23
$\Sigma\cdot$	30
[ ]	42
$A \cdot B$	30
A .^ r	48
A .* B	48
A * B	48
$A \otimes B$	28
A ^ p	48
associative law for addition	35
associative law for multiplication	35
clock	47
colon notation	39
commutative	27
cumprod	39
cumsum	39
diag	37
diff	39
direct product	28
direct sum	29
disp	43
double sum	25
empty matrix	42
eps	47
etime	47
eye	37
floor	45
Hadamard product	30

inf	47
input	41
kron	44
Kronecker product	28
matrix addition	26
matrix product	27
nan	47
ones	37
pi	47
prod	39
rand	45
real quaternions	36
rounding	48
scalar product	26
Schur product	30
size	39
sqrt	39
sum	39
tensor product	28
tril	46
triu	46
while	39
zeros	37

Chapter 2

Complex Numbers

Topics
- *complex number operations*
- *argument, modulus, polar form*
- *De Moivre's Theorem*
- *principal values*
- *elementary functions*
- *complex matrices, inner product*

2.1 Algebra of Complex Numbers

The *real* numbers are the ordinary decimal numbers, terminating or non-terminating, positive, zero, or negative. In analogy with an ordinary measuring ruler, real numbers are identified with points on a line (see Fig. 1).

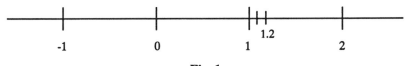

Fig. 1

We know how to combine the points on the real line using addition, subtraction, multiplication, and division. The real line, together with these operations, is usually denoted by $\mathbb{R}$. The basic operations of addition and multiplication (and hence subtraction and division) can be extended to all the points in the cartesian plane in such a way that the original operations for points on the real line $\mathbb{R}$ remain the same.

Addition of points in the plane is simple to define geometrically:

$$(a, b) + (c, d) = (a + c, b + d).$$

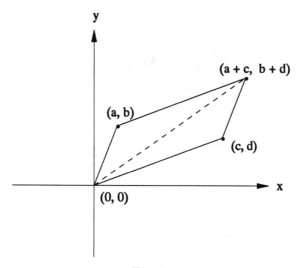

Fig. 2

The *multiplication* of points (a, b) and (c, d) is defined geometrically as follows. First join each of the points to the origin. Let θ and φ be the angles measured counterclockwise from the positive x-axis to (a, b) and (c, d), respectively. Let Δ and δ denote the distances from the origin to (a, b) and (c, d), respectively. Then the *product* of (a, b) and (c, d) is the point (u, v) whose distance to the origin is $\Delta\delta$ and whose angle with the positive x-axis is $\theta + \varphi$.

2.1 Algebra of Complex Numbers

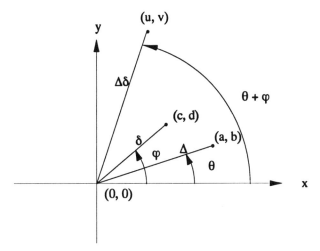

Fig. 3

For example, if (a, b) = (1, 1) and (c, d) = (0, 2) then the angles θ and φ are 45° and 90° respectively, and the distances Δ and δ are $\sqrt{2}$ and 2, respectively. Then the angle for (u, v) is 45° + 90° = 135° and the distance is $2\sqrt{2}$. If either of (a, b) or (c, d) is the origin (0, 0), then we simply define the product (u, v) to be (0, 0). The set of all points in the plane, together with the two operations of addition and multiplication, is denoted by $\mathbb{C}$. A point in $\mathbb{C}$ is called a *complex number* (for historical reasons) and it is customary to denote complex numbers with a single letter, e.g.,

$$z = (1, 1), \quad w = (0, 2).$$

A more or less standard language and notation is used in studying complex numbers. A diagram depicting the positions of complex numbers in the Cartesian plane, such as Fig. 2 or Fig. 3, is called an *Argand* diagram. The point or complex number z = (x, y) is written as

$$z = x + yi. \tag{1}$$

In order to explain the notation in (1), first note that complex numbers on the real line add precisely as before, e.g.,

$$(5, 0) + (3, 0) = (8, 0). \tag{2}$$

We immediately abandon awkward statements such as (2) and simply write (5, 0) as 5 etc., so that (2) becomes the familiar

$$5 + 3 = 8.$$

We next compute the product of the two points (y, 0) and (0, 1). Assume for simplicity that y > 0. Then the angle for (y, 0) is 0 and the angle for (0, 1) is 90°. Hence the angle for (y, 0) · (0, 1) is 90°, and clearly the product of the distances of the two points to the origin is y. Hence (y, 0) · (0, 1) = (0, y). For example,

$$(3, 0)\,(0, 1) = (0, 3).$$

But, writing y instead of (y, 0) and denoting the complex number (0, 1) by i,

$$i = (0, 1), \tag{3}$$

we have

$$\begin{aligned} z &= (x, y) \\ &= (x, 0) + (0, y) \\ &= (x, 0) + (y, 0)\,(0, 1) \\ &= x + yi. \end{aligned}$$

The number i is called the *imaginary unit*. Observe that since the angle for i is 90° and its distance to the origin is 1 we conclude that

$$i^2 = -1. \tag{4}$$

Thus once we know that all the laws of algebra hold for complex numbers, we can easily perform calculations that might not be obvious from the definition alone, e.g.,

$$\begin{aligned} (3 + 7i)\,(2 - 3i) &= 3\,(2 - 3i) + 7i\,(2 - 3i) \\ &= 6 - 9i + 14i - 21i^2 \\ &= 6 + 5i - 21(-1) \\ &= 27 + 5i. \end{aligned}$$

2.1 Algebra of Complex Numbers

In fact, these laws do hold and we list them here:

(commutative law of addition)	$z + w = w + z$	(5)
(commutative law of multiplication)	$zw = wz$	(6)
(additive identity)	$z + 0 = 0 + z = z$	(7)
(multiplicative identity)	$z \cdot 1 = 1 \cdot z = z$	(8)
(associativity of addition)	$z + (w + u) = (z + w) + u$	(9)
(associativity of multiplication)	$(zw)u = z(wu)$	(10)
(distributive law)	$z(w + u) = zw + zu$	(11)

All of these laws are easy to verify. For example, if z, w, and u make angles of θ, φ, and ω with the positive real axis and their distances to the origin are r, ρ, and δ, respectively, then directly from the definition of multiplication, the angle for (zw)u is

$$(\theta + \varphi) + \omega = \theta + (\varphi + \omega), \qquad (12)$$

and the product of the distances of zw and u to the origin is

$$(r\rho)\delta = r(\rho\delta). \qquad (13)$$

But the right sides of (12) and (13) are, respectively, the angle for z(wu) and the distance from the origin to z(wu). Hence (10) is confirmed. Similar geometric arguments can be used to prove the other algebraic laws for $\mathbb{C}$. In view of (6) we can write either x + yi or x + iy to denote a complex number.

The *modulus* or *absolute value* of z = x + yi is the number

$$|z| = \sqrt{x^2 + y^2}.$$

Observe that $|z|$ is the distance from z to the origin.

The *real part* of z = x + yi is the number x. This is written

$$\text{Re}(z) = x.$$

The *imaginary part* of z is the number y. This is written

$$\text{Im}(z) = y.$$

The *complex conjugate* of z = x + yi is the complex number

$$\bar{z} = x - yi.$$

The complex conjugate of z is the mirror image of z across the x - axis.

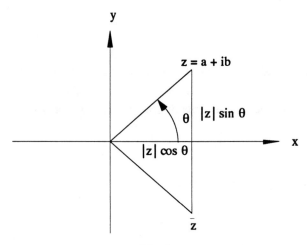

Fig. 4

The angle θ in Fig. 4 is frequently called the *argument* of z. If z is written as

$$z = |z| \cos \theta + i \, |z| \sin \theta = |z| (\cos \theta + i \sin \theta) \qquad (14)$$

then z is said to be in *polar form*.

2.1 Algebra of Complex Numbers

Theorem 1.

Let $z = |z|(\cos\theta + i\sin\theta)$ and $w = |w|(\cos\varphi + i\sin\varphi)$. Then

$$zw = |z||w|(\cos(\theta + \varphi) + i\sin(\theta + \varphi)). \tag{15}$$

Proof.

The formula (15) is simply a restatement of the definition of complex number multiplication: add the angles; multiply the moduli. ∎

The formulas for the sine and cosine of a sum of angles are useful in solving the following problem. Suppose we are given two complex numbers

$$z = a + bi, \quad w = c + di. \tag{16}$$

How do we write the product zw in the form

$$zw = u + vi\,? \tag{17}$$

To answer this, simply use (15) and the trig formulas for $\cos(\theta + \varphi)$ and $\sin(\theta + \varphi)$ to compute that

$$zw = |z||w|(\cos\theta\cos\varphi - \sin\theta\sin\varphi) + i|z||w|(\sin\theta\cos\varphi + \cos\theta\sin\varphi)$$

$$= (|z|\cos\theta)(|w|\cos\varphi) - (|z|\sin\theta)(|w|\sin\varphi)$$

$$+ i\,[(|z|\sin\theta)(|w|\cos\varphi) + (|z|\cos\theta)(|w|\sin\varphi)]. \tag{18}$$

From Fig. 4,

$$a = |z|\cos\theta, \quad b = |z|\sin\theta, \tag{19}$$

and similarly

$$c = |w|\cos\varphi, \quad d = |w|\sin\varphi. \tag{20}$$

Substituting (19) and (20) in (18), we have

$$zw = ac - bd + i(bc + ad),$$

or, referring to (17),

$$u = ac - bd,$$
$$v = bc + ad. \tag{21}$$

If we look back at Fig. 2 and let $z = a + bi$ and $w = c + di$ it is obvious that

$$|z + w| \leq |z| + |w|.$$

This is just another way of saying that the shortest distance between two points in the plane is a straight line. Thus we have:

Theorem 2. (Triangle Inequality)

For any two complex numbers z and w, the following inequality holds:

$$|w + z| \leq |w| + |z|.$$

The triangle inequality can immediately be extended to more than two complex numbers.

Theorem 3.

Let $z_1, z_2, ..., z_n$ be any n complex numbers. Then

$$|z_1 + z_2 + \cdots + z_n| \leq |z_1| + |z_2| + \cdots + |z_n|.$$

Proof.

By repeated use of Theorem 2, we have

2.1 Algebra of Complex Numbers

$$|z_1 + z_2 + z_3 + \cdots + z_n| \leq |z_1| + |z_2 + z_3 + \cdots + z_n|$$
$$\leq |z_1| + |z_2| + |z_3 + \cdots + z_n|$$
$$\vdots$$
$$\leq |z_1| + |z_2| + |z_3| + \cdots + |z_n|. \blacksquare$$

By applying Theorem 1 repeatedly we see that in multiplying several complex numbers z, w, ..., k we simply add the arguments of z, w, ..., k and multiply their moduli to get the angle and modulus, respectively, of the product zw ⋯ k. This leads us to the following important theorem.

Theorem 4. (De Moivre's Theorem)

Let t be an integer and let $z = r(\cos\theta + i\sin\theta)$ where $r > 0$ and θ is real. Then

$$z^t = r^t(\cos t\theta + i\sin t\theta). \tag{22}$$

Proof.

If t is a positive integer then (22) follows from repeatedly applying Theorem 1, taking all the factors to be z. We prove next that (22) holds if $t = -1$: for, the product of the moduli of (22) for $t = 1$ and $t = -1$ is clearly 1 and the sum of the angles is $\theta + (-\theta) = 0$. Thus the expression (22) (for $t = -1$) when multiplied by z produces the product 1. In other words, for $t = -1$ the right side of (22) is the multiplicative inverse of z. If t is a negative integer we can set $t = -n$ where $n > 0$. Then

$$z^t = z^{-n} = (z^n)^{-1}$$
$$= (r^n(\cos n\theta + i\sin n\theta))^{-1}$$
$$= r^{-n}(\cos(-n\theta) + i\sin(-n\theta))$$
$$= r^t(\cos t\theta + i\sin t\theta).$$

If $t = 0$ then $z^t = z^0 = 1$ by convention, and thus

$$r^t(\cos t\theta + i\sin t\theta) = r^0(\cos 0 + i\sin 0) = 1(1 + i0) = 1. \blacksquare$$

As an example, we can use (22) to express $\cos 3\theta$ as a polynomial in $\cos \theta$. Let

$$z = \cos \theta + i \sin \theta .$$

Then from (22),

$$z^3 = \cos 3\theta + i \sin 3\theta . \qquad (23)$$

On the other hand,

$$\begin{aligned} z^3 &= (\cos \theta + i \sin \theta)^3 \\ &= \cos^3 \theta + 3i \cos^2 \theta \sin \theta - 3 \cos \theta \sin^2 \theta - i \sin^3 \theta \\ &= \cos \theta (\cos^2 \theta - 3 \sin^2 \theta) + i (3 \cos^2 \theta \sin \theta - \sin^3 \theta) . \qquad (24) \end{aligned}$$

Hence, equating the real parts of (23) and (24) we have

$$\begin{aligned} \cos 3\theta &= \cos \theta (\cos^2 \theta - 3 \sin^2 \theta) \\ &= 4 \cos^3 \theta - 3 \cos \theta . \end{aligned}$$

2.1 Exercises

1. Find the complex conjugate and the modulus of each of the following complex numbers:

   (a) 2   (b) -2
   (c) $2i$ (d) $-2i$
   (e) $1 + i$ (f) $-3 + i$
   (g) $-3 - i$ (h) i^2
   (i) $i + 2$ (j) $-5i - 7$
   (k) $\sqrt{3} - \sqrt{3} i$ (l) $-1 + \sqrt{3} i$
   (m) $(3 + i) + (2 - i)$ (n) $(i - 1) + (2 - i)$
   (o) $(\sqrt{3} + i) + (2\sqrt{3} - 3i)$ (p) $-\sqrt{3} i / 2$

2.1 Exercises

2. Express each of the following complex numbers in the form $x + yi$, in which x and y are real:

(a) i^{-1}

(b) $(2 + i)(3 - i)$

(c) $(1 + i)^2$

(d) $(3 - 2i)(-4 + i)$

(e) $(1 + i)^{-1}$

(f) $\dfrac{2 + 3i}{2 - 3i}$ (i.e., $(2+3i)(2-3i)^{-1}$)

(g) $\dfrac{3 + 2i}{-1 + 5i}$

(h) $\dfrac{1 - 2i}{-2 + 3i}$

(i) $\dfrac{1 + 2i}{2 - i} + \dfrac{2i}{-3 + i}$

(j) $\dfrac{1 + 2i}{2 - i} \cdot \dfrac{2i}{-3 + i}$

(k) $\dfrac{7 - 6i}{1 + i} - \dfrac{3 - i}{2 - 9i}$

(l) $\dfrac{7 - 6i}{1 + i} \cdot \dfrac{3 - i}{2 - 9i}$

(m) $3 - 2i + \dfrac{4 - 3i}{2 + i}$

(n) $\text{Im}((1 + i)^2)$

(o) $(\text{Re}(i)^2)$

(p) $\dfrac{(2 - i)(1 + 3i)}{(1 + i)(3 + 2i)}$

3. (a) Show that $\text{Re}(z) = 0$ if and only if $\bar{z} = -z$.

(b) Show that $\text{Im}(z) = 0$ if and only if $z = \bar{z}$.

(c) Show that for any two complex numbers z_1 and z_2, $|z_1 z_2| = 0$ if and only if $z_1 = 0$ or $z_2 = 0$.

(d) Show that $|z^{-1}| = |z|^{-1}$, for any nonzero complex number z.

4. Find the inverse of each of the following complex numbers:

   (a) $2i$ (b) $-3i$

   (c) $1 - i$ (d) $3 + 4i$

   (e) $-\frac{1}{2} + \frac{\sqrt{3}}{2} i$ (f) $\frac{1}{\sqrt{2}} + \frac{1}{\sqrt{2}} i$

   (g) $\frac{\sqrt{3} - i}{1 - \sqrt{3} i}$ (h) $\left(-\frac{1}{2} + \frac{\sqrt{3}}{2} i\right)^2$

5. Show that $\left|\frac{z_1}{z_2}\right| = \frac{|z_1|}{|z_2|}$.

6. Show that a complex number is 0 if and only if its modulus is 0.

7. If z and w are two complex numbers, draw an appropriate Argand diagram exhibiting the complex number $w - z$.

8. Illustrate each of the following by means of an Argand diagram:

   (a) $(1 + i) - (2 + 3i)$
   (b) $|z| = |w|$ for two complex numbers z and w
   (c) $\text{Re}(z) = \text{Im}(z)$ (d) $\text{Re}(z) = 0$
   (e) $\text{Im}(z) = 0$ (f) $\text{Re}(z) > 0, \text{Im}(z) > 0$
   (g) $\text{Re}(z) > 0, \text{Im}(z) < 0$ (h) $\text{Re}(z) < 0, \text{Im}(z) < 0$
   (i) $\text{Re}(z) < 0, \text{Im}(z) > 0$

9. Find the rectangular coordinates of each of the following points (r, θ) given in polar coordinates, i.e., r is the modulus, θ is the argument.

   (a) $(4, -90°)$ (b) $(2, 180°)$
   (c) $(0, \pi/6)$ (i.e., $\pi/6$ radians)   (d) (π, π)
   (e) $(1, 30°)$ (f) $(2, 135°)$
   (g) $(1, 210°)$ (h) $(3, 20\pi)$
   (i) $(1, 0)$ (j) $(\pi, 60°)$

2.1 Exercises

10. Express the following complex numbers in polar form.

    (a) $1 - i$ (b) $-i$
    (c) $-\cos 30° - i \sin 30°$ (d) $\cos 35° + i \cos 35°$
    (e) $(\sqrt{3} - i)^8$

11. (a) Let $z_1 = r_1(\cos \theta_1 + i \sin \theta_1)$ and $z_2 = r_2(\cos \theta_2 + i \sin \theta_2)$ be two complex numbers in polar form, $z_2 \neq 0$. Show that

    $$z_1 / z_2 = (r_1 / r_2)(\cos(\theta_1 - \theta_2) + i \sin(\theta_1 - \theta_2)).$$

    (b) Use the above formula to find the polar form of z_1 / z_2 if

    $$z_1 = -1 + i\sqrt{3} \quad \text{and} \quad z_2 = 1 - i.$$

In #12 - 20, use De Moivre's Theorem to compute the indicated complex numbers. Express your answers in the form $a + bi$, where a and b are real numbers.

12. $[3(\cos(45°) + i \sin(45°))]^6$

13. $[2(\cos(15°) + i \sin(15°))]^4$

14. $(\cos(30°) - i \sin(30°))^8$

15. $(\cos(150°) + i \sin(150°))^{10}$

16. $(\cos(30°) + i \sin(30°))^5$

17. $(1 + i)^{10}$

18. $(-2 - 2i)^8$

19. $(1 - i)^{10}$

20. $(1/2 - i\sqrt{3}/2)^9$

21. Show that if z and w are two complex numbers then the distance between z and w is $|z-w|$.

22. Geometrically describe the set of all complex numbers z which satisfy $|z| = 2$.

23. Geometrically describe the set of all complex numbers z which satisfy $1 < |z| < 2$.

24. Geometrically describe the set of all complex numbers z which satisfy $|2z + 1| > 2$.

25. Geometrically describe the set of all complex numbers z which satisfy $|z + i| = |z - 2|$.

26. Geometrically describe the set of all complex numbers z which satisfy $|z - 1| = |z - 3| = |z - i|$.

2.1 MatLab

1. Type help **abs** at the Command prompt (») and read the entry for abs. Use MatLab to compute each of the following.

    (a) abs(- i)
    (b) abs(i * (1 + i))
    (c) abs((1 + i) ^ 2)
    (d) $\text{abs}\begin{pmatrix}\begin{bmatrix} i & i \\ 1+i & 2i \end{bmatrix}\end{pmatrix}$
    (e) abs((1 + i) ^ 3) / (abs(1 + i)) ^ 3

2. Use the Edit window to write a MatLab script that

    • calls for the input of a positive integer n and a complex number z
    • prints out a table of values of $z, z^2, \ldots, z^n$ in the following typical format:

    enter a positive integer N: 4

    enter a complex number z: i

n	z^n
===	===
1.0000	0.0000 + 1.0000i
2.0000	-1.0000 + 0.0000i
3.0000	0.0000 - 1.0000i
4.0000	1.0000 + 0.0000i

Fig. 1

Hint: Use Help to read about **disp, :, ones, ., ^**

```
N = input('enter a positive integer N: ');
z = input('enter a complex number z: ');
disp([' ']);
disp(['     n              z ^ n']);
disp(['    ==             ===']);
power = (z*ones(1,N)).^(1:N);
A = [1:N;power];
disp(A.')
```

In order to understand the last line of the program, recall that A' is the transpose of A when A has real entries. If A has complex entries then A' transposes A, but also replaces each entry with its complex conjugate. To simply transpose a complex matrix, use A.' . Use Help to read about the related **conj**.

3. Type help **angle** at the Command prompt (») and read the entry for angle. Use MatLab to evaluate each of the following:

(a) angle(1 + i)
(b) angle(i)
(c) angle(-1 + i)
(d) angle(-1)
(e) angle(-1 - i)
(f) angle(- i)
(g) angle(1 - i)
(h) angle(1)
(i) angle(-1 + i * 1e-6)
(j) angle(-1 - i * 1e-6)

In mathematical notation what are the precise values of (a) - (h) in radians expressed in terms of π, e.g., angle(1 + i) = $\pi/4$? The angle function in MatLab yields the argument of a complex number. As we see from (i) and (j), MatLab chooses an angle θ for the argument that satisfies

$$-\pi < \theta \leq \pi.$$

This choice of the argument of a complex number $z \neq 0$ is called the *principal value* of the argument and mathematically it is denoted by

$$\arg(z).$$

4. Type help **plot** at the Command prompt and note the statement concerning PLOT(Y) if Y is complex:

    PLOT(Y) is equivalent to PLOT(real(Y), imag(Y)) if Y is complex.

What this means is: if Y is a complex vector, say

$$Y = [a_1 + ib_1, a_2 + ib_2, ..., a_n + ib_n],$$

then the points (a_1, b_1), (a_2, b_2), ..., (a_n, b_n) are plotted, together with line segments joining (a_i, b_i) to (a_{i+1}, b_{i+1}), $i = 1, ..., n-1$. Modify the program in #2 as follows:

```
N = input('enter a positive integer N: ');
z = input('enter a complex number z: ');
power = (z*ones(1,N+1)).^(1:N+1);
axis('square')
plot(power)
```

Run it for $N = 4$, $z = i$ and explain the resulting graphic. Use Help to read about **axis**.

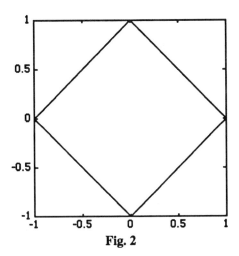

Fig. 2

5. Run the program in #4 for N = 12 and z = cos(π/6) + i sin(π/6). Use De Moivre's Theorem (Theorem 4) to explain the output.

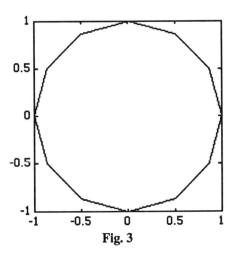

Fig. 3

6. Use the program in #4 to draw a regular pentagon (5 sides).

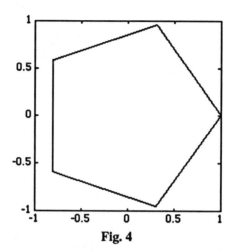

Fig. 4

7. Type help **polar** and read the entry for polar. Thus, to graph a set of complex numbers whose polar coordinates are (t, r), do the following:

- define the vector of t values
- define the corresponding vector of r values
- invoke polar(t, r)

For example, to graph a semicircle of radius 1 we can use the following script:

```
t = 0: .05 * pi : pi;
r = ones(t);
polar(t, r)
```

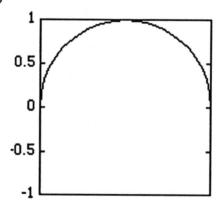

Fig. 5

(a) Modify the preceding script to obtain a polar graph of the unit circle.

(b) Graph the polar equation r = sin (2θ) for 0 ≤ θ ≤ 2π.

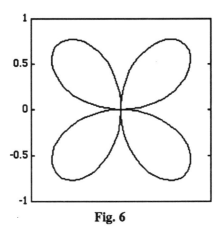

Fig. 6

(c) Write a script to obtain a polar graph of a straight line through the origin whose slope you input. Use Help to read the entry for **atan**, the arctangent function.

Hint:
    m = input ('enter the slope. ')
    th = atan(m);
    r = -1 : .05 : 1;
    t = th * ones(r);
    polar (t, r)

Use Help to read the entries for **axis** and **hold**. Experiment with hold in order to retain several lines at once on the graph.

8. The name usually used for a program in MatLab is **M-file**. There are two kinds of M-files: **Scripts** and **Functions**. M-files are sequences of MatLab commands stored in files. An M-file may include references to other M-files or to itself (a recursive reference). One type of M-file is a script file, or command file; it is a sequence of commands which are performed in order. Another type of M-file is a

function file. A function file is analogous to a Pascal function; it returns one or more values. A function file may be used in the same way as the built-in functions are used. A function file must contain the reserved word **function** as the first word in the file. The word function must be on the first line of the function file. A function file differs from a script file in two ways: arguments may be passed to the function, and a value is returned as the value of the function; variables defined in a function file are local; i.e., the variables cannot be accessed from outside the function.

To define a function and use it on the command line or in a program, do the following:

- Choose New from the File menu to bring up a new Edit window.

- Type in the following (for example):

    function y = cube(x)
    % for a scalar, cube(x) returns the value of x ^ 3
    % for a vector, cube(x) returns a vector with each entry cubed
    % for a matrix, cube(x) returns a matrix with each entry cubed
    % remember that '%' is used for comments in MatLab
    [m,n] = size(x);
    if m == 1 & n == 1   %use Help to read about == and &
        y = x * x * x;
    else
        y = x .* x .* x;   %use Help to read about if
    end

- Choose Save from the File menu.

- You will see a dialog box asking you to name the file. The default name will be cube. Click Save. (If you had wanted to change the name, you could have simply typed another name in the dialog box.)

- Now close the Edit window by clicking in the close box (in the upper left-hand corner), or by choosing Close from the File menu.

- At this point, the Command window should be active. If not, click anywhere in the window to make it active.

2.1 MatLab

- At the prompt in the Command window (»), type in A = [1 2 3; 4 5 6; 7 8 9], then press Return.

- Next, type in Z = cube(A) and press Return. The value of Z should be the 3-square matrix each of whose entries is the cube of the corresponding entry of A. If you now type help cube at the prompt (») in the Command window the comments in the program will be exhibited. Also, if you open the About MacII-MatLab window from the Apple menu, highlight the entry MatLab 1.2(c), and click the Help button, you can find cube in the listing. Simply highlight it and click the Help button to see the comments. Thus you can define and have available your own functions.

To use this function in a program:

- Choose New from the File menu to get a new Edit window.

- Type in the following program:

    ```
    % This program will ask the user for n, the size of an n x n matrix.
    n = input ('Enter n: ');
    A = rand(n)
    B = cube(A)
    ```

- Choose Save And Go from the File menu. When you see the dialog box, type in the name cubeit and click Save. The program will then run. At the prompt, type in 5. You will then see a 5 × 5 matrix A with randomly generated elements, and a matrix with each entry the cube of the corresponding entry in A.

- Type help **function** at the Command prompt (») and read the entry for function.

9. Write a function named quadr that yields the value z .^ 2 + c when evaluated for the complex vector z and the complex number c.

Hint:

```
function w = quadr(z, c)
% QUADR is a quadratic function that
% produces the value z .^ 2 + c
% when passed the complex vector z
% and a complex number c.
w = z .^ 2 + c;
```

10. Let q momentarily denote the function quadr in #9. Write a MatLab script that:

  - calls for complex numbers z and c
  - calls for a positive integer n
  - prints out the values

  $$z_0 = z, \quad z_1 = q(z_0), \quad z_2 = q(z_1), \cdots, \quad z_n = q(z_{n-1})$$

  as in the following typical run:

  enter a complex number z: .5 + .2 * i

  enter a complex number c: -1

  enter a positive integer n: 10

  0.5000 + 0.2000i
  -0.7900 + 0.2000i
  -0.4159 - 0.3160i
  -0.9269 + 0.2628i
  -0.2100 - 0.4873i
  -1.1933 + 0.2046i
  0.3822 - 0.4884i
  -1.0925 - 0.3733i
  0.0541 + 0.8156i
  -1.6623 + 0.0883i
  1.7553 - 0.2935i

Hint:

```
z = input ('enter a complex number z: ');
c = input ('enter a complex number c: ');
n = input ('enter a positive integer n: ');
```

2.1 MatLab

```
        orb = [z];
        for k = 1 : n
            z = quadr(z, c);
            orb = [orb  z];
        end
        disp (' ')
        disp (orb.')
```

11. If f is any complex valued function of a complex variable z, then the *orbit of* z_0 *under* f is the set of complex numbers

$$z_0, \quad z_1 = f(z_0), \quad z_2 = f(z_1), \quad z_3 = f(z_3), \quad \cdots, \quad z_n = f(z_{n-1}), \quad \cdots$$

Such a set is called a *Julia set*, named after the French mathematician Gaston Julia. Julia studied these sets in the 1920's. Julia sets are closely related to sets of complex numbers called *fractals*.

Review the entry for plot by typing help plot at the prompt (») in the Command window. Then study the following script called julia that plots the Julia set for the function $f_c(z) = z^2 + c =$ quadr(z, c) (see #9, #10). The user is prompted to enter the value of c. Then:

- The program computes the first 21 points of the orbits for each of the points $z_0 = x_0 + iy_0$, $-2 \leq x_0 \leq 2$, $-2 \leq y_0 \leq 2$. The points z_0 are defined by the grids $x_0 = -2: .05: 2$ and $y_0 = -2: .05: 2$. For each point in the orbit of z_0 under f_c the program checks whether the point lies outside the circle of radius 2 centered at the origin.

- If any of the first 21 points in the orbit of z_0 under f_c lies outside the circle of radius 2, z_0 is ignored.

- If all 21 points of the orbit of z_0 under f_c are within the circle of radius 2 centered at the origin, then z_0 is plotted.

```
        c = input ('enter a complex number c = c1 + i * c2  ');
        jul = [ ];
```

```
for x = -2: .05: 2
    for y = -2:.05:2
        z = x + i * y;
        b = z;
        n = 1;
        while ((n <= 20) & (abs(b) <= 2) )
            b = quadr(b, c);
            n = n + 1;
        end
        if n == 21
            jul = [jul  z];
        end
    end
end
axis('square')
plot(jul, '+')
```

(a) Run julia for c = 0. (Depending on your computer, this can take as long as 10 minutes).

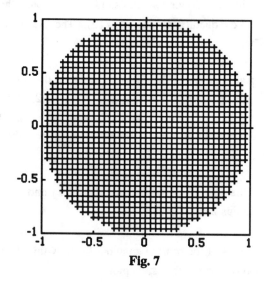

Fig. 7

Explain why the Julia set for c = 0 appears to be the unit disk.

(b) Run julia for c = -1. To compute the length of time that the program

2.1 MatLab

runs, read the Help entries for **clock** and **etime**. Then make the changes necessary in julia to compute the elapsed time.

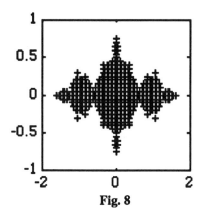

Fig. 8

(c)   Run julia for c = − 0.1 + 0.8i

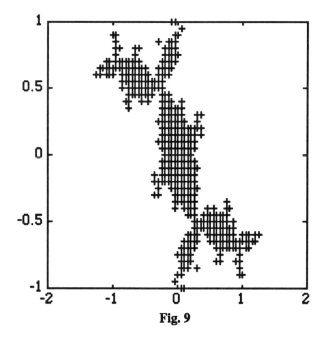

Fig. 9

After the program has run, choose any one point in the set jul (you can examine jul by typing jul at the Command prompt (»)). One such point is z = 1.25 - .6i, but choose any you wish. Run julia for this value of c. Observe that the resulting Julia set appears the same.

2.1 Glossary

```
abs  . . . . . . . . . . . . . . . . . . . . .  64
absolute value . . . . . . . . . . . . . . . .  55
angle  . . . . . . . . . . . . . . . . . . . .  65
arg  . . . . . . . . . . . . . . . . . . . . .  66
Argand diagram . . . . . . . . . . . . . . . .  53
argument . . . . . . . . . . . . . . . . . . .  56
axis . . . . . . . . . . . . . . . . . . . . .  66
ℂ  . . . . . . . . . . . . . . . . . . . . . .  53
clock  . . . . . . . . . . . . . . . . . . . .  75
complex addition . . . . . . . . . . . . . . .  52
complex conjugate  . . . . . . . . . . . . . .  56
complex multiplication . . . . . . . . . . . .  52
De Moivre's Theorem  . . . . . . . . . . . . .  59
etime  . . . . . . . . . . . . . . . . . . . .  75
fractal . . . . . . . . . . . . . . . . . . . . 73
function . . . . . . . . . . . . . . . . . . .  69
hold . . . . . . . . . . . . . . . . . . . . .  69
i  . . . . . . . . . . . . . . . . . . . . . .  54
Im . . . . . . . . . . . . . . . . . . . . . .  56
imaginary part . . . . . . . . . . . . . . . .  56
imaginary unit . . . . . . . . . . . . . . . .  54
Julia set  . . . . . . . . . . . . . . . . . .  73
M-file . . . . . . . . . . . . . . . . . . . .  69
modulus  . . . . . . . . . . . . . . . . . . .  55
orbit  . . . . . . . . . . . . . . . . . . . .  73
plot . . . . . . . . . . . . . . . . . . . . .  66
polar  . . . . . . . . . . . . . . . . . . . .  68
polar form . . . . . . . . . . . . . . . . . .  56
principal value  . . . . . . . . . . . . . . .  66
```

ℝ	51
Re	56
real numbers	51
real part	56
script	69
triangle inequality	58
$\bar{z}$	56

2.2 Elementary Complex Functions

The reader will recall that the quadratic formula for the roots of a quadratic equation

$$ax^2 + bx + c = 0, \quad a \neq 0 \tag{1}$$

can be derived as follows:

- divide through the equation by a to obtain

$$x^2 + \frac{b}{a}x + \frac{c}{a} = 0 \tag{2}$$

- move $\frac{c}{a}$ to the right side of the equation and add $\frac{b^2}{4a^2}$ to both sides of (2),

$$x^2 + \frac{b}{a}x + \frac{b^2}{4a^2} = \frac{b^2 - 4ac}{4a^2} \tag{3}$$

- the left side of (3) is a perfect square so that

$$\left(x + \frac{b}{2a}\right)^2 = \frac{b^2 - 4ac}{4a^2}. \tag{4}$$

At this point, if we can tabulate every complex number α whose square satisfies

$$\alpha^2 = \frac{b^2 - 4ac}{4a^2} \tag{5}$$

then every root of (4), and hence of (1), must have the form

$$x = -\frac{b}{2a} + \alpha.$$

Hence, the problem of evaluating roots, fractional powers, and more general functions of a complex variable is an important one to consider. We begin with a simple example of complex roots.

Let $w_1 = -1 + i\sqrt{3}$ and $w_2 = -1 - i\sqrt{3}$. Putting w_1 and w_2 in polar form we have

$$|w_1| = |w_2| = 2,$$

and

$$w_1 = 2\left(\cos\frac{2\pi}{3} + i\sin\frac{2\pi}{3}\right), \quad w_2 = 2\left(\cos\frac{4\pi}{3} + i\sin\frac{4\pi}{3}\right).$$

Then, by De Moivre's Theorem (§2.1 Theorem 4) with t = 3, we have

$$w_1^3 = 2^3 \left(\cos 3\left(\frac{2\pi}{3}\right) + i\sin 3\left(\frac{2\pi}{3}\right)\right)$$

$$= 8 \left(\cos 2\pi + i\sin 2\pi\right)$$

$$= 8,$$

and similarly

$$w_2^3 = 2^3 \left(\cos 3\left(\frac{4\pi}{3}\right) + i\sin 3\left(\frac{4\pi}{3}\right)\right)$$

$$= 8 \left(\cos 4\pi + i\sin 4\pi\right)$$

$$= 8.$$

We also know that if $w_3 = 2 = 2(\cos 0 + i\sin 0)$ then $w_3^3 = 8$. Thus we see that there are at least three distinct complex numbers whose cubes are equal to 8. This motivates the following definition for the *roots of a complex number*:

2.2 Elementary Complex Functions

Let q be a positive integer. If z is a given complex number and w is a complex number such that $w^q = z$, then w is called a qth root of z. We shall denote any q^{th} root of z by $z^{1/q}$.

Theorem 1.

If q is a positive integer and z is a given nonzero complex number in polar form,

$$z = r(\cos\theta + i\sin\theta),$$

then there are precisely q different complex numbers that are q^{th} roots of z, i.e., whose q^{th} powers equal z. These are given in polar form by

$$w_k = r^{1/q}\left(\cos\frac{\theta + 2\pi k}{q} + i\sin\frac{\theta + 2\pi k}{q}\right), \quad k = 0, \cdots, q-1. \tag{6}$$

Proof.

First observe that by §2.1 Theorem 4,

$$w_k^q = (r^{1/q})^q \left(\cos q\left(\frac{\theta + 2\pi k}{q}\right) + i\sin q\left(\frac{\theta + 2\pi k}{q}\right)\right)$$

$$= r(\cos(\theta + 2\pi k) + i\sin(\theta + 2\pi k))$$

$$= r(\cos\theta + i\sin\theta)$$

$$= z.$$

Note that the q numbers (6) are equally spaced around the circumference of a circle of radius $r^{1/q}$ centered at the origin: the first number (i.e., k = 0) has argument θ/q.

Now suppose $w = \rho(\cos\varphi + i\sin\varphi)$ is a qth root of z. Then

$$w^q = \rho^q(\cos q\varphi + i \sin q\varphi)$$

$$= z$$

$$= r(\cos\theta + i\sin\theta).$$

Since $w^q = z$ we have $\rho^q = |w|^q = |z| = r$ and hence $\rho = r^{1/q}$. Also,

$$\cos(q\varphi) = \cos\theta, \qquad (7)$$

and

$$\sin(q\varphi) = \sin\theta. \qquad (8)$$

The equations (7) and (8) can hold if and only if $q\varphi$ and θ differ by an integral multiple of 2π:

$$q\varphi = \theta + 2\pi k,$$

$$\varphi = \frac{\theta + 2\pi k}{q}. \qquad (9)$$

The angles $\varphi_k = \frac{\theta + 2\pi k}{q}$, $k = 0, \cdots, q-1$, are distinct, and if k is any positive integer not in the set $\{0, \ldots, q-1\}$ then the corresponding value of φ in (9) will differ from one of the φ_k by a multiple of 2π. Also, if u and v are nonnegative integers,

$$0 \le u \le v \le q-1,$$

and

$$\cos\varphi_u = \cos\varphi_v,$$
$$\sin\varphi_u = \sin\varphi_v,$$

then again we know that φ_u and φ_v must differ by an integral multiple of 2π, say $|\varphi_v - \varphi_u| = h2\pi$. But, if $u < v$ then

2.2 Elementary Complex Functions

$$h2\pi = |\varphi_v - \varphi_u|$$

$$= \left|\frac{\theta + 2\pi v}{q} - \frac{\theta + 2\pi u}{q}\right|$$

$$= \frac{2\pi(v-u)}{q}$$

$$< 2\pi.$$

It follows that $h = 0$ and hence that $u = v$. We have confirmed that the q complex numbers on the right in (6) are all the q distinct q^{th} roots of z. ∎

For example, it is simple to find all the nth roots of 1, that is, all complex numbers z such that $z^n = 1$, n a positive integer. According to Theorem 1, the distinct n^{th} roots of 1 are the n complex numbers

$$\cos\left(\frac{2k\pi}{n}\right) + i\sin\left(\frac{2k\pi}{n}\right), \quad k = 0, 1, \ldots, n-1.$$

Note that these numbers lie on the unit circle and that their arguments are $0, 2\pi/n, 4\pi/n, \ldots, 2(n-1)\pi/n$. But these are the vertices of the n-sided regular polygon inscribed in the unit circle with one vertex at $(1, 0)$. For example, the 12th roots of 1 are situated as shown in Fig. 1.

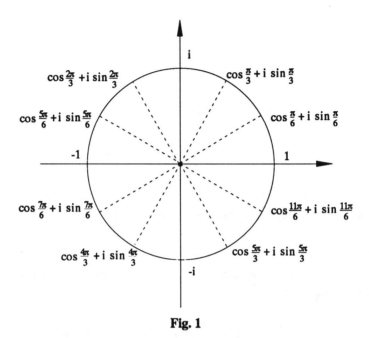

Fig. 1

It is now relatively simple to define what is meant by a *rational power* of a complex number. If p and q are integers, $q > 0$, then by $z^{p/q}$ is meant the pth power of any of the qth roots of z. Any one of these numbers $z^{p/q}$ is called a p/q power of z. We can now give a more complete version of De Moivre's Theorem.

Theorem 2.

If p and q are integers, $q > 0$, and $z = r(\cos\theta + i\sin\theta)$ is a nonzero complex number in polar form, then all of the p/q powers of z are given by

$$r^{p/q}\left(\cos p\left(\frac{\theta + 2\pi k}{q}\right) + i \sin p\left(\frac{\theta + 2\pi k}{q}\right)\right), \quad k = 0, \ldots, q-1. \tag{10}$$

Proof.

By Theorem 1, the numbers $z^{1/q}$ are given by

$$r^{1/q}\left(\cos\left(\frac{\theta + 2\pi k}{q}\right) + i \sin\left(\frac{\theta + 2\pi k}{q}\right)\right), \quad k = 0, \ldots, q-1. \tag{11}$$

2.2 Elementary Complex Functions

We then apply §2.1 Theorem 4 to (11) to immediately obtain the result (10). ∎

As an example, we compute $i^{2/3}$. First note that $i = \cos\frac{\pi}{2} + i\sin\frac{\pi}{2}$ and hence $i^{2/3}$ is any of the three numbers

$$\cos\frac{2}{3}\left(\frac{\pi}{2} + 2\pi k\right) + i\sin\frac{2}{3}\left(\frac{\pi}{2} + 2\pi k\right), \quad k = 0, 1, 2.$$

These are

$$\cos\frac{\pi}{3} + i\sin\frac{\pi}{3} = \frac{1}{2} + i\frac{\sqrt{3}}{2},$$

$$\cos\frac{5\pi}{3} + i\sin\frac{5\pi}{3} = \frac{1}{2} - i\frac{\sqrt{3}}{2},$$

and

$$\cos 3\pi + i\sin 3\pi = -1.$$

If q is a positive integer then a qth root of 1 is usually called a qth *root of unity*. There is a relationship between roots of unity as explained in the following result.

Theorem 3.

If $p > q$ are two positive integers then the number of qth roots of unity that are also p^{th} roots of unity is $d = \gcd(p, q)$.

Proof.

Recall that the *greatest common divisor*, i.e., gcd, of two integers is the largest positive integer that divides both of them. The arguments for the pth roots and the qth roots of 1 are, respectively,

$$\frac{k}{p} 2\pi, \quad k = 1, \cdots, p,$$

$$\frac{s}{q} 2\pi, \quad s = 1, \cdots, q.$$

So the question becomes: how many numbers do the two lists:

$$\frac{k}{p}, \quad k = 1, \cdots, p$$

and

$$\frac{s}{q}, \quad s = 1, \cdots, q$$

have in common? If we multiply both lists by pq the question is: how many integers do the lists

$$qk, \quad k = 1, \ldots, p$$

and

$$ps, \quad s = 1, \ldots, q$$

have in common? Now if $q = q_1 d$ and $p = p_1 d$, where $d = \gcd(p, q)$ (so that q_1 and p_1 have no factors in common), then the problem is: how many integers do the two lists

$$q_1 k, \quad k = 1, \cdots, p \tag{12}$$

$$p_1 s, \quad s = 1, \cdots, q \tag{13}$$

have in common? Suppose then that s is such that there is a k for which

$$q_1 k = p_1 s$$

(i.e., the lists (12) and (13) have a common number). Then $q_1 | p_1 s$ (i.e., q_1 divides $p_1 s$). But q_1 and p_1 have no factors in common so $q_1 | s$. In other words, $s = t\, q_1$, but then

$$q_1 k = p_1 s = p_1 t\, q_1,$$

so

2.2 Elementary Complex Functions

$$k = t\, p_1.$$

Now, s is at most q, so t could only be one of the numbers $t = 1, \ldots, d$ (i.e., $dq_1 = q$). We have established that the two lists

$$q_1 k,\ k = 1, \ldots, p$$

and

$$p_1 s,\ s = 1, \ldots, q$$

can have at most d numbers in common and that these common numbers are obtained by taking

$$s = t\, q_1,\ k = t\, p_1,\ t = 1, \ldots, d.$$

But with these choices of k and s it is indeed true that

$$q_1 k = q_1 t\, p_1$$

and

$$p_1 s = p_1 t\, q_1$$

are equal. Hence the two lists have precisely $d = \gcd(p, q)$ numbers in common. ■

All the standard mathematical operations and elementary functions are implemented in MatLab for complex numbers. In order for us to know what to expect in computations involving complex valued functions of a complex variable we must briefly discuss precisely how these are defined theoretically and how they are implemented in MatLab. The functions in question are: **abs, exp, log, log10, sqrt, sin, cos, tan, asin, acos, atan, atan2, sinh, cosh, tanh, asinh, acosh, atanh.**

We assume that the reader has encountered the Taylor series expansion of e^x in a beginning calculus course:

$$e^x = 1 + x + \frac{x^2}{2!} + \cdots$$
$$= \sum_{k=0}^{\infty} \frac{x^k}{k!}. \tag{14}$$

The definition (14) also makes sense for complex z and thus our starting point is the definition of $e^z = \exp(z)$ as

$$e^z = \sum_{k=0}^{\infty} \frac{z^k}{k!}. \tag{15}$$

If z and w are any two complex numbers then

$$e^{z+w} = e^z \cdot e^w. \tag{16}$$

The formula (16) is easily verified:

$$e^z \cdot e^w = \sum_{k=0}^{\infty} \frac{z^k}{k!} \sum_{k=0}^{\infty} \frac{w^k}{k!}$$
$$= \sum_{k=0}^{\infty} \left(\sum_{p+q=k} \frac{z^p}{p!} \frac{w^q}{q!} \right)$$
$$= \sum_{k=0}^{\infty} \frac{1}{k!} \sum_{p=0}^{k} \frac{k!}{p!(k-p)!} z^p w^{k-p}$$
$$= \sum_{k=0}^{\infty} \frac{1}{k!} \cdot (z+w)^k$$
$$= e^{z+w}.$$

The functions sin z, cos z, tan z, sec z, csc z, cot z are defined in terms of e^z by the formulas:

2.2 Elementary Complex Functions

$$\sin z = \frac{e^{iz} - e^{-iz}}{2i}, \tag{17}$$

$$\cos z = \frac{e^{iz} + e^{-iz}}{2}, \tag{18}$$

$$\tan z = \frac{\sin z}{\cos z}, \tag{19}$$

$$\sec z = \frac{1}{\cos z}, \tag{20}$$

$$\csc z = \frac{1}{\sin z}, \tag{21}$$

$$\cot z = \frac{1}{\tan z}. \tag{22}$$

If we replace z by the real number y in (17) and (18), multiply (17) by i, and then add the two formulas we obtain

$$\cos y + i \sin y = e^{iy}. \tag{23}$$

Hence from (16) (with x and y real)

$$e^{x + iy} = e^x \cdot e^{iy}$$
$$= e^x (\cos y + i \sin y) \tag{24}$$

which in turn implies that

$$\left| e^{x + iy} \right| = e^x. \tag{25}$$

To define the complex logarithm we begin by writing $w = a + ib$ and

$$z = e^w$$
$$= e^{a + ib}$$
$$= e^a e^{ib}. \tag{26}$$

If the polar form of z is $z = r (\cos \theta + i \sin \theta)$ then from (26),

$$r = e^a$$

and

$$\theta = b + k\, 2\pi,$$

where k is any integer. Thus

$$a = \log r,$$

and to specify b uniquely we take k to be 0 and θ to be the principal value of the argument of z, i.e., $-\pi < \theta \leq \pi$. Thus $w = \log z$ has the value

$$\log z = \log r + i\theta \tag{27}$$

and we define the *principal value* of $\log z$ to be (27) in which $z = r\,e^{i\theta}$, $-\pi < \theta \leq \pi$, and log is the natural logarithm, i.e., the base is e. Some examples are:

$$\log i = \frac{i\pi}{2},$$
$$\log (1 + \sqrt{3}\, i) = \log 2 + i\frac{\pi}{3},$$
$$\log (-1) = i\pi,$$
$$\log (1 + i) = \frac{\log 2}{2} + i\frac{\pi}{4}.$$

The hyperbolic functions are defined as follows:

$$\sinh z = \frac{e^z - e^{-z}}{2}, \tag{28}$$

$$\cosh z = \frac{e^z + e^{-z}}{2}, \tag{29}$$

$$\tanh z = \frac{\sinh z}{\cosh z}, \tag{30}$$

$$\operatorname{sech} z = \frac{1}{\cosh z}, \tag{31}$$

2.2 Elementary Complex Functions

$$\operatorname{csch} z = \frac{1}{\sinh z}, \tag{32}$$

$$\operatorname{coth} z = \frac{1}{\tanh z}. \tag{33}$$

There are a large number of identities, analogous to those in ordinary trigonometry, that are easily verified directly from the definitions:

$$\cos^2 z + \sin^2 z = 1, \tag{34}$$

$$\sin 2z = 2 \sin z \cos z, \tag{35}$$

$$\cos 2z = \cos^2 z - \sin^2 z, \tag{36}$$

$$\sin(z \pm w) = \sin z \cos w \pm \cos z \sin w, \tag{37}$$

$$\sin(-z) = -\sin z, \tag{38}$$

$$\cos(-z) = \cos z, \tag{39}$$

$$\cosh(-z) = \cosh z, \tag{40}$$

$$\sinh(-z) = -\sinh z, \tag{41}$$

$$\cosh z + \sinh z = e^z, \tag{42}$$

$$\cosh z - \sinh z = e^{-z}, \tag{43}$$

$$\cos(iz) = \cosh z, \tag{44}$$

$$\cosh(iz) = \cos z, \tag{45}$$

$$\sin(iz) = i \sinh z, \tag{46}$$

$$\sinh(iz) = i \sin z, \tag{47}$$

$$\cosh^2 z - \sinh^2 z = 1, \tag{48}$$

$$\cosh^2 z + \sinh^2 z = \cosh 2z. \tag{49}$$

Remember that for the nonzero complex number $z = x + iy$, the principal value of the argument function, arg z, is the unique θ such that $\sin \theta = y$, $\cos \theta = x$ and $-\pi < \theta \leq \pi$. Moreover, if z is written in polar form, $z = re^{i\theta}$, then $\log z = \log(re^{i\theta}) = \log r + i\theta$. If θ = arg z, i.e., θ is the principal value of the argument function, then any of the numbers

$$w = \log r + i(\theta \pm 2k\pi), \quad k = 0, 1, 2, \ldots$$

satisfies $e^w = z$. We defined the principal value of $\log z$ as

$$\log z = \log r + i\theta. \tag{50}$$

If c is any complex number then the *principal value of z^c* is defined by

$$z^c = \exp(c \log z) \tag{51}$$

where $\log z$ is the principal value of the logarithm of z. For example, we can compute i^i as follows:

$$\begin{align*}
i^i &= \exp(i \log i) & \text{(from (51))} \\
&= \exp(i \, (i \tfrac{\pi}{2})) & \text{(from (50))} \\
&= \exp(-\tfrac{\pi}{2}) \\
&= e^{-\pi/2}.
\end{align*}$$

As another example, consider $z^{1/p}$ where p is a positive integer:

2.2 Elementary Complex Functions

$$z^{1/p} = \exp(\frac{1}{p} \log z)$$

$$= \exp(\frac{1}{p}(\log r + i\theta))$$

$$= e^{(\log r)/p} e^{i\theta/p}$$

$$= r^{1/p} (\cos \frac{\theta}{p} + i \sin \frac{\theta}{p}).$$

This coincides with one of the values of the pth roots of z produced by Theorem 1.

To obtain formulas for evaluating the inverse trigonometric functions we begin by setting (see (17))

$$z = \sin w = \frac{e^{iw} - e^{-iw}}{2i}. \tag{52}$$

Hence

$$e^{2iw} - 2ize^{iw} - 1 = 0,$$

which is a quadratic equation for the determination of e^{iw}. Then

$$e^{iw} = iz + (1 - z^2)^{1/2}$$

and hence

$$w = -i \log(iz + (1 - z^2)^{1/2}). \tag{53}$$

Since $z = \sin w$ we write

$$\sin^{-1} z = -i \log(iz + (1 - z^2)^{1/2}). \tag{54}$$

Similarly, from (18) we have

$$\cos^{-1} z = -i \log(z + (z^2 - 1)^{1/2}), \tag{55}$$

and from (19),

$$\tan^{-1} z = \frac{i}{2} \log\left(\frac{i+z}{i-z}\right). \tag{56}$$

In formulas (54) - (56) the principal values of the square root and the logarithm are used to specify uniquely the values of these functions. As a matter of notation we have used the usual mathematical designations for the exponential, hyperbolic, and trigonometric functions. The following is the notational correspondence between mathematical and MatLab notations:

Mathematical	MatLab
sin	sin
cos	cos
tan	tan
$\sin^{-1}$	asin
$\cos^{-1}$	acos
$\tan^{-1}$	atan
sinh	sinh
cosh	cosh
tanh	tanh
$\sinh^{-1}$	asinh
$\cosh^{-1}$	acosh
$\tanh^{-1}$	atanh

The MatLab functions compute the functions asin, acos, atan in conformity with the principal values of the formulas in (54), (55), and (56), respectively.

As an example of the use of (55), we confirm that $\cos^{-1}(-1/2)$ is $2\pi/3$, as we know from elementary trigonometry:

2.2 Elementary Complex Functions

$$- i \log(-1/2 + ((-1/2)^2 - 1)^{1/2}) = -i \log(-1/2 + i\sqrt{3}/2)$$

$$= -i \log(e^{i2\pi/3})$$

$$= -i (i\, 2\pi/3)$$

$$= 2\pi/3 \ .$$

According to the MatLab documentation, for complex $z = x + iy$ MatLab uses:

$$\sin(z) = \sin(x) \cosh(y) + i \cos(x) \sinh(y) \tag{57}$$

$$\cos(z) = \cos(x) \cosh(y) - i \sin(x) \sinh(y) \tag{58}$$

$$\tan(z) = \frac{\sin(z)}{\cos(z)} \ . \tag{59}$$

These formulas are easily confirmed from the definitions of the indicated functions. The MatLab function **atan2**(y, x) is precisely the same as angle(x + iy). The function atan2(y, x) is called the *four-quadrant arctangent function*.

There are several important classes of matrices with complex number entries that arise in applications. If the entries of an m × n matrix are complex numbers, we can write

$$A = [\, a_{pq} \,] = [\, c_{pq} + i d_{pq} \,],$$

where c_{pq} and d_{pq} are real numbers. The *conjugate* of A is the matrix

$$\overline{A} = [\, \overline{a}_{pq} \,] = [\, c_{pq} - i d_{pq} \,], \tag{60}$$

and the *conjugate transpose* of A is the matrix

$$A^* = \overline{A}^T . \tag{61}$$

For example, the conjugate transpose of

$$D = \begin{bmatrix} \sqrt{2} & \pi & -1 & i \\ 3 & 7 & 0 & e \\ \frac{1}{2} & -1 & 1+3i & 2 \end{bmatrix}$$

is

$$D^* = \begin{bmatrix} \sqrt{2} & 3 & \frac{1}{2} \\ \pi & 7 & -1 \\ -1 & 0 & 1-3i \\ -i & e & 2 \end{bmatrix}.$$

The general formulas

$$(A^*)^* = A, \quad \overline{\overline{A}} = A, \quad A^* = \overline{A^T}, \tag{62}$$

$$(A^*)_{(p)} = (A^{(p)})^*, \quad (A^*)^{(q)} = (A_{(q)})^*$$

are easy to confirm.

If $A = [\, a_{ij}\,]$ is a square matrix and

$$A = A^T,$$

then A is *symmetric*; we emphasize that A may have complex number entries. If

$$A = A^*, \tag{63}$$

then A is *hermitian*. If

$$A^T = -A, \tag{64}$$

2.2 Elementary Complex Functions

then A is *skew-symmetric*. If

$$A^* = -A, \qquad (65)$$

then A is *skew-hermitian*. For example, the matrices

$$\begin{bmatrix} 1 & 1+i \\ 1+i & 2 \end{bmatrix}, \quad \begin{bmatrix} 0 & 1+i \\ 1-i & 2 \end{bmatrix}, \quad \begin{bmatrix} 0 & 1 \\ -1 & 0 \end{bmatrix}, \quad \begin{bmatrix} 0 & 1+i \\ -1+i & i \end{bmatrix}$$

are, respectively, symmetric, hermitian, skew-symmetric, and skew-hermitian.

An n-square matrix A is *orthogonal* if

$$AA^T = A^T A = I_n . \qquad (66)$$

If

$$AA^* = A^* A = I_n \qquad (67)$$

then A is *unitary*. If A is n-square and

$$AA^* = A^* A \qquad (68)$$

then A is a *normal* matrix. Note that hermitian, skew-hermitian and unitary matrices are all normal. However, $A = (1+i) I_n$ is certainly normal but it is not hermitian (why?), nor skew-hermitian (why?), nor unitary (why?).

If u and v are complex column n-tuples, say $u = [u_1 \ldots u_n]^T$ and $v = [v_1 \ldots v_n]^T$, then

$$v^* u = \sum_{k=1}^n u_k \bar{v}_k . \qquad (69)$$

Actually v*u is a 1 × 1 matrix, but a 1 × 1 matrix and its single entry are never distinguished. Thus, with any pair of column vectors u and v in $M_{n,1}(\mathbb{C})$ we can associate the complex number v*u. The value v*u is called the *standard inner product* of the two vectors u and v. Observe that if u, v, and w are in $M_{n,1}(\mathbb{C})$ and α and β are complex numbers then:

$$\overline{v^*u} = u^*v \tag{70}$$

$$v^*(\alpha u + \beta w) = \alpha\, v^*u + \beta\, v^*w \tag{71}$$

$$(\alpha u + \beta w)^*v = \overline{\alpha}\, u^*v + \overline{\beta}\, w^*v \tag{72}$$

$$u^*u \geq 0 \text{ and } u^*u = 0 \text{ if and only if } u = 0. \tag{73}$$

These properties are known as the *conjugate-symmetric property* ((70)), the *conjugate bilinear property* ((71) and (72)), and the *positive-definite property* ((73)). These properties are all very simple to establish. For example,

$$u^*u = \sum_{k=1}^{n} u_k \overline{u_k}$$
$$= \sum_{k=1}^{n} |u_k|^2$$
$$\geq 0$$

and obviously $u^*u = 0$ can hold if and only if $u_1 = \cdots = u_n = 0$. The non-negative number

$$(u^*u)^{1/2}$$

is called the *norm* of u and is denoted by

$$\|u\|. \tag{74}$$

The inner product has its origins in the analytic geometry of the plane. Suppose that $n = 2$ and we think of $u = [\,u_1\ u_2\,]^T$ and $v = [\,v_1\ v_2\,]^T$ as vectors of length 1 in the plane:

2.2 Elementary Complex Functions

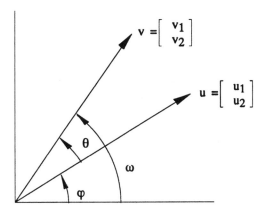

Fig. 2

As we know from elementary trigonometry,

$$\cos \theta = u_1 v_1 + u_2 v_2.$$

To prove this last formula, recall that if φ and ω, respectively, denote the angles that u and v make with the positive horizontal axis then

$$\cos \varphi = \frac{u_1}{\sqrt{u_1^2 + u_2^2}}, \quad \sin \varphi = \frac{u_2}{\sqrt{u_1^2 + u_2^2}},$$

$$\cos \omega = \frac{v_1}{\sqrt{v_1^2 + v_2^2}}, \quad \sin \omega = \frac{v_2}{\sqrt{v_1^2 + v_2^2}}.$$

But u and v are unit vectors so that $u_1^2 + u_2^2 = v_1^2 + v_2^2 = 1$. Now $\theta = \omega - \varphi$ and using the formula for $\cos(\omega - \varphi)$ we have

$$\cos \theta = \cos \omega \cos \varphi + \sin \omega \sin \varphi = u_1 v_1 + u_2 v_2.$$

In general, if u and v are arbitrary nonzero vectors, not necessarily of unit length, then

$$\cos \theta = \frac{u_1 v_1 + u_2 v_2}{\sqrt{u_1^2 + u_2^2} \sqrt{v_1^2 + v_2^2}}. \tag{75}$$

The numerator of the fraction (75) is v^*u. So we can rewrite (75) as

$$\cos\theta = \frac{v^*u}{\|v\|\,\|u\|}. \tag{76}$$

We conclude with the following important result that shows how hermitian matrices can be used to construct any n-square complex matrix.

Theorem 4.

Let A be an n-square complex matrix.

(a) Then

$$A = H + iK \tag{77}$$

where H and K are uniquely determined hermitian matrices.

(b) If $u^*Au = 0$ for all $u \in M_{n,1}(\mathbb{C})$ then $A = 0$.

Proof.

(a) Note the following equality:

$$A = \frac{A+A^*}{2} + i\left(\frac{A-A^*}{2i}\right). \tag{78}$$

Moreover, $H = \frac{A+A^*}{2}$ and $K = \frac{A-A^*}{2i}$ are both hermitian (why?). On the other hand, if H and K are any two hermitian matrices for which (77) holds, then

$$A^* = H^* - iK^*$$
$$= H - iK. \tag{79}$$

If we first add (79) to (77) and then subtract (79) from (77) we see that

$$H = \frac{A+A^*}{2}$$

2.2 Elementary Complex Functions

and

$$K = \frac{A - A^*}{2i}.$$

(b) Let u and v be in $M_{n,1}(\mathbb{C})$. Then by assumption,

$$(u + v)^* A (u + v) = 0,$$

$$u^* A u + v^* A v + u^* A v + v^* A u = 0,$$

$$u^* A v + v^* A u = 0. \tag{80}$$

Now compute that

$$(u + iv)^* A (u + iv) = 0,$$

$$u^* A u + v^* A v + i u^* A v - i v^* A u = 0,$$

$$u^* A v - v^* A u = 0. \tag{81}$$

If we add (80) and (81), we conclude that

$$u^* A v = 0 \tag{82}$$

for all u and v. Take u to be the vector with 1 in position s and 0 elsewhere and v to be the vector with 1 in position t and 0 elsewhere. With these choices of u and v we have

$$u^* A v = a_{st}$$

and from (82), $a_{st} = 0$. Since s and t can be any integers in 1, ..., n we conclude that $A = 0$. ∎

2.2 Exercises

1. Show that $\exp(z + 2\pi i) = \exp(z)$ for all z, i.e., $\exp(z)$ has *period* $2\pi i$.

2. Show that $\log(1 - i) = \dfrac{\log 2}{2} - i\dfrac{\pi}{4}$.

3. Show that for $z = 1 + i$ the value of $\dfrac{i}{2} \log\left(\dfrac{i+z}{i-z}\right)$ is $\dfrac{\pi - \tan^{-1}2}{2} + i\dfrac{\log 5}{4}$.

    Hint: $\dfrac{i+z}{i-z} = -1 - 2i$ is in the 3rd quadrant; $|-1 - 2i| = \sqrt{5}$ and $\arg(-1 - 2i) = \tan^{-1}2 - \pi$. Then $\dfrac{i}{2}\log\left(\dfrac{i+z}{i-z}\right) = \dfrac{i}{2}\log(-1 - 2i) = \dfrac{i}{2}(\log\sqrt{5} + i(\tan^{-1}2 - \pi)) = i\dfrac{\log 5}{4} + \dfrac{(\pi - \tan^{-1}2)}{2}$.

4. Establish the identity (34).

5. Establish the identity (35).

6. Establish the identity (36).

7. Establish the identity (37).

8. Establish the identity (38).

9. Establish the identity (39).

10. Establish the identity (40).

11. Establish the identity (41).

12. Establish the identity (42).

13. Establish the identity (43).

2.2 Exercises

14. Establish the identity (44).

15. Establish the identity (45).

16. Establish the identity (46).

17. Establish the identity (47).

18. Establish the identity (48).

19. Establish the identity (49).

20. Establish the identity (57):

$$\sin z = \sin x \cosh y + i \cos x \sinh y$$

    where $z = x + iy$ and x and y are real.

    Hint: Use (37), (44) and (46).

21. Show that

$$\cos(z \pm w) = \cos(z)\cos(w) \mp \sin(z)\sin(w).$$

    Hint: From (17) and (18), for the "+" alternative we have

$$\cos(z)\cos(w) + \sin(z)\sin(w) = \frac{e^{iz} + e^{-iz}}{2} \cdot \frac{e^{iw} + e^{-iw}}{2} + \frac{e^{iz} - e^{-iz}}{2i} \cdot \frac{e^{iw} - e^{-iw}}{2i}$$

$$= \frac{e^{iz} \cdot e^{-iw} + e^{iw} \cdot e^{-iz}}{2}$$

$$= \frac{e^{i(z-w)} + e^{-i(z-w)}}{2}$$

$$= \cos(z - w).$$

Similarly for the " - " alternative.

22. Establish the identity (58):

$$\cos(z) = \cos(x)\cosh(y) - i\sin(x)\sinh(y)$$

where $z = x + iy$ and x and y are real.

Hint: Use #21, (44), (46).

23. Using the notation of #22, show that

$$|\sin(z)|^2 = \sin^2 x + \sinh^2 y$$

and

$$|\cos(z)|^2 = \cos^2 x + \sinh^2 y .$$

Hint: Use #20 to compute that

$$|\sin(z)|^2 = (\sin(x)\cosh(y))^2 + (\cos(x)\sinh(y))^2 .$$

Replace $\cosh(y)$ by $\dfrac{e^y + e^{-y}}{2}$, $\sinh(y)$ by $\dfrac{e^y - e^{-y}}{2}$, and simplify.

24. Use (56) to confirm that $\tan^{-1}(1) = \dfrac{\pi}{4}$.

Hint:

$$\tan^{-1}(1) = \frac{i}{2}\log\left(\frac{i+1}{i-1}\right)$$

$$= \frac{i}{2}\log(-i)$$

$$= \frac{i}{2}(i\arg(-i))$$

2.2 Exercises

$$= \frac{i}{2}(-i\frac{\pi}{2})$$

$$= \frac{\pi}{4} \ .$$

25. Find the value of $\log(i^{1/2})$.

    Hint: $i^{1/2} = e^{(1/2)\log i} = e^{(1/2) i\pi/2} = e^{i(\pi/4)}$. Thus $\log(i^{1/2}) = i\pi/4$.

26. Show that $(-i)^i = e^{\pi/2}$ and $(-1)^i (i^i) = e^{-3\pi/2}$. Thus, in general it is not true that $(pz)^\alpha = p^\alpha z^\alpha$ for arbitrary complex numbers α, p, and z.

    Hint: $(-i)^i = e^{i \log(-i)} = e^{i(-i\pi/2)} = e^{\pi/2}$;

    $(i^i) = e^{i \log i} = e^{i(i\pi/2)} = e^{-\pi/2}$;

    $(-1)^i = e^{i \log(-1)} = e^{i(i\pi)} = e^{-\pi}$;

    $(-1)^i (i^i) = e^{-\pi} \cdot e^{-\pi/2} = e^{-3\pi/2}$.

27. Show that if p is a nonzero complex number then

    $$(pz)^\alpha = p^\alpha z^\alpha$$

    for all z and α if and only if $p > 0$.

    Hint: $(pz)^\alpha = e^{\alpha(\log pz)} = e^{\alpha(\log|pz| + i \arg(pz))} = e^{\alpha \log|pz|} \cdot e^{i \alpha \arg(pz)}$.
    Also, $p^\alpha z^\alpha = e^{\alpha \log p} \cdot e^{\alpha \log z} = e^{\alpha(\log|p| + i \arg(p))} \cdot e^{\alpha(\log|z| + i \arg(z))}$
    $= e^{\alpha(\log|p| + \log|z|)} \cdot e^{i \alpha(\arg(p) + \arg(z))} = e^{\alpha \log|pz|} \cdot e^{i \alpha(\arg(p) + \arg(z))}$.

    Thus the equality $(pz)^\alpha = p^\alpha z^\alpha$ holds for z and α iff

    $$e^{i\alpha (\arg(pz) - (\arg(p) + \arg(z)))} = 1$$

    for any α. Thus $\arg(pz) = \arg(p) + \arg(z)$ for all z. The reader will confirm that this can happen if and only if $\arg(p) = 0$, i.e., $p > 0$.

28. Show that if $\overline{e^{iz}} = e^{-iz}$ then z is real.

   Hint: Let $z = x + iy$. Then $e^{iz} = e^{-y} e^{ix} = e^{-y} \cos x + i e^{-y} \sin x$. Also, $e^{-iz} = e^{-ix+y} = e^{y} \cos x - i e^{y} \sin x$. Thus $\overline{e^{iz}} = e^{-iz}$ if and only if $e^{-y} \cos x = e^{y} \cos x$ and $e^{y} \sin x = e^{-y} \sin x$. Since not both $\sin x$ and $\cos x$ are 0, it follows that $e^{2y} = 1$ so that $y = 0$, i.e., z is real.

29. Show that if $e^{i \alpha 2\pi} = 1$ then α is an integer.

   Hint: Write $\alpha = x + iy$, x and y real. Then $e^{i(x+iy)2\pi} = 1$ implies that $e^{ix 2\pi} e^{-y 2\pi} = 1$. Taking absolute values implies that $y = 0$ so that $e^{ix 2\pi} = 1$. Then $x 2\pi$ must be an integral multiple of 2π, i.e., $\alpha = x$ must be an integer.

30. Let $-2\pi < w < 2\pi$. Then show that

$$\arg(e^{iw}) = \begin{cases} w & \text{if } |w| < \pi \text{ or } w = \pi \\ w + 2\pi & \text{if } |w| \geq \pi \text{ and } w < 0 \\ w - 2\pi & \text{if } |w| > \pi \text{ and } w > 0 \end{cases}$$

   Hint: Recall that the principal value, $\arg(z)$, is always chosen so that $-\pi < \arg(z) \leq \pi$. If $|w| < \pi$ or $w = \pi$ then clearly $\arg(e^{iw}) = w$. If $|w| \geq \pi$ and $w < 0$ then w is in the interval $-2\pi < w \leq -\pi$. Then $0 < w + 2\pi \leq \pi$ so $\arg(e^{iw}) = w + 2\pi$. If $|w| > \pi$ and $w > 0$ then $\pi < w < 2\pi$ so that $-\pi < w - 2\pi < 0$. Hence $\arg(e^{iw}) = w - 2\pi$.

31. Let $-2\pi < w < 2\pi$ and let $z = r e^{iw}$, $r > 0$. Assume that $\alpha \in \mathbb{C}$ and let $\xi_\alpha = e^{i\alpha 2\pi}$. Define a function $\varepsilon_\alpha(w)$ by

$$\varepsilon_\alpha(w) = \begin{cases} 1 & \text{if } |w| < \pi \text{ or } w = \pi \\ \xi_\alpha & \text{if } |w| \geq \pi \text{ and } w < 0 \\ \xi_\alpha^{-1} & \text{if } |w| > \pi \text{ and } w > 0 \end{cases}$$

   Show that $z^\alpha = r^\alpha e^{iw\alpha} \varepsilon_\alpha(w)$.

   Hint: $z^\alpha = (re^{iw})^\alpha = r^\alpha e^{\alpha i \arg(e^{iw})}$. If $|w| < \pi$ or $w = \pi$ then $\arg(e^{iw}) = w$ (see

#30) so that $z^\alpha = r^\alpha e^{iw\alpha} = r^\alpha e^{iw\alpha} \varepsilon_\alpha(w)$. If $|w| \geq \pi$ and $w < 0$ then $\arg(e^{iw}) = w + 2\pi$ (see #30) and $z^\alpha = r^\alpha e^{\alpha i(w+2\pi)} = r^\alpha e^{iw\alpha} e^{i\alpha 2\pi} = r^\alpha e^{iw\alpha} \varepsilon_\alpha(w)$. If $|w| > \pi$ and $w > 0$, then $\arg(e^{iw}) = w - 2\pi$ (see #30) and $z^\alpha = r^\alpha e^{\alpha i(w-2\pi)} = r^\alpha e^{iw\alpha} e^{-i\alpha 2\pi} = r^\alpha e^{iw\alpha} \varepsilon_\alpha(w)$.

32. Use Theorem 4 to show that if A is an n-square complex matrix then A is hermitian if and only if u^*Au is real for all $u \in M_{n,1}(\mathbb{C})$.

33. Use Theorem 4 to show that if A is an n-square complex matrix then A is unitary if and only if $\|Au\| = \|u\|$ and $\|A^*u\| = \|u\|$ for all $u \in M_{n,1}(\mathbb{C})$.

34. Use Theorem 4 to show that if A is an n-square complex matrix then A is normal if and only if $\|Au\| = \|A^*u\|$ for all $u \in M_{n,1}(\mathbb{C})$.

35. Let A be an m × n complex matrix. Show that both AA^* and A^*A are hermitian.

2.2 MatLab

1. Compute each of the following complex numbers using the exponential operator (^) in MatLab.

    (a) $i^{1/3}$
    (b) $i^{2/3}$
    (c) $(1 + i)^{1/2}$
    (d) $(-1)^{1/3}$
    (e) $(-1 + i)^{1/2}$
    (f) i^i
    (g) $(1 + i)^i$
    (h) i^{i^i}

2. Using the formula $z^\alpha = e^{\alpha \log z}$, obtain the theoretical values of #1 (a) - (h).

3. According to Theorem 2, if p and q are integers, q > 0, then all the p/q powers of $z = re^{i\theta}$ are given by $r^{p/q} e^{ip(\theta + 2\pi k)/q}$, $k = 0, \ldots, q - 1$.

   (a) What is the principal value of $z^{p/q}$ (see (51))?

   (b) Use MatLab to confirm that the exponential operator (^) produces the principal value of z^α for each of the following values of z and $\alpha = p/q$:

   (i) $z = e^{i\pi/4}$, $\alpha = 2/3$
   (ii) $z = e^{i\pi/2}$, $\alpha = 3/4$
   (iii) $z = e^{i3\pi/4}$, $\alpha = 1/2$
   (iv) $z = e^{i\pi}$, $\alpha = 2/5$
   (v) $z = e^{i5\pi/4}$, $\alpha = -1/2$

   (c) Use MatLab to find complex numbers z, α, and β for which $(z\wedge\alpha)\wedge\beta \neq z\wedge(\alpha\beta)$, i.e., show that since MatLab computes the principal value of a power according to the formula (51), the familiar rule of exponents, $(z^\alpha)^\beta = z^{\alpha\beta}$, may fail. Explain your example.

4. Use Help to review **input, axis, plot,** and **hold**. Then write a MatLab script that
   - prompts the user to enter two positive integers p and q
   - plots on the same graph the pth roots of unity and the qth roots of unity using point types '+' and 'x' respectively.

   What values of p and q produced the following outputs? What theorem is used as the basis for your program?

2.2 MatLab

(a)

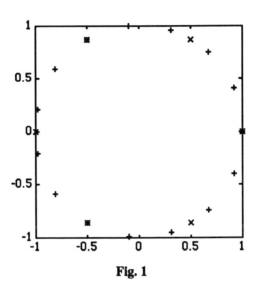

Fig. 1

(b)

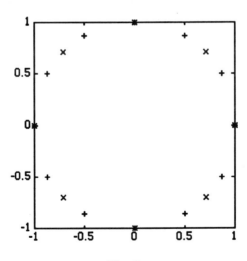

Fig. 2

(c)

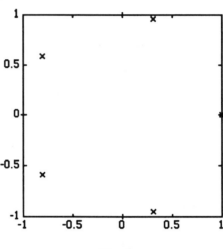

Fig. 3

5. Use Help to read the entry for **rand**. Then write a MatLab script to generate 10 random 2-square matrices of the form u u* where u is a 2×1 complex column vector whose elements are of absolute value at most $\sqrt{2}$. Run the script several times and examine the output.

   (a) To what class of matrices does the output belong: skew-hermitian; unitary; hermitian; symmetric; skew-symmetric; orthogonal?

   (b) Is it possible for any 2-square matrix of the class you identified in (a) to appear as a positive scalar multiple of a matrix in the output? Why?

   (c) Note that the main diagonal entries of every matrix in the output are non-negative. Prove that this is always the case for a matrix of the form uu*.

6. Use Help to review the entries for **if, eps, any, imag, =,** and **&**. Then type the following function declaration into an Edit window:

function p = notneg (A)
% NotNeg returns a 1 if and only if
% no entry of A is 0 or negative,
% otherwise it returns a 0.

2.2 MatLab

```
if any(any(abs(A) < eps))
        p=0;
else
        X=(A < 0);
        Y=(imag (A) == 0);
        p=(any(any(X&Y))) == 0;
end
```

Save the function under the name notneg by choosing **Save** from the File menu. Then at the Command prompt (») enter "type notneg" to see a listing of the program. Next, enter help notneg at the Command prompt to view the comments (i.e., the text following the % lines). To explain the function notneg, note the following:

- the comparison abs(A) < eps generates a matrix of 1's and 0's the same size as A; the i, j entry is 1 if $|A_{ij}| <$ eps ($= 2.204\text{e-}16$)

- any(any(abs(A) < eps)) has the value 1 if and only if some entry of A satisfies $|A_{ij}| <$ eps, otherwise it has value 0

- if any(any(abs(A) < eps)) = 1 then p is assigned the value 0, i.e., p = 0 if some entry of A is so small that its absolute value is below 2.204e-16

- if any(any(abs(A) < eps)) = 0, i.e., if every entry of A is at least eps in absolute value, then the following occurs:

    $X = (A < 0);$ $X_{ij} = 1$ if $\text{Re}(A_{ij}) < 0$, otherwise $X_{ij} = 0$
    $Y = (\text{imag}(A) == 0);$ $Y_{ij} = 1$ if imag $(A_{ij}) = 0$, otherwise $Y_{ij} = 0$.

- $(X \& Y)_{ij} = 1$ if $X_{ij} = Y_{ij} = 1$, otherwise $(X \& Y)_{ij} = 0$

- any(any(X&Y)) is 1 if X&Y contains at least one 1 and is 0 otherwise. Thus p is assigned the value 1 if and only if no entry of A is a negative real.

Use MatLab to evaluate notneg(A) for each of the following A:

(a) $\begin{bmatrix} 0 & 1 \\ 2 & 1+i \end{bmatrix}$

(b) $\begin{bmatrix} 1-i & -1+i \\ 2 & -3i \end{bmatrix}$

(c) $\begin{bmatrix} 1 & -1 \\ 1 & 1 \end{bmatrix}$

7. For $z = 1 + i$ use MatLab to compute each of the following pairs of numbers:

(Except for (e), hints refer to formulas in §2.2 Elementary Complex Functions.)

(a) asin(z), $-i \log (iz + (1-z^2)^{1/2})$

Hint: see (54).

(b) acos(z), $-i \log (z + (z^2 - 1)^{1/2})$

Hint: see (55).

(c) atan(z), $\dfrac{i}{2} \log \left(\dfrac{i+z}{i-z} \right)$

Hint: see (56).

(d) log(z), $\log (|z|) + i \arg(z)$

Hint: see (57).

(e) sin(x + iy),   sin(x) cosh(y) + i cos (x) sinh (y)

Hint: z = x + iy, see §2.2 Exercises, #20.

(f) z^i, $e^{i \log(z)}$

Hint: see (51).

2.2 MatLab

8. Enter the following script in a new Edit window. Then Save it with the name nrange. The numbers at the right following the % signs are for reference only.

```
hold off                                                      %1
disp ('Enter a complex n-square matrix A: ')                  %2
A = input ('A = ');                                           %3
[m,n] = size(A);                                              %4
bool = 1;                                                     %5
while bool                                                    %6
        W = [ ];                                              %7
        for k = 1:500                                         %8
                u = (2*rand(n,1)-1)+(2*rand(n,1)-1)*i;        %9
                u = (1/norm(u))*u;                            %10
                z = u'*A*u;                                   %11
                W = [W z];                                    %12
        end                                                   %13
        axis('square')                                        %14
        if norm(imag(W), inf) < n*eps                         %15
                plot(real(W), zeros(W),'*')                   %16
        else                                                  %17
                plot(W,'*')                                   %18
        end                                                   %19
        hold on                                               %20
        bool = input('Enter 1 for another 500 points, 0 to quit. ');  %21
end                                                           %22
```

(a) Run nrange by typing nrange at the Command window prompt. Describe the output in the Graph window after you have completed the following dialog in the Command window:

»nrange
Enter a complex n-square matrix A:

A = rand(2,2)

Enter 1 for another 500 points, 0 to quit. 0

(b) Run the program again and input A = [1]. Describe the output.

(c) Run the program and input $A = \begin{bmatrix} 0 & 1 \\ 0 & 0 \end{bmatrix}$. Obtain 1000 points. Describe the shape of the output.

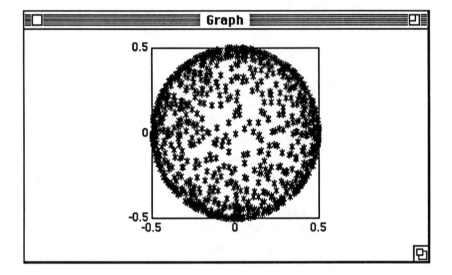

Fig. 4

9. It is evident from Fig. 4 that the shape of the output of nrange for $A = \begin{bmatrix} 0 & 1 \\ 0 & 0 \end{bmatrix}$ is a circular disk, centered at the origin, with radius 0.5. The program nrange does the following:

(i) calls for the input of an n-square matrix A

(ii) obtains the size of A

(iii) generates 500 random 2 × 1 complex column 2-tuples u whose elements have both real and imaginary parts in the interval between -1 and 1

(iv) replaces u by u / ‖ u ‖ so that the resulting vector has norm 1 (see (74))

(v) computes the complex number z = u*Au for each u

2.2 MatLab

(vi) plots the totality of numbers z

(vii) asks the user to enter 1 for another 500 points or 0 to end.

The region plotted by this program, for any square matrix A, is a representation of a set in the complex plane called the *numerical range* or *field of values* of A and is denoted by W(A).

(a) Identify the lines in the program that result in each of the tasks (i) - (vii). Use Help for any unfamiliar commands.

(b) Explain the purpose of the while loop that begins at line 6.

(c) Explain lines 15 - 20 that plot W(A).

Hint: First review the Help entries for **norm** and **plot**. Remember that if W is a complex n-tuple, plot(W) plots the points imag(W) against real(W). If a point type is not specified explicitly, MatLab will connect the points by a line segment. We have specified the point type '*' in lines 16 and 18 so that the points are not connected in the graph.

10. Prove that for the matrix $A = \begin{bmatrix} 0 & 1 \\ 0 & 0 \end{bmatrix}$ in #8(c) the set W(A) is a circular disk, centered at the origin, of radius 0.5.

    Hint: Let $u = [\ s\ t\ ]^T$ where s and t are arbitrary complex numbers that satisfy $\|u\| = 1$, i.e., $|s|^2 + |t|^2 = 1$. Next verify that $u^*Au = \bar{s}\,t$. Write $s = |s|\,e^{i\varphi}$, $t = |t|\,e^{i\theta}$. Then $u^*Au = |s||t|\,e^{i(\theta-\varphi)}$. For a fixed $|s|$ and $|t|$, θ and φ can be chosen arbitrarily so that $\theta - \varphi$ varies over all angles between 0 and 2π. Hence, $|s||t|\,e^{i(\theta-\varphi)}$ describes a circle of radius $|s||t|$ with center at the origin. It only remains to show that $|s||t| \leq 0.5$.

11. Run nrange for several random 2-square matrices of the form $A = vv^*$ where v is 2×1. Describe the outputs.

12. Why do the outputs in #11 have the form that they do?

13. Run nrange for $A = \begin{bmatrix} 0 & 1 \\ 1 & 0 \end{bmatrix}$ (500 points) and describe the output.

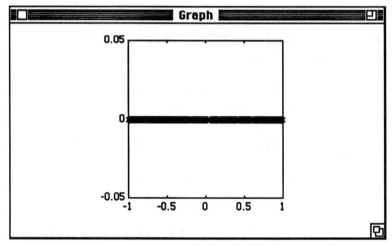

Fig. 5

14. Run nrange for several random 2-square hermitian matrices and describe the outputs.

15. Explain the output in #14.

16. Run nrange for $A = \begin{bmatrix} 1+i & 0 \\ 0 & 2+2i \end{bmatrix}$.

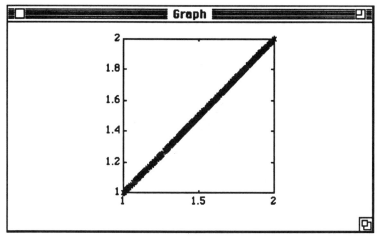

Fig. 6

Explain the output in Fig. 6.

17. Run nrange for A = diag([0 1 i]) for 2000 points.

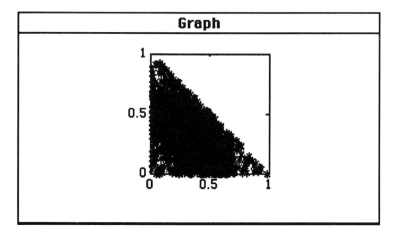

Fig. 7

18. Explain the output in #17.

19. Run nrange for $A = \begin{bmatrix} 0 & 1 & 0 \\ 0 & 0 & 1 \\ 1 & 0 & 0 \end{bmatrix}$ for 1000 points and describe the output.

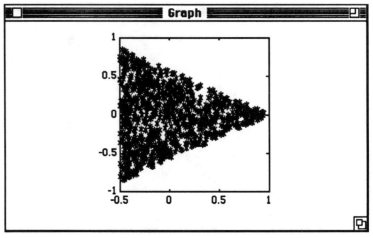

Fig. 8

20. Run nrange for variously sized random A. What common property do all the images W(A) have?

   Hint: It appears that W(A) is always a *convex* set in the complex plane. That is, the line segment joining any two points in W(A) is entirely contained in W(A). In fact, the convexity of W(A) for arbitrary complex n × n matrices A was discovered over 70 years ago by the mathematicians O. Toeplitz and F. Hausdorff.

2.2 Glossary

&	108
+	108
==	108
$\overline{A}$	93
A^*	93
abs	85
acos	85
acosh	85
any	108
arg z	90
arg(z)	104
asin	85
asinh	85
atan	85
atan2	85
atanh	85
axis	106
complex roots	78
conjugate	93
conjugate bilinear	96
conjugate transpose	93
conjugate-symmetric	96
convex set	116
cos	85
cosh	85
eps	108
exp	85
field of values	113
four-quadrant arctangent	93
gcd	83
greatest common divisor	83
Hausdorff, F.	116
hermitian matrix	94

2.2 Glossary

hold	106
hyperbolic functions	88
if	108
imag	108
input	106
log	85
log z	88
log10	85
norm	96
normal matrix	95
numerical range	113
orthogonal matrix	95
plot	106
positive-definite	96
quadratic formula	77
rand	108
rational power	82
sin	85
sinh	85
skew-hermitian matrix	95
skew-symmetric matrix	95
sqrt	85
standard inner product	95
symmetric matrix	94
tan	85
tanh	85
Taylor series	86
Toeplitz, O.	116
unitary matrix	95
W(A)	113
z^c	90
$\|u\|$	96

Chapter 3

Partitioned Matrices

Topics
- *sequence sets*
- *submatrices*
- *block multiplication*
- *conformal partitioning*
- *Toeplitz and Hankel matrices*
- *Kronecker products*

3.1 Sequences and Submatrices

Although matrix multiplication is not commutative, we saw in §1.2 Exercises, #25 that it is *associative*. In fact, if A is m × p, B is p × n, and D is n × q then

$$A(BD) = (AB)D . \quad \text{(associative law)} \tag{1}$$

To confirm (1), simply note that the (i, j) entry of A(BD) is

$$\sum_{k=1}^{p} a_{ik} (BD)_{kj} = \sum_{k=1}^{p} a_{ik} \sum_{s=1}^{n} (b_{ks} d_{sj})$$

$$= \sum_{s=1}^{n} \sum_{k=1}^{p} (a_{ik} b_{ks}) d_{sj}$$

$$= \sum_{s=1}^{n} (AB)_{is} d_{sj}$$

$$= ((AB)D)_{ij} . \tag{2}$$

Note that in confirming (1) we only required that the entries of the matrices satisfy the associative, distributive, and commutative laws of addition and multiplication. Since matrix multiplication is associative, if A is n-square then A(AA) and (AA)A are equal. Similarly, if p is a positive integer, all products of p factors, each equal to A, must be the same. We denote by A^p the common value of all such products. If p = 0, we make the convention $A^0 = I_n$. Similarly, if A(BC) and (AB)C are both defined, then their common value is denoted by ABC. More generally, a product of m matrices, $A_1 \cdots A_m$, is independent of the way in which the factors are associated two at a time for multiplication. Thus MatLab recognizes a command such as A*B*C, without parentheses.

In order to identify matrices constructed from parts of other matrices it is convenient to have a notation for *sequence sets*. Thus, let $n_1, ..., n_p$ be positive integers. Then $\Gamma(n_1, ..., n_p)$ denotes the totality of sequences of positive integers,

$$\omega = (\omega(1), ..., \omega(p)),$$

for which $1 \le \omega(i) \le n_i$, $i = 1, ..., p$. Frequently, the value of $\omega(i)$ is written ω_i; $\Gamma_{r, n}$ is the set of sequences $\Gamma(n, ..., n)$ in which n occurs r times; $G_{r, n}$ is the set of nondecreasing sequences $\omega \in \Gamma_{r, n}$, i.e.,

$$1 \le \omega(1) \le \omega(2) \le \cdots \le \omega(r) \le n.$$

If $1 \le r \le n$ then $Q_{r, n}$ is the set of strictly increasing sequences $\omega \in G_{r,n}$, i.e.,

$$1 \le \omega(1) < \omega(2) < \cdots < \omega(r) \le n.$$

If α and β are in $\Gamma(n_1, ..., n_p)$ then α precedes β in *lexicographic order*, written

$$\alpha \le \beta,$$

if the first nonzero difference $\beta(i) - \alpha(i)$ is positive. If $1 \le c < p$, $\alpha \in \Gamma(n_1, ..., n_c)$, and $\beta \in \Gamma(n_{c+1}, ..., n_p)$ then we can *adjoin* β to α to obtain the sequence

$$\alpha : \beta = (\alpha(1), ..., \alpha(c), \beta(1), ..., \beta(p - c)),$$

3.1 Sequences and Submatrices

a sequence in $\Gamma(n_1, ..., n_p)$.

Let $A = [a_{ij}]$ be an m × n matrix and let $\alpha \in \Gamma_{r,m}$, $\beta \in \Gamma_{s,n}$. Then the r × s matrix $B = [b_{ij}]$ defined by

$$b_{ij} = a_{\alpha(i), \beta(j)}, \quad i = 1, ..., r, \quad j = 1, ..., s,$$

is denoted by

$$A[\alpha | \beta]. \qquad (3)$$

If $r \le m$, $s \le n$, $\alpha \in Q_{r,m}$, $\beta \in Q_{s,n}$, then (3) is called an r × s *submatrix* of A lying in rows α and columns β. If $\alpha \in Q_{r,m}$ (and r < m) then the *sequence complementary* to α is the sequence $\alpha' \in Q_{m-r, m}$ such that $\{\alpha(1), ..., \alpha(r), \alpha'(1), ..., \alpha'(m-r)\}$ is precisely the set of integers $\{1, 2, ..., m\}$. If $\alpha \in Q_{r,m}$ and $\beta \in Q_{s,n}$ then $A[\alpha' | \beta]$ is denoted by $A(\alpha | \beta]$, $A[\alpha | \beta']$ by $A[\alpha | \beta)$, and $A[\alpha' | \beta']$ by $A(\alpha | \beta)$. If A is n-square, $\alpha \in Q_{r,n}$, then $A[\alpha | \alpha]$ is called a *principal submatrix* of A. As examples of the preceding notations let

$$A = \begin{bmatrix} 1 & 2 & 3 & 4 \\ 5 & 6 & 7 & 8 \\ 9 & -8 & -7 & -6 \\ -5 & -4 & -3 & -2 \end{bmatrix}.$$

Then

$$A[1, 2 | 1, 3] = \begin{bmatrix} a_{11} & a_{13} \\ a_{21} & a_{23} \end{bmatrix}$$

$$= \begin{bmatrix} 1 & 3 \\ 5 & 7 \end{bmatrix};$$

$$A[2, 1 \mid 2, 2] = \begin{bmatrix} a_{22} & a_{22} \\ a_{12} & a_{12} \end{bmatrix}$$

$$= \begin{bmatrix} 6 & 6 \\ 2 & 2 \end{bmatrix};$$

$$A(1, 3 \mid 2, 3] = A[2, 4 \mid 2, 3] = \begin{bmatrix} a_{22} & a_{23} \\ a_{42} & a_{43} \end{bmatrix}$$

$$= \begin{bmatrix} 6 & 7 \\ -4 & -3 \end{bmatrix};$$

$$A[1, 1, 1 \mid 2, 2] = \begin{bmatrix} a_{12} & a_{12} \\ a_{12} & a_{12} \\ a_{12} & a_{12} \end{bmatrix}$$

$$= \begin{bmatrix} 2 & 2 \\ 2 & 2 \\ 2 & 2 \end{bmatrix}.$$

If A is an m × n matrix then A can be *partitioned* into nonoverlapping submatrices. For example,

$$A = \left[\begin{array}{ccc|ccc} 1 & 2 & 3 & 4 & 5 & 6 \\ 7 & 8 & 9 & -1 & -2 & -3 \\ \hline -4 & -5 & -6 & -7 & -8 & -9 \\ \hline 2 & 5 & 7 & 4 & 8 & 3 \end{array}\right]$$

is a partitioning of the form

$$A = \left[\begin{array}{c|c|c} A_{11} & A_{12} & A_{13} \\ \hline A_{21} & A_{22} & A_{23} \\ \hline A_{31} & A_{32} & A_{33} \end{array}\right]$$

in which

$$A_{11} = \begin{bmatrix} 1 & 2 \\ 7 & 8 \end{bmatrix}, \quad A_{12} = \begin{bmatrix} 3 \\ 9 \end{bmatrix}, \quad A_{13} = \begin{bmatrix} 4 & 5 & 6 \\ -1 & -2 & -3 \end{bmatrix},$$

$$A_{21} = [-4 \ -5], \quad A_{22} = [-6], \quad A_{23} = [-7 \ -8 \ -9],$$

$$A_{31} = [2 \ 5], \quad A_{32} = [7], \quad A_{33} = [4 \ 8 \ 3].$$

Note that

$$A_{11} = A[1, 2 \mid 1, 2], \quad A_{12} = A[1, 2 \mid 3], \quad A_{13} = A[1, 2 \mid 4, 5, 6],$$

$$A_{21} = A[3 \mid 1, 2], \quad A_{22} = A[3 \mid 3], \quad A_{23} = A[3 \mid 4, 5, 6],$$

$$A_{31} = A[4 \mid 1, 2], \quad A_{32} = [4 \mid 3], \quad A_{33} = A[4 \mid 4, 5, 6].$$

(In this context, A_{ij} is a submatrix of A, not the (i, j) entry of A.)

3.1 Exercises

1. Let A be an n × n matrix and let p and q be nonnegative integers. Prove that $A^p A^q = A^{p+q}$.

2. If $A = \text{diag}(r_1, \ldots, r_n)$, find a formula for A^p.

3. Compute A^p if A is p-square and $a_{12} = a_{23} = \cdots = a_{p-1, p} = 1$, and $a_{ij} = 0$ otherwise.

   Hint: Letting δ_{st} denote the Kronecker delta (i.e., $\delta_{st} = 1$ if $s = t$, 0 if $s \neq t$), A is the matrix $A = [\delta_{i+1, j}]$. By induction on k we show that $A^k = [\delta_{i+k, j}]$. This is clear for $k = 1$. Assume $A^k = [\delta_{i+k, j}]$. Then $A^{k+1} = A^k A = [\delta_{i+k, j}][\delta_{i+1, j}] = \left[\sum_{t=1}^{p} \delta_{i+k, t} \delta_{t+1, j}\right]$. For a given i and j the terms in the indicated sum are 0 unless $t = i + k$ and $t + 1 = j$, i.e., $j = i + k + 1$. Then $A^{k+1} = [\delta_{i+k+1, j}]$. Note that if $k = p$, then $A^p = [\delta_{i+p, j}]$ and $i + p > p$, so that every entry of A^p is 0.

4. Show that the number of sequences in $\Gamma(n_1, \ldots, n_p)$ is $n_1 \cdots n_p$.

5. If $1 \leq r \leq n$, show that the number of sequences in $Q_{r, n}$ is the binomial coefficient $\binom{n}{r}$.

6. Show that the number of sequences in $G_{r, n}$ is the binomial coefficient $\binom{n + r - 1}{r}$.

   Hint: Given $\gamma \in G_{r, n}$ define a sequence $\omega \in Q_{r, n+r-1}$ as follows: $\omega(t) = \gamma(t) + t - 1$, $t = 1, \ldots, r$. Clearly, distinct γ in $G_{r, n}$ produce distinct ω in $Q_{r, n+r-1}$. Conversely, given $\omega \in Q_{r, n+r-1}$ define $\gamma \in G_{r, n}$ by $\gamma(t) = \omega(t) - (t - 1)$, $t = 1, \ldots, r$. Again, note that distinct ω in $Q_{r, n+r-1}$ produce distinct γ in $G_{r, n}$. Hence the number of sequences in $G_{r, n}$ is the same as the number of sequences in $Q_{r, n+r-1}$. See #5 to conclude that there are $\binom{n + r - 1}{r}$ sequences in $Q_{r, n+r-1}$.

7. If $A = [a_{ij}]$ is a 4×5 matrix write out the following: $A[1, 2 \mid 1, 2]$, $A[2, 1 \mid 2, 1]$, $A[1, 1, 2 \mid 1, 1, 3]$, $A[2, 2 \mid 3, 3]$, $A(1, 2 \mid 3)$, $A(3, 4 \mid 2, 3, 5)$, $A[1, 2 \mid 3, 4)$, $A(1, 2 \mid 3, 4]$.

3.1 Exercises

8. If $A = [a_{ij}]$ is a 4×4 matrix, write out all 2×2 principal submatrices of A.

9. If A is an $m \times n$ matrix, show that the number of $k \times k$ submatrices of A is $\binom{m}{k}\binom{n}{k}$.

10. Partition

$$\begin{bmatrix} 1 & 2 & 3 & 4 \\ 5 & 6 & 7 & 8 \\ 9 & -1 & -2 & -3 \\ -4 & -5 & -6 & -7 \end{bmatrix}$$

into nonoverlapping 2×2 submatrices.

11. Partition the matrix in #10 into nonoverlapping submatrices as follows:

$$\begin{array}{c} \\ 2 \text{ rows} \\ \\ 2 \text{ rows} \end{array} \begin{bmatrix} \overset{1 \text{ col.}}{} & \overset{1 \text{ col.}}{} & \overset{2 \text{ cols.}}{} \\ & & \\ \hline & & \end{bmatrix}.$$

12. If $A = [a_{ij}]$ is a 16×16 matrix partitioned into 64 nonoverlapping 2×2 submatrices A_{ij}, write out $A_{12}, A_{24}, A_{73}, A_{51}$.

Hint: $A_{73} = \begin{bmatrix} a_{13,5} & a_{13,6} \\ a_{14,5} & a_{14,6} \end{bmatrix}.$

13. Prove that

$$(A_1 A_2 \cdots A_m)^T = A_m^T A_{m-1}^T \cdots A_1^T$$

and

$$(A_1 \cdots A_m)^* = A_m^* A_{m-1}^* \cdots A_1^*.$$

Hint: Both identities follow from the case m = 2 by a simple induction. For m = 2,

$$((AB)^T)_{ij} = (AB)_{ji} = \sum_k a_{jk} b_{ki} = \sum_k (B^T)_{ik} (A^T)_{kj} = (B^T A^T)_{ij}.$$

Similarly $(AB)^* = B^* A^*$.

14. If A is n-square and satisfies $AA^T = A^T A = I_n$, recall that A is called an *orthogonal* matrix. Prove that if A and B are n × n orthogonal matrices then AB is orthogonal.

15. Exhibit an example of a 2 × 2 orthogonal matrix in which all four entries are non-zero.

16. Prove that if A and B are n × n hermitian matrices and AB = BA then AB is hermitian.

17. If A is an n × n complex matrix (i.e., A has complex number entries), then recall that A is *unitary* if $AA^* = A^*A = I_n$. Prove that if A and B are n × n unitary matrices then AB is unitary.

18. Exhibit an example of a 2 × 2 unitary matrix whose entries are non-zero non-real numbers.

19. Let A be an n × n hermitian matrix. Prove that any principal submatrix of A is hermitian.

Hint: Let $\omega \in Q_{k,n}$ and $B = A[\omega | \omega]$. Then

$$(B^*)_{ij} = \overline{B_{ji}} = \overline{A[\omega | \omega]_{ji}} = \overline{A_{\omega_j \omega_i}} = A_{\omega_i \omega_j} = A[\omega | \omega]_{ij} = B_{ij}.$$

20. Recall that a complex matrix A is *normal* if $AA^* = A^*A$; if A has real entries it is a *real normal* matrix (i.e., $AA^T = A^TA$). Exhibit an example of a 2 × 2 normal matrix with nonzero entries that is not unitary.

3.1 MatLab

1. Enter $A = \begin{bmatrix} 1 & 1 \\ 0 & 1 \end{bmatrix}$ and write a script to print out A^k for $k = 1, 2, \ldots, 10$. Prove that $A^k = \begin{bmatrix} 1 & k \\ 0 & 1 \end{bmatrix}$.

2. Let A be the 8-square matrix with $a_{ii} = 1$, $i = 1, \ldots, 8$, $a_{i,i+1} = 1$, $i = 1, \ldots, 7$, and all other entries 0. Use Help to review diag and then write and run a script to print out the first row of A^k, $k = 1, 2, \ldots, 8$. What are the numbers in $(A^k)_{(1)}$? State and prove a theorem that generalizes the results of this experiment.

3. Let A be an m × n matrix. Then MatLab has the following useful notations:

    - $A(:, j)$ is $A^{(j)}$
    - $A(i, :)$ is $A_{(i)}$
    - $A(:)$ is the *column form* of A, i.e., it is the mn × 1 matrix, denoted mathematically by $c(A)$, and obtained by writing the columns of A in sequence. For example, if

    $$A = \begin{bmatrix} 1 & 2 \\ 3 & 4 \end{bmatrix}$$

    then

    $$A(:) = \begin{bmatrix} 1 \\ 3 \\ 2 \\ 4 \end{bmatrix}.$$

- If $\alpha = (\alpha_1, \alpha_2, ..., \alpha_r)$ is a sequence of length r, then $A([\alpha_1 \; \alpha_2 \; \cdots \; \alpha_r])$ consists of entries numbered $\alpha_1, ..., \alpha_r$ in A(:); these appear as a row.
- A(:, α) is A [1, ..., m | α]
- A(α, :) is A [α | 1, ..., n]
- A(α, β) is A [α | β].

Enter the matrix

$$A = \begin{bmatrix} 1 & 2 & 3 & 4 & 5 \\ 6 & 7 & 8 & 9 & 10 \\ 11 & 12 & 13 & 14 & 15 \\ 16 & 17 & 18 & 19 & 20 \end{bmatrix}.$$

Use MatLab to exhibit A [α | β] for each of the following sequences α, β:

(a) $\alpha = \beta = (1 \; 2)$ Hint: A([1 2], [1 2]).

(b) $\alpha = (1 \; 4)$, $\beta = (2 \; 4)$

(c) $\alpha = (1 \; 1)$, $\beta = (2 \; 1)$

(d) $\alpha = (1 \; 1)$, $\beta = (1 \; 1)$

(e)   Use MatLab to print out $[a_{41} \; a_{22} \; a_{22} \; a_{11} \; a_{45}]$ as a row.

Hint:
       A([4 6 6 1 20]).

4. If $\alpha \in Q_{r,n}$ then α', the complementary subsequence in $Q_{n-r,n}$, is obtained by collapsing the entries in [1 ... n] corresponding to α into the *empty matrix* [ ]. For example, if u = [1 2 3 4 5], α = [2 4], then the commands

       u([2 4]) = [ ];
       u

will produce [1 3 5] = α' as output. Write a MatLab script that calls for the input of $\alpha \in Q_{r,4}$ and $\beta \in Q_{s,5}$ and then exhibits each of the following submatrices of the matrix A in #3: A[α | β], A(α | β], A[α | β), A(α | β).

5. Use Help to review the entry for **for..end**. Then write a script that generates a 3-square random matrix A and exhibits all 2-square submatrices of A.

   Hint:
   ```
   A = rand(3)
   for a = 1:3
           for b = 1:3
                   h = 1:3;
                   v = h;
                   h(a) = [ ];
                   v(b) = [ ];
                   A(h,v)
           end
   end
   ```

6. Write a script that calls for the input of integers m and n and then generates an mn × 1 matrix X whose entries are the integers 1, ..., mn. Then the program is to exhibit the unique m × n matrix A that satisfies c(A) = X.

   Hint:
   ```
   m = input('Enter an integer m: ');
   n = input('Enter an integer n: ');
   X = (1:m*n)'
   A = [ ];
   for k = 1:n
           A = [A X((k-1)*m+1 : k*m)];
   end
   A
   ```

7. Write a script that exhibits all 2-square submatrices of the 3 × 4 matrix

$$A = \begin{bmatrix} 1 & 2 & 3 & 4 \\ 5 & 6 & 7 & 8 \\ 9 & 10 & 11 & 12 \end{bmatrix}.$$

8. *Horner's method* for computing $f(x) = c_n x^n + c_{n-1} x^{n-1} + \cdots + c_1 x + c_0$ is based on the following idea (n = 4):

$$f(x) = c_4 x^4 + c_3 x^3 + c_2 x^2 + c_1 x + c_0$$
$$= x(c_4 x^3 + c_3 x^2 + c_2 x + c_1) + c_0$$
$$= x(x(c_4 x^2 + c_3 x + c_2) + c_1) + c_0$$
$$= x(x(x(c_4 x + c_3) + c_2) + c_1) + c_0$$
$$= x(x(x(x(c_4) + c_3) + c_2) + c_1) + c_0.$$

Write a script that calls for the input of $c = [c_n\ c_{n-1}\ c_{n-2}\ \ldots\ c_1\ c_0]$, calls for the input of an arbitrary complex number x, and then exhibits f(x). Review the Help entries for **length, sum,** and **.^** . Use the program to compute

$$\sum_{k=1}^{10} k \cdot 2^k.$$

Hint: (This solution does not directly use Horner's method.)

disp('Enter the coefficient matrix of the polynomial ')
c = input('in decreasing powers of the variable: ');
x = input('enter the value of x: ');
L = length(c);
f = sum(c.*((x*ones(1, L)).^(L-1:-1:0)))

Enter 10:-1:0 for c and 2 for x. The value of the sum is 18434.

9. If w is a non-zero complex column vector of length n, write a MatLab formula for a column vector v that is a multiple of w and satisfies

$$\sum_{k=1}^{n} |v_k|^2 = 1.$$

10. Write a script that calls for the input of a nonzero complex column vector w, replaces w by the vector v in #9, exhibits the matrix

$$Q = I_n - 2vv^*,$$

and exhibits the two products QQ* and Q*Q. A matrix A constructed in this way is called a *Householder matrix*. Show that in general a Householder matrix is both unitary and hermitian.

Hint:
```
w = input('enter a complex column n-vector: ');
v = w/sqrt(w'*w);
n = length(v);
Q = eye(n)-2*v*v'
Q*Q'
Q'*Q
```

11. In (a) - (c), without using either while..end or for..end loops, write scripts that:

(a) calls for the input of the integer n and the number w and exhibits the n–square matrix $A = [w^{ij}]$;

(b) calls for the input of the integer n and exhibits the n-square *Hilbert matrix*

$$A = \left[\frac{1}{i+j-1}\right];$$

(c) calls for the input of the integer n and the n-vector $x = (x_1, ..., x_n)$ and exhibits the n-square matrix

$$A = [(x_i)^j].$$

Hint:
(a)
```
n = input('enter an integer n: ');
w = input('enter a number w: ');
A = (w*ones(n)).^((1:n)'*(1:n))
```

(b)
```
n = input('enter an integer n: ');
A = ((1:n)'*ones(1,n)+ones(n,1)*(1:n)-1).^(-1)
```

(c)
```
n = input('enter an integer n: ');
x = input ('enter an n-vector x: ');
A = ((x.')*ones(1,n)).^(ones(n,1)*(1:n))
```

12. Write a script that calls for an integer n, calls for a complex n-vector p, and then exhibits the n-square matrix

$$A = \begin{bmatrix} 0 & 0 & \cdots & 0 & -p(1) \\ 1 & 0 & \cdots & 0 & -p(2) \\ 0 & 1 & 0 & \cdots & 0 & -p(3) \\ 0 & & & & & \vdots \\ \vdots & & & & \vdots \\ 0 & & \cdots & 0 & 1 & -p(n) \end{bmatrix}$$

The matrix A is called the *companion matrix* of p.

13. Use Help to review **if**, **&**, **|**, and **~**. Use for..end loops to generate and exhibit the 6 *permutations* of 1 2 3. (For the purposes of this exercise, a permutation of 1, 2, 3 is simply a rearrangement of the numbers 1, 2, 3. Permutations will be covered in detail in Chapter 5.) The output should have the following appearance.

```
        1       2       3
        1       3       2
        2       1       3
        2       3       1
        3       1       2
        3       2       1
```

Hint:
```
for i = 1:3
    for j = 1:3
        for k = 1:3
            if (j ~= i)&(k ~= i)&(k ~= j)
                disp([i j k])
            end
        end
    end
end
```

14. If p and q are permutations of 1, ..., n they may be represented as row matrices, e.g., p = [2 1 3] is the permutation p(1) = 2, p(2) = 1, p(3) = 3. The *product* pq is defined as the permutation [p(q(1)) p(q(2)) p(q(3))]. Write a script that calls for the input of an integer n, calls for the input of a permutation p of 1, ..., n, and then produces as output the least integer k for which p^k = [1 2 3 ... n], where p^k is the product of p with itself k times. Use Help to review while and any.

   Hint:
   ```
   n = input('enter an integer n: ');
   p = input('enter a permutation of 1...n: ');
   k = 1;
   id = 1:n;
   q = p;
   while any(p ~= id)
           p = q(p);
           k = k+1;
   end
   disp(k)
   ```

15. If p is a permutation of 1, ..., n then the n-square matrix A(p) whose i, j entry is $\delta_{i,p(j)}$ (*Kronecker delta*) is called the *permutation matrix* corresponding to p (recall that the Kronecker delta is defined by δ_{st} = 1 if s = t and δ_{st} = 0 if s ≠ t). Modify the script for #13 in order to exhibit the 6 permutation matrices of order 3. The output should have the following form.

    ```
    1  0  0
    0  1  0
    0  0  1

    1  0  0
    0  0  1
    0  1  0

    0  1  0
    1  0  0
    0  0  1
    ```

$$\begin{matrix} 0 & 1 & 0 \\ 0 & 0 & 1 \\ 1 & 0 & 0 \end{matrix}$$

$$\begin{matrix} 0 & 0 & 1 \\ 1 & 0 & 0 \\ 0 & 1 & 0 \end{matrix}$$

$$\begin{matrix} 0 & 0 & 1 \\ 0 & 1 & 0 \\ 1 & 0 & 0 \end{matrix}$$

Hint:

```
A = zeros(3);
for i = 1:3
    for j = 1:3
        for k = 1:3
            if (j ~= i)&(k ~= i)&(k ~= j)
                p = [i j k];
                for s = 1:3
                    for t = 1:3
                        if t == p(s)
                            A(s, t) = 1;
                        else
                            A(s, t) = 0;
                        end
                    end
                end
                disp(A)
            end
        end
    end
end
```

3.1 Glossary

A(:)	127
A(:, j)	127
A($\alpha \mid \beta$)	121
A($\alpha \mid \beta$]	121
A(i, :)	127
A[$\alpha \mid \beta$)	121
A[$\alpha \mid \beta$]	121
associative law	119
α(A)	127
column form	127
companion matrix	132
complementary sequence	121
$G_{r,n}$	120
$\Gamma(n_1, \cdots, n_p)$	120
$\Gamma_{r,n}$	120
Hilbert matrix	131
Horner's method	130
Householder matrix	130
Kronecker delta	133
lexicographic order	120
normal matrix	127
orthogonal matrix	126
permutation	132
permutation matrix	133
principal submatrix	121
$Q_{r,n}$	120
sequence sets	120
submatrix	121
unitary matrix	126

3.2 Conformal Partitioning

Let A be m × n, b be n × r and let m_i, n_j, r_k be positive integers, i = 1, ..., d, j = 1, ..., e, k = 1, ..., f, for which

$$m_1 + \cdots + m_d = m, \ n_1 + \cdots + n_e = n, \ r_1 + \cdots + r_f = r.$$

Partition both A and B into nonoverlapping submatrices according to the following schemes:

$$A = \begin{array}{c} \\ m_1 \\ m_2 \\ \vdots \\ m_d \end{array} \begin{array}{c} n_1 \ \ n_2 \ \ \cdots \ \ n_e \\ \left[\begin{array}{cccc} A_{11} & A_{12} & \cdots & A_{1e} \\ A_{21} & A_{22} & \cdots & A_{2e} \\ \vdots & \vdots & \vdots & \vdots \\ A_{d1} & A_{d2} & \cdots & A_{de} \end{array} \right] \end{array}, \quad B = \begin{array}{c} \\ n_1 \\ n_2 \\ \vdots \\ n_e \end{array} \begin{array}{c} r_1 \ \ r_2 \ \ \cdots \ \ r_f \\ \left[\begin{array}{cccc} B_{11} & B_{12} & \cdots & B_{1f} \\ B_{21} & B_{22} & \cdots & B_{2f} \\ \vdots & \vdots & \vdots & \vdots \\ B_{e1} & B_{e2} & \cdots & B_{ef} \end{array} \right] \end{array}. \quad (1)$$

Then A and B are said to be *conformally partitioned* for multiplication. Let C = AB, an m × r matrix, partition the rows of C in the same way as the rows of A, and partition the columns of C in the same way as the columns of B. Let C_{ik} be the (i, k) block (i.e., submatrix) in this partitioning of C.

Theorem 1.

The block C_{ik} is obtained from the blocks in (1) according to the formula

$$C_{ik} = \sum_{j=1}^{e} A_{ij} B_{jk}, \ i = 1, \ldots, d, \ k = 1, \ldots, f. \quad (2)$$

Proof.

First note that since A_{ij} is $m_i \times n_j$, and B_{jk} is $n_j \times r_k$, each matrix product $A_{ij} B_{jk}$ is $m_i \times r_k$ so that the sum on the right in (2) is an $m_i \times r_k$ matrix. Moreover, C was partitioned according to the row partitioning in A and the column partitioning in B, so that C_{ik} is $m_i \times r_k$ as well. The (p, q) entry of C is the product of row p of A and column q of B:

3.2 Conformal Partitioning

$$c_{pq} = A_{(p)} B^{(q)}. \tag{3}$$

Suppose row p of A lies in the ith block row of A, say it is row s of this block, and column q of B lies in the kth block column of B, say it is in column t of this block. This means that

$$A_{(p)} = [\overset{n_1}{A_{i1}} | \overset{n_2}{A_{i2}} | \cdots | \overset{n_e}{A_{ie}}]_{(s)} \tag{4}$$

and

$$B^{(q)} = \begin{bmatrix} n_1 & \overline{B_{1k}} \\ n_2 & \overline{B_{2k}} \\ & \vdots \\ n_e & \overline{B_{ek}} \end{bmatrix}^{(t)}. \tag{5}$$

From the definition of matrix multiplication,

$$\begin{aligned} c_{pq} &= A_{(p)} B^{(q)} \\ &= (A_{i1})_{(s)} (B_{1k})^{(t)} + (A_{i2})_{(s)} (B_{2k})^{(t)} + \cdots + (A_{ie})_{(s)} (B_{ek})^{(t)} \\ &= (A_{i1} B_{1k})_{st} + (A_{i2} B_{2k})_{st} + \cdots + (A_{ie} B_{ek})_{st} \\ &= \left(\sum_{j=1}^{e} A_{ij} B_{jk} \right)_{st}. \end{aligned} \tag{6}$$

If we now construct a partitioned matrix M with the same row partitioning as A, the same column partitioning as B, and whose (i, k) block is the matrix on the right in (2), then it is clear that the (p, q) entry of M is precisely (6). In other words, M = C, or stated otherwise, the i, k block of C is the right side of (2). ∎

Partitioning for block multiplication is particularly useful if one of the factors in a matrix product contains a block of zero entries. For example, let

$$A = \begin{bmatrix} 1 & 2 & 0 & 0 \\ 0 & 1 & 0 & 0 \\ 1 & -1 & 2 & 3 \\ 1 & -1 & 0 & 1 \end{bmatrix}. \tag{7}$$

In computing A^2 we may partition A as follows:

$$\left[\begin{array}{cc|cc} 1 & 2 & 0 & 0 \\ 0 & 1 & 0 & 0 \\ \hline 1 & -1 & 2 & 3 \\ 1 & -1 & 0 & 1 \end{array}\right] \left[\begin{array}{cc|cc} 1 & 2 & 0 & 0 \\ 0 & 1 & 0 & 0 \\ \hline 1 & -1 & 2 & 3 \\ 1 & -1 & 0 & 1 \end{array}\right] = \begin{bmatrix} C_{11} & C_{12} \\ C_{21} & C_{22} \end{bmatrix}$$

where

$$C_{11} = \begin{bmatrix} 1 & 2 \\ 0 & 1 \end{bmatrix} \begin{bmatrix} 1 & 2 \\ 0 & 1 \end{bmatrix} + 0_{2,2} \begin{bmatrix} 1 & -1 \\ 1 & -1 \end{bmatrix} = \begin{bmatrix} 1 & 4 \\ 0 & 1 \end{bmatrix},$$

$$C_{12} = \begin{bmatrix} 1 & 2 \\ 0 & 1 \end{bmatrix} 0_{2,2} + 0_{2,2} \begin{bmatrix} 2 & 3 \\ 0 & 1 \end{bmatrix} = 0,$$

$$C_{21} = \begin{bmatrix} 1 & -1 \\ 1 & -1 \end{bmatrix} \begin{bmatrix} 1 & 2 \\ 0 & 1 \end{bmatrix} + \begin{bmatrix} 2 & 3 \\ 0 & 1 \end{bmatrix} \begin{bmatrix} 1 & -1 \\ 1 & -1 \end{bmatrix}$$

$$= \begin{bmatrix} 1 & 1 \\ 1 & 1 \end{bmatrix} + \begin{bmatrix} 5 & -5 \\ 1 & -1 \end{bmatrix} = \begin{bmatrix} 6 & -4 \\ 2 & 0 \end{bmatrix},$$

$$C_{22} = \begin{bmatrix} 1 & -1 \\ 1 & -1 \end{bmatrix} 0_{2,2} + \begin{bmatrix} 2 & 3 \\ 0 & 1 \end{bmatrix} \begin{bmatrix} 2 & 3 \\ 0 & 1 \end{bmatrix} = \begin{bmatrix} 4 & 9 \\ 0 & 1 \end{bmatrix}.$$

Thus

$$A^2 = \begin{bmatrix} 1 & 4 & 0 & 0 \\ 0 & 1 & 0 & 0 \\ 6 & -4 & 4 & 9 \\ 2 & 0 & 0 & 1 \end{bmatrix}.$$

Note that $A^2[1, 2 \mid 3, 4] = 0_{2,2}$.

3.2 Exercises

1. For the matrix A in (7), show that if p is any positive integer then the (1, 2) block of A^p is 0 (where the partitioning is into four 2 × 2 blocks).

2. Let A be m × n, B be n × r and let k be a positive integer, $1 \leq k \leq m$. Show that if $A[1, 2, ..., k \mid 1, 2, ..., n] = 0_{k,n}$ then $(AB)[1, 2, ..., k \mid 1, 2, ..., r] = 0_{k,r}$.

3. Let A and B be n × n matrices and define a 2n × 2n matrix M by

$$M = \left[\begin{array}{c|c} 0_n & A \\ \hline B & 0_n \end{array} \right].$$

Find M^3.

4. If A is the matrix in #3 and $B = A^*$ find M^2, M^3, M^4.

3.2 MatLab

1. Write a script that calls for the input of two 4-square matrices A and B and then computes $(AB)[1, 2 \mid 1, 2]$ by conformally partitioning A and B into 2-square blocks as in Theorem 1.

2. Review the Help entries for **length** and **floor**. If A is an m × n matrix then length(A) is the larger of m and n, so that if A is n-square, length(A) = n. Write a script that generates 3 random square matrices and exhibits their direct sum. Each of the matrices is to be of random dimension not exceeding 3.

    Hint:
    A = [ ];
    p = 1;
    while p <= 3

```
            n = floor(1 + 3*rand);
            B = rand(n);
            s = length(A);
            A = [A   zeros(s, n); zeros(n, s)   B];
            p = p + 1;
      end
      disp(A)
```

3. Review the Help entry for ^. Then compute $A^{1/2}$ where

    $$A = \begin{bmatrix} 1 & 1 \\ 0 & 1 \end{bmatrix}.$$

    Check whether the square of the answer yields A. Does there exist an X such that $X^2 = A$?

4. Read the Help entry for **toeplitz**. An n-square Toeplitz matrix has all elements on each superdiagonal and all elements on each subdiagonal equal. For example, if c = [6 4 5 0 0] and r = [0 2 3 0 0] then A = toeplitz(c, r) generates the matrix

    $$A = \begin{bmatrix} 6 & 2 & 3 & 0 & 0 \\ 4 & 6 & 2 & 3 & 0 \\ 5 & 4 & 6 & 2 & 3 \\ 0 & 5 & 4 & 6 & 2 \\ 0 & 0 & 5 & 4 & 6 \end{bmatrix}.$$

    Note that the first entry of c is always the main diagonal entry. Review the Help entry for diag. Then write a MatLab script that calls for the input of c and r and generates toeplitz(c, r) without using the toeplitz command.

    Hint:
```
            c = input('enter the first column: ');
            r = input('enter the first row: ');
            n = length(c);
            T = diag(c(1)*ones(1, n), 0);
            for k = 2:n
```

```
        T = T + diag(c(k)*ones(1,n-k+1), 1-k) + diag(r(k)*ones(1, n-k+1), k-1);
     end
     disp(T)
```

5. Write a MatLab script that calls for the input of an n-vector d and an (n-1)-vector u and then exhibits the matrix

$$A = \begin{bmatrix} d(1) & u(1) & & & \\ & d(2) & u(2) & & 0 \\ & & \ddots & \ddots & \\ 0 & & & \ddots & u(n-1) \\ & & & & d(n) \end{bmatrix}.$$

6. Recall that if v is an n-vector then the MatLab script

    ```
    for p = v
        statements
    end
    ```

    has the effect of p successively taking on the values v(1), v(2), ..., v(n), each time executing the statements. Predict the output of the following MatLab script.

    ```
    x = zeros(1,4);
    y = (1:7).^2;
    v = [2 4 6];
    for p = v
        x(p) = y(p);
    end
    x
    ```

7. Recall that if A is an m × n matrix then the MatLab script

    ```
    for p = A
        statements
    end
    ```

has the effect of p successively taking on the values A(:,1), A(:,2), ..., A(:,n), each time executing statements. Read the Help entries for **any** and **all**. Then predict the output of the following MatLab script for the matrix

$$A = \begin{bmatrix} 0 & 1 & 4 \\ 0 & 2 & 5 \\ 0 & 3 & 0 \end{bmatrix}.$$

```
r = [ ];
for p = A
    r = [r any(p) - all(p)];
end
r
```

8. Read the Help entry for **find**. If A is a matrix then find(A) returns the column whose entries are the indices in A(:) at which nonzero entries occur. Devise a one-line MatLab command that returns the number of nonzero entries in a matrix A.

9. Read the Help entries for **rem** and **fix**. Then write a MatLab script that solves the following problem: given positive integers m, n and p, $1 \le p \le mn$, find the pair of integers i, j such that the (i, j) entry in an m × n matrix A is the pth entry in its column form A(:).

Hint:
```
m = input('enter m: ');
n = input('enter n: ');
p = input('enter p: ');
if rem(p, m) == 0
    i = m;
    j = fix(p/m);
else
    i = rem(p, m);
    j = fix(p/m) + 1;
end
i, j
```

10. Write a script that calls for the input of an m × n matrix A and then prints out the set of positions i, j for which $a_{ij} \neq 0$.

   Hint:
   ```
   A = input('enter a matrix A: ');
   disp([ ])
   j = 1;
   disp([' column    rows']);
   for p = A
       f = find(p);
       if any(f ~=0)
           disp([j f' ])
       end
       j = j+1;
   end
   ```

11. Test the random number generator in MatLab by generating 1000 random numbers and counting how many lie in each of the half-open intervals [k·0.1, (k + 1)·0.1), k = 0, ..., 9.

12. Write a MatLab script that calls for the input of a matrix A and then replaces A by PA where P is a permutation matrix so chosen that all zero rows of PA are at the bottom of the matrix.

   Hint:
   ```
   A = input('enter a matrix A');
   [m, n] = size(A);
   v = find(any(A'));
   p = length(v);
   A = [A(v,:); zeros(m - p, n)]
   ```

13. Read the Help entries for **max** and **min**. Write a MatLab script that calls for the input of a matrix A and then exhibits the set of all pairs of integers i, j such that a_{ij} is the entry in A of largest absolute value.

   Hint:
   ```
   A = input('enter a matrix A: ');
   mm = max(max(abs(A)));
   M = (abs(A) == mm*ones(A));        %M is the matrix of 0's
   ```

```
            disp([ ]);                    %and 1's in which a 1 appears
            j = 1;                        %if the corresponding entry in
            disp(['  column    rows'])    %A has max absolute value.
            for p = M
                f = find(p);
                if any(f ~= 0)
                    disp([j f' ])
                end
                j = j+1;
            end
```

14. Review the Help entries for **any, all, real,** and **imag**. Then devise one-line commands to test for each of the following:

    (a) every entry of a matrix A is less than 1 in absolute value
    (b) a square matrix A is hermitian
    (c) a square matrix A is real symmetric
    (d) a square matrix A is normal
    (e) a square matrix A is unitary   Hint: all(all(A*A'==eye(A*A')))
    (f) a matrix A is square

    In solving these problems keep in mind that if ρ is a relational operator (<, <=, ==, ~=, >=, >) and A and B are the same size then A ρ B returns a 0, 1 matrix of the same size as A and B whose i, j entry is 0 if $a_{ij} \rho b_{ij}$ is false and 1 if $a_{ij} \rho b_{ij}$ is true.

    Similarly, if ρ is a logical operator (&, |, ~) then: A&B has 0's where either A or B does and 1's otherwise; A|B has 0's where both A and B do and 1's otherwise; ~A has 0's where A has nonzero elements and 1's where A has 0's.

15. Review the Help entries for **any, all, tril,** and **triu**. Then devise one-line commands as follows:

    (a) Use **any** to determine if a square matrix A is upper triangular

    Hint: any(any(tril(A, -1))). The explanation for this is the following: tril(A, k) is the matrix which agrees with A in positions (i, j) for which i + k ≥ j and is 0 otherwise. If k = -1 then tril(A, -1) has the following form:

$$\begin{array}{c} & \begin{array}{cccc} 1 & 2 & 3 & 4 \end{array} \\ \begin{array}{c} 1 \\ 2 \\ 3 \\ 4 \end{array} & \left[\begin{array}{cccc} 0 & 0 & 0 & 0 \\ x & 0 & 0 & 0 \\ x & x & 0 & 0 \\ x & x & x & 0 \end{array} \right] \end{array} \text{, i.e., } i-1 \geq j \text{ or } i \geq j+1.$$

Then if any(any(tril(A, -1))) is 0 it means that all the lower triangular entries are 0. If it is 1 then there is a nonzero entry in A below the main diagonal.

(b) Use **any** to determine if a square matrix A is lower triangular
(c) Use **all** to determine if a square matrix A is upper triangular
(d) Use **all** to determine if a square matrix A is lower triangular

16. Read the Help entry for **hankel**. An n-square Hankel matrix H has all elements on each diagonal parallel to the **sinister diagonal** equal. The sinister diagonal of H is the diagonal consisting of $h_{1,n}, h_{2,n-1}, h_{3,n-2}, \ldots, h_{n,1}$. For example, if c = [6 4 5 0 1] and r = [0 2 3 0 0] then A = hankel(c, r) generates the matrix

$$A = \begin{bmatrix} 6 & 4 & 5 & 0 & 1 \\ 4 & 5 & 0 & 1 & 2 \\ 5 & 0 & 1 & 2 & 3 \\ 0 & 1 & 2 & 3 & 0 \\ 1 & 2 & 3 & 0 & 0 \end{bmatrix}.$$

Note that the last entry of c is always the sinister diagonal entry. Incidentally hankel(c) produces the same matrix as hankel(c, zeros(1, n)), i.e., it is 0 below the sinister diagonal.

(a) The **sinister diagonal matrix** S is the matrix whose sinister diagonal consists of 1's and whose other entries are 0. Devise a MatLab function named sinis whose value for any positive integer n is the n-square sinister diagonal matrix. Save function sinis.

Hint:

```
function S = SINIS(n)
%Sinis(n) is the n-square
%matrix with 1's on the
%sinister diagonal, 0's elsewhere.
X = eye(n);
S = X(:,n:-1:1);
```

(b) Prove: T is a Toeplitz matrix if and only if either of ST or TS is a Hankel matrix.

Hint: Let H = ST and assume T is a Toeplitz matrix. Then

$$T_{nj} = T_{n-1, j-1} = T_{n-2, j-2},$$

etc. Then

$$H_{ij} = (ST)_{ij} = \sum_{k=1}^{n} S_{ik} T_{kj} = \sum_{k=1}^{n} (\delta_{n-i+1, k}) T_{kj} = T_{n-i+1, j}.$$

Thus

$$H_{1j} = T_{nj} = T_{n-1, j-1} = T_{n-2, j-2},$$

etc. But

$$H_{2, j-1} = T_{n-1, j-1}, \quad H_{3, j-2} = T_{n-2, j-2},$$

etc. Hence

$$H_{1j} = H_{2, j-1} = H_{3, j-2} \dots$$

and thus H is a Hankel matrix. The rest of the argument is similar.

(c) Write a MatLab function named hankint whose value for any positive integer n is a random n-square Hankel matrix with integer entries in the range -10, ..., 10.

Hint:

function H = hankint(n)
%Hankint(n) delivers a random n-square Hankel matrix with
%integer entries in the range -10:10
c = floor(21*rand(1,n) - 10);
r = floor(21*rand(1,n) - 10);
H = hankel(c, r);

3.2 Glossary

```
~  . . . . . . . . . . . . . . . . . . . . . . . .  144
&  . . . . . . . . . . . . . . . . . . . . . . . .  144
<  . . . . . . . . . . . . . . . . . . . . . . . .  144
<= . . . . . . . . . . . . . . . . . . . . . . . .  144
~= . . . . . . . . . . . . . . . . . . . . . . . .  144
== . . . . . . . . . . . . . . . . . . . . . . . .  144
>  . . . . . . . . . . . . . . . . . . . . . . . .  144
>= . . . . . . . . . . . . . . . . . . . . . . . .  144
all . . . . . . . . . . . . . . . . . . . . . . .  142
any . . . . . . . . . . . . . . . . . . . . . . .  142
conformal partitioning . . . . . . . . . . . . . .  136
find . . . . . . . . . . . . . . . . . . . . . . .  142
fix  . . . . . . . . . . . . . . . . . . . . . . .  142
floor . . . . . . . . . . . . . . . . . . . . . .  139
hankel . . . . . . . . . . . . . . . . . . . . . .  145
imag . . . . . . . . . . . . . . . . . . . . . . .  144
length . . . . . . . . . . . . . . . . . . . . . .  139
logical operator . . . . . . . . . . . . . . . . .  144
max . . . . . . . . . . . . . . . . . . . . . . .  143
min . . . . . . . . . . . . . . . . . . . . . . .  143
real . . . . . . . . . . . . . . . . . . . . . . .  144
relational operator . . . . . . . . . . . . . . .  144
rem . . . . . . . . . . . . . . . . . . . . . . .  142
sinister diagonal . . . . . . . . . . . . . . . .  145
```

toeplitz 140
tril   144
triu 144
| 144

3.3 Kronecker Products

Partitioned (block) matrix multiplication is useful in examining products of Kronecker products.

Theorem 1.

Let A, B, C, and D be matrices for which the two products AC and BD are defined. Then

$$(A \otimes B)(C \otimes D) = AC \otimes BD. \qquad (1)$$

Proof.

Suppose A is $m \times n$, C is $n \times r$, B is $p \times q$, and D is $q \times s$. Then $A \otimes B$ can be partitioned into mn blocks in which the (i, j) block is $a_{ij}B$, $i = 1, ..., m$, $j = 1, ..., n$. It is convenient to denote this partitioning by

$$A \otimes B = [a_{ij} B]. \qquad (2)$$

Similarly partition $C \otimes D$:

$$C \otimes D = [c_{ij} D]. \qquad (3)$$

The partitionings (2) and (3) are conformal for block multiplication; i.e., the ith block row in (2) is

$$[a_{i1} B \mid a_{i2} B \mid \cdots \mid a_{in} B]$$

and the jth block column in (3) is

3.3 Kronecker Products

$$\begin{bmatrix} c_{1j} D \\ \hline c_{2j} D \\ \hline \vdots \\ \hline c_{nj} D \end{bmatrix}. \tag{4}$$

Thus applying §3.2 Theorem 1, we see that the (i, j) block in $(A \otimes B)(C \otimes D)$ is

$$\sum_{k=1}^{n} (a_{ik} B)(c_{kj} D)$$
$$= \left(\sum_{k=1}^{n} a_{ik} c_{kj} \right) BD$$
$$= (AC)_{ij} BD. \tag{5}$$

But (5) is precisely the (i, j) block in the partitioning of $(AC) \otimes (BD)$. ∎

Let X be an m × n matrix. We can associate with X a unique mn × 1 matrix by writing the columns of X in succession in a single column matrix which we denoted earlier by c(X), the *column form* of X:

$$c(X) = \begin{bmatrix} X^{(1)} \\ \hline X^{(2)} \\ \hline \vdots \\ \hline X^{(n)} \end{bmatrix}. \tag{6}$$

Recall the Kronecker delta:

$$\delta_{ij} = \begin{cases} 0 & \text{if } i \neq j \\ 1 & \text{if } i = j. \end{cases}$$

If A is a p × m matrix then $I_n \otimes A$ can be partitioned into n^2 blocks of size p × m in which each block is $\delta_{ij}A$.

$$I_n \otimes A = \begin{array}{c} \\ p \\ p \\ \\ p \end{array} \begin{bmatrix} \overset{m}{A} & \overset{m}{0_{p,m}} & \cdots & \overset{m}{0_{p,m}} \\ 0_{p,m} & A & & 0_{p,m} \\ \hline & & & \\ 0_{p,m} & & 0_{p,m} & A \end{bmatrix} \qquad (7)$$

The matrices (6) and (7) are conformally partitioned for multiplication and thus by §3.2 Theorem 1,

$$(I_n \otimes A)\, c(X) = \begin{bmatrix} AX^{(1)} \\ \hline AX^{(2)} \\ \hline \vdots \\ \hline AX^{(n)} \end{bmatrix}$$

$$= \begin{bmatrix} (AX)^{(1)} \\ \hline (AX)^{(2)} \\ \hline \vdots \\ \hline (AX)^{(n)} \end{bmatrix}$$

$$= c(AX). \qquad (8)$$

3.3 Kronecker Products

Suppose now that D is an n × q matrix. Then

$$c(XD) = \begin{bmatrix} (XD)^{(1)} \\ \hline (XD)^{(2)} \\ \hline \vdots \\ \hline (XD)^{(q)} \end{bmatrix}$$

$$= \begin{bmatrix} \sum_{t=1}^{n} X^{(t)} d_{t1} \\ \hline \sum_{t=1}^{n} X^{(t)} d_{t2} \\ \hline \vdots \\ \hline \sum_{t=1}^{n} X^{(t)} d_{tq} \end{bmatrix}$$

$$= \begin{bmatrix} d_{11}I_m & d_{21}I_m & \cdots & d_{n1}I_m \\ d_{12}I_m & d_{22}I_m & \cdots & d_{n2}I_m \\ \vdots & \vdots & \vdots & \vdots \\ d_{1q}I_m & d_{2q}I_m & \cdots & d_{nq}I_m \end{bmatrix} \begin{bmatrix} X^{(1)} \\ \hline X^{(2)} \\ \hline \vdots \\ \hline X^{(n)} \end{bmatrix}$$

$$= (D^T \otimes I_m) \, c(X). \tag{9}$$

Note that in (9) the identity matrix I_m has the same dimensions as the number of rows of X. If we combine (8) and (9), we have

$$c(AXD) = c((AX)D)$$

$$= (D^T \otimes I_p) c(AX) \quad \text{(i.e. the number of rows of AX is p)}$$

$$= (D^T \otimes I_p)(I_n \otimes A) c(X)$$

$$= (D^T \otimes A) c(X). \tag{10}$$

The formula (10) shows that ordinary matrix multiplication AXD can be effected by multiplying $c(X)$ by the Kronecker product $D^T \otimes A$.

As an example of the use of formula (10) suppose we want to solve an equation of the form

$$AXD = B \tag{11}$$

where A is $p \times m$, the required matrix X is $m \times n$, D is $n \times q$, and B is $p \times q$. Then (11) is equivalent to solving the equation

$$(D^T \otimes A) c(X) = c(B). \tag{12}$$

For the purposes of illustration let

$$A = \begin{bmatrix} 1 & -1 \\ 0 & 1 \end{bmatrix}, \quad D = \begin{bmatrix} 0 & -1 \\ 1 & 0 \end{bmatrix},$$

$$B = \begin{bmatrix} 0 & 1 \\ 0 & 0 \end{bmatrix}, \quad X = \begin{bmatrix} x_{11} & x_{12} \\ x_{21} & x_{22} \end{bmatrix}.$$

Then (12) becomes

3.3 Kronecker Products

$$\begin{bmatrix} 0 & 0 & 1 & -1 \\ 0 & 0 & 0 & 1 \\ -1 & 1 & 0 & 0 \\ 0 & -1 & 0 & 0 \end{bmatrix} \begin{bmatrix} x_{11} \\ x_{21} \\ x_{12} \\ x_{22} \end{bmatrix} = \begin{bmatrix} 0 \\ 0 \\ 1 \\ 0 \end{bmatrix},$$

or

$$x_{12} - x_{22} = 0$$
$$x_{22} = 0$$
$$-x_{11} + x_{21} = 1$$
$$-x_{21} = 0.$$

Hence

$$X = \begin{bmatrix} -1 & 0 \\ 0 & 0 \end{bmatrix}.$$

It is simple to check that (11) is satisfied for this value of X.

An important result holds for Kronecker product multiplication of matrices, namely that it is **associative**. Let A be $m \times n$, B be $p \times q$, and C be $e \times f$.

Theorem 2.

Kronecker product multiplication is associative:

$$A \otimes (B \otimes C) = (A \otimes B) \otimes C. \tag{13}$$

Proof.

Let e_i^n denote the single column matrix with n rows, with a 1 in row position i, and 0 elsewhere. (If the n is understood, we sometimes omit it and write e_i.) Then note that

$$A e_i^n = A^{(i)}.$$

Suppose for the moment we know that for single column matrices, associativity of Kronecker product multiplication is true:

$$(x \otimes y) \otimes z = x \otimes (y \otimes z). \tag{14}$$

Then using Theorem 1 we compute that

$$(A \otimes (B \otimes C))(e_i^n \otimes (e_j^q \otimes e_k^f)) = Ae_i^n \otimes (B \otimes C)(e_j^q \otimes e_k^f)$$
$$= Ae_i^n \otimes (Be_j^q \otimes Ce_k^f)$$
$$= A^{(i)} \otimes (B^{(j)} \otimes C^{(k)}).$$

By (14),

$$A^{(i)} \otimes (B^{(j)} \otimes C^{(k)}) = (A^{(i)} \otimes B^{(j)}) \otimes C^{(k)}$$
$$= (Ae_i^n \otimes Be_j^q) \otimes Ce_k^f$$
$$= (A \otimes B)(e_i^n \otimes e_j^q) \otimes Ce_k^f$$
$$= ((A \otimes B) \otimes C)((e_i^n \otimes e_j^q) \otimes e_k^f). \tag{15}$$

Now observe that

$$e_i^n \otimes e_j^q = \begin{bmatrix} 0 \\ \hline 0 \\ \hline \vdots \\ \hline e_j^q \\ \hline \vdots \\ \hline 0 \\ \hline 0 \end{bmatrix} \begin{matrix} \\ \\ \\ \text{block } i \rightarrow \\ \\ \\ \end{matrix} \quad \text{(n blocks of size } q \times 1\text{)} \tag{16}$$

3.3 Kronecker Products

In (16) the 1 appears in row position $(i-1)q+j$. Thus

$$e_i^n \otimes e_j^q = e_{(i-1)q+j}^{nq}, \quad i=1,\cdots,n, \; j=1,\cdots,q. \tag{17}$$

Distinct i and j produce distinct $e_i^n \otimes e_j^q$ and every e_t^{nq}, $t = 1, \ldots, nq$, appears in the form (17). Next apply (17) to obtain

$$(e_i^n \otimes e_j^q) \otimes e_k^f = e_{(i-1)q+j}^{nq} \otimes e_k^f = e_{((i-1)q+j-1)f+k}^{nqf} \tag{18}$$

and

$$e_i^n \otimes (e_j^q \otimes e_k^f) = e_i^n \otimes e_{(j-1)f+k}^{qf} = e_{(i-1)qf+(j-1)f+k}^{nqf}. \tag{19}$$

Note that the subscripts on the right in (18) and (19) are the same. Hence

$$(e_i^n \otimes e_j^q) \otimes e_k^f = e_{(i-1)qf+(j-1)f+k}^{nqf} = e_i^n \otimes (e_j^q \otimes e_k^f). \tag{20}$$

We will denote this common value as

$$e_i^n \otimes e_j^q \otimes e_k^f.$$

Finally, we confirm that single column Kronecker product multiplication is associative, i.e., that (14) holds.

Suppose that

$$x = \sum_{i=1}^{n} x_i \, e_i^n, \quad y = \sum_{j=1}^{q} y_j \, e_j^q, \quad z = \sum_{k=1}^{f} z_k \, e_k^f.$$

Then, as we will confirm below in §3.3 Exercises, #4, we have

$$x \otimes y = \sum_{i=1, j=1}^{n,q} x_i \, y_j \, e_i^n \otimes e_j^q$$

so that

$$(x \otimes y) \otimes z = \left(\sum_{i=1, j=1}^{n,q} x_i y_j e_i^n \otimes e_j^q \right) \otimes \sum_{k=1}^{f} z_k e_k^f$$

$$= \sum_{i=1, j=1, k=1}^{n,q,f} x_i y_j z_k (e_i^n \otimes e_j^q) \otimes e_k^f \qquad \text{(by \#4)}$$

$$= \sum_{i=1, j=1, k=1}^{n,q,f} x_i y_j z_k e_i^n \otimes (e_j^q \otimes e_k^f) \qquad \text{(by (20))}$$

$$= \sum_{i=1}^{n} x_i e_i^n \otimes \left(\sum_{j=1}^{q} y_j e_j^q \otimes \sum_{k=1}^{f} z_k e_k^f \right) \qquad \text{(by \#4)}$$

$$= x \otimes (y \otimes z).$$

The common value of $(x \otimes y) \otimes z = x \otimes (y \otimes z)$ will be denoted by

$$x \otimes y \otimes z.$$

From (15) we have

$$(A \otimes (B \otimes C))(e_i^n \otimes (e_j^q \otimes e_k^f)) = ((A \otimes B) \otimes C)((e_i^n \otimes e_j^q) \otimes e_k^f) \qquad (21)$$

and as we observed in (17) and (20), the $nqf \times 1$ matrices $e_i^n \otimes (e_j^q \otimes e_k^f)$ and $(e_i^n \otimes e_j^q) \otimes e_k^f$ are equal, and moreover their common value is an $nqf \times 1$ matrix of the form

$$e_t^{nqf}, \quad t = 1, \cdots, nqf.$$

This means that $A \otimes (B \otimes C)$ and $(A \otimes B) \otimes C$ match up column by column. ∎

The common value of $A \otimes (B \otimes C)$ and $(A \otimes B) \otimes C$ will be denoted by

3.3 Kronecker Products

$$A \otimes B \otimes C. \tag{22}$$

As with ordinary matrix multiplication, associativity allows us to unambiguously define Kronecker products of more than 3 matrices. For example,

$$A \otimes B \otimes C \otimes D = (A \otimes B) \otimes (C \otimes D).$$

Associativity can be used to answer such questions as the following: if A is 2×3, B is 2×2, and C is 5×7, what is the (17, 23) entry of $A \otimes B \otimes C$ in terms of entries of A, B, and C? Write

$$(A \otimes B \otimes C)_{17,\,23} = (A \otimes (B \otimes C))_{17,\,23}$$
$$= a_{22} (B \otimes C)_{7,\,9}$$

because $B \otimes C$ is 10×14. The (7, 9) entry of $B \otimes C$ is $b_{22} c_{22}$ (why?). Thus

$$(A \otimes B \otimes C)_{17,\,23} = a_{22} b_{22} c_{22}.$$

The kth *Kronecker power* of an $m \times n$ matrix is

$$A \otimes A \otimes \cdots \otimes A \tag{23}$$

in which A appears k times in (23), $k \geq 1$. The matrix (23) is usually denoted by

$$A^{[k]}.$$

3.3 Exercises

1. Prove that $(A \otimes B)^T = A^T \otimes B^T$, and if A and B are complex matrices then $(A \otimes B)^* = A^* \otimes B^*$.

2. Prove that if A_i and B_i are n_i-square matrices, $i = 1, \ldots, p$, then
$$\sum_{i=1}^{p} \oplus A_i \sum_{i=1}^{p} \oplus B_i = \sum_{i=1}^{p} \oplus A_i B_i .$$

   Hint: Use block multiplication.

3. Write $c\left(\begin{bmatrix} 2 & 1 \\ 3 & 5 \end{bmatrix} \begin{bmatrix} x_{11} & x_{12} \\ x_{21} & x_{22} \end{bmatrix} \begin{bmatrix} -1 & 2 \\ 0 & 1 \end{bmatrix}\right)$ as a Kronecker product times $c(X)$.

   Hint: From formula (10), $c(AXD) = (D^T \otimes A)c(X)$. In this case
$$A = \begin{bmatrix} 2 & 1 \\ 3 & 5 \end{bmatrix}, \quad D = \begin{bmatrix} -1 & 2 \\ 0 & 1 \end{bmatrix}$$

   so that
$$D^T \otimes A = \begin{bmatrix} -1 & 0 \\ 2 & 1 \end{bmatrix} \otimes \begin{bmatrix} 2 & 1 \\ 3 & 5 \end{bmatrix} = \begin{bmatrix} -2 & -1 & 0 & 0 \\ -3 & -5 & 0 & 0 \\ 4 & 2 & 2 & 1 \\ 6 & 10 & 3 & 5 \end{bmatrix}.$$

4. Prove that
$$(A + B) \otimes C = A \otimes C + B \otimes C$$

   and
$$C \otimes (A + B) = C \otimes A + C \otimes B.$$

   Hint: $(A + B) \otimes C = [(A + B)_{ij}C] = [a_{ij}C + b_{ij}C] = [a_{ij}C] + [b_{ij}C] = A \otimes C + B \otimes C$. The other equality is similarly proved.

3.3 Exercises

5. If A is p × q and B is m × n, devise an algorithm for computing the (s, t) entry of $A \otimes B$.

   Hint: Let $C = A \otimes B$. Then $c_{st} = a_{ij}b_{kw}$. The integers, i, j, k, w are computed by the following algorithm:

   Step 1: divide s by m, s = mu + h, $0 \le h < m$. Then

   $$i = \begin{cases} u + 1 & \text{if } h > 0 \\ u & \text{if } h = 0 \end{cases}$$

   and

   $$k = \begin{cases} h & \text{if } h > 0 \\ m & \text{if } h = 0 \end{cases}$$

   Step 2: divide t by n, t = nv + r, $0 \le r < n$. Then

   $$j = \begin{cases} v + 1 & \text{if } r > 0 \\ v & \text{if } r = 0 \end{cases}$$

   and

   $$w = \begin{cases} r & \text{if } r > 0 \\ n & \text{if } r = 0. \end{cases}$$

6. Let A be an m × n matrix with m, n ≥ 2. Show that $(A^{[k]})_{22} = a_{11}^{k-1} a_{22}$.

   Hint: Use Theorem 2 to write $A^{[k]} = A \otimes A^{[k-1]}$.

7. Prove that $(A \otimes B) \otimes (C \otimes D) = (A \otimes (B \otimes C)) \otimes D = A \otimes ((B \otimes C) \otimes D)$.

8. Show that the Kronecker product of $A_1, A_2, \ldots, A_m$ is well defined, i.e., use the associative law to show that the way in which parentheses are inserted in order to compute $A_1 \otimes A_2 \otimes \cdots \otimes A_m$ is irrelevant. As indicated earlier, the Kronecker product of $A_1, \ldots, A_m$ is denoted by $A_1 \otimes \cdots \otimes A_m$ or $\overset{m}{\underset{1}{\otimes}} A_i$.

   Hint: Use Theorem 2 and induction on m.

9. Prove that if the products $A_i B_i$ are defined, $i = 1, \ldots, m$, then
$$\left(\overset{m}{\underset{i=1}{\otimes}} A_i \right) \left(\overset{m}{\underset{i=1}{\otimes}} B_i \right) = \overset{m}{\underset{i=1}{\otimes}} A_i B_i.$$

   Hint: Use Theorem 1 and induction on m.

10. Let $A = \begin{bmatrix} 1 & 1 \\ 0 & 1 \end{bmatrix}$, $B = [1]$, $C = [1]$. Show that $A \otimes (B \oplus C) \neq A \otimes B \oplus A \otimes C$.

11. Let B be $p \times p$, C be $q \times q$, and A be $n \times n$. Prove that $(B \oplus C) \otimes A = B \otimes A \oplus C \otimes A$.

12. Prove: If A_i is hermitian, $i = 1, \ldots, m$, then $A_1 \otimes \cdots \otimes A_m$ is hermitian.

    Hint: Use #1 and induction on m.

13. Prove: If I_{n_i} is the $n_i \times n_i$ identity matrix, $i = 1, \ldots, m$, and $n = n_1 \ldots n_m$ then $I_n = I_{n_1} \otimes \cdots \otimes I_{n_m}$.

    Hint: Use Theorem 2 and induction on m.

14. Let A_t be a diagonal matrix, $t = 1, \ldots, m$. Prove that $A_1 \otimes \cdots \otimes A_m$ is a diagonal matrix.

    Hint: Use Theorem 2 and induction on m.

15. Let A_t be $n_t \times n_t$, $t = 1, \ldots, m$. Prove that the main diagonal entries of $A_1 \otimes \cdots \otimes A_m$ are $A_1(\alpha(1), \alpha(1)) \cdots A_m(\alpha(m), \alpha(m))$, $\alpha \in \Gamma(n_1, \ldots, n_m)$.

Hint: By definition, the main diagonal entries of the Kronecker product of two square matrices A and B are all possible products of the form $a_{ii}b_{jj}$. By induction on m, any diagonal entry of $A_1 \otimes \cdots \otimes A_{m-1}$ is of the form $A_1(\beta(1), \beta(1)) \cdots A_{m-1}(\beta(m-1), \beta(m-1))$, $\beta \in \Gamma(n_1, \ldots, n_{m-1})$. Thus any main diagonal entry of $A_1 \otimes \cdots \otimes A_m$ is of the form

$$A_1(\beta(1), \beta(1)) \cdots A_{m-1}(\beta(m-1), \beta(m-1)) A_m(\alpha(m), \alpha(m))$$

where $\alpha(m)$ is any integer, $1 \leq \alpha(m) \leq n_m$. Clearly, this product is of the form $A_1(\alpha(1), \alpha(1)) \cdots A_m(\alpha(m), \alpha(m))$ where $\alpha = \beta{:}\alpha(m)$ (i.e., $\alpha(m)$ is adjoined to β).

16. Let A and B be $p \times p$ matrices with $AB = BA$. Show that the binomial theorem holds: $(A + B)^n = A^n + \binom{n}{1} A^{n-1}B + \binom{n}{2} A^{n-2}B^2 + \cdots + B^n$.

    Hint: Induction on n.

17. Let A and B be square matrices of the same dimensions, say $p \times p$. Prove that the Schur (or Hadamard) product is a principal submatrix of the Kronecker product $A \otimes B$.

    Hint: By #5, $(A \otimes B)_{(i-1)p+i,\, (j-1)p+j} = a_{ij}b_{ij}$.

18. Let A and B be $n \times n$ matrices and let C be the $2n \times 2n$ partitioned matrix $C = \begin{bmatrix} A & B \\ B & -A \end{bmatrix}$. Find C^2.

19. Let A be $m \times n$, B be $p \times q$. For α in $\Gamma(m, p)$ and β in $\Gamma(n, q)$, define $k_{\alpha, \beta} = a_{\alpha(1), \beta(1)} b_{\alpha(2), \beta(2)}$ and arrange the numbers $k_{\alpha, \beta}$ in an $mp \times nq$ matrix K, lexicographically in α and β, i.e., the "(α, β) entry" of K is $k_{\alpha, \beta}$ where the α are ordered lexicographically and the β are ordered lexicographically. Show that $K = A \otimes B$.

    Hint: A is $m \times n$ and B is $p \times q$. Write the sequences α in $\Gamma(m, p)$ in lexicographic order to index the rows in a matrix K with mp rows. Similarly, write the sequences β in $\Gamma(n, q)$ in lexicographic order to index the columns of K with nq columns. Schematically K has the following appearance:

$\Gamma(n, q)$

t^{th} block

	(1, 1) ... (1, q)	...	(t, 1) ... (t, q)	...	(n, 1) ... (n, q)
(1, 1)					
(1, p)					
(s, 1)					
(s, p)			$a_{st} B$		
(m, 1)					
(m, p)					

with $\Gamma(m, p)$ indexing rows and s^{th} block indicated.

Each block in the indicated partitioning of K is p × q. By definition, the element $k_{\alpha, \beta}$ is equal to $a_{\alpha(1), \beta(1)} b_{\alpha(2), \beta(2)}$. If α is in the sth block and β is in the tth block then α has the form $\alpha = (s, i)$ and β has the form $\beta = (t, j)$. Then $k_{\alpha, \beta} = a_{st} b_{ij}$ where i runs from 1, ..., p and j runs from 1, ..., q. That is, the entries of K in block row s and block column t precisely comprise the matrix $a_{st} B$. Thus $K = A \otimes B$.

20. Let A_i be $m_i \times n_i$ matrix, i = 1, ..., p. For α in $\Gamma(m_1, ..., m_p)$ and β in $\Gamma(n_1, ..., n_p)$, define $k_{\alpha, \beta} = A_1(\alpha(1), \beta(1)) \cdots A_p(\alpha(p), \beta(p))$ and let $K = [k_{\alpha, \beta}]$, with the $k_{\alpha, \beta}$ arranged lexicographically in α and β. Prove that $K = A_1 \otimes \cdots \otimes A_p$.

Hint: The sequences in $\Gamma(m_1, ..., m_p)$ can be arranged in lexicographic order as follows. Let $\Gamma(m_2, ..., m_p)$ be arranged in lexicographic order. Then let $s:\Gamma(m_2, ..., m_p)$ denote the sequences in $\Gamma(m_1, ..., m_p)$ obtained by prefixing each sequence in $\Gamma(m_2, ..., m_p)$ by s. Then $\Gamma(m_1, ..., m_p)$ in lexicographic order is

$$1:\Gamma(m_2, ..., m_p); \; 2:\Gamma(m_2, ..., m_p); \; ... \; ; \; m_1:\Gamma(m_2, ..., m_p).$$

Similarly, $\Gamma(n_1, ..., n_p)$ is arranged lexicographically by

$$1:\Gamma(n_2, ..., n_p); \; ... \; ; \; n_1:\Gamma(n_2, ..., n_p).$$

By Theorem 2, write the matrix $A_1 \otimes A_2 \otimes \cdots \otimes A_p$ as $A_1 \otimes B$ where $B = A_2 \otimes \cdots \otimes A_p$. Let $1 \leq \alpha(1) \leq m_1$ and $1 \leq \beta(1) \leq n_1$. Then the $\alpha(1)$, $\beta(1)$ block in $A_1 \otimes B$ is $A_1(\alpha(1), \beta(1))B$. But, by induction on p, the entries of B are indexed rowwise by the sequences $(\alpha(2), ..., \alpha(p)) \in \Gamma(m_2, ..., m_p)$ and columnwise by the sequences $(\beta(2), ..., \beta(p)) \in \Gamma(n_2, ..., n_p)$. Thus, by #19, the entry in row position $\alpha = (\alpha(1), \alpha(2), ..., \alpha(p)) = \alpha(1):(\alpha(2), ..., \alpha(p))$ and column position $\beta = (\beta(1), \beta(2), ..., \beta(p)) = \beta(1):(\beta(2), ..., \beta(p))$ is $A_1(\alpha(1), \beta(1)) \, B((\alpha(2), ..., \alpha(p)), (\beta(2), ..., \beta(p)))$. But, by induction on p, the second factor in the preceding product is $A_2(\alpha(2), \beta(2)) \cdots A_p(\alpha(p), \beta(p))$. In other words, the α, β entry of $A_1 \otimes \cdots \otimes A_p$ is

$$A_1(\alpha(1), \beta(1))A_2(\alpha(2), \beta(2)) \cdots A_p(\alpha(p), \beta(p)).$$

21. If A is n × n then recall that A is *lower (upper) triangular* if $a_{ij} = 0$ whenever $j > i$ ($i > j$). Prove that if A and B are lower (upper) triangular and AB is defined, then AB is lower (upper) triangular.

22. Prove: If $A_1, ..., A_p$ are lower (upper) triangular then so is $K = A_1 \otimes \cdots \otimes A_p$.

Hint: By #20, if $\alpha \in \Gamma(m_1, ..., m_p)$ and $\beta \in \Gamma(n_1, ..., n_p)$ then $K_{\alpha, \beta} = \prod_{t=1}^{p} A_t(\alpha(t), \beta(t))$. If $\alpha > \beta$ in lexicographic order then there is an integer t, $1 \leq t \leq p$, such that $\alpha(t) > \beta(t)$. But then $A_t(\alpha(t), \beta(t)) = 0$.

23. Prove: If $A_1, ..., A_p$ are square matrices, and $A_1 \otimes \cdots \otimes A_p$ has at least one nonzero main diagonal entry and is lower (upper) triangular, then every A_i is lower (upper) triangular.

Hint: Suppose there exists t, $1 \leq t \leq p$, such that A_t is not lower triangular. Then there are integers $\alpha(t), \beta(t)$ such that $\alpha(t) < \beta(t)$, and $A_t(\alpha(t), \beta(t)) \neq 0$. Since $A_1 \otimes \cdots \otimes A_p$ has a nonzero diagonal entry there exists a sequence γ such that $\prod_{i=1}^{p} A_i(\gamma(i), \gamma(i)) \neq 0$. Consider the sequences

$$\alpha = (\gamma(1), ..., \gamma(t-1), \alpha(t), \gamma(t+1), ..., \gamma(p))$$

and

$$\beta = (\gamma(1), ..., \gamma(t-1), \beta(t), \gamma(t+1), ..., \gamma(p)).$$

Note that $\alpha < \beta$. Then

$$(A_1 \otimes \cdots \otimes A_p)_{\alpha, \beta} = (\prod_{i \neq t} A_i(\gamma(i), \gamma(i))) \cdot A_t(\alpha(t), \beta(t)) \neq 0.$$

Hence $A_1 \otimes \cdots \otimes A_p$ is not lower triangular, a contradiction.

24. Prove: If $A_1, ..., A_p$ are square matrices and $A_1 \otimes \cdots \otimes A_p \neq 0$ is a diagonal matrix, then every A_i is a diagonal matrix.

Hint: By #23, all of $A_1, ..., A_p$ must be both lower and upper triangular.

25. Prove: If $A_1, ..., A_p$ are complex normal matrices then so is $A_1 \otimes \cdots \otimes A_p$.

Hint: Use #1 and Theorem 1.

3.3 MatLab

1. If A is m × n and v is mn × 1 then the MatLab command A(:) = v is equivalent to the loop

    ```
    for j = 1:n
        A(:, j) = v(((j-1)*m+1):j*m))
    end
    ```

    Write a MatLab script that calls for the entry of a pair of integers m and n, generates a random mn × 1 vector, and reconstructs the unique m × n matrix A for which c(A) = v. Use the above idea.

    Hint:
    ```
    m = input('enter an integer m: ');
    n = input('enter an integer n: ');
    A = zeros(m, n);
    v = rand(m*n, 1);
    A(:) = v;
    disp(A)
    ```

2. Use Help to review **floor** and **rand**. Then write a Matlab script that calls for the input of three integers m, a, b, m > 0, a < b, and then outputs an m-square matrix A whose entries are random integers in [a, b].

    Hint:
    ```
    m = input('enter m: ');
    a = input('enter a: ');
    b = input('enter b: ');
    b = b + 1;
    A = floor(a + (b - a) * rand(m))
    ```

3. Write a MatLab script that calls for the input of three integers m, a, and b. The script then outputs m-square matrices A and B, both with entries that are random integers in [a, b], as well as the Hadamard and Kronecker products H = A · B and K = A ⊗ B. Run the script several times. Is there a relationship between H and K?

Hint:
```
m = input('enter m: ');
a = input('enter a: ');
b = input('enter b: ');
b = b + 1;
A = floor(a+(b-a)*rand(m));
B = floor(a+(b-a)*rand(m));
H = A.*B
K = kron(A, B)
```

4. Read the Help entry for **ceil** and **rem**. Then write a MatLab script that calls for the input of an m × n matrix A, a p × q matrix B, and two integers r and s, $1 \leq r \leq mp$, $1 \leq s \leq nq$. The outputs of the program are the pairs (k, i) and (t, j) for which $(A \otimes B)_{r,\,s} = a_{ki} b_{tj}$ as well as the elements a_{ki} and b_{tj}.

Hint:
```
A = input('Enter an m x n matrix A: ');
[m, n] = size(A);
B = input('Enter a p x q matrix B: ');
[p, q] = size(B);
r = input('Enter r: ');
s = input('Enter s: ');
k = ceil(r/p);
t = rem(r, p);
if t==0
        t = p;
end
i = ceil(s/q);
j = rem(s, q);
if j == 0
        j = q;
end
disp(['   k   ','   i   ','   t   ','   j   '])
disp([k i t j])
disp([A(k, i), B(t, j) ])
```

3.3 MatLab

5. Write a MatLab command that exhibits the Hadamard product, A.*B as a submatrix of the Kronecker product K = kron(A, B). (The matrices A and B are both assumed to be p-square.)

   Hint: $a_{ki}b_{ki} = (A \otimes B)_{(k-1)p+k,\ (i-1)p+i}$

6. Write a MatLab script that calls for the input of two matrices A and B and outputs K = A ⊗ B. Do not use the **kron** command.

7. Write a MatLab function called kronpwr that computes the kth Kronecker power of a matrix A. Use a recursive definition. What theorem does kronpwr depend on?

   Hint:
   ```
   function y = kronpwr(A, k)
   %KRONPWR computes the kth Kronecker power of A
   if k==1
       y = A;
   else
       y = kron(A, kronpwr(A, k - 1));
   end
   ```

8. There are built-in MatLab functions that return more than a single value, e.g., [m, n] = size(A). Write a MatLab function called sroot that does the following: for a positive number a the function computes the sequence of values

$$x_0 = 1, \quad x_{n+1} = \frac{x_n + ax_n^{-1}}{2}, \quad n = 0, 1, 2, \ldots$$

until $|x_{n+1} - x_n| < 10^{-6}$. Then the value of sroot is the value of $n+1$ and the value of x_{n+1}. The function sroot is called at the Command prompt (>>) by a statement such as [n, r] = sroot(2).

   Hint:
   ```
   function [n, r] = sroot(a)
   %SROOT uses the recursion x <-- (x + a /x)/2 to compute
   ```

```
%the square root of the positive number a.
x = 1;
y = (x + a/x)/ 2;
k = 1;
while abs(y - x) >= 1e-6
        x = y;
        y = (x + a/x)/ 2;
        k = k + 1;
end
n = k;
r = y;
```

9. Modify the function sroot in #8 so that it will compute the square root of a matrix a. Use the notation a/x to denote ax^{-1}. Use Help to read the entry for /. We have not yet discussed inverses of square matrices, but for the purposes of this problem assume that the inverse is definable and that it satisfies $xx^{-1} = x^{-1}x = I_n$. Call the function matroot and invoke it precisely in the same way as sroot.

Hint:
```
function [n, r] = matroot(a)
%MATROOT uses the recursion x <-- (x + a/x)/ 2 to compute
%the square root of the matrix a.
x = eye(a);
y = (x + a/x)/2;
k = 1;
while max(max(abs(y - x))) >= 1e-6
        x = y;
        y = (x + a/x)/ 2;
        k = k + 1;
end
n = k;
r = y;
```

10. Use the function matroot to compute the square root of each of the following matrices. Verify your answers by squaring the output of matroot.

(a) $\begin{bmatrix} 5 & 3 \\ 3 & 2 \end{bmatrix}$

(b) $\begin{bmatrix} 3 & 1 & 1 \\ 1 & 3 & 1 \\ 1 & 1 & 3 \end{bmatrix}$

(c) I_4

(d) $\begin{bmatrix} 1 & 1 & 0 \\ 0 & 1 & 1 \\ 0 & 0 & 1 \end{bmatrix}$

(e) $\begin{bmatrix} 0 & 1 & 0 \\ -1 & 0 & 1 \\ 0 & -1 & 0 \end{bmatrix}$

3.3 Glossary

/	168
$A^{[k]}$	157
$c(X)$	149
ceil	166
column form	149
e_i^n	153
floor	165
Kronecker delta	149
Kronecker power	157
lower triangular	163
rand	165
rem	166
upper triangular	163

Chapter 4

Elementary Matrices and Rank

Topics
- *inverses*
- *elementary operations and matrices*
- *Hermite Normal Form*
- *canonical form*
- *rank*
- *Moore-Penrose inverse*
- *least squares*

4.1 Inverses

We begin with a careful description of the sets from which the entries in a matrix are chosen. For example, A may have integer entries, rational number entries, entries which are polynomials in x with real or, perhaps, complex coefficients. The entries of a matrix can themselves be matrices, e.g., the entries of A might be real quaternions (see §1.2 Exercises, #26). In each of these sets, i.e., integers, rational numbers, polynomials in x, real quaternions, there is an *identity* for multiplication, call it 1, the multiplication is associative, and, in general, some but not necessarily all entries have multiplicative inverses. For example, 1 and -1 have multiplicative inverses (i.e., themselves) in the integers, but 2 does not, i.e., 1/2 is not an integer. In the set of all polynomials in x with real coefficients, the polynomial x + 1 does not have a multiplicative inverse, but the constant polynomial 3 does have the inverse 1/3. In general, if r is an element of the set R from which the entries of the matrices are being taken, then r is a *unit*, or equivalently, r has an *inverse*, if there exists an element s ∈ R such that rs = sr = 1.

Recall that $M_n(R)$ is the set of n-square matrices whose entries are in the set R. If $A \in M_n(R)$ and there exists $B \in M_n(R)$ such that $AB = BA = I_n$ then we say that A itself is a *unit matrix*. Other terms in common use are: A is *nonsingular;* A is *regular;* A has an *inverse;* A is *invertible*. Observe that if A has an inverse B, so that $AB = BA = I_n$, and $C \in M_n(R)$ satisfies $AC = CA = I_n$, then we have $C = CI_n = C(AB) = (CA)B = I_nB = B$. In other words, there is at most one matrix B for which

$$AB = BA = I_n . \qquad (1)$$

The matrix B is called the *inverse* of A and is denoted by A^{-1}. If A does not have an inverse in $M_n(R)$ then A is *singular*. Note that a matrix may not have an inverse in $M_n(R)$ but be nonsingular in $M_n(S)$ where $R \subset S$. For example, [2] has no inverse in the set of 1×1 matrices with integer entries, whereas it has the inverse [1/2] in the set of 1×1 matrices with rational number entries.

The standard notations for the sets R we generally use are:

- $\mathbb{Z}$ integers

- $\mathbb{Q}$ rational numbers

- $\mathbb{R}$ real numbers

- $\mathbb{C}$ complex numbers

- $\mathbb{Z}[x]$ polynomials in x with coefficients in $\mathbb{Z}$

- $\mathbb{Q}[x]$ polynomials in x with coefficients in $\mathbb{Q}$

- $\mathbb{R}[x]$ polynomials in x with coefficients in $\mathbb{R}$

- $\mathbb{C}[x]$ polynomials in x with coefficients in $\mathbb{C}$

The quotient of two polynomials in $\mathbb{Z}[x]$ is called a *rational function* in x over $\mathbb{Z}$. Similar language is used for quotients of polynomials in $\mathbb{Q}[x]$, $\mathbb{R}[x]$, $\mathbb{C}[x]$. The set of all rational functions in x over $\mathbb{Z}$ is denoted by $\mathbb{Z}(x)$. A similar meaning is given to $\mathbb{Q}(x)$, $\mathbb{R}(x)$, $\mathbb{C}(x)$. We assume that the reader is acquainted with the usual rules for adding and multiplying rational functions:

$$\frac{p(x)}{q(x)} + \frac{r(x)}{s(x)} = \frac{p(x)\,s(x) + q(x)\,r(x)}{q(x)\,s(x)},$$

$$\frac{p(x)}{q(x)} \cdot \frac{r(x)}{s(x)} = \frac{p(x)\,r(x)}{q(x)\,s(x)}.$$

Unless we specifically indicate otherwise, the matrices appearing henceforth will have entries from one (or more) of the sets just described. Thus, a phrase such as "let A be an m × n matrix over R" will mean simply that $A \in M_{m,n}(R)$ and that the entries of A come from a set listed above.

4.1 Exercises

In exercises #1 - 15, decide whether the matrix is nonsingular or singular.

1. $\begin{bmatrix} 1 & 2 \\ 0 & 1 \end{bmatrix} \in M_2(\mathbb{Z})$
2. $\begin{bmatrix} 1 & 0 \\ -1 & 1 \end{bmatrix} \in M_2(\mathbb{Z})$
3. $\begin{bmatrix} 1 & 0 \\ 0 & -1 \end{bmatrix} \in M_2(\mathbb{Z})$
4. $\begin{bmatrix} 1 & 0 \\ 1 & 0 \end{bmatrix} \in M_2(\mathbb{Z})$
5. $\begin{bmatrix} 1 & 1 \\ 0 & 0 \end{bmatrix} \in M_2(\mathbb{Z})$
6. $\begin{bmatrix} 1 & 1 \\ 1 & 1 \end{bmatrix} \in M_2(\mathbb{Z})$
7. $\begin{bmatrix} 0 & 0 \\ 0 & 1 \end{bmatrix} \in M_2(\mathbb{Z})$
8. $\begin{bmatrix} 1 & 0 \\ 0 & 2 \end{bmatrix} \in M_2(\mathbb{Z})$

9. $\begin{bmatrix} 1 & 0 \\ 0 & 1 \end{bmatrix} \in M_2(\mathbb{Z})$ 10. $\begin{bmatrix} 1 & x \\ 0 & 1 \end{bmatrix} \in M_2(\mathbb{Z}[x])$

11. $\begin{bmatrix} 1 & 2x+1 \\ 0 & -1 \end{bmatrix} \in M_2(\mathbb{Q}[x])$ 12. $\begin{bmatrix} x & x \\ 1 & 1 \end{bmatrix} \in M_2(\mathbb{Q}[x])$

13. $\begin{bmatrix} 2+i & 1 \\ 0 & i \end{bmatrix} \in M_2(\mathbb{C})$ 14. $\begin{bmatrix} ix+1 & 1 \\ 1 & 0 \end{bmatrix} \in M_2(\mathbb{C}[x])$

15. $\begin{bmatrix} x-1 & x^2 \\ 1 & x+1 \end{bmatrix} \in M_2(\mathbb{Q}[x])$

16. Prove: if A and B are $n \times n$ nonsingular matrices, then AB is non-singular and $(AB)^{-1} = B^{-1}A^{-1}$.

17. Prove: if $A_1, \ldots, A_p$ are $n \times n$ nonsingular matrices, then $A_1 \cdots A_p$ is nonsingular and $(A_1 \cdots A_p)^{-1} = A_p^{-1} A_{p-1}^{-1} \cdots A_1^{-1}$.

18. Prove: if A is nonsingular then A^T is nonsingular and $(A^T)^{-1} = (A^{-1})^T$.

19. Prove: if $A \in M_n(\mathbb{C})$ and A is nonsingular then A^* is nonsingular and $(A^*)^{-1} = (A^{-1})^*$.

20. Prove: if A is nonsingular then A^{-1} is nonsingular and $(A^{-1})^{-1} = A$.

21. Prove: if A is symmetric and nonsingular then A^{-1} is symmetric.

22. Prove: if $A \in M_n(\mathbb{C})$ is unitary then $A^* = A^{-1}$.

23. Prove: if A is orthogonal then $A^T = A^{-1}$.

24. Prove: if $A_1, \ldots, A_p$ are nonsingular then $A_1 \otimes \cdots \otimes A_p$ is nonsingular.

    Hint: Use §3.3 Exercises, #9.

25. Prove: if $A \in M_n(\mathbb{C})$ is normal and nonsingular then A^{-1} is normal.

26. Prove: if A is nonsingular then A^p is nonsingular for every positive integer p.

27. If A is nonsingular and p is a positive integer then define A^{-p} to be $(A^{-1})^p$. As before, A^0 will denote the identity matrix. Prove that if s and t are arbitrary integers then:

$$A^s A^t = A^{s+t}; \quad (A^s)^t = A^{st}; \quad (A^s)^{-1} = (A^{-1})^s = A^{-s}.$$

28. Find a 2×2 matrix $A \in M_2(\mathbb{Z})$ with 0's on the main diagonal satisfying $A^2 = I_2$.

29. Find a 3×3 matrix $A \in M_3(\mathbb{Z})$ with 0's on the main diagonal satisfying $A^3 = I_3$.

30. Prove: if A has a row (column) of 0 entries then A is singular.

4.1 MatLab

1. Read the Help entries for \, /, **inv**, **eye**, and **ones**. Let A be the matrix

$$A = \begin{bmatrix} 2 & -1 & 0 & 0 \\ -1 & 2 & -1 & 0 \\ 0 & -1 & 2 & -1 \\ 0 & 0 & -1 & 2 \end{bmatrix}$$

and B be the matrix

$$B = \begin{bmatrix} 3 & 1 & 1 & 1 \\ 1 & 3 & 1 & 1 \\ 1 & 1 & 3 & 1 \\ 1 & 1 & 1 & 3 \end{bmatrix}.$$

Use MatLab to compute

(a) A^{-1}

(b) $A^{-1}B$

(c) AB^{-1}

(d) $A^{-1}v$, $v = [1\ 0\ 0\ 0]'$

(e) uB^{-1}, $u = [0\ 1\ 0\ 0]$

2. Let A be an n-square nonsingular matrix and let x and b be n × 1 matrices. Clearly, if Ax = b and A is nonsingular, then $x = A^{-1}b$. Solve each of the following systems of equations by entering A and b and computing A\b.

(a) $x_2 + x_3 + x_4 = 4$
$3x_1 + 3x_3 - 4x_4 = 7$
$x_1 + x_2 + x_3 + 2x_4 = 6$
$2x_1 + 3x_2 + x_3 + 3x_4 = 6$

(b) $2x_1 + 3x_2 + 2x_3 = 9$
$x_1 + 2x_2 - 3x_3 = 14$
$3x_1 + 4x_2 + x_3 = 16$

(c) $36.47x_1 + 5.28x_2 + 6.34x_3 = 12.26$
     $7.33x_1 + 28.74x_2 + 5.86x_3 = 15.15$
     $4.63x_1 + 6.31x_2 + 26.17x_3 = 25.22$

3. The *division algorithm for integers* states the following: Let a and b be integers, $b \neq 0$. Then there exist unique integers q and r that satisfy

$$a = bq + r, \; 0 \leq r < |b|.$$

The integer q is called the *quotient* and the integer r is called the *remainder*. This result is also called the *Euclidean algorithm*. (If $a = 0$, then $q = 0$ and $r = 0$.)

Read the Help entries for **round, fix, floor, ceil,** and **rem**. Let $\rho = \text{rem}(\text{abs}(a), \text{abs}(b))$ and $Q = \text{fix}(\text{abs}(a)/\text{abs}(b))$. Show that the values of r and q in the division algorithm are given according to the following table:

a	b	ρ	r	q
+, 0	+	*	ρ	Q
-	+	0	0	-Q
-	+	$\neq 0$	$b - \rho$	$-(Q + 1)$
+	-	*	ρ	-Q
-	-	0	0	Q
-	-	$\neq 0$	$-(b+\rho)$	$Q + 1$

(The * entry in the ρ column means that whatever the value of ρ, r has the indicated value.)

Hint: We confirm the first row and last two rows of the table. The other parts are handled similarly. MatLab computes $\text{fix}(a/b)$ as the nearest integer to $\frac{a}{b}$ towards 0, i.e., if we set $q = \text{fix}(a/b)$ then $q \leq \frac{a}{b} < q + 1$, so that $bq \leq a < bq + b$, $0 \leq a - bq < b$. Also, MatLab computes $\text{rem}(a, b)$ as $a - \text{fix}(a/b) * b = a - qb$, so that if we set $r = a - qb$ we have $r = \text{rem}(a, b) = \rho$. This takes care of the first row of the table. We proceed to the last two rows.

Write $|a| = |b|Q + \rho$, $0 \le \rho < |b|$. If $\rho = 0$ then $|a| = |b|Q$; $-a = (-b)Q$, $a = bQ$. Thus set $r = \rho = 0$, $q = Q$. If $\rho \ne 0$ (i.e., $\rho > 0$) then from $|a| = |b|Q + \rho$ we have $-a = (-b)Q + \rho$, $a = bQ - \rho = bQ - |b| + |b| - \rho = bQ + b + (-b - \rho)$ $= b(Q + 1) + (-(b + \rho))$. Also, $-(b + \rho) = -b - \rho = |b| - \rho$. So (since $0 < \rho < |b|$), $0 < |b| - \rho < |b|$. Thus $q = Q + 1$, $r = -(b + \rho)$.

4. Write a MatLab function called euclid such that the value of euclid(a, b) is the pair q, r as given by the division algorithm described in #3; a and b are integers and $b \ne 0$.

   Hint:
   ```
   function [q, r] = euclid(a, b)
   %EUCLID delivers the quotient and remainder
   %in the division algorithm.
   if b == 0
           disp('divisor cannot be 0');
   else
           rho = rem(abs(a), abs(b));
           que = fix(abs(a)/abs(b));
           if (a >= 0) & (b > 0)
                   r = rho;
                   q = que;
           elseif (a < 0) & (b > 0) & (rho == 0)
                   r = rho;
                   q = - que;
           elseif (a < 0) & (b > 0) & (rho ~= 0)
                   r = b-rho;
                   q = -(que+1);
           elseif (a > 0) & (b < 0)
                   r = rho;
                   q = - que;
           elseif (a < 0) & (b < 0) & (rho == 0)
                   r = rho;
                   q = que;
           elseif (a < 0) & (b < 0) & (rho ~= 0)
                   r = - (b+rho);
                   q = que+1;
           end
   end
   ```

5. The computation of the *greatest common divisor* of two integers a and b is based on the Euclidean algorithm. Note that if $a = bq + r$, $0 \leq r < |b|$ then gcd(a, b) = gcd(b, r). Write a MatLab function called gcd that computes the greatest common positive integer divisor of a and b. Use the function euclid in #4.

   Hint:
   ```
   function d = gcd(a, b)
   %GCD implements the Euclidean Algorithm; gcd(a,b) is the
   % greatest common integer divisor of a and b.
   a = abs(a);
   b = abs(b);
   if (a == 0) | (b == 0)
           d = a+b;
   else
           [q, r] = euclid(a, b);
           if r ==0
                   d = b;
           else
                   while r ~= 0
                           z = b;
                           [q, r] = euclid(a, b);
                           b = r;
                           a = z;
                   end
                   d = a;
           end
   end
   ```

6. Note that if a and b are integers then gcd(a, b) = gcd(a - b, b). Thus to compute the greatest common divisor of 24 and 18 we can use the following scheme:

24	18
6	18
6	12
6	6

The smaller entry in a column is subtracted from the larger until the last pair of entries are equal. The last number is the gcd. Use this idea to write a MatLab function called gcdd to compute the gcd of two integers. Review the Help entry for ~.

Hint:

```
function d = gcdd(a, b)
%GCDD computes gcd(a, b) by successive differencing.
a = abs(a);
b = abs(b);
if (a == 0) | (b == 0)
    d = a+b;
else
    while a ~= b
        if a > b
            a = a-b;
        else
            b = b-a;
        end
    end
    d = a;
end
```

7. Write a MatLab function called ratadd that computes the sum of two rational fractions, $\frac{a}{b} + \frac{c}{d}$, and expresses the answer in lowest terms, i.e., $\frac{1}{8} + \frac{5}{8} = \frac{6}{8}$, however, ratadd should output the answer as $\frac{3}{4}$.

Hint:

```
function [u, v] = ratadd(f, g)
%RATADD computes the sum of two rational fractions
%u/v = a/b + c/d. The answer is expressed in lowest terms.
%The fractions are represented in the form [a b].
a = f(1);
b = f(2);
c = g(1);
d = g(2);
```

```
            if b*d == 0
                    disp('denominator is 0 ')
            else
                    u = a*d+b*c;
                    v = b*d;
                    d = gcd(u, v);
                    u = u/d;
                    v = v/d;
            end
```

8. Modify ratadd in #7 appropriately and write MatLab functions called ratdiff, ratmult, and ratdiv to do differences, products, and quotients of rational fractions, expressing the answers in lowest terms. Create a "library" (i.e., folder) called Arithmetic that contains euclid, gcd, gcdd, ratadd, ratdiff, ratmult, and ratdiv.

    Hint:
```
            function [u, v] = ratdiff(f, g)
            %RATDIFF computes the difference of two rational fractions
            %u/v = a/b - c/d. The answer is expressed in lowest terms.
            %The fractions are represented in the form [a b].
            a = f(1);
            b = f(2);
            c = g(1);
            d = g(2);
            if b*d == 0
                    disp('denominator is 0 ')
            else
                    u = a*d-b*c;
                    v = b*d;
                    d = gcd(u, v);
                    u = u/d;
                    v = v/d;
            end
```

```
function [u, v] = ratmult(f, g)
%RATMULT computes the product of two rational fractions
%u/v = a/b * c/d.  The answer is expressed in lowest terms.
%The fractions are represented in the form [a b].
a = f(1);
b = f(2);
c = g(1);
d = g(2);
if b*d == 0
        disp('denominator is 0 ')
else
        u = a*c;
        v = b*d;
        d = gcd(u, v);
        u = u/d;
        v = v/d;
end


function [u, v] = ratdiv(f, g)
%RATDIV computes the quotient of two rational fractions
%u/v = (a/b) / (c/d).  The answer is expressed in lowest terms.
%The fractions are represented in the form [a b].
a = f(1);
b = f(2);
c = g(1);
d = g(2);
if (b*d == 0) | (b*c == 0)
        disp('denominator is 0 ')
else
        u = a*d;
        v = b*c;
        d = gcd(u, v);
        u = u/d;
        v = v/d;
end
```

9. The greatest common divisor of two integers can also be found by a simple search process; e.g., in looking for gcd(693, 126) simply test the integers from 126 down to 1 to determine if they are divisors of 693. Write a MatLab function called gcds based on this idea and save it in Arithmetic.

10. Write a MatLab script called commondiv that prints out all the positive common integer divisors of two integers a and b. Review the Help entries for **fix**, **==**, and **&**. Save commondiv in Arithmetic.

    Hint:
    ```
    a = input('Enter a: ');
    b = input('Enter b: ');
    a = abs(a);
    b = abs(b);
    m = min([a b]);
    D = [ ];
    for k = m:-1:1
            x = a / k;
            y = b / k;
            if (x == fix(x)) & (y == fix(y))
                    D = [D k];
            end
    end
    disp(D);
    ```

11. Write a MatLab function called prime such that prime(n) is 0 or 1 according as n is a prime integer or not. Save prime in Arithmetic.

    Hint:
    ```
    function p = prime(n)
    %PRIME yields the value 0 if the argument is prime, 1 otherwise.
    if (n == 2) | (n == 3) | (n == 5) | (n == 7)
            p = 0;
    elseif n/2 == fix(n/2)
            p = 1;
    else
            m = fix(sqrt(n));
            k = 3;
            while (n/k ~= fix(n/k)) & (k <= m)
    ```

```
            k = k + 2;
        end
        if k > m
            p = 0;
        else
            p = 1;
        end
end
```

12. Write a MatLab script called primelist that calls for the input of a positive integer N and the prints out a table of all the primes not exceeding N. Use the function prime in #11. Save primelist in Arithmetic.

13. Read the Help entry for **poly**. The important part of the Help description for poly in the present context can be described as follows. Let v be a vector, $v = [v_1\ v_2\ ...\ v_n]$. Then poly(v) is a vector whose elements are the coefficients of the polynomial whose roots are $v_1, v_2, ..., v_n$. For example, if v = 1:3 and poly(v) are entered at two successive prompts (>>) in the Command window then the output will be

$$1\ -6\ 11\ -6$$

This means that the polynomial with roots 1, 2, 3 is $\lambda^3 - 6\lambda^2 + 11\lambda - 6$. Use **poly** to compute the polynomials whose roots are:

(a)   1, 2, 3

(b) $e^{i2\pi k/5}$, k = 0, 1, 2, 3, 4

(c)   -1 with multiplicity 5

Hint:

(a)   poly([1 2 3])

(b)   z = exp(i * 2 * pi / 5);
       k = 0:4;
       r = (z * ones(1, 5)).^k;
       poly(r)

(c)   r = -ones(1, 5);
       poly(r)

14. Write a one line MatLab command that generates the binomial coefficients
$$\binom{n}{0}, \binom{n}{1}, \ldots, \binom{n}{n}.$$

15. If $r_1, \ldots, r_n$ are numbers then the polynomial

$$p(\lambda) = \prod_{t=1}^{n} (\lambda - r_t)$$

can be expanded into

$$\lambda^n - E_1(r)\lambda^{n-1} + E_2(r)\lambda^{n-2} + \cdots + (-1)^{n-k}E_k(r)\lambda^{n-k} + \cdots + (-1)^n E_n(r).$$

The coefficient $E_k(r)$ is called the kth *elementary symmetric function* of $r_1, \ldots, r_n$. Write a MatLab function called esf whose value on the vector $r = [r_1 \ldots r_n]$ is the vector $e = [E_1(r)\ E_2(r)\ \ldots\ E_n(r)]$.

Hint:
>   function e = esf(r);
>   %ESF returns the elementary symmetric
>   %functions of the numbers r = [r1, r2, ..., r n].
>   p = poly(r);
>   n = length(r);
>   p = p(2:(n + 1));   %The first coefficient of p is just 1.
>   s = - ones(1, n);
>   e = p.*(s.^(1:n));

16. (a) What must p and q be so that the roots of $x^2 + px + q$ are p and q?

    (b) Solve the cubic equation $x^3 + 2x^2 + 3x + 2 = 0$, given that the roots satisfy $r_1 = r_2 + r_3$.

17. Read the Help entry for **roots**, ****, and **for**. Use roots to find the roots of each of the following polynomials.

(a) $x^3 + 6x^2 - 72x + 27 = 0$

(b) $2x^4 + x^3 - 2x - 8 = 0$

18. Let x_i, $i = 0, \ldots, n$, be a set of $n + 1$ distinct numbers. Let y_i and y_i', respectively, denote the numerical values of a polynomial p and its derivative p' at x_i, $i = 0, \ldots, n$. There are $2n + 2$ linear conditions on the coefficients of p imposed by the $2n + 2$ equations

$$p(x_i) = y_i$$
$$p'(x_i) = y_i', \quad i = 0, \ldots, n .$$

If p has degree $2n + 1$ then it has $2n + 2$ coefficients which can be determined by the above equations. Such a polynomial is called a *Hermite interpolating polynomial*. Find a Hermite interpolating polynomial for the data

$$\{-2, -63, 16\}$$

$$\{0, 1, 0\}$$

$$\{3, 82, 189\}$$

$$\{4, 513, 768\}$$

where it is meant that $x_0 = -2$, $y_0 = -63$, $y_0' = 16$, etc. Also evaluate the polynomial for $x = 1$.

Hint: In this case $n = 3$, so that $2n + 1 = 7$.

$$p(t) = c_0 + c_1 t + c_2 t^2 + c_3 t^3 + c_4 t^4 \text{ b } + c_5 t^5 + c_6 t^6 + c_7 t^7$$

$$p'(t) = \quad c_1 + c_2(2t) + c_3(3t^2) + c_4(4t^3) + c_5(5t^4) + c_6(6t^5) + c_7(7t^6)$$

To compute the rows of the coefficient matrix we need to evaluate

$$u = [1 \ t \ t^2 \ t^3 \ t^4 \ t^5 \ t^6 \ t^7]$$

at -2, 0, 3, 4 and also

$$v = [0 \ 1 \ 2t \ 3t^2 \ 4t^3 \ 5t^4 \ 6t^5 \ 7t^6]$$

at -2, 0, 3, 4. Complete this problem by writing MatLab functions called herm and dherm to compute the vectors u and v, respectively. Then write a script called Hermite to compute the coefficients $c_0, \ldots, c_{2n+1}$. Read the Help entry for **sum**. Save herm, dherm, and Hermite in a library (i.e., folder) called "Hermite Polynomials."

Hint:

The vector u can be generated by a function as follows:

function u = herm(n, t)
%HERM computes [1 t t^2 ... t^(2n+1)].
u = (t*ones(1, 2*n+2)).^(0:(2*n+1));

We can then generate v by another function:

function v = dherm(n, t)
%DHERM computes [0 1 2*t 3*(t^2) ... (2*n+1)*t^(2*n)].
w = [0 herm(n, t)];
w = w(1:(2*n+2));
m = 0:(2*n+1);
v = m.*w;

Then a MatLab script to obtain the Hermite interpolating polynomial is

n = input('Enter n: ');   %n+1 is the number of points.
x = input('Enter [x0, ..., xn] ');   %x0, ..., xn are the points.
y = input('Enter [y0, ..., yn] ');   %y0, ..., yn are the
                                     %corresponding y values.
dy = input('Enter [dy0, ..., dyn] ');%dy0, ..., dyn are the
                                     %corresponding derivative values.

```
A = [ ];
b = [ ];
for k = 1:(n+1)
        A = [A; herm(n, x(k)); dherm(n, x(k))];
        b = [b  y(k)  dy(k)];
end
c = A\b.';                          %The coefficients.
sum(c)                              %The value of a polynomial at 1
                                    %is the sum of the coefficients.
```

Then $-0.0356t^7 + 0.4267t^6 - 0.6000t^5 - 1.2178t^4 + 6.8267t^3 - 10.2400t^2 + 1$ is the required polynomial.

19. Read the Help entry for **polyval**. Then modify the MatLab script in #18 to compute the value of the Hermite interpolating polynomial in #18 for any value t.

    Hint: In the previous script add a new input line after the line dy = input('Enter [dy0 ... dyn]'):

    t = input('enter a value of t ');

    remove the sum(c) line at the end of the script and replace it with

    ```
    c = c((2*n+2):-1:1);        %This changes c to c = [c(2n+1) ... c(0)]
    polyval(c,t)                %Computes c(2n+1)*t^(2n+1)+ ... + c(1)*t + c(0)
    ```

    To compute several values of the Hermite interpolating polynomial, use the following script:

    ```
    n = input('Enter n ');                  %n+1 is the number of points
    x = input('Enter [x0 ... xn] ');        %x0, ..., xn are the points
    y = input('Enter [y0 ... yn] ');        %y0, ..., yn are the
                                            %corresponding y values.
    dy = input('Enter [dy0 ... dyn] ');     %dy0, ..., dyn are the corresponding
                                            %derivative values.
    A = [ ];
    b = [ ];
    for k = 1:(n+1)
    ```

```
            A = [A; herm(n, x(k)); dherm(n, x(k))];
            b = [b y(k) dy(k)];
end
c = A\b.';                      %The coefficients.
rev = (2*n+2):-1:1;
c = c(rev);                     %This changes c to c = [c(2n+1) ... c0].
go = 1;
while go ==1
    t = input('Enter a value of t:');
    polyval(c,t)                %Computes c(2n+1)*t^(2n+1) + ... + c0
    go = input('Enter 1 to continue, 0 to quit: ');
end
```

20. Let $p(t) = c_n t^n + c_{n-1} t^{n-1} + \cdots + c_1 t + c_0$ and let A be a square matrix. The notation $p(A)$ denotes the matrix

$$p(A) = c_n A^n + c_{n-1} A^{n-1} + \cdots + c_1 A + c_0 I_n.$$

In MatLab this is implemented with the command **polyvalm**. Read the Help entry for polyvalm. Thus if $c = [c_n \; c_{n-1} \; ... \; c_0]$ then

$$\text{polyvalm}(c, A)$$

computes $p(A)$.

Evaluate $p(A)$ for each of the following:

(a) $A = \text{ones}(3)$, $p(t) = t^3 - 3t^2$

(b) $A = \text{eye}(3)$, $p(t) = (t - 1)^3$

(c) $A = \text{rand}(2)$, $p(t) = t^2 - (a_{11} + a_{22})t + (a_{11} a_{22} - a_{12} a_{21})$

Hint: The answer is the zero matrix for (a), (b), and (c).

21. Read the Help entries for **conv** and **deconv**. Use conv to write a MatLab function named coef that computes the coefficient of t^k in the product of the polynomials $f(t) = c_n t^n + \cdots + c_0$ and $g(t) = d_m t^m + \cdots + d_0$. Save coef in a library called "Polynomial."

Hint:
```
function co = coef(f, g, k)
%COEF(f, g, k) computes the coefficient of t^k
%in the product of the polynomials f(t) and g(t).
r = conv(f, g);
L = length(r);
if k > L - 1          %length(r) - 1 = deg f(t)g(t)
    co = 0;
else
    co = r(L - k);
end
```

22. The MatLab function **deconv** acts for polynomials in much the same way as the function euclid in #4 does for integers. In using deconv(f,g) it is necessary that the leading coefficient of g is not 0. Write a MatLab function called gcdp that computes the monic greatest common polynomial divisor of f and g by using the Euclidean algorithm for polynomials. Save gcdp in Polynomial. Use gcdp on:

(a) $f = t^6 + 2t^5 + t^3 + 3t^2 + 3t + 2$, $g = t^4 + 4t^3 + 4t^2 - t - 2$

(b) $f = t^5 - t^4 - 2t^3 + 2t^2 + t - 1$, $g = 5t^4 - 4t^3 - 6t^2 + 4t + 1$

Hint:
```
function u = strip(r, tol)
%STRIP sets to 0 every component of r with modulus < tol.
%Then the initial segment of 0's is removed from r.
w = find(abs(r) < tol);
if length(w) == 1
    r(w) = 0;
else
    r(w) = zeros(w);
end
if r == 0
    u = 0;                              %If abs(r) < tol then strip = 0.
else
    v = find(abs(r) > tol);
    u = r(v(1):length(r));
end
```

4.1 MatLab

```
function dp = gcdp(f, g)
%GCDP uses the Euclidean algorithm to compute
%the monic gcd of polynomials f = c(m)*t^m + ··· + c(0) and
%g = g(n)*t^n + ··· + g(0).
tol = 1e-6;
f = strip(f, tol);
g = strip(g, tol);
if length(g) > length(f)
        temp = f;
        f = g;
        g = temp;
end
if f == 0
        dp = g/g(1);
elseif g == 0
        dp = f/f(1);
else
        [q r] = deconv(f, g);
        r = strip(r, tol);
        if r == 0
                dp = g/g(1);
        else
                while max(abs(r)) > tol
                        z = g;
                        [q r] = deconv(f, g);
                        r = strip(r, tol);
                        g = r;
                        f = z;
                end
                dp = f /f(1);
        end
end
```

(a)   dp = gcdp([1 2 0 1 3 3 2], [1 4 4 -1 -2])
The monic greatest common divisor is $t^2 + 3t + 2$.

(b)   dp = gcdp ([1 -1 -2 2 1 -1], [5 -4 -6 4 1])
The monic greatest common divisor is $t^3 - t^2 - t + 1$.

4.1 Glossary

~	180
/	175
\	175
A\b	176
$\mathbb{C}$	172
$\mathbb{C}(x)$	173
$\mathbb{C}[x]$	172
ceil	177
conv	189
deconv	189
division algorithm for integers	177
$E_k(r)$	185
elementary symmetric function	185
Euclidean algorithm	177
eye	175
fix	177
floor	177
for	185
greatest common divisor	179
Hermite interpolating polynomial	186
identity	171
inv	175
inverse	171
invertible	172
monic	190
nonsingular	172
ones	175
poly	184
polyval	188
polyvalm	189
$\mathbb{Q}$	172
$\mathbb{Q}(x)$	173
$\mathbb{Q}[x]$	172
quotient	177

$\mathbb{R}$	172
$\mathbb{R}(x)$	173
$\mathbb{R}[x]$	172
rational function	173
regular	172
rem	177
remainder	177
roots	185
round	177
singular	172
sum	187
unit	171
unit matrix	172
$\mathbb{Z}$	172
$\mathbb{Z}(x)$	173
$\mathbb{Z}[x]$	172

4.2 Elementary Matrices

There are three types of elementary row operations on m × n matrices in $M_{m,n}(R)$ that are important in solving linear equations, computing rank, and finding inverses:

- Type I: interchange two rows;
- Type II: add to a row a multiple of another row;
- Type III: multiply a row by a unit in R.

The result of multiplying $A_{(i)}$ by $r \in R$ is the matrix $rA_{(i)}$, i.e., the row multiplication is written on the left in case the elements of R do not commute multiplicatively (they do for the sets R in this book). The notation describing these operations is:

- $I_{(i),(j)}$ interchange rows i and j;
- $II_{(i)+c(j)}$ to row i add c times row j (i ≠ j);
- $III_{r(i)}$ multiply row i by the unit $r \in R$.

Corresponding to the three types of elementary row operations on m × n matrices are the three types of m × m matrices called *elementary row matrices*:

- $E_{(i),(j)}$ results from I_m by performing $I_{(i),(j)}$;
- $E_{(i)+c(j)}$ results from I_m by performing $II_{(i)+c(j)}$;
- $E_{r(i)}$ results from I_m by performing $III_{r(i)}$.

For example, if m = 3 then

$$E_{(1)+2(3)} = \begin{bmatrix} 1 & 0 & 2 \\ 0 & 1 & 0 \\ 0 & 0 & 1 \end{bmatrix}.$$

Observe that

$$E_{(1)+2(3)} E_{(1)-2(3)} = \begin{bmatrix} 1 & 0 & 2 \\ 0 & 1 & 0 \\ 0 & 0 & 1 \end{bmatrix} \begin{bmatrix} 1 & 0 & -2 \\ 0 & 1 & 0 \\ 0 & 0 & 1 \end{bmatrix} = \begin{bmatrix} 1 & 0 & 0 \\ 0 & 1 & 0 \\ 0 & 0 & 1 \end{bmatrix} = I_3,$$

and similarly,

$$E_{(1)-2(3)} E_{(1)+2(3)} = I_3.$$

Thus $E_{(1)+2(3)}$ is nonsingular in $M_3(R)$. Also observe that

$$E_{(1)+2(3)} \begin{bmatrix} a_{11} & a_{12} & a_{13} & a_{14} \\ a_{21} & a_{22} & a_{23} & a_{24} \\ a_{31} & a_{32} & a_{33} & a_{34} \end{bmatrix} = \begin{bmatrix} 1 & 0 & 2 \\ 0 & 1 & 0 \\ 0 & 0 & 1 \end{bmatrix} \begin{bmatrix} A_{(1)} \\ A_{(2)} \\ A_{(3)} \end{bmatrix}$$

$$= \begin{bmatrix} A_{(1)}+2A_{(3)} \\ \hline A_{(2)} \\ \hline A_{(3)} \end{bmatrix}$$

$$= \begin{bmatrix} a_{11}+2a_{31} & a_{12}+2a_{32} & a_{13}+2a_{33} & a_{14}+2a_{34} \\ a_{21} & a_{22} & a_{23} & a_{24} \\ a_{31} & a_{32} & a_{33} & a_{34} \end{bmatrix}.$$

In other words, left multiplication of A by $E_{(1)+2(3)}$ effects $II_{(1)+2(3)}$ on A.

The simple analogues of these results for Type I and Type III operations are found in §4.2 Exercises.

4.2 Exercises

1. Prove: $E^{-1}_{(i),(j)} = E_{(i),(j)}$ ($E^{-1}_{(i),(j)}$ means $(E_{(i),(j)})^{-1}$).

2. Prove: $E^{-1}_{(i)+c(j)} = E_{(i)-c(j)}$ ($E^{-1}_{(i)+c(j)}$ means $(E_{(i)+c(j)})^{-1}$).

3. Prove: $E^{-1}_{r(i)} = E_{r^{-1}(i)}$ ($E^{-1}_{r(i)}$ means $(E_{r(i)})^{-1}$).

4. Prove: left multiplication of a matrix A by $E_{(i),(j)}$ effects $I_{(i),(j)}$ on A.

5. Prove: left multiplication of a matrix A by $E_{(i)+c(j)}$ effects $II_{(i)+c(j)}$ on A.

6. Prove: left multiplication of a matrix A by $E_{r(i)}$ effects $III_{r(i)}$ on A.

7. Find a sequence of elementary row operations that will reduce

$$\begin{bmatrix} 1 & 1 & 1 \\ 1 & 2 & 1 \\ 3 & 3 & 0 \end{bmatrix} \in M_3(\mathbb{Z})$$

to a matrix with only 0's below the main diagonal. Remember that the units in $\mathbb{Z}$ are 1 and -1.

8. The *elementary column operations* are defined by replacing the word "row" by the word "column" in the definitions of the elementary row operations, with the following alteration: the scalar multiplication of columns is done on the right in case the entries of R are not multiplicatively commutative. The notations for the three types are: $I^{(i),(j)}$, $II^{(i)+(j)c}$, $III^{(i)r}$. The corresponding *elementary column matrices* are $E^{(i),(j)}$, $E^{(i)+(j)c}$, $E^{(i)r}$ and are obtained from the identity matrix by performing the corresponding column operations. Write out the following 3 × 3 matrices: $E^{(1),(2)}$; $E^{(2)+(3)7}$; $E^{(1)4}$; $E^{(1),(3)} E^{(2)+(3)(-1)}$; $E^{(2),(3)} E^{(2),(3)}$.

9. Prove: $\left(E^{(i),(j)}\right)^{-1} = E^{(i),(j)}$.

10. Prove: $\left(E^{(i)+(j)c}\right)^{-1} = E^{(i)+(j)(-c)}$.

11. Prove: $\left(E^{(i)r}\right)^{-1} = E^{(i)r^{-1}}$.

12. Prove: right multiplication of a matrix A by $E^{(i),(j)}$ effects $I^{(i),(j)}$ on A.

13. Prove: right multiplication of a matrix A by $E^{(i)+(j)c}$ effects $II^{(i)+(j)c}$ on A.

14. Prove: right multiplication of a matrix A by $E^{(i)r}$ effects $III^{(i)r}$ on A.

15. Let $1 \le j < k \le n$, $\alpha, \beta \in \mathbb{R}$, $0 \le \theta < 2\pi$, and define $U(j, k) \in M_n(\mathbb{C})$ by

$$U[j, k \mid j, k] = \begin{bmatrix} e^{i\alpha} \cos \theta & e^{i\beta} \sin \theta \\ -e^{-i\beta} \sin \theta & e^{-i\alpha} \cos \theta \end{bmatrix},$$

$$U(j, k \mid j, k) = I_{n-2}.$$

Then $U(j, k)$ is called a *plane rotation in the plane* (j, k). Prove: $U(j, k)$ is unitary and $U(j, k)^*$ is a plane rotation in the plane (j, k) in which α is replaced by $-\alpha$ and β by $\beta + \pi$.

Exercises #16 - 20 use a common notation.

16. Let $w \in M_{n,1}(\mathbb{C})$, $w^*w = 1$ (i.e., [1]) and define $P_\mu = I_n - \mu w w^*$ where μ is a complex number. Prove if $\mu + \overline{\mu} = |\mu|^2$ then P_μ is unitary, and if $\mu = 2$ then P_μ is also hermitian. Matrices of the form P_2 are called *Householder matrices*, after the distinguished American mathematician Alston S. Householder.

17. Prove: $P_2^2 = I_n$.

18. Prove: If P_μ is unitary then $\mu + \overline{\mu} = |\mu|^2$.

19. Prove: $P_2 w = -w$.

20. Prove: If $v \in M_{n,1}(\mathbb{C})$ and $w^*v = 0$, then $P_2 v = v$.

4.2 MatLab

1. Read the Help entries for :, |, if, and size. Write a MatLab function called rtyp1 such that rtyp1(A, s, t) replaces the m × n matrix A by the matrix obtained from A by performing $I_{(s),(t)}$. Create a library (i.e., folder) called Elemops and put rtyp1 in it.

   Hint:
   ```
   function B = rtyp1(A, s, t)
   %RTYP1
   %A is an m × n matrix; B(s, :) = A(t, :) & B(t, :) = A(s, :).
   %If s or t is out of the integer range 1, ..., m then
   %rtyp1(A, s, t) returns A.
   [m n] = size(A);
   if (s < 1) | (s > m) | (t < 1) | (t > m)
        B = A;
   else
        temp = A(s,:);
        A(s,:) = A(t,:);
        A(t,:) = temp;
   end
   B = A;
   ```

2. Write a MatLab function called rtyp2 such that rtyp2(A, s, t, c) replaces the m × n matrix A by the matrix obtained from A by performing $II_{(s) + c(t)}$. Put rtyp2 in Elemops.

   Hint:
   ```
   function B = rtyp2(A, s, t, c)
   %RTYP2
   %A is an m x n matrix; B(s, :) = A(s, :) + c*A(t, :).
   %If s or t is out of the integer range 1, ..., m or s = t
   %then rtyp2(A, s, t, c) returns A.
   [m n] = size(A);
   if (s < 1) | (s > m) | (t < 1) | (t > m) | (s == t)
       B = A;
   else
       A(s, :) = A(s, :) + c*A(t, :);
   end
   B = A;
   ```

3. Write a MatLab function called rtyp3 such that rtyp3(A, s, c) replaces the m × n matrix A by the matrix obtained from A by performing $III_{c(s)}$. Put rtyp3 in Elemops.

   Hint:
   ```
   function B = rtype3(A, s, c)
   %RTYP3
   %A is an m x n matrix, B(s, :) = c*A(s, :).
   %If s is out of the integer range 1, ..., m
   %or abs(c)<eps, then rtyp3(A, s, c) returns A.
   [m n] = size(A);
   if (s < 1) | (s > m) | (abs(c) < eps)
       B = A;
   else
       A(s, :) = c*A(s, :);
   end
   B = A;
   ```

4. Write three MatLab functions called ctyp1, ctyp2, and ctyp3 that replace the m × n matrix A by the matrix obtained from A by performing the corresponding elementary column operations. Put these three functions in Elemops.

5. The m × n matrix A is said to be in *Hermite normal form (HNF)* if:

   - all the nonzero rows appear contiguously in the first r rows ($1 \leq r \leq m$)
   - for $i = 1, \ldots, r$, if $a_{i\,n_i}$ is the first nonzero entry in row i then $a_{i\,n_i} = 1$, $n_1 < \ldots < n_r$, and $a_{t\,n_i} = 0, t = 1, \ldots, m, t \neq i$.

   In other words: all the nonzero rows appear in the first r row positions; the first nonzero entry in row i appears in column n_i, $n_1 < n_2 < \ldots < n_r$; the 1 in column n_i is the only nonzero entry in that column.

   Use rtyp1, rtyp2, and rtyp3 to reduce the following matrix to HNF.

   $$A = \begin{bmatrix} 0 & 2 & 0 & -1 & 3 & 0 & 0 & 1 & -2 \\ 0 & -1 & 0 & 2 & 0 & 3 & 0 & -1 & 2 \\ 0 & 1 & 0 & -1 & 1 & -1 & 0 & 2 & -4 \\ 0 & 1 & 0 & -1 & 1 & -1 & 0 & 1 & -2 \end{bmatrix}$$

   Then read the Help entry for **rref**. The *reduced row echelon form* is another name used for the Hermite normal form.

6. Using rtytp1, rtyp2, rtyp3:

   (a) reduce

   $$A = \begin{bmatrix} 2 & 2 & 1 & 1 \\ 1 & 1 & 0 & 1 \\ 0 & 1 & 1 & 1 \\ 2 & 1 & 0 & 0 \end{bmatrix}$$

   to HNF;

(b) reduce

$$A = \begin{bmatrix} 3 & 3 & -1 & 4 & -2 \\ 1 & -1 & 7 & -1 & 0 \\ 5 & 1 & 13 & 2 & -2 \\ 2 & 4 & -8 & 5 & -2 \end{bmatrix}$$

to HNF;

(c) solve the system of linear equations

$$Ax = b$$

where A is the matrix in (b) and b = [14 -2 10 16]'.

Reducing a matrix A to HNF using elementary row operations is also called *Gaussian elimination*.

4.2 Glossary

```
:  . . . . . . . . . . . . . . . . . . . . . . .  197
E(i) + c(j)  . . . . . . . . . . . . . . . . . .  194
E(i), (j)  . . . . . . . . . . . . . . . . . . .  194
elementary column matrices  . . . . . . . . . .  196
elementary column operations . . . . . . . . . . 196
elementary row matrices . . . . . . . . . . . .  194
elementary row operations . . . . . . . . . . .  193
Er(i) . . . . . . . . . . . . . . . . . . . . .  194
Gaussian elimination  . . . . . . . . . . . . .  200
Hermite normal form . . . . . . . . . . . . . .  199
HNF . . . . . . . . . . . . . . . . . . . . . .  199
Householder matrix  . . . . . . . . . . . . . .  197
I(i), (j)  . . . . . . . . . . . . . . . . . . .  193
if  . . . . . . . . . . . . . . . . . . . . . .  197
```

II$_{(i) + c(j)}$ 193
III$_{r(i)}$ 193
plane rotation 196
reduced row echelon form 199
rref 199
size 197
Type I 193
Type II 193
Type III 193
| 197

4.3 Canonical Forms and Generalized Inverses

If B is an m × n matrix, then B is in *canonical form* if it is the zero matrix or if it has the form

$$B = \left[\begin{array}{c|c} I_r & 0_{r, n-r} \\ \hline 0_{m-r, r} & 0_{m-r, n-r} \end{array}\right]. \tag{1}$$

Some of the zero blocks may be missing. For example,

$$\begin{bmatrix} 1 & 0 \\ 0 & 1 \\ 0 & 0 \\ 0 & 0 \end{bmatrix}, \quad \begin{bmatrix} 1 & 0 \\ 0 & 0 \end{bmatrix}, \quad \begin{bmatrix} 1 & 0 & 0 & 0 \\ 0 & 1 & 0 & 0 \end{bmatrix}, \quad \begin{bmatrix} 1 \\ 0 \\ 0 \end{bmatrix}, \quad [1 \ 0 \ 0]$$

are all in canonical form. The following theorem is important.

Theorem 1.

Let R be a set in which every nonzero element is a unit. If A is an m × n matrix over R, then there exist an m-square matrix P and an n-square matrix Q such that PAQ = B is in canonical form. Moreover, P and Q are products of elementary matrices.

Proof.

Elementary row and column operations on A can be performed by pre- and post-multiplication, respectively, by elementary matrices. If $A \neq 0$ then it has a nonzero entry, and this entry can be brought to the (1,1) position by type I elementary row and column operations. The (1, 1) entry in the resulting matrix can be made equal to 1 with a type III elementary row operation. Finally, by type II elementary row and column operations, the entries in the first row and column, other than the (1, 1) entry, may be reduced to 0. The matrix on hand at this stage has the form

$$\begin{bmatrix} 1 & 0 & \cdots & 0 \\ 0 & & & \\ \vdots & & A_1 & \\ 0 & & & \end{bmatrix} \qquad (2)$$

where A_1 is $(m - 1) \times (n - 1)$. Note that elementary operations on A_1 can be effected by elementary operations on the matrix (2) without altering either the first row or first column of (2). The proof is completed by a straightforward induction on $m + n$. ∎

Henceforth in this chapter we shall assume that the set R satisfies the hypothesis in Theorem 1.

At this point we require a somewhat special result that will be subsumed by theorems that are proved subsequently. But Theorem 2 is simple and makes the proof of Theorem 3 attractively short.

Theorem 2.

If X and Y are matrices in $M_n(R)$ and $XY = I_n$ then neither the last column of X nor the last row of Y can be 0.

Proof.

To show that the last column of X is not 0 write $X = UZV$ where

$$Z = \begin{bmatrix} I_r & 0 \\ 0 & 0 \end{bmatrix}$$

is in canonical form and U and V are products of elementary matrices. From §4.1 Exercises, #17 and §4.2 Exercises, #1, #2, #3, #9, #10, #11, products of elementary matrices have inverses. The equality $XY = I_n$ implies that

$$UZVY = I_n$$

and hence

$$ZW = T$$

where $W = VY$ and T is the nonsingular matrix U^{-1}. If we partition W and T conformally with Z, we have

$$ZW = \begin{bmatrix} I_r & 0 \\ 0 & 0 \end{bmatrix} \begin{bmatrix} W_{11} & W_{12} \\ W_{21} & W_{22} \end{bmatrix} = \begin{bmatrix} T_{11} & T_{12} \\ T_{21} & T_{22} \end{bmatrix}.$$

Hence

$$\begin{bmatrix} W_{11} & W_{12} \\ 0 & 0 \end{bmatrix} = \begin{bmatrix} T_{11} & T_{12} \\ T_{21} & T_{22} \end{bmatrix}.$$

If r were strictly less than n, it would follow that T has a zero last row. But then $TT^{-1} = I_n$ would have a zero last row, a contradiction. Hence $r = n$, $Z = I_n$ and $X = UZV = UV$ is a product of elementary matrices and hence has an inverse. It follows that $X^{-1}X = I_n$, so that X could not have a zero last column. To prove that Y cannot have a zero last row simply take transposes in $XY = I_n$ to obtain $Y^T X^T = I_n$. Then apply what was just proved to conclude that Y^T cannot have a zero last column. ∎

Theorem 3.

Let A be an m × n matrix as in Theorem 1. If A is reduced to canonical form B as defined in (1), then r is uniquely determined.

Proof.

Suppose $PAQ = B$, $P_1 A Q_1 = D$ are both in canonical form and P, P_1, Q, Q_1 are all products of elementary matrices and hence have inverses. Suppose that D has s 1's in it and B has r 1's in it. If $s \neq r$, we can assume $s > r$. Then

$$UB = DV \tag{3}$$

where U and V are products of elementary matrices, i.e., $A = P^{-1} B Q^{-1} = P_1^{-1} D Q_1^{-1}$ and $P_1 P^{-1} B = D Q_1^{-1} Q$. Now partition V conformally with D so that

$$DV = \begin{bmatrix} I_s & 0_{s,\,n-s} \\ \hline 0_{m-s,\,s} & 0_{m-s,\,n-s} \end{bmatrix} \begin{bmatrix} V_{11} & V_{12} \\ \hline V_{21} & V_{22} \end{bmatrix}$$

$$= \begin{array}{c} \\ s \\ \\ \end{array} \begin{bmatrix} \overset{s}{V_{11}} & V_{12} \\ \hline 0_{m-s,\,s} & 0_{m-s,\,n-s} \end{bmatrix}. \tag{4}$$

The last $n - r$ columns of $UB = DV$ are 0 because the last $n - r$ columns of B are 0. Since $r < s$ it follows that $n - r \geq n - s + 1$. Hence from (4), V_{12} is 0 and $(V_{11})^{(s)}$ is also 0. Thus V has the following form:

$$V = \begin{array}{c} \\ s \\ \\ \end{array} \begin{bmatrix} \overset{s}{V_{11}} & 0 \\ \hline V_{21} & V_{22} \end{bmatrix} \tag{5}$$

4.3 Canonical Forms and Generalized Inverses

in which V_{11} has a 0 last column and hence must be singular (why?). Let $W = V^{-1}$ and partition W precisely as in (5):

$$W = \begin{bmatrix} s & W_{11} & W_{12} \\ \hline & W_{21} & W_{22} \end{bmatrix}.$$

By block multiplication we have $I_n = VW$ and hence $V_{11}W_{11} = I_s$. But $(V_{11})^{(s)} = 0$. Thus by Theorem 2, $s > r$ is impossible and we conclude that $s \leq r$. The argument is symmetric in r and s and hence $r \leq s$. This proves that $r = s$. ∎

Note that if $m = n$ in the canonical form matrix B in (1), then $r < n$ implies that $(XB)^{(n)}$ and $(BY)_{(n)}$ are both 0 matrices for any $n \times n$ matrices X and Y. Hence, in particular, neither XB nor BY can be I_n. If $r = n$, note that $B = I_n$. In general, if X and A are $n \times n$ and $XA = I_n$ then we say X is a *left inverse* for A and A is a *right inverse* for X. These simple observations lead to the following interesting result.

Theorem 4.

Let A be an n-square matrix over R. Then precisely one of the following alternatives is true for A:

(i) A has neither a right nor a left inverse;
(ii) A has both a right and a left inverse, A is nonsingular, and any right or left inverse of A must be A^{-1}.

Proof.

First recall that if P is a product of elementary matrices then P^{-1} exists and is a product of elementary matrices. By Theorem 1, let

$$PAQ = I_r \oplus 0_{n-r} \tag{6}$$

be canonical where P and Q are products of elementary matrices. Then there are two mutually exclusive cases: $r < n$; $r = n$.

(i) $r < n$:

Suppose A has a right inverse Y, $AY = I_n$. Then $P = PI_n = P(AY) = (PA)Y = (I_r \oplus 0_{n-r})Q^{-1}Y$. Since $r < n$, this last product has a 0 last row. But then $P_{(n)} = [0 \ldots 0]$, so that certainly there is no matrix P^{-1} for which $PP^{-1} = I_n$ (why?). Hence A has no right inverse, and similarly, A has no left inverse. In other words, if $r < n$ then A has neither a left nor a right inverse.

(ii) $r = n$:

We observed above that $PAQ = I_n$, so that $A = P^{-1}Q^{-1}$ and hence A is nonsingular, i.e., it is a product of elementary matrices. Hence A has the left and right inverse A^{-1}. If $XA = I_n$ then $X = A^{-1}$ and if $AY = I_n$ then $Y = A^{-1}$ so that any left or right inverse must be the inverse of A. ∎

Theorem 4 tells us that if A is $n \times n$ and $XA = I_n$ or $AY = I_n$ then $r = n$ and thus $X = A^{-1}$ and $Y = A^{-1}$. Also note that if A is nonsingular then r must equal n (as we saw in the above proof), so that $A = P^{-1}I_nQ^{-1}$ is a product of elementary matrices. In other words, A is nonsingular if and only if A is a product of elementary matrices.

The unique integer r in the canonical form for A in Theorem 1 is called the *rank* of A and is denoted by $\rho(A)$. Thus Theorem 4 implies that an $n \times n$ matrix A has an inverse if and only if $\rho(A) = n$. Notation: the phrase "if and only if" is frequently abbreviated to "iff."

The result in Theorem 1 shows us how to construct right and left inverses for a matrix A satisfying the conditions of Theorem 1. However, we can extend the notions of right and left inverses to not necessarily square matrices. If A is an $m \times n$ matrix as in Theorem 1 then X is a *right inverse* of A if X is an $n \times m$ matrix that satisfies

4.3 Canonical Forms and Generalized Inverses

$$AX = I_m. \tag{7}$$

If Y is an $n \times m$ matrix then Y is a *left inverse* of A if

$$YA = I_n. \tag{8}$$

Theorem 5.

Let A be an $m \times n$ matrix as in Theorem 1. Then A has a right inverse iff $\rho(A) = m$, and A has a left inverse iff $\rho(A) = n$.

Proof.

Let P and Q be products of elementary matrices for which $PAQ = B$ is in canonical form (1). Suppose first that

$$AX = I_m \tag{9}$$

for some $n \times m$ matrix X. Then

$$P^{-1}BQ^{-1}X = I_m, \tag{10}$$

or

$$BQ^{-1}X = P \tag{11}$$

and P is an $m \times m$ nonsingular matrix. If $r = \rho(A) < m$ then B has a 0 last row and hence, from (11), so does P. But we know that a nonsingular matrix cannot have a 0 last row. Thus $r = m$.

Conversely, suppose $r = m$. Then from $PAQ = B$ we have

$$\begin{aligned} AQ &= P^{-1}B \\ &= P^{-1}[\,I_m \mid O_{m,\,n-m}\,] \\ &= [\,P^{-1} \mid O_{m,\,n-m}\,]. \end{aligned} \tag{12}$$

Now let M be the n × m matrix

$$M = \left[\begin{array}{c} P \\ \hline 0_{n-m,\,m} \end{array}\right]. \tag{13}$$

Then, using block multiplication, we multiply both sides of (12) on the right by M to obtain

$$A(QM) = I_m$$

and $X = QM$ is the required right inverse. Note that if $n > m$, the 0 block in (13) could be replaced with any $(n - m) \times m$ matrix. Thus a right inverse is not uniquely determined. If $n = m = r$ then Theorem 4 tells us that any right inverse of A must be A^{-1}. The arguments for a left inverse of A are analogous and will be left to the reader. ∎

If A is an m × n matrix and A^+ is an n × m matrix for which

$$AA^+A = A \tag{14}$$

and

$$A^+AA^+ = A^+, \tag{15}$$

then A^+ is called a *generalized inverse* of A. Note that if B is the matrix (1) in canonical form then

$$BB^TB = \left[\begin{array}{c|c} I_r & 0_{r,\,n-r} \\ \hline 0_{m-r,\,r} & 0_{m-r,\,n-r} \end{array}\right] \left[\begin{array}{c|c} I_r & 0_{r,\,m-r} \\ \hline 0_{n-r,\,r} & 0_{n-r,\,m-r} \end{array}\right] \left[\begin{array}{c|c} I_r & 0_{r,\,n} \\ \hline 0_{m-r,\,r} & 0_{m-r,\,n} \end{array}\right]$$

4.3 Canonical Forms and Generalized Inverses

$$= \begin{bmatrix} I_r & 0_{r,\,m-r} \\ \hline 0_{m-r,\,r} & 0_{m-r,\,m-r} \end{bmatrix} \begin{bmatrix} I_r & 0_{r,\,n-r} \\ \hline 0_{m-r,\,r} & 0_{m-r,\,n-r} \end{bmatrix}$$

$$= \begin{bmatrix} I_r & 0_{r,\,n-r} \\ \hline 0_{m-r,\,r} & 0_{m-r,\,n-r} \end{bmatrix}$$

$$= B,$$

and similarly, it is easy to confirm that

$$B^T B B^T = B^T.$$

Thus B^T is a generalized inverse of B. Now suppose that A is an $m \times n$ matrix of rank r so that by Theorem 1, $A = P^{-1}BQ^{-1}$, where P and Q are products of elementary matrices. Let $X = QB^T P$ and compute directly that

$$AXA = (P^{-1}BQ^{-1})(QB^T P)(P^{-1}BQ^{-1})$$

$$= P^{-1} B B^T B Q^{-1}$$

$$= P^{-1} B Q^{-1}$$

$$= A,$$

and similarly

$$XAX = X.$$

The matrix X satisfies the same defining equations (14) and (15) as A^+ does and hence is a generalized inverse of A. We conclude that if $PAQ = B$, where B is in canonical form and P and Q are nonsingular, then

$$A^+ = QB^TP \qquad (16)$$

is a generalized inverse of A. In other words, any matrix over R has a generalized inverse.

If the m × n matrix A^+ is a generalized inverse of the m × n matrix A and if A is a matrix over $\mathbb{C}$ (or $\mathbb{R}$) then A^+ is called a *Moore-Penrose inverse* of A if it satisfies

$$AA^+A = A, \qquad (17)$$

$$A^+AA^+ = A^+, \qquad (18)$$

$$(AA^+)^* = AA^+, \qquad (19)$$

$$(A^+A)^* = A^+A. \qquad (20)$$

Thus, in addition to conditions (14) and (15) listed earlier, A^+ must satisfy (19) and (20). It is interesting to note that the four conditions (17) - (20) are enough to specify A^+ uniquely. To see this, suppose that both X and Y are n × m matrices that satisfy the equations (17) - (20) for A^+. Then

$$\begin{aligned}
X &= XAX & &\text{(from (18))} \\
&= X(AX)^* & &\text{(from (19))} \\
&= XX^*A^* & &(21) \\
&= XX^*(AYA)^* & &\text{(from (17))} \\
&= XX^*A^*(AY)^* & & \\
&= X(AY)^* & &\text{(from (21))} \\
&= XAY & &\text{(from (19))}
\end{aligned}$$

4.3 Canonical Forms and Generalized Inverses

$$= (XA)^* Y \qquad \text{(from (20))}$$

$$= A^* X^* Y$$

$$= (AYA)^* X^* Y \qquad \text{(from (17))}$$

$$= (YA)^* A^* X^* Y$$

$$= YAA^* X^* Y \qquad \text{(from (20))}$$

$$= YA(XA)^* Y$$

$$= YAXAY \qquad \text{(from (20))}$$

$$= Y(AXA)Y$$

$$= YAY \qquad \text{(from (17))}$$

$$= Y. \qquad \text{(from (18))}$$

The question remains whether a Moore-Penrose inverse exists at all. We will defer the proof of its existence until after we study unitary and hermitian matrices in more detail. However, suppose that $A \in M_{m,n}(\mathbb{C})$ can be written in the form

$$A = PDQ \qquad (22)$$

where $P \in M_m(\mathbb{C})$ is unitary, $Q \in M_n(\mathbb{C})$ is unitary, and $D \in M_{m,n}(\mathbb{C})$ is a matrix of the form

$$\left[\begin{array}{c|c} S & 0_{r,\,n-r} \\ \hline 0_{m-r,\,r} & 0_{m-r,\,n-r} \end{array} \right] \qquad (23)$$

in which

$$S = \text{diag}(\alpha_1, \cdots, \alpha_r),$$

$\alpha_t > 0$, $t = 1, \ldots, r$. (We will prove in the sequel that such a decomposition is possible for any $A \in M_{m,n}(\mathbb{C})$.) Then set

$$X = Q^* \left[\begin{array}{c|c} S^{-1} & 0_{r,\,m-r} \\ \hline 0_{n-r,\,r} & 0_{n-r,\,m-r} \end{array} \right] P^* \tag{24}$$

and denote the middle $n \times m$ matrix on the right in (24) by Δ. It is obvious by block multiplication that

$$AX = PDQQ^*\Delta P^*$$

$$= PD\Delta P^* \quad (QQ^* = I_n)$$

$$= P(I_r \oplus 0_{m-r})\, P^* \tag{25}$$

and (25) is hermitian. Similarly,

$$XA = Q^*\Delta P^* PDQ$$

$$= Q^*\Delta DQ \quad (P^*P = I_n)$$

$$= Q(I_r \oplus 0_{n-r})\, Q^* \tag{26}$$

and (26) is hermitian. Also,

$$AXA = PDQQ^*\Delta P^* PDQ$$

$$= PD\Delta DQ. \tag{27}$$

But, by block multiplication, it is simple to confirm that

$$D\Delta D = D \tag{28}$$

and also that

$$\Delta D\Delta = \Delta. \tag{29}$$

Hence, combining (27) and (28),

$$AXA = PDQ$$

$$= A.$$

Similarly, using (29),

$$XAX = Q^*\Delta P^* PDQQ^*\Delta P^*$$

$$= Q^*\Delta D\Delta P^*$$

$$= Q^*\Delta P^*$$

$$= X. \tag{30}$$

Thus matrix X in (24) satisfies the conditions (17) - (20) and must therefore be the Moore-Penrose inverse of A.

4.3 Exercises

1. Write out all 3×4 matrices in canonical form.

2. Write out all 4×2 matrices in canonical form.

In #3 - 20 reduce the indicated matrix over $\mathbb{Q}$ (i.e., the rational numbers) to canonical form.

3. $\begin{bmatrix} 1 & 1 & 0 \\ 0 & 0 & 0 \\ 0 & 0 & 0 \end{bmatrix}$ 4. $\begin{bmatrix} 1 & 0 & 0 \\ 1 & 0 & 0 \\ 0 & 0 & 0 \end{bmatrix}$

5. $\begin{bmatrix} 1 & 1 & 0 \\ 1 & 1 & 0 \\ 0 & 0 & 0 \end{bmatrix}$ 6. $\begin{bmatrix} 1 & 1 & 0 \\ 1 & -1 & 0 \\ 0 & 0 & 0 \end{bmatrix}$

7. $\begin{bmatrix} 1 & 1 & 1 \\ 1 & 1 & 1 \\ 1 & 1 & 1 \end{bmatrix}$ 8. $\begin{bmatrix} 1 & 2 & 3 & 4 \\ 0 & 0 & 0 & 0 \\ 1 & 2 & 3 & 4 \end{bmatrix}$

9. $\begin{bmatrix} 1 & -1 & 2 \\ 2 & 0 & 1 \end{bmatrix}$ 10. $\begin{bmatrix} 1 & 2 \\ -1 & 0 \\ 2 & 1 \end{bmatrix}$

11. $\begin{bmatrix} 1 & 1 & 1 \\ 0 & 1 & 1 \\ 0 & 0 & 1 \end{bmatrix}$ 12. $\begin{bmatrix} 1 \\ 1 \\ 1 \end{bmatrix}$

13. $\begin{bmatrix} 1 & 1 & 1 \end{bmatrix}$ 14. $\begin{bmatrix} 1 & 2 & 3 & 4 \\ 5 & 6 & 7 & 8 \end{bmatrix}$

15. $\begin{bmatrix} 5 & 1 & 2 \\ 0 & 0 & 1 \\ 0 & 0 & 1 \\ 0 & 0 & 0 \end{bmatrix}$ 16. $\begin{bmatrix} 7 & 5 & 1 \\ 1 & 1 & 0 \\ 0 & 1 & 0 \\ 1 & 0 & 0 \\ 0 & 0 & 1 \\ 0 & 0 & 0 \end{bmatrix}$

17. $\begin{bmatrix} 0 \end{bmatrix}$ 18. $\begin{bmatrix} 0 & 0 & 1 \\ 0 & 0 & 1 \end{bmatrix}$

4.3 Exercises

19. $\begin{bmatrix} 1 & 2 & 3 \\ 4 & 5 & 6 \\ 7 & 8 & 9 \end{bmatrix}$ 20. $\begin{bmatrix} 3 & 1 & 1 \\ 1 & 3 & 1 \\ 1 & 1 & 3 \end{bmatrix}$

21. Prove: if A is m × n then $\rho(A) \leq \min\{m, n\}$.

22. Prove: $\rho(A) = \rho(A^T)$.

23. Prove: Any nonsingular matrix is a product of elementary matrices.

    Hint: By Theorem 1 there exist P and Q, products of elementary matrices, such that $PAQ = I_r \oplus 0_{n-r}$. By Theorem 4, A is nonsingular iff r = n. Thus $A = P^{-1}Q^{-1}$ and both P^{-1} and Q^{-1} are products of elementary matrices.

24. Prove: if C and D are nonsingular then for any matrix A, $\rho(CAD) = \rho(A)$.

    Hint: Let $A = P\Delta Q$ where P and Q are products of elementary matrices and Δ is in canonical form. Then $CAD = CP\Delta QD$. Since C and D are nonsingular they are products of elementary matrices so the canonical form of CAD is also Δ.

25. Prove: if $A = [\,B\,|\,0\,]$ or $A = \begin{bmatrix} B \\ 0 \end{bmatrix}$ then $\rho(A) = \rho(B)$.

    Hint: Elementary row and column operations on B can be performed by doing them on A. Thus there exist P and Q, both nonsingular, such that

    $$PAQ = \begin{bmatrix} I_r & 0 \\ 0 & 0 \end{bmatrix}$$

    is in canonical form where $\rho(B) = r$. But then $\rho(A) = r$.

26. Prove: $\rho(AB) \leq \min\{\rho(A), \rho(B)\}$.

    Hint: Let Δ be the canonical form for B and secure nonsingular P and Q such that $B = P\Delta Q$. Then by #24, $\rho(AB) = \rho(AP\Delta Q) = \rho(AP\Delta)$. Let $X = AP$ and suppose

$$\Delta = \left[\begin{array}{c|c} I_r & 0 \\ \hline 0 & 0 \end{array}\right],$$

$r = \rho(B)$. Then partition X conformally with Δ and compute

$$\left[\begin{array}{c|c} X_{11} & X_{12} \\ \hline X_{21} & X_{22} \end{array}\right] \left[\begin{array}{c|c} I_r & 0 \\ \hline 0 & 0 \end{array}\right] = \left[\begin{array}{c|c} \overset{r}{X_{11}} & 0 \\ \hline X_{21} & 0 \end{array}\right].$$

By #21 and #25, $\rho(X\Delta) \le r = \rho(B)$, and hence $\rho(AB) \le \rho(B)$. Now, by #22, $\rho(AB) = \rho(B^T A^T) \le \rho(A^T) = \rho(A)$.

27. Prove: if $A \in M_{m,n}(\mathbb{C})$ then $\rho(A) = \rho(A^*)$.

28. Let A be $m \times n$, u be $m \times 1$, and let $C = [\,A \mid u\,]$, be an $m \times (n+1)$ matrix. Prove: $\rho(A) \le \rho(C) \le \rho(A) + 1$. In other words, adjoining a column to A either leaves the rank unchanged or increases it by 1.

Hint: Perform elementary row and column operations on C to bring A to canonical form:

$$D = PCQ = \left[\begin{array}{c|c|c} I_r & 0 & v_1 \\ & & \vdots \\ \hline 0 & 0 & v_m \end{array}\right].$$

By elementary column operations on D, make $v_1, \ldots, v_r$ into 0 elements. If $v_{r+1} = \cdots = v_m = 0$ then $\rho(C) = \rho(D) = r = \rho(A)$. If some v_t, $t \ge r+1$, is not 0, make it 1 and reduce the rest of the v_j's to 0. Then put v_t ($= 1$) in row $r+1$ and column $r+1$. Then we have produced nonsingular P_1 and Q_1 such that

$$P_1 C Q_1 = \left[\begin{array}{c|c} I_{r+1} & 0 \\ \hline 0 & 0 \end{array}\right]$$

and $\rho(C) = r + 1$.

4.3 Exercises

29. Prove: if

$$\begin{array}{c} \overset{n}{} \\ \begin{array}{c} r \\ n-r \end{array}\left[\begin{array}{c} S_1 \\ \hline S_2 \end{array}\right] \end{array}$$

is nonsingular then $\rho(S_1) = r$.

Hint: Suppose $\rho(S_1) = p < r$ and perform elementary row and column operations on S to bring S_1 to canonical form Δ_1. Clearly, $p < r$ implies Δ_1 has a zero last row. Then by an elementary row operation make this zero row the last row of the entire matrix. Thus there are P and Q, nonsingular, such that PSQ has a zero last row. But PSQ is nonsingular since S is, and a nonsingular matrix cannot have a zero last row.

30. Let A be $m \times n$ and B be $n \times p$. Prove: $\rho(AB) \geq \rho(A) + \rho(B) - n$.

Hint: Secure P, an $m \times m$ matrix, Q an $n \times n$ matrix, P_1 an $n \times n$ matrix, and Q_1 a $p \times p$ matrix, all non-singular, so that $A = P\Delta Q$, $B = P_1 \Gamma Q_1$, and both Δ and Γ are canonical. Let $\rho(A) = r$, $\rho(B) = s$ and set $QP_1 = S$, S non-singular. Then by #24, $\rho(AB) = \rho(P\Delta QP_1 \Gamma Q_1) = \rho(\Delta S \Gamma)$. Write

$$\Delta = \begin{array}{c} \overset{r\ \ n-r}{} \\ \begin{array}{c} r \\ m-r \end{array}\left[\begin{array}{c|c} I_r & 0 \\ \hline 0 & 0 \end{array}\right] \end{array}, \quad S = \begin{array}{c} \overset{n}{} \\ \begin{array}{c} r \\ n-r \end{array}\left[\begin{array}{c} S_1 \\ \hline S_2 \end{array}\right] \end{array}$$

so that $\Delta S = \begin{bmatrix} S_1 \\ 0 \end{bmatrix}$. Then partition again so that

$$(\Delta S)\Gamma = \begin{array}{c} \overset{s\ \ \ n-s}{} \\ \begin{array}{c} r \\ m-r \end{array}\left[\begin{array}{c|c} S_{11} & S_{12} \\ \hline 0 & 0 \end{array}\right] \end{array} \begin{array}{c} \overset{s\ \ \ p-s}{} \\ \begin{array}{c} s \\ n-s \end{array}\left[\begin{array}{c|c} I_s & 0 \\ \hline 0 & 0 \end{array}\right] \end{array} = \begin{array}{c} \overset{s}{} \\ r\left[\begin{array}{c|c} S_{11} & 0 \\ \hline 0 & 0 \end{array}\right] \end{array}.$$

Now S is non-singular, and thus #29 implies that $\rho(S_1) = r$. Also S_{11} is obtained from S_1 by deleting $n - s$ columns. By #28, each such column deletion reduces the rank by at most 1. Hence $\rho(S_{11}) \geq \rho(S_1) - (n - s) = r - (n - s) = r + s - n$. However, it is obvious that $\rho(S_{11}) = \rho(\Delta S \Gamma) = \rho(AB)$.

31. Prove: if $\rho(A) = r$ then A is a sum of r rank 1 matrices.

32. Prove: if A is m × n, and $\rho(A) = 1$, then $A = xy$ where x is m × 1 and y is 1 × n.

33. Prove: if A is n-square, and $\rho(A) < n$ then there exists $x \neq 0$, n × 1, such that $Ax = 0$. Also prove: if B is m × n, m ≥ n, then $\rho(B) < n$ iff there exists $y \neq 0$, n × 1, such that $By = 0$.

    Hint: Let $A = P\Delta Q$, Δ canonical. Then the equation $Ax = 0$ becomes $P\Delta Qx = 0$, $\Delta Qx = 0$, $\Delta y = 0$ where $y = Qx$. It is simple to see that if $\rho(A) = r$ then e.g., $y = e^n_{r+1}$ is a solution (see §3.3 Kronecker Products, Theorem 2 for notation). By the same argument, if $\rho(B) = r < n$ then $B = P\Delta Q$, and $By = 0$ becomes $\Delta z = 0$, where $z = Qy$. Now note that $z = e^n_{r+1}$ satisfies $\Delta z = 0$, so that $y = Q^{-1}z \neq 0$ satisfies $By = P\Delta Qy = P\Delta QQ^{-1}z = P\Delta z = 0$. Conversely, suppose $By = 0$, $y \neq 0$. Write $B = P\Delta Q$ as before so that $By = 0$ becomes $\Delta z = 0$, $z = Qy \neq 0$. If $\rho(B) = n$ then Δ has the form

    $$\Delta = m \begin{bmatrix} I_n \\ 0 \end{bmatrix}^n$$

    so that $\Delta z = 0$ implies $z = 0$, a contradiction. Hence $\rho(B) < n$.

34. Prove: if $X \in M_{n,r}(\mathbb{C})$, $r \leq n$, and $\rho(X) = r$, then $\rho(X^*X) = r$.

    Hint: Since X is n × r it follows that X^*X is r × r. So suppose $\rho(X^*X) < r$. By #33, obtain $z = [z_1, ..., z_r]^T$ such that $z \neq 0$ and $X^*Xz = 0$. Then $z^*X^*Xz = 0$, $(Xz)^*(Xz) = 0$. But then $Xz = 0$ (why?). Let $X = P\Delta Q$, where P is n × n, Q is r × r, both are nonsingular, and, moreover, Δ is an n × r matrix in canonical form. Since $\rho(X) = r$, it follows that

    $$\Delta = \begin{bmatrix} I_r \\ 0_{n-r,r} \end{bmatrix}.$$

Then $0 = Xz = P\Delta Qz$. Thus $\Delta y = 0$ where $y = Qz \neq 0$ (i.e., $z \neq 0$). However, $\Delta y = y$ as seen by block multiplication. This contradiction implies that $\rho(X^*X)$ must be r.

35. Prove: if $A \in M_{m,n}(\mathbb{C})$, then $\rho(A) = \rho(AA^*) = \rho(A^*A)$.

    Hint: Let $\rho(A) = r$ and write $A = P\Delta Q$ where Δ is in canonical form, P is $m \times m$, and Q is $n \times n$, both nonsingular. Then

    $$\rho(A^*A) = \rho(Q^*\Delta^* P^*P\Delta Q) = \rho(\Delta^* P^*P\Delta).$$

    It is easy to see by block multiplication that

    $$\Delta^*(P^*P)\Delta = (P^*P)[1, ..., r \mid 1, ..., r] \oplus 0_{n-r}$$

    $$= P[1, ..., m \mid 1, ..., r]^* P[1, ..., m \mid 1, ..., r] \oplus 0_{n-r}.$$

    By #29, $\rho(P[1, ..., m \mid 1, ..., r]) = r$ because P is $m \times m$ non-singular. By #34, $\rho(P[1, ..., m \mid 1, ..., r]^* P[1, ..., m \mid 1, ..., r]) = r$. Hence $\rho(A^*A) = r$. Now $\rho(A^*) = r$ so $\rho((A^*)^*A^*) = r$ by the first part of the proof. Hence $\rho(AA^*) = r$.

36. Prove: If A is $m \times n$ and $m < n$ then there exists $x \neq 0$, $n \times 1$, such that $Ax = 0$. In words, a homogeneous system of equations with more unknowns than equations always has a nonzero solution. (Remember that *homogeneous* means the right hand side is 0.)

37. Prove: If A is $m \times n$, then there exists x, $n \times 1$, such that $Ax = b$, if and only if $\rho(A) = \rho([A \mid b])$.

    Hint: Write $A = P\Delta Q$, Δ a canonical $m \times n$ matrix, so that $Ax = b$ becomes $P\Delta Qx = b$, $\Delta y = P^{-1}b = c$, where $y = Qx$. Suppose $\rho(A) = r (\leq m)$. Then $\Delta y = c$ becomes $y_1 = c_1, ..., y_r = c_r$, $0 = c_{r+1} = \cdots = c_m$. Thus there is a solution iff $c_{r+1} = \cdots = c_m = 0$. Consider the matrix $[\Delta \mid c] = [P^{-1}AQ^{-1} \mid P^{-1}b] = P^{-1}[AQ^{-1} \mid b]$. Now the condition $c_{r+1} = \cdots = c_m = 0$ is precisely the condition that $\rho([\,\Delta\,|\,c\,]) = r$, or equivalently, $\rho(P^{-1}[\,AQ^{-1}\,|\,b\,]) = \rho([\,AQ^{-1}\,|\,b\,]) = r$. But

$$[\,A\,|\,b\,]\begin{bmatrix} Q^{-1} & 0_{n,1} \\ 0_{1,n} & 1 \end{bmatrix} = [\,AQ^{-1}\,|\,b\,].$$

Obviously,

$$\begin{bmatrix} Q^{-1} & 0 \\ 0 & 1 \end{bmatrix}$$

is nonsingular, so

$$\rho([\,A\,|\,b\,]) = \rho([AQ^{-1}\,|\,b\,]).$$

Thus $\rho([\,A\,|\,b\,]) = r = \rho(A)$ is a necessary and sufficient condition for the existence of a solution. Note that if this condition holds then $y_{r+1}, \ldots, y_n$ can be chosen arbitrarily so that the general solution to $Ax = b$ contains $n - r$ arbitrary parameters.

In # 38 - 40, the generalized inverse A^+ is defined as in (16).

38. Let the notation be the same as in #37 and assume $Ax = b$ is solvable. Prove: $x = A^+b$ is a solution.

Hint: By definition (see(16)) $A^+ = QB^TP$ where $PAQ = B$ is the canonical form of A. We compute that

$$A(A^+b) = P^{-1}BQ^{-1}QB^TPb$$
$$= P^{-1}BB^TPb.$$

Thus $AA^+b = b$ iff $BB^T(Pb) = Pb$. Let $c = Pb$ and assume $\rho(A) = \rho(B) = r$. Then it is easy to check that $BB^Tc = c$ holds iff $c_{r+1} = \cdots = c_m = 0$. On the other hand, the original equation $Ax = b$ is equivalent to $P^{-1}BQ^{-1}x = b$, or $By = c$, where $y = Q^{-1}x$ and $Pb = c$. The equation $By = c$ has a solution iff $c_{r+1} = \cdots = c_m = 0$. Thus we have proved that if there is a solution to $Ax = b$, then $c_{r+1} = \cdots = c_m = 0$ and it follows that A^+b satisfies $Ax = b$.

39. Devise a method of using the canonical form of A for obtaining all solutions to the *homogeneous equation* $Ax = 0$.

   Hint: Let $A = P\Delta Q$, Δ in canonical form, so that $Ax = 0$ becomes $P\Delta Qx = 0$, or $\Delta y = 0$ where $y = Qx$. If $\rho(A) = r$ then $y_1 = y_2 = \cdots = y_r = 0$ and $y_{r+1}, \ldots, y_n$ can be chosen arbitrarily. Thus any solution to $Ax = 0$ has the form $Q^{-1}y$ where $y = [\, 0_{1,r} \mid y_{r+1}, \ldots, y_n \,]^T$.

40. Prove: (i) Any two solutions to $Ax = b$ differ by a solution to $Ax = 0$. (ii) If $Ax = b$ has a solution then any solution is given by $x = A^+b + Q^{-1}(I_n - \Delta^+\Delta)v$ where $A = P\Delta Q$, Δ is canonical, and v is an arbitrary $n \times 1$ matrix.

   Hint: (i) is simple.

   (ii) By #38, $A^+ b$ is a particular solution to $Ax = b$. Thus by part (i) we need only exhibit the general form of a solution to $Ax = 0$. By the Hint for #39, any solution to $Ax = 0$ has the form $Q^{-1}y$ where $y = [\, 0_{1,r} \mid y_{r+1}, \ldots, y_n \,]^T$, $P\Delta Q = A$, $\rho(A) = \rho(\Delta) = r$ and Δ is in canonical form. We can write y in the form $y = (I_n - \Delta^+\Delta)v$, where v is any $n \times 1$ matrix. But then from part (i), the general solution to $Ax = b$ becomes

   $$x = A^+b + Q^{-1}y$$
   $$= A^+b + Q^{-1}(I_n - \Delta^+\Delta)v .$$

4.3 Matlab

1. Using the functions in the library Elemops created in §4.2 MatLab, reduce the matrices in §4.3 Exercises, #3 - 12 to canonical form.

2. Use rref to write a MatLab function called canf that returns the canonical form of an m × n matrix A. Save canf in Elemops.

    Hint:
    > function f = canf(A)
    > %CANF(A) returns the canonical form of A
    > B = rref(A);
    > C = rref(B.');
    > f = C.';

3. Use canf to write a MatLab function called rk that returns the rank of an m × n matrix A. Read the Help entry for **round** and use it to ensure that rk(A) is an integer. Save rk in Elemops.

    Hint:
    > function r = rk(A)
    > %RK(A) returns the rank of A.
    > A = canf(A);
    > b = sum(sum(A));
    > r = round(b);

4. Read the Help entry for **rank**. Write a MatLab script called rankola that generates 20 random matrices A of dimensions at most 10 and compares the values of rk(A) and rank(A). Use round and rand to generate random integers.

5. Use MatLab to show that

$$A = \begin{bmatrix} 1 & -1 & 2 \\ 2 & 0 & 1 \end{bmatrix}$$

4.3 MatLab

has a right inverse X. Find X using the method explained in formulas (12) and (13).

6. Write a MatLab script called ranktest to generate 20 random m × n complex matrices A of dimensions at most 10 and compare the ranks of A, AA* and A*A.

7. Recall that if $u \in M_{n,1}(\mathbb{C})$ then $(u^*u)^{1/2}$ is called the *2-norm* of u and is frequently written as $\| u \|$. Read the Help entry for norm, in particular the example for norm(x,2). Write a MatLab function called smax that accepts a 2-square complex matrix A and returns the maximum value of $\| Au \|$ as u runs over 100 random complex 2-vectors of unit norm, i.e., $\| u \| = 1$. Save smax in a library called "Norms."

   Hint:
   ```
   function f = smax(A)
   %SMAX(A) returns the max of the 2-norm of Au as u runs
   %over 100 random vectors u of 2-norm 1.
   a = [ ];
   big = 0;
   for k = 1:100
           u = (2*rand(2,1)-1)+i*(2*rand(2,1)-1);
           u = u/norm(u,2);
           m = norm(A*u,2);
           if m > big
                   big = m;
           end
   end
   f = big;
   ```

8. Write a MatLab function called lsqrs (for "least squares") that

   - accepts a real m × n matrix A and an m × 1 matrix b
   - generates 100 random n × 1 real matrices x
   - returns an x generated in the previous step that minimizes $\| Ax - b \|$.

   Save lsqrs in Norms.

Hint: Suppose a vector x_0 is generated with the command $x_0 = 2*\text{rand}(n,1) -1$. To find the minimum value of $\|A(tx_0) - b\|$ as a function of t, note that

$$\varphi(t) = \|A(tx_0) - b\|^2 = t^2 \|Ax_0\|^2 - 2tb^T Ax_0 + \|b\|^2.$$

If $Ax_0 \neq 0$, the minimum value of $\varphi(t)$ is taken on for

$$t = \frac{b^T Ax_0}{\|Ax_0\|^2}.$$

If $Ax_0 = 0$ then $\varphi(t)$ is a constant, namely $\|b\|^2$.

9. Run lsqrs on the pair

$$A = \begin{bmatrix} 1 & 2 \\ -1 & -2 \end{bmatrix}, \qquad b = \begin{bmatrix} 6 \\ -4 \end{bmatrix}.$$

10. It is not difficult to prove that an x which minimizes $\|Ax - b\|$ is given by $A^+ b$ where A^+ is the Moore-Penrose inverse of A defined in formula (24). (This result will be found in §8.3 Exercises, #4, #5.) Read the Help entry for **pinv**. The Moore–Penrose inverse of A is denoted by pinv(A). Use pinv to solve $Ax = b$ for the matrices in #9 above, in the sense that the solution minimizes the 2-norm of $Ax - b$. Such a solution is called a *least squares* solution to $Ax = b$. For, if A is $m \times n$ and b is $m \times 1$ then

$$\|Ax - b\|^2 = (Ax - b)^*(Ax - b) = \sum_{k=1}^{m} |(Ax)_k - b_k|^2.$$

Thus a solution that minimizes $\|Ax - b\|$ minimizes the sum of the squares of the absolute values of the differences of the components of Ax and b. Note that if A is nonsingular then $A^+ = A^{-1}$. To see this we use the definitions in formulas (22) - (24): $A = PDQ$, $A^{-1} = Q^{-1}D^{-1}P^{-1} = Q^*D^{-1}P^* = A^+$ (from (26) and the fact that P and Q are unitary matrices.). Thus if A is nonsingular in the equation $Ax = b$, the least squares solution coincides with the usual solution $x = A^{-1}b$.

11. Use pinv to find A^+ for each of the following matrices A:

(a) $A = \begin{bmatrix} -3 & 1 \\ -2 & 1 \\ -1 & 1 \\ 0 & 1 \\ 1 & 1 \\ 2 & 1 \\ 3 & 1 \end{bmatrix}$ (b) $A = \begin{bmatrix} 0 & 0 & 1 & 2 \\ 1 & 2 & 2 & 3 \end{bmatrix}$

12. Find the least squares solution to the system

$$x_1 + 2x_2 + 2x_3 + 3x_4 = 2$$
$$x_3 + 2x_4 = 1$$

13. Find the line $y = at + c$ that in the least squares sense best fits the points { (–3, 10), (-2, 15), (-1, 19), (0, 27), (1, 28), (2, 34), (3, 42) }.

Hint: if the line $y = at + c$ were actually to go through the points we would have

$$-3a + c = 10$$

$$-2a + c = 15$$

$$-a + c = 19$$

$$c = 27$$

$$a + c = 28$$

$$2a + c = 34$$

$$3a + c = 42.$$

If we set

$$A = \begin{bmatrix} -3 & 1 \\ -2 & 1 \\ -1 & 1 \\ 0 & 1 \\ 1 & 1 \\ 2 & 1 \\ 3 & 1 \end{bmatrix}, \quad x = \begin{bmatrix} a \\ c \end{bmatrix}, \quad b = \begin{bmatrix} 10 \\ 15 \\ 19 \\ 27 \\ 28 \\ 34 \\ 42 \end{bmatrix}$$

then the equations become $Ax = b$. The least squares solutions to this system is $x = A^+ b$.

14. Read the Help entry for **norm(A, 'fro')** (i.e., the Frobenius norm). Find a way of implementing norm(A, 'fro') using the commands, abs, sqrt, sum, and diag. The Frobenius norm is frequently designated mathematically as $\|A\|_F$.

15. Write a MatLab script called ptrans that generates 20 random complex matrices A of dimensions not exceeding 5 and prints out a list of the numbers $\|(A^*)^+ - (A^+)^*\|_F$. Save ptrans in norms. What do you conjecture about the relationship between $(A^*)^+$ and $(A^+)^*$? Prove your conjecture.

Hint:
```
d = [ ];
for k = 1: 20
        m = floor(1+5*rand);
        n = floor(1+5*rand);
        C = 2*rand(m, n)-1;
        D = 2*rand(m, n)-1;
        A = C + i*D;
        f = norm(pinv(A')-pinv(A)','fro');
        d = [d f];
end
disp(d');
```

It appears that $(A^*)^+ = (A^+)^*$. To prove this equality we use the fact that equations (17) - (20) uniquely specify the Moore-Penrose inverse of a matrix. Let $B = A^*$. Then $B^+ = (A^*)^+$ is characterized uniquely by the equations

$$BB^+B = B,$$

$$B^+BB^+ = B^+,$$

$$(BB^+)^* = BB^+,$$

$$(B^+B)^* = B^+B.$$

Consider the following computations in which we replace B by A^* and B^+ by $(A^+)^*$ in the above equations:

$$BB^+B = A^*(A^+)^*A^* = (AA^+A)^* = A^* = B,$$

$$B^+BB^+ = (A^+)^*A^*(A^+)^* = (A^+AA^+)^* = (A^+)^* = B^+,$$

$$(BB^+)^* = (A^*(A^+)^*)^* = ((A^+A)^*)^* = A^+A = (A^+A)^* = A^*(A^+)^* = BB^+,$$

$$(B^+B)^* = ((A^+)^*A^*)^* = (AA^+)^{**} = AA^+ = (AA^+)^* = (A^+)^*A^* = B^+B.$$

Thus $(A^+)^*$ satisfies the defining equations for the Moore-Penrose inverse of A^*.

4.3 Glossary

2-norm	223
A^+	208
canonical form	201
generalized inverse	208
homogeneous	219

iff	206
least squares	224
left inverse	205
Moore-Penrose inverse	210
norm (A,'fro')	226
$\rho(A)$	206
rank	206
right inverse	205
$\|A\|_F$	226

Chapter 5

Permutations

Topics
- *composition of permutations*
- *disjoint cycles, orbits*
- *conjugacy, cycle structure*
- *inversions, interchanges*

5.1 Introduction to Permutations

Permutations and permutation groups are important throughout applied and pure mathematics. At an elementary level, permutations usually appear in a beginning course on discrete mathematics. In this chapter we study permutations in terms of their applications in matrix theory.

A *permutation* on n objects, labeled 1, ..., n, is a one-one mapping (i.e., function) of the set {1, ..., n} onto itself. In conformity with standard notational conventions, we denote the image of i under a permutation σ by σ(i). Sometimes it is convenient to denote a permutation as a 2 × n array:

$$\sigma = \begin{pmatrix} 1 & 2 & \cdots & n \\ \sigma(1) & \sigma(2) & \cdots & \sigma(n) \end{pmatrix} \quad (1)$$

in which the value of σ at i appears directly below i. Clearly, the order in which the pairs "σ(i) below i" appear is immaterial. Thus, if σ is defined as

$$\sigma = \begin{pmatrix} 1 & 2 & 3 & 4 \\ 3 & 2 & 4 & 1 \end{pmatrix} = \begin{pmatrix} 2 & 4 & 3 & 1 \\ 2 & 1 & 4 & 3 \end{pmatrix}$$

then $\sigma(1) = 3$, $\sigma(2) = 2$, $\sigma(3) = 4$, and $\sigma(4) = 1$. It is customary to say "σ maps 1 into 3, 2 into 2, 3 into 4, and 4 into 1."

The *product* of two permutations σ and φ is defined in terms of function composition:

$$(\sigma\varphi)(i) = \sigma(\varphi(i)), \quad i = 1, \ldots, n.$$

The set of all permutations on n objects, together with the operation of multiplication defined in terms of function composition, is called the *symmetric group of degree n*, and is denoted by S_n.

The following are examples of permutation multiplication:

$$\begin{pmatrix} 1 & 2 & 3 & 4 & 5 \\ 2 & 4 & 1 & 5 & 3 \end{pmatrix} \begin{pmatrix} 1 & 2 & 3 & 4 & 5 \\ 4 & 1 & 2 & 5 & 3 \end{pmatrix} = \begin{pmatrix} 1 & 2 & 3 & 4 & 5 \\ 5 & 2 & 4 & 3 & 1 \end{pmatrix} \quad (2)$$

$$\begin{pmatrix} 1 & 2 & 3 & 4 & 5 \\ 4 & 1 & 2 & 5 & 3 \end{pmatrix} \begin{pmatrix} 4 & 1 & 2 & 5 & 3 \\ 1 & 2 & 3 & 4 & 5 \end{pmatrix} = \begin{pmatrix} 1 & 2 & 3 & 4 & 5 \\ 1 & 2 & 3 & 4 & 5 \end{pmatrix} \quad (3)$$

$$\begin{pmatrix} 1 & 2 & 3 & 4 & 5 \\ 5 & 2 & 4 & 3 & 1 \end{pmatrix} \begin{pmatrix} 4 & 1 & 2 & 5 & 3 \\ 1 & 2 & 3 & 4 & 5 \end{pmatrix} = \begin{pmatrix} 1 & 2 & 3 & 4 & 5 \\ 2 & 4 & 1 & 5 & 3 \end{pmatrix} \quad (4)$$

$$\begin{pmatrix} 1 & 2 & 3 & 4 & 5 \\ 4 & 1 & 2 & 5 & 3 \end{pmatrix} \begin{pmatrix} 1 & 2 & 3 & 4 & 5 \\ 2 & 4 & 1 & 5 & 3 \end{pmatrix} = \begin{pmatrix} 1 & 2 & 3 & 4 & 5 \\ 1 & 5 & 4 & 3 & 2 \end{pmatrix}. \quad (5)$$

For example, in formula (2), 3 is mapped into 2 by the second factor on the left and then 2 is mapped into 4 by the first factor. Thus the product maps 3 into 4 as indicated. Examples (2) and (5) show that multiplication of permutations is not always commutative, i.e., it is not always the case that $\sigma\varphi = \varphi\sigma$.

Theorem 1.

(a)   A product of two permutations in S_n is a permutation in S_n.

5.1 Introduction to Permutations

(b) Multiplication of permutations is associative.

(c) If $e = \begin{pmatrix} 1 & 2 & \cdots & n \\ 1 & 2 & \cdots & n \end{pmatrix}$ and σ is any permutation in S_n, then

$$e\sigma = \sigma e = \sigma. \tag{6}$$

(d) For every $\sigma \in S_n$ there exists a unique permutation $\sigma^{-1} \in S_n$ such that $\sigma\sigma^{-1} = \sigma^{-1}\sigma = e$.

(e) There are n! distinct permutations in S_n.

(f) If $S_n = \{\sigma_1, \sigma_2, ..., \sigma_M\}$, $M = n!$, and φ is a fixed permutation in S_n, then

$$\{\varphi\sigma_1, \varphi\sigma_2, ..., \varphi\sigma_M\} = \{\sigma_1\varphi, \sigma_2\varphi, ..., \sigma_M\varphi\} = S_n. \tag{7}$$

Proof.

Parts (a) and (c) follow immediately from the definition. However, we note that if both e and e' satisfy (6) then $e = ee' = e'$. The unique permutation e that satisfies (6) is called the *identity*.

The statement in part (b) is true for function composition in general: if $\sigma, \varphi, \tau \in S_n$, then using the definition of multiplication, we have

$$((\sigma\varphi)\tau)(i) = (\sigma\varphi)(\tau(i)) = \sigma(\varphi(\tau(i)))$$

and

$$(\sigma(\varphi\tau))(i) = \sigma((\varphi\tau)(i)) = \sigma(\varphi(\tau(i)))$$

for any i. Because the multiplication in S_n is associative, we can write the product of several permutations without parentheses and, in particular, we can inductively define the kth power of σ, k a positive integer, by

$$\sigma^k = \sigma(\sigma^{k-1}) \quad \text{for } k > 1.$$

To prove (d), set

$$\sigma^{-1} = \begin{pmatrix} \sigma(1) & \sigma(2) & \cdots & \sigma(n) \\ 1 & 2 & \cdots & n \end{pmatrix}.$$

Then $\sigma\sigma^{-1} = \sigma^{-1}\sigma = e$. Suppose φ is a second permutation that satisfies $\sigma\varphi = \varphi\sigma = e$. Then, using associativity,

$$\sigma^{-1} = \sigma^{-1}e = \sigma^{-1}(\sigma\varphi) = (\sigma^{-1}\sigma)\varphi = e\varphi = \varphi.$$

The permutation of σ^{-1} is called the *inverse* of σ.

Note that the customary laws of exponents hold for powers of a permutation σ. As usual, if k is a negative integer then

$$\sigma^k = (\sigma^{-1})^p \tag{8}$$

where $p = |k|$. Then the reader will easily confirm that

$$\sigma^{r+s} = \sigma^r \sigma^s \tag{9}$$

and

$$(\sigma^r)^s = \sigma^{rs} \tag{10}$$

are valid for any permutation σ and any integers r and s (convention: $\sigma^0 = e$).

We prove part (e) by counting the number of ways in which the second line of the array

$$\begin{pmatrix} 1 & 2 & \cdots & n \\ \sigma(1) & \sigma(2) & \cdots & \sigma(n) \end{pmatrix}$$

can be written. The value of $\sigma(1)$ can be any of the numbers 1, ..., n. Once $\sigma(1)$ is chosen, $\sigma(2)$ can be chosen to be any of the remaining n - 1 numbers. After $\sigma(1)$ and $\sigma(2)$ are chosen, $\sigma(3)$ can be chosen to be any of the remaining n - 2 numbers, and so on.

5.1 Introduction to Permutations

The total number of choices is $n(n-1)(n-2) \cdots 2 \cdot 1 = n!$. Hence the number of permutations in S_n is $n!$.

To prove (f) we first show that if $\sigma_p \neq \sigma_q$ then $\varphi\sigma_p \neq \varphi\sigma_q$. For, by part (d), there exists φ^{-1} such that $\varphi^{-1}\varphi = e$ and therefore $\varphi\sigma_p = \varphi\sigma_q$ implies $\varphi^{-1}\varphi\sigma_p = \varphi^{-1}\varphi\sigma_q$, and hence $\sigma_p = \sigma_q$. Thus all permutations in $\{\varphi\sigma_1, \varphi\sigma_2, \ldots, \varphi\sigma_M\}$ are distinct and, since there are $M = n!$ of them, they must, by part (e), be the permutations of S_n in some order. Similarly,

$$\{\sigma_1\varphi, \ldots, \sigma_M\varphi\} = S_n . \quad \blacksquare$$

Let $N = \{1, 2, \ldots, n\}$. A permutation $\sigma \in S_n$ is called a *k-cycle* if there exists a k element subset of N, $\{i_1, i_2, \ldots, i_k\}$, such that

$$\sigma(i_t) = i_{t+1}, \quad t = 1, \ldots, k-1, \tag{11}$$

$$\sigma(i_k) = i_1, \tag{12}$$

and

$$\sigma(j) = j, \quad j \notin \{i_1, i_2, \ldots, i_k\} . \tag{13}$$

The *cycle notation* abbreviates the 2-row notation to

$$\sigma = (i_1 \ i_2 \ \ldots \ i_k) .$$

As an example of the cycle notation,

$$\begin{pmatrix} 1 & 2 & 3 & 4 & 5 & 6 & 7 & 8 \\ 1 & 4 & 5 & 7 & 2 & 6 & 3 & 8 \end{pmatrix} = (2 \ 4 \ 7 \ 3 \ 5) .$$

Thus, 2 maps into 4, 4 into 7, 7 into 3, 3 into 5, 5 into 2, and 1, 6 and 8 are left fixed. Clearly any 1-cycle must be the identity permutation. We also observe that

$$(i_1 \ i_2 \ \ldots \ i_k) = (i_2 \ i_3 \ \ldots \ i_k \ i_1) = (i_3 \ i_4 \ \ldots \ i_k \ i_1 \ i_2) = \cdots = (i_k \ i_1 \ \ldots \ i_{k-1}) .$$

The simplest nontrivial cycles, the 2-cycles, play a special role in the theory of permutations. A 2-cycle is called a *transposition* or an *interchange*. For example, the cycle (1 2) = (2 1) is a transposition.

Let $i \in N = \{1, ..., n\}$ and let σ be any permutation in S_n. Then the set

$$\sigma[i] = \{ i, \sigma(i), \sigma^2(i), ..., \sigma^{k-1}(i) \}, \qquad (14)$$

where k is the least positive integer such that $\sigma^k(i) = i$, is called the *orbit of i under σ*. Observe that if t is any integer, then $\sigma^t(i)$ is in the orbit of i. For, let $t = kq + r$, where $0 \le r < k$, and use (8), (9), and (10) to compute

$$\sigma^t(i) = \sigma^{kq+r}(i) = \sigma^{r+kq}(i) = \sigma^r(\sigma^{kq}(i)) = \sigma^r(i).$$

The last equality follows from the fact that $\sigma^k(i) = i$ and hence $\sigma^{kq}(i) = (\sigma^k)^q(i) = i$. Moreover, the orbit of i has exactly k distinct elements. For, if p and q, $p < q$, are any two numbers in $\{1, ..., k\}$ then $\sigma^p(i) = \sigma^q(i)$ implies that $\sigma^{q-p}(i) = i$ and, since $0 < q - p < k$, this contradicts our assumption about the minimality of k. Note that this argument also shows that the first integer that is repeated in the sequence $i, \sigma(i), \sigma^2(i), ..., \sigma^{k-1}(i)$ must be i itself. The number k is called the *length* of the orbit.

If $\sigma = (i_1 i_2 ... i_k)$ is a cycle of length k then it is clear that

$$\sigma[i_1] = \{i_1, i_2, ..., i_k\}.$$

We say that two cycles, $\sigma = (i_1 i_2 ... i_k)$ and $\varphi = (j_1 j_2 ... j_r)$, are *disjoint* if $\sigma[i_1] \cap \varphi[j_1] = \emptyset$. A set of cycles is said to be disjoint if they are pairwise disjoint. The reader should confirm the easy assertion that disjoint cycles commute.

Theorem 2.

(a) The orbits under a permutation $\sigma \in S_n$ partition the set $N = \{1, ..., n\}$ into disjoint subsets whose union is N.

(b) Every permutation is a product of disjoint cycles. This decomposition into disjoint cycles is unique except for the order in which the cycles appear.

5.1 Introduction to Permutations

(c) Every permutation is a product of transpositions.

Proof.

(a) Since each integer in N belongs to at least one orbit, it suffices to prove that if two orbits have an element in common then they are equal. Suppose then that $\sigma^s(i) = \sigma^t(j)$. We first show that i belongs to the orbit of j,

$$i \in \sigma[j], \tag{15}$$

and also that

$$j \in \sigma[i]. \tag{16}$$

Let k be the length of the orbit of i. If $s = k$, there is nothing to prove. If $s < k$ then

$$i = \sigma^k(i) = \sigma^{k-s+s}(i) = \sigma^{k-s}(\sigma^s(i)) = \sigma^{k-s}(\sigma^t(j)),$$

or

$$i = \sigma^{k-s+t}(j).$$

Similarly, j belongs to the orbit of i and (15) and (16) are established. But then (15) implies that

$$\sigma[i] \subset \sigma[j] \tag{17}$$

and (16) implies that

$$\sigma[j] \subset \sigma[i], \tag{18}$$

and we conclude that $\sigma[i] = \sigma[j]$.

(b) Let $\sigma \in S_n$. Let $i_1 \in N$ and let k_1 be the length of the orbit of i_1 under σ. Let ψ_1 be the cycle $(i_1 \; \sigma(i_1) \; \ldots \; \sigma^{k_1-1}(i_1))$. If $\psi_1 = \sigma$, the proof is complete.

Otherwise, there exists $i_2 \in N$, not in the orbit of i_1, such that $\sigma(i_2) \neq i_2$. For, if no such i_2 exists, we would have $\psi_1 = \sigma$. Let $\psi_2 = (i_2 \, \sigma(i_2) \, \ldots \, \sigma^{k_2-1}(i_2))$ where k_2 is the length of the orbit of i_2 under σ. Note that, by part (a), the cycles ψ_1 and ψ_2 are disjoint. Again, either $\sigma = \psi_2 \psi_1$ and the proof is complete or there exists $i_3 \in N$, not in the union of the orbits of i_1 and i_2 under σ, such that $\sigma(i_3) \neq i_3$, and so on. Since N is finite, the process will terminate after a finite number of steps, say m. Then

$$\sigma = \psi_m \cdots \psi_1 . \tag{19}$$

Clearly, in (19) we can omit any occurrences of 1-cycles, i.e., omit the identity permutation as one of the factors ψ_j, $j = 1, \ldots, m$.

To prove the uniqueness of the decomposition (19) of σ into a product of disjoint cycles, suppose that θ_1 is a cycle of length at least 2 that appears in some factorization of σ into disjoint cycles:

$$\theta_1 = (i \, \sigma(i) \, \ldots \, \sigma^{k-1}(i)).$$

Since i is not a *fixed point* of θ_1 (i.e., $k > 1$, so that $\theta_1(i) \neq i$), it follows that i must appear in the orbit determined by one of the ψ_t (which we may take as ψ_1 by the commutativity of disjoint cycles):

$$i = \psi_1^p (j).$$

Then

$$\theta_1(i) = \sigma(i) = \sigma(\psi_1^p(j)) = \psi_1(\psi_1^p(j)) = \psi_1(i). \tag{20}$$

But then $\theta_1 = \psi_1$. Hence any equality of the form

$$\theta_r \theta_{r-1} \cdots \theta_1 = \sigma = \psi_m \psi_{m-1} \cdots \psi_1$$

implies that

$$\theta_r \theta_{r-1} \cdots \theta_2 = \psi_m \psi_{m-1} \cdots \psi_2.$$

The argument is easily completed by an induction on m. Finally, we note that since the product of disjoint cycles is commutative, the decomposition (19) can only be unique to within the order in which the factors on the right occur.

(c) Clearly, by part (b), it suffices to prove that every cycle is a product of transpositions:

$$\sigma = (i_1 \; i_2 \; \ldots \; i_k) = (i_1 \; i_k)(i_1 \; i_{k-1}) \cdots (i_1 \; i_2). \blacksquare$$

5.1 Exercises

1. Write each of the following permutations as a product of disjoint cycles:

   (a) (1 2 3)(4 1 3)

   Hint: Let $\sigma = \theta\varphi$, $\theta = (1\;2\;3)$, $\varphi = (4\;1\;3)$. Then $\sigma(1) = \theta(\varphi(1)) = \theta(3) = 1$, $\sigma(2) = \theta(\varphi(2)) = \theta(2) = 3$, $\sigma(3) = \theta(\varphi(3)) = \theta(4) = 4$, $\sigma(4) = \theta(\varphi(4)) = \theta(1) = 2$. Thus $\sigma = (2\;3\;4)$.

   (b) (1 2)(3 4)(1 2)

   (c) (1 2 3)(1 4 3)

   (d) (1 2 3 4)(5 6 4 3 2 1)

   (e) (1 2)(1 3 4 5)(2 6 7 8)

   (f) (1 2 3 4)(2 3 4 5)(3 4 5 6)

   (g) $[\,(1\;2\;3\;4)(4\;3\;1\;2)(7\;1\;3\;5)(2\;6\;1\;3)\,]^{-1}$

   Hint: $(1\;2\;3\;4)^{-1} = (4\;3\;2\;1)$. Also, for arbitrary permutations,

$$(\sigma_1 \sigma_2 \cdots \sigma_m)^{-1} = \sigma_m^{-1} \sigma_{m-1}^{-1} \cdots \sigma_2^{-1} \sigma_1^{-1}.$$

(h) $\begin{pmatrix} 1 & 2 & 3 & 4 & 5 & 6 \\ 1 & 4 & 2 & 3 & 6 & 5 \end{pmatrix}^{-1}$

Hint: First write the permutation itself as a product of disjoint cycles. Then use the hint in part (g).

(i) $\begin{pmatrix} 1 & 2 & 3 & 4 & 5 & 6 & 7 & 8 & 9 \\ 2 & 3 & 4 & 5 & 1 & 6 & 7 & 8 & 9 \end{pmatrix}$

(j) $(1\ 2\ 3)^2 (1\ 3\ 4)^3 (1\ 2\ 3\ 4\ 5)^{-2} (1\ 5\ 7)^5$

2. Prove: The number of distinct k-cycles in S_n is $n!/((n-k)!\, k)$.

Hint: An ordered sequence $i_1\ i_2 \cdots i_k$ can be chosen in $n!/(n-k)!$ ways. For each such choice there corresponds precisely k cyclic rearrangements that yield the same k-cycle: $(i_1\ i_2 \cdots i_k) = (i_2\ i_3 \cdots i_k\ i_1) = \cdots = (i_k\ i_1 \cdots i_{k-1})$. Thus the total number is $n!/((n-k)!\, k)$.

3. A *permutation matrix* $A(\sigma)$ corresponding to $\sigma \in S_n$ is the *incidence matrix* for σ, i.e., $A(\sigma)$ is the n-square matrix whose (i, j) entry is $\delta_{i,\, \sigma(j)}$. Find $A(\sigma)$ for each $\sigma \in S_3$. Recall the Kronecker delta notation:

$$\delta_{p,q} = \begin{cases} 1 & \text{if } p = q \\ 0 & \text{if } p \neq q \end{cases}.$$

4. Prove: If $A(\sigma)$ is an $n \times n$ permutation matrix then each row and each column of $A(\sigma)$ contains precisely one 1 and $n-1$ 0's.

5. Prove: If $\sigma, \varphi \in S_n$ then $A(\sigma\varphi) = A(\sigma)A(\varphi)$.

6. Prove: If $\sigma \in S_n$ then $A(\sigma)^{-1} = A(\sigma^{-1})$.

7. Find $\sigma \in S_5$ such that the least positive integer k for which $A(\sigma)^k = I_5$ is k = 6.

8. Prove: $A(\sigma^{-1}) = A(\sigma)^T$, i.e., permutation matrices are orthogonal.

9. If $\sigma \in S_m$, $\varphi \in S_n$, X is m × n, and $Y = A(\sigma)XA(\varphi)$, find a formula for Y_{ij} in terms of σ, φ and the entries of X.

Hint:

$$Y_{ij} = (A(\sigma)XA(\varphi))_{ij} = \sum_{k,r} (A(\sigma))_{ik} X_{kr} (A(\varphi))_{rj} = \sum_{k,r} \delta_{i,\sigma(k)} X_{kr} \delta_{r,\varphi(j)}$$

$$= \sum_r X_{\sigma^{-1}(i), r} \delta_{r, \varphi(j)} = X_{\sigma^{-1}(i), \varphi(j)}.$$

10. Recall that the notation $\prod_{i=1}^n x_i$ means the product of $x_1, \ldots, x_n$. If $A = [a_{ij}]$ is 2 × 2, write out

$$\sum_{\sigma \in S_2} \prod_{i=1}^2 a_{i\,\sigma(i)}.$$

If $A = [a_{ij}]$ is 3 × 3, write out

$$\sum_{\sigma \in S_3} \prod_{i=1}^3 a_{i\,\sigma(i)}.$$

11. Show that if A is a 3 × 3 permutation matrix then

$$\sum_{\sigma \in S_3} \prod_{i=1}^3 a_{i\,\sigma(i)}$$

is 1.

12. Let J be the n × n matrix each of whose entries is 1. Prove:

$$\sum_{\sigma \in S_n} \prod_{i=1}^n a_{i\,\sigma(i)} = n!.$$

13. Find a 4 × 4 matrix A with every entry either 0 or 1, with at least one 1 in each row and column, and for which

$$\sum_{\sigma \in S_4} \prod_{i=1}^{4} a_{i\,\sigma(i)} = 0.$$

5.1 MatLab

1. Write a MatLab function called pmtly that accepts two permutations x and y in S_n and returns the product xy. Save pmtly in a library (folder) called Permute.

   Hint: A permutation

   $$x = \begin{pmatrix} 1 & 2 & 3 & \cdots & n \\ x(1) & x(2) & x(3) & \cdots & x(n) \end{pmatrix}$$

   may be represented in MatLab as the row matrix

   $$x = [\, x(1) \ \ x(2) \ \ \cdots \ \ x(n) \,].$$

   Remember that MatLab computes x(y) as the row matrix $[\,(xy)(1) \cdots (xy)(n)\,]$.

2. Write a function called ppwr that accepts a permutation x in S_n and a nonnegative integer k and returns the permutation x^k. Save ppwr in Permute.

3. Write a function called cmpr that accepts two permutations x and y in S_n and returns 0 if they are unequal, 1 if they are equal. Save cmpr in Permute.

4. Write a function called pmat that accepts a permutation x in S_n and returns the n–square incidence matrix $A(x) = [\,\delta_{s,x(t)}\,]$. Save pmat in Permute.

5.1 MatLab

Hint:

```
function A = pmat(x)
%PMAT returns the incidence matrix A(x)
%corresponding to a permutation x.
n = length(x);
A = eye(n);
A = A(:, x);
```

5. Write a function called pnv that accepts a permutation x in S_n and returns x^{-1}, the inverse of x. Save pnv in Permute.

Hint:

```
function u = pnv(x)
%PNV returns the inverse of
%a permutation x.
n = length(x);
u(x) = 1:n;
```

6. Write a function called powr that accepts a permutation x in S_n and any integer k and returns x^k. Save powr in Permute.

Hint:

```
function u = powr(x, k)
%POWR returns the kth power of a permutation x.
n = length(x);
if k >= 0
    u = ppwr(x, k);
else
    u = pnv(ppwr(x, abs(k)));
end
```

7. One method of generating a random permutation x of 1, ..., n is to first choose x(1) from 1, ..., n. Then choose one of the remaining numbers for x(2), etc. Thus choosing x involves n choices with probabilities $\frac{1}{n}$, $\frac{1}{n-1}$, ..., 1, so that the probability of any particular permutation being chosen is $\frac{1}{n!}$. Write a function

called prand that accepts an integer n and returns a random permutation x in S_n. Save prand in Permute.

Hint: Suppose x = 1:4. We trace the results of the following for..end loop

```
for m = 1:4
    L = floor(m + rand * (5 - m));
    L1 = X(L);
    X(L) = X(m);
    X(m) = L1;
end
```

Obviously, the last 3 statements inside the for..end loop simply interchange the contents (i.e., values) of X(L) and X(m). The following table shows the exit values from the loop for each of the values of m. The values of L are chosen randomly depending on the formula L = floor(m + rand * (5 - m)). The initial value of X is [1 2 3 4].

m	L = floor(m + rand * (5 - m))	X(m)	X(L)	X
1	3	3	1	3 2 1 4
2	2	2	2	3 2 1 4
3	4	4	1	3 2 4 1
4	4	1	1	3 2 4 1

Note that: for m = 1, L is one of the integers 1, 2, 3, 4
for m = 2, L is one of the integers 2, 3, 4
for m = 3, L is one of the integers 3, 4
for m = 4, L is the integer 4

8. In §5.1 Exercises, #5, the assertion is made that the equation A(xy) = A(x)A(y) holds for permutation matrices. Write a MatLab script called Incidence that calls for the input of an integer n and then prints out the values of A(x), A(y), A(x)A(y) and A(xy) for 10 randomly chosen pairs of permutations x and y in S_n. Save Incidence in Permute.

9. Consider the following algorithm for listing the permutations x in S_n in lexicographical order.

Start with x = e:
In the current x do the following
- Step 1: find the largest t such that $x(t-1) < x(t)$
- Step 2: find the largest r such that $x(t-1) < x(r)$
- Step 3: switch $x(t-1)$ and $x(r)$ to obtain a new permutation, also denoted by x
- Step 4: in the current x reverse the order of

$$x(t), x(t+1), \ldots, x(n)$$

and call the new permutation x
- Step 5: if x is the permutation [n n-1 ... 1] then stop; otherwise, go to Step 1.

(a) Run through the algorithm manually for n = 3.

(b) Write a function called Lt that accepts a permutation x and finds the largest t such that $x(t-1) < x(t)$. Save Lt in Permute.

(c) Write a function called Lr that accepts an integer n and a permutation x and returns the largest r such that $x(t-1) < x(r)$ where t is the integer produced by Lt. Save Lr in Permute.

(d) Write a function called swap that accepts a pair of integers, a and b, and a permutation x, and returns the permutation y that results from x by interchanging $x(a)$ and $x(b)$. Save swap in Permute.

(e) Write a function called rev that accepts an integer t and a permutation x and returns a permutation obtained from x by reversing

$$x(t), x(t+1), \ldots, x(n).$$

Save rev in Permute.

(f) Write a MatLab script called Allperms that calls for the input of a positive integer n and then prints out S_n in lexicographical order. Save Allperms in Permute.

Hint:

(b) function t = Lt(x)
%LT returns the largest t such
%that x(t-1) < x(t) in the permutation x.
n = length(x);
t = n;
if x == (n:-1:1)
    t = 0;
else
    while (x(t-1) >= x(t)) & (t >= 2)
        t = t-1;
    end
end

(c) function r = Lr(x, t)
%LR returns the largest r such that
%x(t - 1) < x(r) in the permutation x.
n = length(x);
r = n;
while(x(t - 1) >= x(r)) & (r >= t)
    r = r - 1;
end

(d) function y = swap(x, a, b)
%SWAP returns y, obtained from x with
%x(a) and x(b) exchanged.
temp = x(a);
x(a) = x(b);
x(b) = temp;
y = x;

(e) function y = rev(x, t)
%REV returns x with
%x(t), ..., x(n) reversed.
n = length(x);
u = n:-1:t;
y = [x(1:(t - 1)) x(u)];

(f) n = input('Enter n: ');
x = 1:n;
u = x;
p = Lt(x);   %p is the largest t such that x(t-1)<x(t)
while p > 1
    q = Lr(x, p);   %q is the largest r such that x(t-1) < x(r)
    x = swap(x, p-1, q);   %swaps x(t-1) & x(r)
    x = rev(x, p);   %reverses x(t), ..., x(n)
    u = [u; x];
    p = Lt(x);
end
disp(u)

10. (a) Write a function called fr that accepts a 1 × n matrix x and returns the least j for which x(j) ≠ -1. If no such j exists, i.e., all entries are -1, then fr(x) = n + 1. Save fr in Permute.

(b) Write a function called cycle that accepts a permutation x in S_n and an integer s and returns the cycle containing s in the disjoint cycle decomposition of x. Save cycle in Permute.

(c) Write a function called disj that accepts a permutation x in S_n and then returns the disjoint cycle decomposition of x as rows of a k × n matrix. Save disj in Permute.

Hint:

(a) function f = fr(x)
%FR accepts a 1 x n matrix x
%and returns the least j
%for which x(j) ≠ -1. If no
%such j exists, n + 1 is returned.
n = length(x);
v = find(x ~= -1);
if length(v) == 0

```
                f = n + 1;
        else
                f = v(1);
        end

(b)     function y = cycle(x, s)
        %CYCLE accepts a permutation x and an
        %integer s and returns the cycle in which s lies.
        n = length(x);
        y = s;
        if x(s) ~= s
                j = s;
                while x(j) ~= s
                        j = x(j);
                        y = [y  j];
                end
        end

(c)     function A = disj(x)
        %DISJ accepts a permutation x
        %and returns the disjoint
        %cycle decomposition of x in Sn
        %as rows of a k x n matrix filled out with zeros.
        n = length(x);
        A = zeros(n);
        k = 0;
        s = fr(x);
        while s ~= (n + 1)
                y = cycle(x, s);
                m = length(y);
                k = k + 1;
                A(k, :) = [y   zeros(1, n - m) ];
                x(y) = -ones(1, m);
                s = fr(x);
        end
        A = A(1:k, :);
```

5.1 Glossary

cycle notation	233
disjoint cycles	234
fixed point	236
identity permutation	231
incidence matrix	238
interchange	234
inverse	232
k-cycle	233
length of orbit	234
orbit of i under σ	234
permutation	229
permutation matrix	238
permutation product	230
S_n	230
$\sigma[i]$	234
symmetric group of degree n	230
transposition	234

5.2 The Cauchy Index and Conjugacy

Let $\sigma = \sigma_m \cdots \sigma_1 \in S_n$ be the decomposition of σ into disjoint cycles, as guaranteed by §5.1, Theorem 2. Let σ_j be a k_j - cycle, j = 1, ..., m. The integer

$$\nu(\sigma) = \sum_{j=1}^{m} (k_j - 1) \qquad (1)$$

is called the *Cauchy index of* σ. The permutation σ is said to be *even* or *odd* according as its Cauchy index is even or odd. The *sign of* σ, denoted by $\varepsilon(\sigma)$, is 1 or -1 according to the formula

$$\varepsilon(\sigma) = (-1)^{\nu(\sigma)}. \tag{2}$$

If $\sigma = e$ then $\nu(\sigma)$ is defined to be 0 and $\varepsilon(e) = 1$.

There are several simple formulas for multiplying permutations that are important to know. To state these, let a, b, $x_1, \ldots, x_r, y_1, \ldots, y_s$ be $r + s + 2$ distinct integers in $N = \{1, 2, \ldots, n\}$. Then

$$(a\ b)(a\ x_1\ x_2\ \ldots\ x_r\ b\ y_1\ y_2\ \ldots\ y_s) = (a\ x_1\ x_2\ \ldots\ x_r)(b\ y_1\ y_2\ \ldots\ y_s), \tag{3}$$

$$(a\ b)(a\ x_1\ x_2\ \ldots\ x_r) = (a\ x_1\ x_2\ \ldots\ x_r\ b), \tag{4}$$

and

$$(a\ b)(a\ x_1\ x_2\ \ldots\ x_r)(b\ y_1\ y_2\ \ldots\ y_s) = (a\ x_1\ x_2\ \ldots\ x_r\ b\ y_1\ y_2\ \ldots\ y_s). \tag{5}$$

We leave the easy verifications of (3), (4) and (5) to the reader and simply remark that (5) follows from (3) by multiplying both sides of (3) on the left by the transposition (a b).

Theorem 1.

If σ is a product of k transpositions, then

$$\varepsilon(\sigma) = (-1)^k, \tag{6}$$

i.e., σ is even or odd according as k is.

Proof.

We repeatedly use the fact that disjoint cycles commute. We first prove that if $\tau = (a\ b)$ is a transposition and σ is any permutation in S_n then

$$\nu(\tau\sigma) = \nu(\sigma) \pm 1. \tag{7}$$

If $\sigma = e$ then (7) is evident. Next assume that σ has a decomposition into a product of disjoint cycles of lengths at least 2. There are 4 possibilities:

5.2 The Cauchy Index and Conjugacy

(i) a and b do not appear in any of the cycles. Then obviously

$$v(\tau\sigma) = v(\sigma) + 1$$

because $\tau\sigma$ has one more cycle of length 2, namely τ, in its disjoint cycle decomposition than σ does.

(ii) Both a and b occur in the same cycle in the disjoint cycle decomposition of σ. Then we can assume that the cycle in which a and b occur is the first cycle (i.e., disjoint cycles commute). Hence $\tau\sigma$ begins with the product on the left side of (3). The contribution of

$$(a\ x_1\ x_2\ \ldots\ x_r\ b\ y_1\ y_2\ \ldots\ y_s)$$

to $v(\sigma)$ is $r + s + 2 - 1$. But the contribution of the right side of (3) to $v(\tau\sigma)$ is $(r + 1 - 1) + (s + 1 - 1)$. Hence

$$v(\tau\sigma) = v(\sigma) - 1. \tag{8}$$

(iii) a and b both occur in separate cycles. Using (5), the reader will readily confirm as in (ii) that

$$v(\tau\sigma) = v(\sigma) + 1. \tag{9}$$

(iv) Precisely one of a or b appears in a cycle. Then (4) implies that

$$v(\tau\sigma) = v(\sigma) + 1. \tag{10}$$

To complete the proof, suppose that

$$\sigma = \tau_1 \tau_2 \cdots \tau_k, \tag{11}$$

where $\tau_1, \ldots, \tau_k$ are transpositions. Since $\tau^{-1} = \tau$ for any transposition, we have

$$\tau_k \tau_{k-1} \cdots \tau_2 \tau_1 \sigma = e$$

and then

$$\begin{aligned} 0 &= \nu(e) \\ &= \nu(\tau_k \tau_{k-1} \cdots \tau_1 \sigma) \\ &= \nu(\tau_{k-1} \cdots \tau_1 \sigma) \pm 1 \qquad \text{(from (7))} \\ &\vdots \\ &= \nu(\sigma) \pm 1 \pm 1 \pm \cdots \pm 1 \qquad (12) \end{aligned}$$

in which ± 1 occurs k times in (12). Suppose -1 occurs q times and $+1$ occurs p times in (12), $p + q = k$. Then

$$0 = \nu(\sigma) + p - q,$$

$$\nu(\sigma) = q - p,$$

and hence

$$\nu(\sigma) - k = q - p - (p + q)$$

$$= -2p.$$

It follows that

$$\varepsilon(\sigma) = (-1)^{\nu(\sigma)} = (-1)^{k-2p} = (-1)^k. \quad \blacksquare$$

For example, in S_7 we can use Theorem 1 to compute $\varepsilon(\sigma)$ for the product

$$\sigma = (1\ 2\ 3\ 5)(2\ 4\ 1\ 7)(1\ 2\ 3)(6\ 7)(4\ 5\ 7).$$

We write σ as a product of transpositions according to §5.1 Theorem 2(c):

$$\sigma = (1\ 5)(1\ 3)(1\ 2)(2\ 7)(2\ 1)(2\ 4)(1\ 3)(1\ 2)(6\ 7)(4\ 7)(4\ 5).$$

There are 11 transpositions in this factorization and hence from (6)

5.2 The Cauchy Index and Conjugacy

$$\varepsilon(\sigma) = (-1)^{11} = -1.$$

Although a permutation σ can be written in many ways as a product of transpositions, Theorem 1 shows that any two such products must both involve an even number of transpositions or both involve an odd number of transpositions.

Theorem 2.

If $\sigma, \varphi \in S_n$, then

$$\varepsilon(\sigma\varphi) = \varepsilon(\sigma)\varepsilon(\varphi)$$

and

$$\varepsilon(\sigma^{-1}) = \varepsilon(\sigma).$$

Proof.

If σ is a product of h transpositions and φ is a product of k transpositions, then $\sigma\varphi$ is a product of h + k transpositions and therefore

$$\varepsilon(\sigma\varphi) = (-1)^{h+k} = (-1)^h(-1)^k = \varepsilon(\sigma)\varepsilon(\varphi).$$

Also,

$$1 = \varepsilon(e) = \varepsilon(\sigma\sigma^{-1}) = \varepsilon(\sigma)\varepsilon(\sigma^{-1})$$

and thus

$$\varepsilon(\sigma^{-1}) = 1/\varepsilon(\sigma) = \varepsilon(\sigma). \blacksquare$$

There is an important equivalence relation that can be defined on S_n. This relation is called *conjugacy* and two permutations σ and φ are said to be *conjugate* if there is a permutation θ such that

$$\sigma = \theta\varphi\theta^{-1}. \tag{13}$$

It is easy to confirm that conjugacy is an equivalence relation. We shall denote the fact that σ and φ are conjugate, i.e., (13) holds, with the notation

$$\sigma \sim \varphi. \tag{14}$$

Obviously, $\sigma \sim \sigma$ by taking $\theta = e$. Suppose that (14) holds. Then, from (13),

$$\theta^{-1}\sigma\theta = \varphi$$

so that

$$\theta^{-1}\sigma(\theta^{-1})^{-1} = \varphi$$

and it follows that $\varphi \sim \sigma$. Finally, (13) and

$$\varphi = \omega\mu\omega^{-1}$$

imply that

$$\sigma = \theta\varphi\theta^{-1}$$

$$= \theta\omega\,\mu\omega^{-1}\theta^{-1}$$

$$= (\theta\omega)\,\mu(\theta\omega)^{-1}$$

so that $\sigma \sim \mu$. Thus "$\sim$" is *reflexive, symmetric,* and *transitive*.

Let $\sigma \in S_n$ be a product of disjoint cycles in which there are λ_t cycles of length t, $t = 2, ..., n$. We permit some of the λ_t to be 0. We also define λ_1 to be the number of integers held fixed by σ, i.e., a cycle of length 1 can be interpreted as an integer held fixed by σ. Note that

$$\lambda_1 \cdot 1 + \lambda_2 \cdot 2 + \cdots + \lambda_n \cdot n = n. \tag{15}$$

5.2 The Cauchy Index and Conjugacy

The *cycle structure of* σ is simply the data just described. A common notation to indicate that σ has the preceding cycle structure is to say that σ has cycle structure

$$s_\lambda = [\, 1^{\lambda_1} 2^{\lambda_2} 3^{\lambda_3} \ldots n^{\lambda_n} \,], \tag{16}$$

i.e., σ has λ_t cycles of length t, t = 1, ..., n, in which we incorporate the fixed integers as 1-cycles.

There is an intimate relationship between the cycle structure of a permutation and the conjugacy class in which it resides.

Theorem 3.

Two permutations σ and φ in S_n are conjugate if and only if they have the same cycle structure.

Proof.

Assume that σ has the cycle structure s_λ in (16). Assume $\varphi \sim \sigma$, so that

$$\varphi = \theta \sigma \theta^{-1}. \tag{17}$$

Let $s_\beta = [\, 1^{\beta_1} 2^{\beta_2} 3^{\beta_3} \ldots n^{\beta_n} \,]$ be the cycle structure of φ. Note that if j is fixed by σ then

$$\begin{aligned} \varphi(\theta(j)) &= \theta \sigma \theta^{-1}(\theta(j)) \\ &= \theta \sigma \theta^{-1} \theta(j) \\ &= \theta \sigma(j) \\ &= \theta(j). \end{aligned}$$

In other words, the number of fixed points of φ, namely β_1, is at least λ_1, i.e., $\beta_1 \geq \lambda_1$. But since conjugacy is symmetric, $\lambda_1 \geq \beta_1$, and hence $\lambda_1 = \beta_1$. In general, if

$$\sigma = \sigma_1 \sigma_2 \cdots \sigma_k$$

is the disjoint cycle factorization of σ then

$$\varphi = \theta\sigma\theta^{-1} = (\theta\sigma_1\theta^{-1})(\theta\sigma_2\theta^{-1})\cdots(\theta\sigma_k\theta^{-1}). \tag{18}$$

It is also easy to confirm that if

$$\sigma_m = (i_1\ i_2\ \cdots\ i_p)$$

then

$$\theta\sigma_m\theta^{-1} = (\theta(i_1)\ \theta(i_2)\cdots\theta(i_p)). \tag{19}$$

For, consider the value of both sides of (19) on $\theta(j)$, $j = 1, \ldots, n$. If j is some i_t then the value of the right side on $\theta(j) = \theta(i_t)$ is $\theta(i_{t+1})$. The value of the left side is

$$\theta\sigma_m\theta^{-1}\theta(i_t) = \theta\sigma_m(i_t) = \theta(i_{t+1}).$$

If j is none of $i_1, \ldots, i_p$ then $\sigma_m(j) = j$ and $\theta(j)$ cannot be any of the $\theta(i_1), \ldots, \theta(i_p)$. Thus the value of the right side of (19) on $\theta(j)$ is $\theta(j)$. The value of the left side of (19) on $\theta(j)$ is

$$\theta\sigma_m\theta^{-1}\theta(j) = \theta\sigma_m(j) = \theta(j).$$

Thus (19) is valid. From (18) and (19) it follows that each σ_m in the disjoint cycle decomposition of σ can be paired with a corresponding cycle $\theta\sigma_m\theta^{-1}$ in the disjoint cycle decomposition of φ. It follows that $s_\lambda = s_\beta$ and thus σ and φ have the same cycle structure.

Conversely, suppose σ and φ have the same cycle structure s_λ. Write out the disjoint cycle decompositions of both σ and φ, but include the fixed points:

$$\sigma = (a_1)\cdots(a_{\lambda_1})\cdots(b_{11}\cdots b_{1k})\cdots(b_{\lambda_k 1}\cdots b_{\lambda_k k})\cdots \tag{20}$$

$$\varphi = (c_1)\cdots(c_{\lambda_1})\cdots(d_{11}\cdots d_{1k})\cdots(d_{\lambda_k 1}\cdots d_{\lambda_k k})\cdots \tag{21}$$

in which corresponding cycles of the same length in φ are directly below those in σ.

5.2 The Cauchy Index and Conjugacy

Define a permutation

$$\theta = \begin{pmatrix} a_1 & a_2 & \cdots & a_{\lambda_1} & b_{11} & \cdots & b_{1k} & \cdots \\ c_1 & c_2 & \cdots & c_{\lambda_1} & d_{11} & \cdots & d_{1k} & \cdots \end{pmatrix}. \quad (22)$$

Now we compute that

$$\theta\sigma\theta^{-1}(c_t) = \theta\sigma(a_t) = \theta(a_t) = c_t = \varphi(c_t), \quad t = 1, 2, \cdots, \lambda_1.$$

Next

$$\theta\sigma\theta^{-1}(d_{1t}) = \theta\sigma(b_{1t}) = \theta(b_{1,t+1}) = d_{1,t+1} = \varphi(d_{1t}).$$

Continuing in this way, we easily conclude that

$$\varphi = \theta\sigma\theta^{-1},$$

and hence $\varphi \sim \sigma$. ∎

As an example of Theorem 3, suppose $\varphi \sim \sigma \in S_7$ and that

$$\sigma = (2)\,(3)\,(1\ 4)\,(5\ 6\ 7),$$

$$\varphi = (7)\,(4)\,(3\ 1)\,(6\ 2\ 5).$$

As in the proof of Theorem 3, define θ to be

$$\theta = \begin{pmatrix} 2 & 3 & 1 & 4 & 5 & 6 & 7 \\ 7 & 4 & 3 & 1 & 6 & 2 & 5 \end{pmatrix}.$$

Note that

$$\theta\sigma\theta^{-1}(7) = \theta\sigma\theta^{-1}(\theta(2)) = \theta\sigma(2) = \theta(2) = 7 = \varphi(7)$$

and similarly,

$$\theta\sigma\theta^{-1}(4) = 4 = \varphi(4).$$

Also,

$$\theta\sigma\theta^{-1}(2) = \theta\sigma(6) = \theta(7) = 5 = \varphi(2).$$

Thus we see that

$$\varphi = \theta\sigma\theta^{-1}$$

by checking the value of both sides on each of 1, ..., 7.

5.2 Exercises

1. Prove: Every even permutation in S_n, $n > 2$, is a product of 3-cycles.

    Hint: $(1\ 2)(3\ 4) = (1\ 2\ 3)(2\ 3\ 4)$, $(1\ 2)(1\ 3) = (1\ 3\ 2)$, $(1\ 2)^2 = e = (1\ 2\ 3)(1\ 3\ 2)$.

2. Prove: The number of even permutations in S_n is $n!/2$.

3. A $k \times k$ *full cycle permutation matrix* is the incidence matrix of $(1\ 2\ ...\ k)$, e.g.,

    $$\begin{bmatrix} 0 & 0 & 1 \\ 1 & 0 & 0 \\ 0 & 1 & 0 \end{bmatrix}$$

    is a full cycle permutation matrix. Prove: If $\sigma \in S_n$ then there exists $\theta \in S_n$ such that $A(\theta)A(\sigma)A(\theta)^T$ is a direct sum of full cycle permutation matrices and (possibly) an identity matrix. A permutation matrix of this type is said to be in *normal form*.

    Hint: Suppose σ has cycle structure $s_\lambda = [\ 1^{\lambda_1}\ 2^{\lambda_2}\ 3^{\lambda_3}\ ...\ n^{\lambda_n}\ ]$. For sake of definiteness suppose $n = 15$, $\lambda_1 = 2$, $\lambda_2 = 3$, $\lambda_3 = 1$, $\lambda_4 = 1$ and the rest of the λ_j are 0. Then σ has the same cycle structure as the permutation

$$\varphi = (1)(2)(3\ 4)(5\ 6)(7\ 8)(9\ 10\ 11)(12\ 13\ 14\ 15).$$

Clearly the incidence matrix $A(\varphi)$ is a direct sum of I_2, the matrix

$$\begin{bmatrix} 0 & 1 \\ 1 & 0 \end{bmatrix},$$

(3 times),

$$\begin{bmatrix} 0 & 0 & 1 \\ 1 & 0 & 0 \\ 0 & 1 & 0 \end{bmatrix},$$

and the matrix

$$\begin{bmatrix} 0 & 0 & 0 & 1 \\ 1 & 0 & 0 & 0 \\ 0 & 1 & 0 & 0 \\ 0 & 0 & 1 & 0 \end{bmatrix}.$$

However, by Theorem 3 there exists θ such that $\varphi = \theta\sigma\theta^{-1}$ and by §5.1 Exercises, #5, #8,

$$A(\varphi) = A(\theta\sigma\theta^{-1}) = A(\theta)A(\sigma)A(\theta)^T.$$

4. Find the normal form of $A(\sigma)$ for each of the permutations σ in §5.1 Exercises, #1.

5. Prove: If A and B are permutation matrices then $A \otimes B$ is a permutation matrix.

6. Identify which of the following pairs of permutations have the same cycle structure:

   (a) $(1\ 2\ 3)(4\ 5); (1\ 2\ 3)(3\ 4)$

   (b) $(1\ 2)(3\ 4)(5\ 6); (1\ 3)(2\ 4)(6\ 5)$

(c)   (1 2 3) (3 4) (5 6); (1 2 3) (4 5 6)

(d)   (1 2 3) (1 2 4) (1 2 5); (1 2 3)^3

(e)   (1 2 3 4 5) (5 3); (1 3 2 4 5) (2 3)

7. Let $X = [x_{ij}]$ be an $n \times n$ matrix. Show that the $n^2 \times n^2$ permutation matrix Q for which $Qc(X) = c(X^T)$, where $c(X)$ is the column form of X, (see formula (10) in §3.3 Kronecker Products), can be described as follows: let E_{ij} be the $n \times n$ matrix with 1 as its (i, j) entry, all other entries 0. If Q is partitioned into n^2, $n \times n$ non-overlapping submatrices $Q = [Q_{ij}]$, then $Q_{ij} = E_{ji}$.

Hint: Note that by block multiplication

$$Qc(X) = \begin{bmatrix} \sum_{k=1}^{n} E_{k1} X^{(k)} \\ \sum_{k=1}^{n} E_{k2} X^{(k)} \\ \vdots \\ \sum_{k=1}^{n} E_{kn} X^{(k)} \end{bmatrix}.$$

But $E_{kt} X^{(k)} = [0, ..., 0, x_{tk}, 0, ..., 0]^T$, where x_{tk} appears in the kth position, and hence $\sum_{k=1}^{n} E_{kt} X^{(k)} = [x_{t1}, x_{t2}, ..., x_{tn}]^T = (X^T)^{(t)}$. Thus

$$Qc(X) = \begin{bmatrix} (X^T)^{(1)} \\ (X^T)^{(2)} \\ \vdots \\ (X^T)^{(n)} \end{bmatrix} = c(X^T).$$

8. Show that the matrix equation $Y = AX - X^T A$ can be written in terms of Kronecker products and the column forms of X and Y as follows:

5.2 Exercises

$$c(Y) = [I_n \otimes A - (A^T \otimes I_n) Q] c(X).$$

Hint: See formula (10) in §3.3 Kronecker Products.

9. Prove that the matrix Q in #7 is not a Kronecker product of two n × n matrices.

10. Prove: $Q^2 = I_{n^2}$.

11. Prove: There does not exist a fixed pair of n × n matrices A and B such that $X^T = AXB$ holds for all n × n matrices X.

    Hint: $c(X^T) = Qc(X)$ from #7. From formula (10) in §3.3 Kronecker Products, $c(AXB) = (B^T \otimes A)c(X)$. Thus if the equation $X^T = AXB$ were to hold, it would require that $(B^T \otimes A)c(X) = Qc(X)$ for all X. But then $Q = B^T \otimes A$, contradicting #9.

12. Analogously to #8, write the matrix equation $Y = BX^T - XB$ in terms of Kronecker products and the column forms of X and Y.

13. Let B be a 2 × 2 nonzero matrix and let Q be the 4 × 4 matrix described in #7. Prove

    $$\rho((I_2 \otimes B)Q - B^T \otimes I_2) = \begin{cases} 1 & \text{if B is symmetric} \\ 3 & \text{if B is not symmetric} \end{cases}$$

14. Let A be m × m, B be n × n. Prove: $A \otimes B = I_{mn}$ iff $A = \alpha I_m$, $B = \beta I_n$ and $\alpha\beta = 1$.

15. Let A_t be an $m_t \times m_t$ matrix and let P_t be an $m_t \times m_t$ permutation matrix, $t = 1, \ldots, k$. Assume $A_1 \otimes \cdots \otimes A_k = P_1 \otimes \cdots \otimes P_k$. Prove $A_t = \alpha_t P_t$, $t = 1, \ldots, k$, and $\alpha_1 \cdots \alpha_k = 1$.

    Hint: $P_1^T A_1 \otimes \cdots \otimes P_k^T A_k = I_{m_1} \otimes \cdots \otimes I_{m_k}$. From #14, $P_1^T A_1 = \alpha_1 I_{m_1}$ and $P_2^T A_2 \otimes \cdots \otimes P_k^T A_k = I_{m_2} \otimes \cdots \otimes I_{m_k}$. Use induction.

16. List all even permutations in S_4.

5.2 MatLab

1. Use the function A = disj(x) in §5.1 MatLab, #10 to write a function called cyst that accepts a permutation $x \in S_n$ and returns the cycle structure of x as an n × 2 matrix of the following form (see (16)):

$$\begin{bmatrix} 1 & \lambda_1 \\ 2 & \lambda_2 \\ \vdots & \\ n & \lambda_n \end{bmatrix}$$

Save disj in Permute.

Hint: The value of A = disj(x) is a k × n matrix whose rows are the disjoint cycles of x, each filled out with 0's to make n columns:

$$A = \begin{bmatrix} * & * & \cdots & * & 0 & \cdots & 0 \\ * & * & \cdots & & * & 0 & \cdots & 0 \\ \vdots & & & & & & \\ * & * & \cdots & & * & 0 & \cdots & 0 \end{bmatrix}$$

To obtain the cycle structure it is necessary to count the number of cycles of each length.

```
function L = cyst(x)
%CYST accepts a permutation x in Sn
%and returns the cycle structure L.
A = disj(x);
[k n] = size(A);
z = zeros(1, n);
for t = 1:k
        j = length(find(A(t, :)));
        z(j) = z(j) + 1;
end
L = [1:n;  z]';
```

2. The *permanent* of an n-square matrix $A = [a_{ij}]$ is the sum

$$\text{per}(A) = \sum_{\sigma \in S_n} \prod_{i=1}^{n} a_{i\sigma(i)} .$$

Write a function called perm that accepts an n-square matrix A and returns the value per(A).

Hint: Modify the script Allperms in §5.1 MatLab, #9(f). Also read the Help entry for **prod**. Save perm in Permute.

3. Use the function perm in #2 to compute the permanents of each of the following matrices:

(a) $\begin{bmatrix} 0 & 1 \\ 1 & 0 \end{bmatrix}$

(b) $\begin{bmatrix} 0 & 1 & 1 \\ 1 & 0 & 1 \\ 1 & 1 & 0 \end{bmatrix}$

(c) $\begin{bmatrix} 0 & 1 & 1 & 1 \\ 1 & 0 & 1 & 1 \\ 1 & 1 & 0 & 1 \\ 1 & 1 & 1 & 0 \end{bmatrix}$

4. Read the Help entry for **flops**. Compute the number of flops performed in computing perm(A) for the matrices in #3 above. The program perm is definitely not the most efficient way to compute per(A), and indeed, better methods are available. However, their exposition would take an excessive amount of time.

5. A *convex combination* of permutation matrices is a sum of the form

$$A = \sum_{x \in S_n} c_x A(x)$$

where A(x) is the permutation matrix corresponding to the permutation $x \in S_n$ and the c_x are nonnegative real numbers that sum to 1.

(a) Show that every row and column sum of A is 1. Such a matrix is called *doubly stochastic* (abbreviated *d.s.*).

(b) Write a MatLab function called rdsm (i.e., random d.s. matrix) that accepts an integer n and returns a random n-square d.s. matrix A. Save rdsm in Permute. Hint: Modify the script of Allperms in §5.1 MatLab, #9(f) and also use the function pmat that appears in §5.1 MatLab, #4.

Hint:

```
function S = rdsm(n)
%RDSM(n) generates a random n-square doubly stochastic matrix
x = 1:n;                    %initialize first permutation as id
S = pmat(x);                %constructs incidence matrix for id
cx = rand;                  %first random coefficient
S = cs*S;                   %multiplies current S by cx
c = cx;                     %c sums the values of the cx
p = Lt(x);                  %This segment generates Sn
while p > 1                 %*
    q = Lr(x, p);           %*
    x = swap(x, p - 1, q);  %*
    x = rev(x, p);          %as in Allperms.
    cx = rand;              %generates a random cx in [0 1)
    c = c + cx;             %accumulates sum of cx
    A = pmat(x);            %generates permutation matrix A(x)
    S = S + cs*A            %accumulates sum of cx*A(x)
    p = Lt(x);              %When Lt(x) drops below 2
end                         %all of Sn has been generated
S = (1/c)*S;                %makes coefficients sum to 1
```

6. Write a MatLab script called vdw that generates 20 random 3-square d.s. matrices S and lists their permanents. Compare each such permanent with the value $3!/3^3$. Save vdw in Permute. This script confirms the so-called *van der Waerden* conjecture for n = 3. The conjecture stated that if S is an n-square d.s. matrix then

5.2 MatLab

per(S) $\geq$ n!/n^n with equality iff S = (1/n) * ones(n, n). The conjecture was finally resolved for general n more than 50 years after its original publication.

7. Write a MatLab script called cnj that

   - calls for a positive integer n
   - generates two random permutations u and v in S_n
   - prints out the cycle structure of u and x = v^{-1}uv

   Save cnj in Permute.

   Hint:
   ```
   n = input('enter a positive integer n: ');
   u = prand(n);              %see §5.1 MatLab, #7
   v = prand(n);
   w = pnv(v);                %see §5.1 MatLab, #5
   x = pmtly(w, pmtly(u, v)); %see §5.1 MatLab, #1
   L = cyst(u);
   Lc = cyst(x);
   disp('     u     x     ');
   disp('_____');
   disp([L'; Lc(:, 2)' ]' )   %L is an n x 2 matrix
   ```

8. A cycle of length m can always be expressed as a product of m - 1 transpositions: (1 2 3 ... m) = (1 m) (1 m - 1) (1 m - 2) ··· (1 2). Use the function cyst in #1 and Theorem 1 to write a MatLab function called sn that returns $\varepsilon(x)$ for any permutation x $\in S_n$. Save sn in Permute.

   Hint:
   ```
   function e = sn(x)
   %SN(x) returns the sign
   %of a permutation x in Sn.
   n = length(x);
   L = cyst(x);               %see #1
   z = L(:, 2) ';
   c = 0:(n - 1);
   ```

```
k = sum(c.*z);              %k = 0*z(1) + ··· + (n - 1)*z(n)
e = (-1)^k;
```

9. The r element subsets of an n element set (i.e., r-combinations of an n-set) can be represented as strictly increasing sequences of length r chosen from 1, 2, ..., n, i.e., as elements in the sequence set $Q_{r,n}$. For example, if n = 7 and r = 4 then the first twelve sequences in $Q_{r,n}$ in lexicographic order are

1.	1 2 3 4	5.	1 2 4 5	9.	1 2 5 7
2.	1 2 3 5	6.	1 2 4 6	10.	1 2 6 7
3.	1 2 3 6	7.	1 2 4 7	11.	1 3 4 5
4.	1 2 3 7	8.	1 2 5 6	12.	1 3 4 6

Note that if $k_1 k_2 k_3 k_4$ is a particular sequence then $k_4 \leq 7$, $k_3 \leq 6$, $k_2 \leq 5$, $k_1 \leq 4$. In fact, the last sequence in lexicographic order is 4 5 6 7. In sequence 10, $k_1 = 1$, $k_2 = 2$, $k_3 = 6$, $k_4 = 7$ and neither k_3 nor k_4 can be further increased to obtain sequence 11. Thus, $k_2 = 2$ is increased by 1 to obtain 1 3 6 7. Then k_3 is decreased to 4 and k_4 is decreased to 5. Thus the changes are

```
1 2 6 7
1 3 6 7
1 3 4 7
1 3 4 5
```

and 1 3 4 5 is the next sequence. In general, here is the algorithm for going from the sequence $a = k_1 \ldots k_r$ in $Q_{r,n}$ to the next sequence.

- find the largest t such that $k_t < n - r + t$
- replace k_t by $k_t + 1$
- replace k_j by $k_{j-1} + 1$, $j = t + 1, \ldots, r$. (if t = r, this step is omitted)

(a) Write a MatLab function named Lxc such that if $a = k_1 \ldots k_r$ is in $Q_{r,n}$ then Lxc(a, n) returns the largest t such that $k_t < n - r + t$; if no such t exists, i.e., $a = (n - r + 1) \ldots n$, then Lxc(a, n) returns 0. Save Lxc in Permute.

(b) Write a MatLab function named cmb that returns an $\binom{n}{r} \times r$ matrix whose rows are the sequences in $Q_{r,\,n}$ in lexicographic order. Save cmb in Permute.

(c) Read the Help entry for gamma. Then use the function in part (b) of this exercise to write a MatLab function named rcmb that returns a random r-combination of 1, ..., n. Save rcmb in Permute.

Hint:

(a) function t = Lxc(a, n);
%LXC(a, n) returns the largest t such that
%a(t) < n - r + t; if no such t exists then
%Lxc(a, n) returns 0
r = length(a);
t = r;
if a == ((n - r + 1):n)
    t = 0;
else
    while a(t) >= n - r + t
        t = t - 1;
    end
end

(b) function C = cmb(r, n)
%CMB(r, n) returns
%a matrix whose rows are
%the r-combinations of 1, ..., n.
a = 1:r;
C = a;
t = Lxc(a, n);
while t ~= 0
    a(t) = a(t) + 1;
    if t < r
        for j = (t + 1):r
            a(j) = a(j - 1) + 1;

```
                    end
                end
                C = [C; a];
                t = Lxc(a, n);
            end
```

(c)   function a = rcmb(r, n);
        %RCMB(r, n) returns a random r-combination of 1, ..., n.
        b = gamma(n + 1)/(gamma(n - r + 1) * gamma(r + 1));
        b = round(b);
        C = cmb(r, n);
        j = floor(1 + b*rand);
        a = C(j, :);

10. Write a MatLab script to test that rcmb is randomly choosing r-combinations of 1, ..., n. Save the script with the name rcmbtest in Permute.

Hint:
```
z = zeros(1, 10);
C = cmb(2, 5);
for j = 1:500
        u = rcmb(2, 5);
        for t = 1:10
            if u == C(t, :);
                z(t) = z(t) + 1;
            end
        end
end
z
```

5.2 Glossary

Cauchy index	247
conjugacy	251
conjugate permutations	251
convex combination	261
cycle structure	253
doubly stochastic matrix	262
d.s.	262
$\varepsilon(\sigma)$	247
equivalence relation	251
even permutation	247
flops	261
full cycle permutation matrix	256
$\nu(\sigma)$	247
normal form	256
odd permutation	247
per(A)	261
permanent	261
prod	261
reflexive	252
$\sigma \sim \varphi$	252
sign of a permutation	247
symmetric	252
transitive	252
van der Waerden conjecture	262

5.3 Some Special Results

In this section we assemble several results about permutations that we will need in the next chapter on determinants.

Theorem 1.

Let $1 \le \alpha_1 < ... < \alpha_k \le n$ and $1 \le \beta_1 < ... < \beta_{n-k} \le n$ be complementary sets of integers in the set $\{1, ..., n\}$, i.e., the two sets are disjoint and

$$\{\alpha_1, \ldots, \alpha_k\} \cup \{\beta_1, \ldots, \beta_{n-k}\} = \{1, \ldots, n\}.$$

Let σ be the permutation in S_n defined by

$$\sigma = \begin{pmatrix} 1 & \cdots & k & k+1 & \cdots & n \\ \alpha_1 & \cdots & \alpha_k & \beta_1 & \cdots & \beta_{n-k} \end{pmatrix}. \tag{1}$$

Then

$$\varepsilon(\sigma) = (-1)^{s(\alpha) + k(k+1)/2}$$

where

$$s(\alpha) = \sum_{t=1}^{k} \alpha_t.$$

Proof.

We use induction on n. If $n = 1$, there is nothing to prove. Assume that $n > 1$ and that the theorem holds for any permutations in S_{n-1} of the type (1). There are two possibilities: either $\beta_{n-k} = n$ or $\alpha_k = n$. If $\beta_{n-k} = n$ so that

$$\sigma = \begin{pmatrix} 1 & \cdots & k & k+1 & \cdots & n-1 & n \\ \alpha_1 & \cdots & \alpha_k & \beta_1 & \cdots & \beta_{n-k-1} & n \end{pmatrix},$$

we define $\tau \in S_{n-1}$ by

$$\tau(i) = \sigma(i), \quad i = 1, \ldots, n-1.$$

Clearly, τ and σ have the same Cauchy index and therefore $\varepsilon(\tau) = \varepsilon(\sigma)$. Also, by the induction hypothesis,

$$\varepsilon(\tau) = (-1)^{s(\alpha) + k(k+1)/2}.$$

5.3 Some Special Results

The other alternative is $\alpha_k = n$. Then

$$\sigma = \begin{pmatrix} 1 & \cdots & k-1 & k & k+1 & \cdots & n \\ \alpha_1 & \cdots & \alpha_{k-1} & n & \beta_1 & \cdots & \beta_{n-k} \end{pmatrix}.$$

Let

$$\theta = \begin{pmatrix} 1 & \cdots & k-1 & k & k+1 & \cdots & n-1 & n \\ \alpha_1 & \cdots & \alpha_{k-1} & \beta_1 & \beta_2 & \cdots & \beta_{n-k} & n \end{pmatrix}.$$

Then, as before, we can apply the induction hypothesis to obtain

$$\varepsilon(\theta) = (-1)^{\alpha_1 + \cdots + \alpha_{k-1} + (k-1)k/2}.$$

Moreover,

$$\theta^{-1}\sigma$$

$$= \begin{pmatrix} \alpha_1 & \cdots & \alpha_{k-1} & \beta_1 & \beta_2 & \cdots & \beta_{n-k} & n \\ 1 & \cdots & k-1 & k & k+1 & \cdots & n-1 & n \end{pmatrix} \begin{pmatrix} 1 & \cdots & k-1 & k & k+1 & \cdots & n \\ \alpha_1 & \cdots & \alpha_{k-1} & n & \beta_1 & \cdots & \beta_{n-k} \end{pmatrix}$$

$$= \begin{pmatrix} 1 & 2 & \cdots & k-1 & k & k+1 & k+2 & \cdots & n-1 & n \\ 1 & 2 & \cdots & k-1 & n & k & k+1 & \cdots & n-2 & n-1 \end{pmatrix}$$

$$= \begin{pmatrix} n & n-1 & n-2 & \cdots & k+1 & k \end{pmatrix}.$$

Then since $\alpha_k = n$, $(-1)^{2k} = 1$, and the length of the preceding cycle, $\theta^{-1}\sigma$, is $n - k + 1$ we have

$$\varepsilon(\theta^{-1}\sigma) = (-1)^{n-k} = (-1)^{\alpha_k - k}(-1)^{2k} = (-1)^{\alpha_k + k}.$$

It follows that

$$\varepsilon(\sigma) = \varepsilon(\theta\theta^{-1}\sigma)$$

$$= \varepsilon(\theta)\varepsilon(\theta^{-1}\sigma)$$

$$= (-1)^{\alpha_1 + \cdots + \alpha_{k-1} + (k-1)k/2} (-1)^{\alpha_k + k}$$

$$= (-1)^{\alpha_1 + \cdots + \alpha_{k-1} + \alpha_k + k + (k-1)k/2}$$

$$= (-1)^{s(\alpha) + k(k+1)/2}. \qquad \blacksquare$$

As an example of Theorem 1, suppose $\sigma \in S_8$,

$$\sigma = \begin{pmatrix} 1 & 2 & 3 & 4 & 5 & 6 & 7 & 8 \\ 4 & 7 & 8 & 1 & 2 & 3 & 5 & 6 \end{pmatrix}.$$

Then $k = 3$, $\alpha_1 = 4$, $\alpha_2 = 7$, $\alpha_3 = 8$ and $\beta_1 = 1$, $\beta_2 = 2$, $\beta_3 = 3$, $\beta_4 = 5$, $\beta_5 = 6$. Hence, $s(\alpha) = 4 + 7 + 8 = 19$ and $s(\alpha) + k(k+1)/2 = 19 + 6 = 25$. Thus $\varepsilon(\sigma) = (-1)^{25} = -1$.

Let

$$\sigma = \begin{pmatrix} 1 & 2 & 3 & \cdots & n \\ \sigma(1) & \sigma(2) & \sigma(3) & \cdots & \sigma(n) \end{pmatrix}. \qquad (2)$$

An *inversion in* σ is simply a pair of integers $i < j$ for which

$$\sigma(i) > \sigma(j).$$

For example, in

$$\sigma = \begin{pmatrix} 1 & 2 & 3 & 4 & 5 & 6 \\ 4 & 5 & 1 & 2 & 3 & 6 \end{pmatrix}$$

we see that the inversions occur for the following pairs: (1, 3), (1, 4), (1, 5), (2, 3), (2, 4), (2, 5). Hence there are 6 inversions in σ. Note that Theorem 1 implies that σ is an even permutation. Examine the second row of σ for the first integer a that is preceded by a larger integer b. Notice that a = 1 and the closest such b is 5. In general, for any σ,

5.3 Some Special Results

the closest such b must always immediately precede a. For, if x is closer to a than b is, then the second row of σ has the appearance

$$\ldots b \ldots x \ldots a \ldots .$$

Since x is closer to a than b, x < a. Also b < x otherwise a would not be the first integer preceded by a larger integer. But then b < x < a and b does not exceed a. This contradiction shows that b immediately precedes a.

If we multiply σ on the right by the transposition (2 3) we obtain

$$\sigma(2\ 3) = \begin{pmatrix} 1 & 2 & 3 & 4 & 5 & 6 \\ 4 & 5 & 1 & 2 & 3 & 6 \end{pmatrix}(2\ 3)$$

$$= \begin{pmatrix} 1 & 2 & 3 & 4 & 5 & 6 \\ 4 & 1 & 5 & 2 & 3 & 6 \end{pmatrix}. \tag{3}$$

The second row of σ(2 3) differs from the second row of σ in that the entries in positions 2 and 3 have been interchanged. Again examine the second row of (3) for the first integer preceded by a larger integer and find the closest such integer that exceeds it. This occurs in positions 1 and 2, i.e. 4 > 1. Multiply (3) on the right by (1 2) to obtain

$$\sigma(2\ 3)(1\ 2) = \begin{pmatrix} 1 & 2 & 3 & 4 & 5 & 6 \\ 1 & 4 & 5 & 2 & 3 & 6 \end{pmatrix}.$$

We can continue until ultimately the second row is the same as the first. Transpositions of the form (2 3), (1 2), i.e., (i i + 1), are called *adjacent interchanges*. The preceding argument shows that any permutation σ can be reduced to the identity by multiplying on the right by adjacent interchanges:

$$\sigma \tau_1 \tau_2 \cdots \tau_k = e, \tag{4}$$

or

$$\sigma = \tau_k \tau_{k-1} \cdots \tau_1 .$$

In other words, any permutation is a product of adjacent interchanges. Also note that the process described above diminishes the number of inversions in the second row by 1 at each step and, in fact, restores the leftmost adjacent inversion to natural order. From (4) the number of inversions in σ is k and hence

$$\epsilon(\sigma) = (-1)^k.$$

We have established the following result:

Theorem 2.

If σ has k inversions in the second row of its 2-row notation then

$$\epsilon(\sigma) = (-1)^k.$$

The discussion leading to Theorem 2, in particular (4), is constructive. We can obtain the transpositions τ which appear as factors of σ. Starting at the left end of σ, record the transposition (i i + 1) where $\sigma(i + 1)$ is the first integer from the left preceded by a larger integer. Interchange $\sigma(i)$ and $\sigma(i + 1)$ in the second row of the permutation and repeat the process. Continue until there are no inversions remaining in the second row.

We have covered three general methods for computing the sign of a permutation: compute the Cauchy index; factor the permutation into a product of transpositions; count the inversions in the second row of the 2-row notation for the permutation. There is an additional method based on computing the product

$$D = \prod_{1 \le i < j \le n} \sigma(j) - \sigma(i). \tag{5}$$

In other words, D is the product of all pairs $\sigma(j) - \sigma(i)$ in which $\sigma(j)$ is to the right of $\sigma(i)$ in the 2-row notation for σ.

Theorem 3.

If $\sigma \in S_n$ then σ is even or odd according as D is positive or negative.

Proof.

Each inversion contributes a negative factor to D. Thus if D > 0 then the number of inversions is even, and if D < 0, the number of inversions is odd. Apply Theorem 2. ∎

For example, if

$$\sigma = \begin{pmatrix} 1 & 2 & 3 & 4 \\ 4 & 1 & 2 & 3 \end{pmatrix}$$

then

$$D = (\sigma(2) - \sigma(1))(\sigma(3) - \sigma(1))(\sigma(4) - \sigma(1))(\sigma(3) - \sigma(2))(\sigma(4) - \sigma(2))(\sigma(4) - \sigma(3))$$

$$= (1 - 4)(2 - 4)(3 - 4)(2 - 1)(3 - 1)(3 - 2)$$

$$= -12$$

Hence σ is odd.

5.3 Exercises

1. Compute

$$\varepsilon \begin{pmatrix} 1 & 2 & 3 & 4 & 5 & 6 & 7 & 8 & 9 \\ 6 & 7 & 8 & 1 & 2 & 3 & 4 & 5 & 9 \end{pmatrix}$$

   using Theorem 1.

2. Let $\sigma = S_n$ and let $\sigma(i) = i$, $i = 1, ..., k$. Define $\varphi \in S_{n-k}$ by $\varphi(t) = \sigma(t + k) - k$, $t = 1, ..., n - k$. Prove: $\varepsilon(\sigma) = \varepsilon(\varphi)$.

3. Prove: Any permutation in S_n is a product of permutations of the form (1 j).

4. Prove: Any even permutation in S_n ($n \geq 3$) is a product of 3-cycles of the form $(1\ 2\ k)$, $k = 3, \ldots, n$.

5. Construct a multiplication table for the permutations

$$K = [e, (1\ 2)(3\ 4), (2\ 3)(1\ 4), (1\ 3)(2\ 4)].$$

This set is called the *Klein four group*.

6. Show that any $\sigma \in S_n$ is a product of the transpositions

$$(1\ 2), (2\ 3), (3\ 4), \ldots, (n-1\ n).$$

7. Write

$$\sigma = \begin{pmatrix} 1 & 2 & 3 & 4 & 5 & 6 & 7 & 8 \\ 3 & 4 & 1 & 2 & 5 & 7 & 8 & 6 \end{pmatrix}$$

as a product of transpositions of the form $(i\ i+1)$.

8. Let G be a subset of S_n *closed under multiplication*, i.e., if σ and φ are in G then $\sigma\varphi \in G$. Prove:

$$\sum_{\sigma \in G} \varepsilon(\sigma) = \begin{cases} |G| \\ 0 \end{cases}$$

where $|G|$ denotes the number of permutations in G.

5.3 MatLab

1. In Theorem 3 it was established that if $\sigma \in S_n$ and

$$D = \prod_{1 \leq i < j \leq n} \sigma(j) - \sigma(i)$$

5.3 MatLab

then σ is even or odd according as D > 0 or D < 0. Write a MatLab function named psn that accepts a permutation σ ∈ S_n and returns ε(σ) using Theorem 3. Read the Help entry for **sign**. Save psn in Permute.

Hint:
```
function e = psn(x)
%PSN(x) is the sign
%of x in Sn.
n = length(x);
d = 1;
for s = 1:(n - 1)
        for t = (s+1):n
                d = d*(x(t) - x(s));
        end
end
e = sign(d);
```

2. As we saw in §5.3 Some Special Results, any permutation σ ∈ S_n is a product of adjacent interchanges. Write a MatLab function named chg that accepts a permutation σ and returns a matrix whose rows, in succession, are adjacent interchanges τ_1, ..., τ_p such that σ = τ_p ... $\tau_2 \tau_1$. If σ is the identity permutation then chg returns [ ] (i.e., the empty matrix) to signal this fact. Save chg in Permute.

Hint:
```
function T = chg(x)
%CHG(x) is a matrix whose
%rows are the adjacent interchanges
%t1, ..., tp such that x = tp ... t1.
%chg(id) = [ ], the empty matrix.
n = length(x);
A = [ ];
id = 1:n;
if x == id
        T = [ ];
else
        while sum(x~=id) ~= 0
```

```
                    for k = 2:n
                        if x(k-1) > x(k)
                            x = swap(x, k-1, k);   %see §5.1 MatLab, #9(d)
                            A = [A; k-1, k];
                        end
                    end
                end
                T = A;
            end
```

3. Use the function chg to write a two line sequence of commands that produces the sign of a permutation $x \in S_n$.

5.3 Glossary

adjacent interchanges 271
inversion 270
Klein four group 274

Chapter 6

Determinants

Topics
- *permanents, determinants*
- *Cauchy-Binet Theorem*
- *Laplace Expansion Theorem*
- *adjugate, inverses*
- *subdeterminants, rank*
- *positive-definite matrices*
- *Cholesky factorization*

6.1 Generalized Matrix Functions

In this section the general concept of a numerical valued function of a square matrix is introduced. The two most important examples of these functions are the determinant and the permanent. In §6.2 Two Classical Determinant Theorems, the expansion theorems of Laplace and Cauchy-Binet are proved. The relationship between the subdeterminants of a matrix and its rank is also investigated. In §6.3 Compound Matrices, there are a number of interesting and important results that describe matrices that are built from the subdeterminants of a matrix. Although compound matrices are not directly applicable to numerical work, they are powerful tools in analyzing the foundations of numerical matrix theory.

If $\sigma \in S_n$ then the *diagonal of A corresponding to* σ is the n-tuple of numbers $(a_{1\sigma(1)}, a_{2\sigma(2)}, \ldots, a_{n\sigma(n)})$. The *diagonal product corresponding to* σ is just the product

$$\prod_{i=1}^{n} a_{i\,\sigma(i)} = a_{1\,\sigma(1)} a_{2\,\sigma(2)} \cdots a_{n\,\sigma(n)}. \tag{1}$$

The matrix functions considered in this chapter are of the following type. Let χ be a function on S_n with values in a set of numbers R, i.e., a function that associates an element of R with each permutation in S_n. Then define a number associated with $A \in M_n(R)$ by the formula

$$d_\chi(A) = \sum_{\sigma \in S_n} \chi(\sigma) \prod_{i=1}^{n} a_{i\,\sigma(i)}. \tag{2}$$

The function $d_\chi: M_n(R) \to R$ is called a *generalized matrix function*. The function d_χ that is obtained by setting $\chi(\sigma) = 1$ for all $\sigma \in S_n$ in (2) is called the *permanent* of A and is denoted by per(A). Thus

$$\text{per}(A) = \sum_{\sigma \in S_n} \prod_{i=1}^{n} a_{i\,\sigma(i)}. \tag{3}$$

The function d_χ obtained by setting $\chi(\sigma) = \varepsilon(\sigma)$ for all $\sigma \in S_n$ in (2) is called the *determinant* of A and is denoted by det(A). Thus

$$\det(A) = \sum_{\sigma \in S_n} \varepsilon(\sigma) \prod_{i=1}^{n} a_{i\,\sigma(i)}. \tag{4}$$

Before we describe the special properties of these functions, we want to consider them in a somewhat different way that facilitates computing their values. Thus, let $v_i = [a_{i1}, \ldots, a_{in}]$, $i = i, \ldots, n$, be a set of $1 \times n$ matrices. Let A be the matrix in $M_n(R)$ whose i^{th} row is v_i, $i = 1, \ldots, n$. Then define a function d_χ on the ordered n-tuple of vectors $v_1, \ldots, v_n$ by the formula

$$d_\chi(v_1, \ldots, v_n) = d_\chi(A). \tag{5}$$

As examples of (5) we can write

$$\det(A_{(1)}, \ldots, A_{(n)}) = \det(A), \quad \text{per}(A_{(1)}, \ldots, A_{(n)}) = \text{per}(A).$$

6.1 Generalized Matrix Functions

Theorem 1.

The generalized matrix function d_χ is linear in each vector v_i separately. That is, for each $i = 1, \ldots, n$,

$$d_\chi(v_1, \ldots, v_{i-1}, cv_i + hv_i', v_{i+1}, \ldots, v_n)$$
$$= cd_\chi(v_1, \ldots, v_{i-1}, v_i, v_{i+1}, \ldots, v_n)$$
$$+ hd_\chi(v_1, \ldots, v_{i-1}, v_i', v_{i+1}, \ldots, v_n)$$

for any vectors $v_1, \ldots, v_i, v_i', v_{i+1}, \ldots, v_n$ in $M_{1,n}(R)$ and all c and h in R.

Proof.

Let $v_k = [a_{k1}, \ldots, a_{kn}]$, $k = 1, \ldots, n$, and set $v_i' = [a_{i1}', \ldots, a_{in}']$. In what follows, the notation $\prod_{t \neq i}$ denotes the product of the n - 1 indicated numbers obtained by setting $t = 1, \ldots, n$, with the exception of $t = i$. Then

$$d_\chi(v_1, \ldots, v_{i-1}, cv_i + hv_i', v_{i+1}, \ldots, v_n)$$

$$= \sum_{\sigma \in S_n} \chi(\sigma) \left(\prod_{t \neq i} a_{t\sigma(t)} \right) (ca_{i\sigma(i)} + ha_{i\sigma(i)}')$$

$$= c \sum_{\sigma \in S_n} \chi(\sigma) \prod_{t=1}^{n} a_{t\sigma(t)} + h \sum_{\sigma \in S_n} \chi(\sigma) \left(\prod_{t \neq i} a_{t\sigma(t)} \right) a_{i\sigma(i)}'$$

$$= cd_\chi(v_1, \ldots, v_{i-1}, v_i, v_{i+1}, \ldots, v_n) + hd_\chi(v_1, \ldots, v_{i-1}, v_i', v_{i+1}, \ldots, v_n). \blacksquare$$

As an immediate consequence of Theorem 1, for any generalized matrix function d_χ we can write

$$d_\chi\left(\sum_{t=1}^k c_{1t}u_t, \sum_{t=1}^k c_{2t}u_t, \ldots, \sum_{t=1}^k c_{nt}u_t\right)$$

$$= \sum_{t_1,\ldots,t_n=1}^k c_{1t_1} c_{2t_2} \cdots c_{nt_n} d_\chi(u_{t_1}, \ldots, u_{t_n}) \tag{6}$$

where $u_t \in M_{1,n}(R)$ and $c_{it} \in R$, $t = 1, \ldots, k$, $i = 1, \ldots, n$. In formula (6) the summation indices $t_1, \ldots, t_n$ each run independently from 1 to k, so that formally the sum involves k^n terms.

We observe an elementary notational convention used in many computations involving products of the form

$$c = \prod_{t=1}^n c_t.$$

Make the substitution $t = \varphi(i)$, $\varphi \in S_n$, and note that

$$c = \prod_{i=1}^n c_{\varphi(i)}.$$

We remark that this is nothing more nor less than the commutativity of multiplication, e.g., if $n = 3$ and $\varphi = (1\ 2\ 3)$, then $\varphi(1) = 2$, $\varphi(2) = 3$, and $\varphi(3) = 1$ so that

$$\prod_{i=1}^3 c_{\varphi(i)} = c_{\varphi(1)} c_{\varphi(2)} c_{\varphi(3)} = c_2 c_3 c_1 = c_1 c_2 c_3 = \prod_{t=1}^3 c_t.$$

Another important general property of the d_χ function is contained in the next theorem.

Theorem 2.

If the function χ has the property that

$$\chi(\theta\varphi) = \chi(\theta)\chi(\varphi) \tag{7}$$

for all θ and φ in S_n, then

6.1 Generalized Matrix Functions

$$d_\chi(v_{\varphi(1)}, \ldots, v_{\varphi(n)}) = \chi(\varphi) d_\chi(v_1, \ldots, v_n) \qquad (8)$$

for any $\varphi \in S_n$.

Proof.

Once again we set $v_i = [a_{i1}, \ldots, a_{in}]$, $i = 1, \ldots, n$, so that

$$d_\chi(v_{\varphi(1)}, \ldots, v_{\varphi(n)}) = \sum_{\sigma \in S_n} \chi(\sigma) \prod_{i=1}^n a_{\varphi(i), \sigma(i)}. \qquad (9)$$

If we set $t = \varphi(i)$ in

$$\prod_{i=1}^n a_{\varphi(i), \sigma(i)},$$

then this term becomes

$$\prod_{t=1}^n a_{t, \sigma\varphi^{-1}(t)}.$$

Thus from (9)

$$d_\chi(v_{\varphi(1)}, \ldots, v_{\varphi(n)}) = \sum_{\sigma \in S_n} \chi(\sigma) \prod_{t=1}^n a_{t, \sigma\varphi^{-1}(t)}. \qquad (10)$$

We know from §5.1 Theorem 1(f), that as σ runs through S_n so does $\sigma\varphi^{-1}$, where φ is any fixed permutation in S_n. Thus in (10) we can set $\theta = \sigma\varphi^{-1}$, i.e., $\sigma = \theta\varphi$, and sum over θ instead of σ. We have

$$d_\chi(v_{\varphi(1)}, \ldots, v_{\varphi(n)}) = \sum_{\theta \in S_n} \chi(\theta\varphi) \prod_{t=1}^n a_{t, \theta(t)}. \qquad (11)$$

But, by (7), $\chi(\theta\varphi) = \chi(\theta)\chi(\varphi)$ so that (8) follows from (11). ∎

We can conclude from Theorem 2 that

$$\operatorname{per}(E_{(i),(j)}A) = \operatorname{per}(A), \quad \det(E_{(i),(j)}A) = -\det(A). \tag{12}$$

For, let φ be the transposition (i j). Then, since the value of $\chi(\varphi)$ is 1 in the permanent and -1 in the determinant, (12) follows immediately from (8). In fact, if $\varphi \in S_n$ then

$$\det(A_{(\varphi(1))}, \ldots, A_{(\varphi(n))}) = \varepsilon(\varphi) \det(A_{(1)}, \ldots, A_{(n)}). \tag{13}$$

We can use Theorem 1 with $\chi(\sigma) = \varepsilon(\sigma)$ to confirm that

$$\det(E_{(i)+c(j)}A) = \det(A_{(1)}, \ldots, A_{(i-1)}, A_{(i)} + cA_{(j)}, A_{(i+1)}, \ldots, A_{(j)}, \ldots, A_{(n)})$$

$$= \det(A_{(1)}, \ldots, A_{(i-1)}, A_{(i)}, A_{(i+1)}, \ldots, A_{(n)})$$

$$+ c \det(A_{(1)}, \ldots, A_{(i-1)}, A_{(j)}, A_{(i+1)}, \ldots, A_{(j)}, \ldots, A_{(n)})$$

$$= \det(A) + c \det(A_{(1)}, \ldots, A_{(i-1)}, A_{(j)}, A_{(i+1)}, \ldots, A_{(j)}, \ldots, A_{(n)}).$$

The last determinant on the right must be zero because the interchange of two equal rows $A_{(j)}$ appearing in the ith and j^{th} positions changes the sign of the determinant, as noted in (12). On the other hand, the determinant obviously does not change when these two equal rows are interchanged. Thus it must be zero. We then have

$$\det(E_{(i)+c(j)}A) = \det(A). \tag{14}$$

In the above calculation we have assumed that $j > i$. If $i > j$, the proof of (14) is essentially the same.

By precisely the same kind of calculation as above, the reader may verify that

$$\det(E_{c(i)}A) = c \det(A). \tag{15}$$

If we take $A = I_n$ in (12), (14), and (15) and observe that $\det(I_n) = 1$ (why?), we have

6.1 Generalized Matrix Functions

$$\det(E_{(i),(j)}) = -1, \quad \det(E_{(i)+c(j)}) = 1, \quad \det(E_{c(i)}) = c. \tag{16}$$

Moreover, if E is any of the three types of elementary matrices, then it follows from (12), (14), (15), and (16) that

$$\det(EA) = \det(E)\det(A). \tag{17}$$

As we shall see momentarily (i.e., formula (20)), $\det(A^T) = \det(A)$, so that also

$$\det(AE) = \det((AE)^T) = \det(E^T A^T) = \det(E^T)\det(A^T) = \det(E)\det(A). \tag{18}$$

We are now able to prove an important and classical result.

Theorem 3.

Let R be a set of numbers in which every nonzero element is a unit. If $A \in M_n(R)$, then A is singular if and only if $\det(A) = 0$.

Proof.

Let $PAQ = I_r \oplus 0_{n-r}$ be the canonical form of A, $\rho(A) = r$, and P and Q products of elementary matrices. As we saw following §4.3 Theorem 3, A is singular iff $r < n$. The formulas (16), (17), and (18) imply that

$$\det(A) = c \det(I_r \oplus 0_{n-r}), \quad c \neq 0,$$

so that obviously $r < n$ iff $\det(A) = 0$. ∎

As an example of the use of Theorem 3, consider the homogeneous system of linear equations

$$x_1 + 2x_2 + x_3 = 0$$

$$x_2 - 3x_3 = 0 \qquad (19)$$

$$-x_1 + x_2 - x_3 = 0.$$

In matrix notation this becomes

$$Ax = 0$$

where $x = [x_1 \; x_2 \; x_3]^T$ and

$$A = \begin{bmatrix} 1 & 2 & 1 \\ 0 & 1 & -3 \\ -1 & 1 & -1 \end{bmatrix}.$$

Now $\det(A) = (-1 + 6 + 0) - (-1 - 3 + 0) = 9$. Hence, A is nonsingular so that A^{-1} exists and $0 = A^{-1}Ax = x$. Thus the only solution to the system (19) is $x_1 = x_2 = x_3 = 0$.

If $A \in M_n(R)$, then we confirm that

$$\det(A^T) = \det(A), \quad \operatorname{per}(A^T) = \operatorname{per}(A). \qquad (20)$$

To verify (20), compute that

$$\det(A^T) = \sum_{\sigma \in S_n} \varepsilon(\sigma) \prod_{i=1}^{n} (A^T)_{i\sigma(i)}$$

$$= \sum_{\sigma \in S_n} \varepsilon(\sigma) \prod_{i=1}^{n} a_{\sigma(i)i} \, .$$

In $\prod_{i=1}^{n} a_{\sigma(i)i}$ the order of the factors is immaterial. Thus set $j = \sigma(i)$, so that $i = \sigma^{-1}(j)$ and then

$$\prod_{i=1}^{n} a_{\sigma(i)i} = \prod_{j=1}^{n} a_{j\sigma^{-1}(j)} .$$

Hence

$$\det(A^T) = \sum_{\sigma \in S_n} \varepsilon(\sigma) \prod_{j=1}^{n} a_{j\sigma^{-1}(j)} .$$

Now let $\varphi = \sigma^{-1}$ and recall that $\varepsilon(\sigma) = \varepsilon(\sigma^{-1}) = \varepsilon(\varphi)$, so that

$$\det(A^T) = \sum_{\varphi \in S_n} \varepsilon(\varphi) \prod_{j=1}^{n} a_{j\varphi(j)}$$

$$= \det(A) .$$

The verification that $\text{per}(A^T) = \text{per}(A)$ is similar.

We will shortly consider determinants of submatrices of a matrix. The determinant of a square submatrix of A is called a *subdeterminant* of A. The determinant of a principal submatrix of a square matrix called a *principal subdeterminant*.

6.1 Exercises

1. Compute det(A) directly from formula (4) for each of the following matrices:

    (a) $\begin{bmatrix} 1 & 2 & 3 \\ 3 & 1 & 2 \\ 0 & 0 & 1 \end{bmatrix}$

    (b) $\begin{bmatrix} 1 & 1 & a \\ 1 & 1 & b \\ 1 & 1 & c \end{bmatrix}$

(c) $\begin{bmatrix} 1 & x & x^2 \\ 1 & y & y^2 \\ 1 & z & z^2 \end{bmatrix}$ 	(d) $E_{(i)} + c_{(j)}$

(e) $E_{c(i)} E_{d(j)}$ 	(f) $(E^{(i)} + (j)c)^T$

2. Compute the sum of all principal 2 × 2 subdeterminants of

$$\begin{bmatrix} 1 & 2 & 3 \\ 1 & -1 & 2 \\ 0 & 1 & 4 \end{bmatrix}.$$

3. Prove: if $A \in M_n(\mathbb{R})$ then $\det(A)\det(A^T) \geq 0$.

4. Find $\text{per}(J_n)$ where J_n is the n × n matrix with every entry $1/n$.

5. Let

$$A = \begin{bmatrix} a_{11} & a_{12} & a_{13} & a_{14} \\ a_{21} & a_{22} & a_{23} & a_{24} \end{bmatrix}$$

and set $a(\omega) = \det(A[1,2 \mid \omega])$ for $\omega \in Q_{2,4}$. Show by direct calculation that

$$a(12)a(34) - a(13)a(24) + a(14)a(23) = 0.$$

This equation is an example of what are called *quadratic Plücker relations*.

6. Show that the system of equations

$$x_1 - x_2 = 0$$
$$x_2 + x_3 = 0$$
$$x_3 + x_4 = 0$$
$$x_4 + x_1 = 0$$

6.1 Exercises

has only the trivial solution $x = 0 = [0\ 0\ 0\ 0]^T$ by writing the system in the form $Ax = 0$ and computing that $\det(A) \neq 0$.

7. Find the number of permutations σ in S_3 satisfying $\sigma(i) \neq i$, $i = 1, 2, 3$.

8. Prove: $\text{per}(A) = \text{per}(A^T)$.

9. Prove: if A is an n-square complex matrix and $b_{ij} = |a_{ij}|$, $i, j = 1, \ldots, n$, then $|\text{per}(A)| \leq \text{per}(B)$.

10. Prove: $\det(\overline{A}) = \overline{\det(A)}$ for any $A \in M_n(\mathbb{C})$.

11. Prove: if A is an $n \times n$ complex hermitian matrix then $\det(A)$ is a real number.

12. Recall that an *upper (lower) triangular* $n \times n$ matrix has zero entries below (above) the main diagonal. Prove: if A is upper (lower) triangular then

$$\det(A) = \prod_{i=1}^{n} a_{ii}.$$

13. Prove: if A is upper (lower) triangular then

$$\text{per}(A) = \prod_{i=1}^{n} a_{ii}.$$

14. Let

$$A = \begin{bmatrix} a_{11} & a_{12} \\ a_{21} & a_{22} \end{bmatrix},$$

and assume $\det(A) \neq 0$. Prove:

$$A^{-1} = \frac{1}{\det(A)} \begin{bmatrix} a_{22} & -a_{12} \\ -a_{21} & a_{11} \end{bmatrix}.$$

15. Prove: if A and B are $n \times n$ matrices then $\det(AB) = \det(A)\det(B)$.

Hint: If either A or B is singular then AB is also singular (i.e., $\rho(AB)$ is at most $\min\{\rho(A), \rho(B)\}$). Hence, in this case, both $\det(AB) = 0$ and $\det(A)\det(B) = 0$. If both A and B are nonsingular then both are products of elementary matrices, say $A = E_1 \cdots E_p$, $B = E_{p+1} \cdots E_q$. Then

$$AB = E_1 \cdots E_p E_{p+1} \cdots E_q,$$

so that by formula (17),

$$\det(AB) = \det(E_1 \cdots E_p E_{p+1} \cdots E_q)$$

$$= \det(E_1) \cdots \det(E_p) \det(E_{p+1}) \cdots \det(E_q)$$

$$= \det(A)\det(B).$$

16. Let X be $p \times q$. Prove:

$$\begin{bmatrix} I_p & X \\ 0 & I_q \end{bmatrix}^{-1} = \begin{bmatrix} I_p & -X \\ 0 & I_q \end{bmatrix}.$$

17. Let A be an upper triangular matrix and B be a lower triangular matrix, p-square and q-square respectively. Assume that the main diagonals of A and B have no values in common. Let C be a $p \times q$ matrix and set

$$M = \begin{bmatrix} A & C \\ 0 & B \end{bmatrix}.$$

Prove: there exists a non-singular $(p+q)$-square matrix S such that $SMS^{-1} = A \oplus B$.

Hint: Set

$$S = \begin{bmatrix} I_p & X \\ 0 & I_q \end{bmatrix}$$

so that by #16,

$$SMS^{-1} = \begin{bmatrix} I_p & X \\ 0 & I_q \end{bmatrix} \begin{bmatrix} A & C \\ 0 & B \end{bmatrix} \begin{bmatrix} I_p & -X \\ 0 & I_q \end{bmatrix}$$

$$= \begin{bmatrix} A & C + XB \\ 0 & B \end{bmatrix} \begin{bmatrix} I_p & -X \\ 0 & I_q \end{bmatrix}$$

$$= \begin{bmatrix} A & -AX + XB + C \\ 0 & B \end{bmatrix}.$$

We want to choose X so that $AX - XB = C$. The column form of this equation is $(I_q \otimes A - B^T \otimes I_p)c(X) = c(C)$. Now $I_q \otimes A - B^T \otimes I_p$ is upper triangular with differences of the form $a_{ii} - b_{jj}$ along the main diagonal. Hence

$$\det(I_q \otimes A - B^T \otimes I_p) \neq 0$$

and thus $c(X)$ and X can be determined.

18. If $A \in M_n(\mathbb{C})$ then $r \in \mathbb{C}$ is an *eigenvalue* of A if r satisfies the equation $\det(rI_n - A) = 0$. We study eigenvalues in detail in Chapter 7.

    Prove: r is an eigenvalue of A iff there is an $x \in M_{n,1}(\mathbb{C})$, $x \neq 0$, such that $Ax = rx$; x is called an *eigenvector* of A corresponding to r (see §4.3 Exercises, #33).

    Hint: By Theorem 3, $\det(rI_n - A) = 0$ iff $rI_n - A$ is singular. By §4.3 Exercises, #33 there is an $x \neq 0$ such that $(rI_n - A)x = 0$, i.e., $Ax = rx$. Conversely, if such a non-zero x exists then $(rI_n - A)x = 0$. Unless r satisfies $\det(rI_n - A) = 0$, $(rI_n - A)^{-1}$ exists by Theorem 3 and hence we could conclude $x = 0$, a contradiction.

19. Let x_1 be a nonzero matrix in $M_{n,1}(R)$. Prove that there exist $n - 1$ column vectors $x_2, ..., x_n$ in $M_{n,1}(R)$ such that the $n \times n$ matrix X whose columns are $x_1, x_2, ..., x_n$ is nonsingular.

20. Prove: if $A \in M_n(R)$ then there exists a nonsingular matrix $P \in M_n(R)$ such that $P^{-1}AP$ is upper (or lower) triangular. Assume that all of the eigenvalues of A are in R.

Hint: Find an eigenvalue of r such that $\det(rI_n - A) = 0$. Then by #18 obtain an $n \times 1$ nonzero x such that $Ax = rx$. By #19, let X be a non-singular matrix whose first column is x. Note that $(X^{-1}AX)^{(1)} = X^{-1}(AX)^{(1)} = X^{-1}(AX^{(1)}) = X^{-1}Ax = X^{-1}rx = rX^{-1}x = rX^{-1}X^{(1)} = r(X^{-1}X)^{(1)} = r(I_n)^{(1)} = [r, 0, ..., 0]^T$. Thus

$$X^{-1}AX = \begin{bmatrix} r & \cdots \\ 0 & \\ \vdots & B \\ 0 & \end{bmatrix},$$

where B is $(n-1)$-square. Proceed by induction to obtain Y such that $Y^{-1}BY$ is upper triangular, and then set $P = X([1] \oplus Y)$. To get the lower triangular result apply the result to A^T.

21. Let $X \in M_n(R)$. Assume that P is a nonsingular matrix in $M_n(R)$ for which $P^{-1}XP = T$ is upper (lower) triangular. Prove: the main diagonal entries of T are the eigenvalues of X.

22. Prove: if $P^{-1}XP$ is upper triangular and $Q^{-1}XQ$ is lower triangular then except possibly for the order in which they appear, the entries in the main diagonal of $P^{-1}XP$ are the same as the entries in the main diagonal of $Q^{-1}XQ$.

23. Let A be a p-square matrix and B be a q-square matrix, both with entries in a set R that contains all the eigenvalues of both A and B. Assume that A and B have no eigenvalues in common. Let C be a $p \times q$ matrix with entries in R and set

$$M = \begin{bmatrix} A & C \\ 0 & B \end{bmatrix}.$$

Prove: there exists a nonsingular $(p + q)$-square matrix S such that $S^{-1}MS = A \oplus B$.

Hint: As in the Hint for #17, the problem reduces to proving the nonsingularity of

$$I_q \otimes A - B^T \otimes I_p.$$

By #20 obtain P and Q nonsingular such that $P^{-1}B^TP$ and $Q^{-1}AQ$ are upper triangular. Then

$$(P^{-1} \otimes Q^{-1})(I_q \otimes A - B^T \otimes I_p)(P \otimes Q) = I_q \otimes (Q^{-1}AQ) - (P^{-1}B^TP) \otimes I_p.$$

The main diagonal entries of $Q^{-1}AQ$ are eigenvalues of A, the main diagonal entries of $P^{-1}B^TP$ are eigenvalues of B^T (and hence of B) so that the main diagonal entries of $L = I_q \otimes (Q^{-1}AQ) - (P^{-1} B^TP) \otimes I_p$ are differences of eigenvalues of A and B. But L is upper triangular, so $\det(L) \neq 0$ and the proof proceeds as before.

6.1 MatLab

1. Use Help to read the entry for **det**. Then use det to compute the determinant of each of the following matrices:

    (a) I_5

    (b)   the 5×5 matrix $E_{(2),(3)}$

    (c)   the 5×5 matrix $E_{(3) + 2(4)}$

    (d)   the 5×5 matrix $E_{3(2)}$

    (e)   the 5×5 matrix $\left[\dfrac{1}{i + j - 1}\right]$.

    Hint: Use Help to read the entry for **hilb**.

2. Write a MatLab function called invs that

    - accepts a permutation x
    - returns the value invs, the number of inversions in x

Save invs in Permute.

Hint:
```
function f=invs(x)
%INVS returns the number of inversions
%in the permutation x.
n=length(x);
c=0;
for i=1:(n-1)
        for j=(i+1):n
                if x(i) > x(j)
                        c=c+1;
                end
        end
end
f=c;
```

3. Use pmat and prand (§5.1 MatLab, #4, #7) to compute $\det(A(\sigma))$ for various permutations σ in S_n. Use invs in #2 to compare $\det(A(\sigma))$ with $\varepsilon(\sigma)$. What appears to be the relationship between these numbers? Prove your conjecture.

Hint: $\det(A(\sigma)) = \varepsilon(\sigma)$. To prove this refer to §5.3 MatLab, #2 for notation. Note that

$$A(\sigma) = A(\tau_p) \cdots A(\tau_1)$$

where $\sigma = \tau_p \cdots \tau_1$ and the τ_j are adjacent interchanges. Then

$$\det(A(\sigma)) = \prod_{j=1}^{p} \det(A(\tau_j))$$

$$= (-1)^p$$

$$= \varepsilon(\sigma).$$

4. Use Help to review the entries for **prod** and **diag**. Then write a MatLab function called dprd that accepts an n-square matrix A and a permutation $p \in S_n$ and returns the diagonal product

6.1 MatLab

$$\prod_{i=1}^{n} a_{i\,p(i)}.$$

Recall that a permutation p is represented as the 1 × n matrix

$$p = [p(1)\ p(2)\ \ldots\ p(n)].$$

Save dprd in a library called determ.

Hint:

    function d=dprd(A,p)
    %DPRD accepts an n-square matrix A
    %and permutation p and returns the diagonal
    %product in A corresponding to p.
    d = prod(diag(A(:,p)));

5. Write a MatLab function called ddiag that accepts an n-square matrix X and two permutations p and q in S_n and returns as a column the main diagonal of A(p)XA(q). Recall that A(σ) is the mathematical notation for the incidence matrix for σ. Save ddiag in determ.

    Hint: See §5.1 Exercises, #9.

    function dd=ddiag(X,p,q)
    %DDIAG accepts an n-square matrix X
    %and two permutations p and q in Sn
    %and returns as a column the
    %main diagonal of A(p)XA(q).
    n=length(X);
    p1(p) = 1:n;   %p1 is the inverse of p.
    dd=diag(X(p1,q));

6. Evaluate each of the following permanents using the function perm in §5.2 MatLab.

    (a)   per(ones(5))

(b) per(ones(5) - eye(5))

(c) per(A), where A is the direct sum of ones(2) and ones(3)

7. Use perm to determine the number of permutations σ in S_5 for which σ has no fixed points, i.e., $\sigma(t) \neq t$, $t = 1, \ldots, 5$.

Hint: The answer is the same as the answer to #6(b). Why?

8. Write a MatLab script call permpos that:

- generates 20 pairs of random positive integers m and n, $m \leq 5$, $n \leq 5$
- generates 20 corresponding m × n random complex matrices A, $|a_{ij}| \leq 1$, $i = 1, \ldots, m$, $j = 1, \ldots, n$
- prints out a table listing m, n, and per(A^*A).

On the basis of this experiment, make an educated guess about the values per(A^*A) for any complex matrix A. Prove that your guess is correct for rank 1 matrices. Save permpos in the determ folder.

Note: The guess is that per(A^*A) ≥ 0 for any $A \in M_{m,n}(\mathbb{C})$. If A is m × n of rank 1 then $A = x^*y$ where x is 1 × m and y is 1 × n. Then $A^*A = (x^*y)^* x^*y = y^*xx^*y = (xx^*)y^*y = xx^* [\bar{y}_i y_j]$. It follows that

$$\text{per}(A^*A) = \text{per}(xx^* [\bar{y}_i y_j])$$

$$= (xx^*)^n \sum_{\sigma \in S_n} \prod_{i=1}^{n} \bar{y}_i y_{\sigma(i)}$$

$$= (xx^*)^n \sum_{\sigma \in S_n} \prod_{i=1}^{n} |y_i|^2$$

$$= n! (\sum_{i=1}^{m} |x_i|^2)^n \prod_{i=1}^{n} |y_i|^2$$

$$\geq 0.$$

6.1 MatLab

Hint:

```
%PERMPOS generates 20 random m × n
%complex matrices, m and n random integers
%at most 5.  Also, every entry of any
%generated matrix is at most 1 in modulus.
%The program then exhibits m, n and per(A*A);
D=[ ];
disp('    m              n              per(A*A)     ');
disp('===================================================');
for j=1:20
    m = floor(5*rand+1); n = floor(5*rand+1);
    theta = (2*pi*rand(m,n));
    r=rand(m, n); A = r.*exp(i*theta);
    per = perm(A'*A);
    D=[D; m n per];
end
disp(D);
```

9. Write a function called direc that accepts two square matrices and returns their direct sum. Save direc in determ.

Hint:

```
function d=direc(A,B)
%DIREC accepts two square matrices
%A and B and returns their
%direct sum.
m=length(A);
n=length(B);
d=[A zeros(m,n); zeros(n,m) B];
```

10. Write a MatLab script called perdir that

- generates 20 pairs of random positive integers m and n, $m \leq 5$, $n \leq 5$
- generates 20 corresponding pairs of random matrices $A \in M_m(R)$, $B \in M_n(R)$ where $R = \{-1, 0, 1\}$

- prints out a table listing per(A), per(B), per(direc(A,B)), and per(A)*per(B).

Save perdir in the folder called determ.

Hint:
```
%PERDIR compares a = per(A), b = per(B),
%d = per(A direct sum B), and p=a*b for
%randomly generated matrices A and B
%with entries -1, 0, 1 and size at most 5.
disp('    a    b    d    p');
disp('   ============');
D=[ ];
for j = 1:20
    m=floor(5*rand + 1);
    n=floor(5*rand + 1);
    A=floor(3*rand(m) - ones(m));
    B=floor(3*rand(n) - ones(n));
    a=perm(A);
    b=perm(B);
    c=perm(direc(A,B));
    D=[D; a b c a*b];
end
disp(D);
```

11. On the basis of the output of the script perdir in #10, make an educated guess about the relationship among per(A), per(B) and per(A ⊕ B). Prove that your guess is correct.

Hint: per(A ⊕ B) = per(A)per(B). To see this, let $\sigma \in S_{m+n}$ and let $X = A \oplus B$. Clearly, the product $\prod_{i=1}^{m+n} X_{i\,\sigma(i)}$ is automatically 0 unless $\sigma\{1, ..., m\} = \{1, ..., m\}$ and $\sigma\{m+1, ..., n\} = \{m+1, ..., n\}$. Thus, the nonzero terms in per(X) arise only from $\sigma = \theta\varphi$, where θ is a permutation of $1, ..., m$, and φ is a permutation of $m+1, ..., n$. Moreover, this representation for σ is unique. Hence,

$$\text{per}(A \oplus B) = \text{per}(X)$$

$$= \sum_{\sigma \in S_{m+n}} \prod_{i=1}^{m+n} X_{i\,\sigma(i)}$$

$$= \sum_{\theta,\varphi} \prod_{i=1}^{m+n} x_{i,\,\theta\varphi(i)}$$

$$= \sum_{\theta,\varphi} \prod_{i=1}^{m} x_{i\,\theta(i)} \prod_{i=m+1}^{n} x_{i\,\varphi(i)}$$

$$= \sum_{\sigma \in S_m} \prod_{i=1}^{m} x_{i\,\theta(i)} \sum_{\varphi} \prod_{i=m+1}^{n} x_{i\,\varphi(i)}$$

$$= \sum_{\sigma \in S_m} \prod_{i=1}^{m} a_{i\,\theta(i)} \sum_{\omega \in S_n} \prod_{i=1}^{n} b_{i\,\omega(i)}$$

$$= \text{per}(A)\,\text{per}(B).$$

12. Review the function cmb in §5.2 MatLab, #9(b). Then write a MatLab function named psdet that

   - accepts an n-square matrix A and a positive integer r, $1 \le r \le n$
   - returns psdet(r,A), the sum of all r-square principal subdeterminants of A.

   Save psdet in determ.

   Hint:

   function s=psdet(r,A)
   %PSDET returns the sum of all r-square
   %principal subdeterminants of A.
   n=length(A);
   C=cmb(r,n);
   add=0;
   cnr=size(C);
   for t=1:cnr(1)   %cnr(1) is binomial coefficient n over r.
       add=add + det(A(C(t,:),C(t,:)));
   end
   s=add;

13. Write a MatLab function named psper that

   - accepts an n-square matrix A and a positive integer r, $1 \le r \le n$
   - returns psper(r, A), the sum of all r-square principal subpermanents of A.

Save psper in determ.

Hint: Use perm in §5.2 MatLab, #2.

14. Write a MatLab function named sumdet that

    - accepts an n-square matrix A and a positive integer r, $1 \leq r \leq n$
    - returns sumdet(r,A), the sum of all r-square subdeterminants of A.

15. Write a MatLab function named cmpnd that

    - accepts an m × n matrix A and a positive integer r, $1 \leq r \leq \min\{m, n\}$
    - returns an $\binom{m}{r} \times \binom{n}{r}$ matrix cmpnd(r,A) whose entries are the r-square subdeterminants of A arranged in doubly lexicographic order, e.g., if

    $$A = \begin{bmatrix} a_{11} & a_{12} & a_{13} \\ a_{21} & a_{22} & a_{23} \\ a_{31} & a_{32} & a_{33} \end{bmatrix}$$

    then

    $$\text{cmpnd}(2,A) = \begin{bmatrix} \det(A[1,2\,|\,1,2]) & \det(A[1,2\,|\,1,3]) & \det(A[1,2\,|\,2,3]) \\ \det(A[1,3\,|\,1,2]) & \det(A[1,3\,|\,1,3]) & \det(A[1,3\,|\,2,3]) \\ \det(A[2,3\,|\,1,2]) & \det(A[2,3\,|\,1,3]) & \det(A[2,3\,|\,2,3]) \end{bmatrix}.$$

    Save cmpnd in the folder determ.

    Hint:
    ```
    function CA = cmpnd(r, A)
    %CMPND(r,A) returns a matrix
    %whose entries are the r-square subdeterminants
    %of A arranged in doubly lexicographic order.
    [m n]=size(A);
    C=cmb(r,n);
    D=cmb(r,m);
    cnr=size(C);
    dnr=size(D);
    for s=1:dnr(1)
        for t=1:cnr(1)
    ```

```
                    B(s,t) = det(A(D(s,:), C(t,:)));
            end
    end
    CA = B;
```

6.1 Glossary

det	278
determinant	278
diag	292
diagonal	277
diagonal product	277
eigenvalue	289
eigenvector	289
generalized matrix function	278
hilb	291
lower triangular	287
per	278
permanent	278
principal subdeterminant	285
prod	292
quadratic Plücker relations	286
subdeterminant	285
upper triangular	287

6.2 Two Classical Determinant Theorems

We next discuss two major classical results about determinants and exploit some of their consequences. The first result, known as the *Cauchy-Binet Theorem*, shows how to calculate a subdeterminant of a product of matrices AB from subdeterminants of the individual factors A and B.

Theorem 1 (Cauchy-Binet Theorem).

Let $A \in M_{p,q}(R)$, $B \in M_{q,r}(R)$ and assume that the integer m satisfies

$$1 \leq m \leq \min(p, q, r).$$

If $\alpha \in Q_{m,p}$, $\beta \in Q_{m,r}$, then

$$\det((AB)[\,\alpha\,|\,\beta\,]) = \sum_{\omega \in Q_{m,q}} \det(A[\,\alpha\,|\,\omega\,]) \det(B[\,\omega\,|\,\beta\,]). \tag{1}$$

In particular, if $p = q = r = n$, it follows that

$$\det(AB) = \det(A)\det(B). \tag{2}$$

Proof.

Let $C = (AB)[\,\alpha\,|\,\beta\,]$. The (i, j) entry of C is the (α_i, β_j) entry of AB, i.e., it is the product of row α_i of A and column β_j of B:

$$c_{ij} = A_{(\alpha_i)} B^{(\beta_j)}, \quad i, j, = 1, \ldots, m.$$

Thus the ith row of C is just

$$C_{(i)} = [A_{(\alpha_i)}B^{(\beta_1)}, A_{(\alpha_i)}B^{(\beta_2)}, \ldots, A_{(\alpha_i)}B^{(\beta_m)}]$$

$$= \left[\sum_{t=1}^{q} a_{\alpha_i t} b_{t\beta_1}, \sum_{t=1}^{q} a_{\alpha_i t} b_{t\beta_2}, \ldots, \sum_{t=1}^{q} a_{\alpha_i t} b_{t\beta_m}\right]$$

$$= \sum_{t=1}^{q} a_{\alpha_i t} [b_{t\beta_1}, \ldots, b_{t\beta_m}].$$

If we let $D = B[1, \ldots, q\,|\,\beta\,]$, then the preceding calculation shows that

$$C_{(i)} = \sum_{t=1}^{q} a_{\alpha_i t} D_{(t)}.$$

Hence, using §6.1, formula (6), we have

$$\det(C) = \det\left(\sum_{t=1}^{q} a_{\alpha_1 t} D_{(t)}, \ldots, \sum_{t=1}^{q} a_{\alpha_m t} D_{(t)}\right)$$

$$= \sum_{(t_1,\ldots,t_m) \in \Gamma_{m,q}} (\prod_{j=1}^{m} a_{\alpha_j,t_j}) \det(D_{(t_1)}, \ldots, D_{(t_m)}) . \quad (3)$$

The terms in the sum (3) arising from choices of $(t_1, \ldots, t_m)$ in which $t_i = t_j$ for some $i \neq j$ drop out because the determinant of a matrix with two rows the same is zero. We can obtain all sequences of distinct integers $(t_1, \ldots, t_m)$ by choosing first an increasing sequence, i.e., $(t_1, \ldots, t_m) \in Q_{m,q}$, and then taking all rearrangements of each such increasing sequence. In other words, by §6.1 formula (13), we conclude from (3) that

$$\det(C) = \sum_{(t_1,\ldots,t_m) \in Q_{m,q}} \sum_{\sigma \in S_m} (\prod_{j=1}^{m} a_{\alpha_j, t_{\sigma(j)}}) \det(D_{(t_{\sigma(1)})}, \ldots, D_{(t_{\sigma(m)})})$$

$$= \sum_{(t_1,\ldots,t_m) \in Q_{m,q}} \left(\sum_{\sigma \in S_m} \varepsilon(\sigma) \prod_{j=1}^{m} a_{\alpha_j t_{\sigma(j)}}\right) \det(D_{(t_1)}, \ldots, D_{(t_m)}). \quad (4)$$

Recall that $D = B[1, \ldots, q \mid \beta]$ and hence

$$\det(D_{(t_1)}, \ldots, D_{(t_m)}) = \det(B[t_1, \ldots, t_m \mid \beta]).$$

By the definition of a determinant we have

$$\sum_{\sigma \in S_m} \varepsilon(\sigma) \prod_{j=1}^{m} a_{\alpha_j t_{\sigma(j)}} = \det(A[\alpha \mid t_1, \ldots, t_m]).$$

Hence, upon setting $\omega = (t_1, \ldots, t_m) \in Q_{m,q}$ in (4) we have

$$\det((AB)[\alpha \mid \beta]) = \det(C) = \sum_{\omega \in Q_{m,q}} \det(A[\alpha \mid \omega]) \det(B[\omega \mid \beta]).$$

If $p = q = r = m$ then there are only the choices $(1, \ldots, m) = \alpha = \beta = \omega$ in the formula (1) and we immediately obtain (2). ∎

We can illustrate Theorem 1 as follows. Let

$$A = \begin{bmatrix} 1 & 0 & 3 & 4 \\ 2 & 0 & -1 & 2 \end{bmatrix}, \quad B = \begin{bmatrix} 1 & 1 \\ 1 & -1 \\ 2 & 2 \\ 0 & 0 \end{bmatrix}.$$

Then AB is a 2×2 matrix and we can compute its determinant by

$$\det(AB) = \sum_{\omega \in Q_{2,4}} \det(A[\,1, 2\,|\,\omega\,])\det(B[\,\omega\,|\,1, 2]).$$

Now

$$\det(A[\,1, 2\,|\,\omega\,])\det(B[\,\omega\,|\,1, 2]) = 0$$

whenever ω involves either a 2 or a 4, i.e., $A[\,1, 2\,|\,\omega\,]$ has a zero column or $B[\,\omega\,|\,1, 2\,]$ has a zero row. The only term that survives is the one obtained by setting $\omega = (1, 3)$. Hence

$$\det(AB) = \det(A[\,1, 2\,|\,1, 3])\det(B[\,1, 3\,|\,1, 2\,])$$

$$= (-7)(0) = 0.$$

Our next result is preliminary to Theorem 3, the *Laplace Expansion Theorem*. In fact, it is a special case of Theorem 3.

Theorem 2.

Let $A \in M_n(R)$ be a matrix of the following form:

$$A = I_k \oplus B \tag{5}$$

where $B \in M_{n-k}(R)$. Then $\det(A) = \det(B)$.

6.2 Two Classical Determinant Theorems

Proof.

Since $A = I_k \oplus B$ and

$$\det(A) = \sum_{\sigma \in S_n} \varepsilon(\sigma) \prod_{i=1}^{n} a_{i\sigma(i)}$$

it follows that

$$\prod_{i=1}^{n} a_{i\sigma(i)} = 0$$

unless $\sigma(i) = i$, $i = 1, \ldots, k$, i.e., $a_{i\sigma(i)} = 0$ if $1 \leq i \leq k$ and $\sigma(i) \neq i$. Consider a permutation of $1, \ldots, n - k$, say φ. Define $\sigma \in S_n$ by

$$\sigma = \begin{pmatrix} 1 & \cdots & k & k+1 & k+2 & \cdots & k+(n-k) \\ 1 & \cdots & k & k+\varphi(1) & k+\varphi(2) & \cdots & k+\varphi(n-k) \end{pmatrix}.$$

Clearly, the number of interchanges necessary to restore $k + \varphi(1), \ldots, k + \varphi(n - k)$ to natural order is the same as the number of interchanges necessary to restore $\varphi(1), \ldots, \varphi(n - k)$ to natural order. Hence $\varepsilon(\sigma) = \varepsilon(\varphi)$. But

$$\det(A) = \sum_{\sigma \in S_n} \varepsilon(\sigma) \prod_{i=1}^{n} a_{i\sigma(i)} = \sum_{\varphi \in S_{n-k}} \varepsilon(\varphi) \prod_{i=1}^{n-k} b_{i\varphi(i)} = \det(B). \blacksquare$$

Theorem 3 (Laplace Expansion Theorem).

Let $A \in M_n(R)$, and let α be a fixed sequence in $Q_{k,n}$, $1 \leq k \leq n$. Then

$$\det(A) = \sum_{\gamma \in Q_{k,n}} (-1)^{s(\alpha) + s(\gamma)} \det(A[\,\alpha\,|\,\gamma\,]) \det(A(\,\alpha\,|\,\gamma\,)) \qquad (6)$$

where, for any $\gamma \in Q_{k,n}$,

$$s(\gamma) = \sum_{j=1}^{k} \gamma_j.$$

Proof.

We write $\det(A) = \det(A_{(1)}, \ldots, A_{(n)})$, and then set

$$A_{(t)} = \sum_{j=1}^{n} a_{tj} e_j, \quad t = 1, \ldots, n,$$

where, in this argument, the e_j are the $1 \times n$ row matrices with 1 in position j, and zero elsewhere. Replacing $A_{(\alpha_i)}$ by

$$\sum_{j=1}^{n} a_{\alpha_i j} e_j$$

for $i = 1, \ldots, k$, we have

$$\det(A) = \det(A_{(1)}, \ldots, A_{(\alpha_1)}, \ldots, A_{(\alpha_k)}, \ldots, A_{(n)})$$

$$= \det(A_{(1)}, \ldots, \sum_{j=1}^{n} a_{\alpha_1 j} e_j, \ldots, \sum_{j=1}^{n} a_{\alpha_k j} e_j, \ldots, A_{(n)})$$

$$= \sum_{(j_1, \ldots, j_k) \in \Gamma_{k,n}} a_{\alpha_1 j_1} \cdots a_{\alpha_k j_k} \det(A_{(1)}, \ldots, e_{j_1}, \ldots, e_{j_k}, \ldots, A_{(n)}). \qquad (7)$$

In (7) we have only used rows $\alpha_1, \ldots, \alpha_k$ to expand. All others are left unaltered. In the last summation in (7), if any two of $j_1, \ldots, j_k$ are equal, then, as we have seen previously,

$$\det(A_{(1)}, \ldots, e_{j_1}, \ldots, e_{j_k}, \ldots, A_{(n)}) = 0.$$

Hence, we need only sum over sets of distinct $j_1, \ldots, j_k$. Moreover, we can do this by choosing increasing sequences, $j_1 < \cdots < j_k$, and then summing separately over all $k!$ rearrangements of each such sequence:

$$(j_{\sigma(1)}, \ldots, j_{\sigma(k)}), \quad \sigma \in S_k.$$

6.2 Two Classical Determinant Theorems

Clearly, the number of inversions in the second row of

$$\sigma = \begin{pmatrix} 1 & 2 & \cdots & k \\ \sigma(1) & \sigma(2) & \cdots & \sigma(k) \end{pmatrix}$$

is the same as the number of inversions in the second row of

$$\begin{pmatrix} j_1 & \cdots & j_k \\ j_{\sigma(1)} & \cdots & j_{\sigma(k)} \end{pmatrix}$$

and hence

$$\det(A_{(1)}, \ldots, e_{j_{\sigma(1)}}, \ldots, e_{j_{\sigma(k)}}, \ldots, A_{(n)})$$

$$= \varepsilon(\sigma) \det(A_{(1)}, \ldots, e_{j_1}, \ldots, e_{j_k}, \ldots, A_{(n)}). \tag{8}$$

Using (7) and (8) and setting $\gamma = (j_1, \ldots, j_k) \in Q_{k,n}$ we can write

$$\det(A) = \sum_{\gamma \in Q_{k,n}} \sum_{\sigma \in S_k} a_{\alpha_1 j_{\sigma(1)}} \cdots a_{\alpha_k j_{\sigma(k)}} \det(A_{(1)}, \ldots, e_{j_{\sigma(1)}}, \ldots, e_{j_{\sigma(k)}}, \ldots, A_{(n)})$$

$$= \sum_{\gamma \in Q_{k,n}} \left(\sum_{\sigma \in S_k} \varepsilon(\sigma) a_{\alpha_1 j_{\sigma(1)}} \cdots a_{\alpha_k j_{\sigma(k)}} \right) \det(A_{(1)}, \ldots, e_{j_1}, \ldots, e_{j_k}, \ldots, A_{(n)})$$

$$= \sum_{\gamma \in Q_{k,n}} \det(A[\alpha \mid j_1, \ldots, j_k]) \det(A_{(1)}, \ldots, e_{j_1}, \ldots, e_{j_k}, \ldots, A_{(n)}). \tag{9}$$

We now examine the determinant

$$\det(A_{(1)}, \ldots, e_{j_1}, \ldots, e_{j_k}, \ldots, A_{(n)}) \tag{10}$$

appearing in the last summation in (9). Rows numbered $\alpha_1, \ldots, \alpha_k$ may be brought into positions $1, \ldots, k$ by a row permutation whose sign is $(-1)^{s(\alpha)-k(k+1)/2}$. That is, $\alpha_1 - 1$ adjacent row interchanges will bring row α_1 to the first row position. Then $\alpha_2 - 2$ adjacent interchanges will bring row α_2 to the second row position, etc. But e_{j_t} appears in row numbered α_t of (10), $t = 1, \ldots, k$. Thus we must examine the value of

$$(-1)^{s(\alpha)-k(k+1)/2} \det(e_{j_1}, \ldots, e_{j_k}, A_{(\beta_1)}, \ldots, A_{(\beta_{n-k})}), \qquad (11)$$

where the rows following the kth row in (11) are precisely those rows $A_{(t)}$ of A for which t is in the set complementary to $\{\alpha_1, \ldots, \alpha_k\}$ in $\{1, \ldots, n\}$; we have called this set $\{\beta_1, \ldots, \beta_{n-k}\}$ in (11).

Write out the matrix in (11) in terms of its rows:

$$\begin{bmatrix} e_{j_1} \\ e_{j_2} \\ \vdots \\ e_{j_k} \\ A_{(\beta_1)} \\ \vdots \\ A_{(\beta_{n-k})} \end{bmatrix}. \qquad (12)$$

By performing elementary row operations of type II on the matrix in (12), it is clear that without altering the value of the determinant we can annihilate the entries appearing in columns $j_1, \ldots, j_k$ and in rows $k+1, \ldots, n$. In the resulting matrix we can then move columns $j_1, \ldots, j_k$ into column positions $1, \ldots, k$, thereby introducing a total of

$$\sum_{t=1}^{k} j_t - k(k+1)/2 = s(\gamma) - k(k+1)/2$$

sign changes in the determinant. The resulting matrix has the form

$$\begin{bmatrix} I_k & 0 \\ 0 & A(\alpha \mid j_1, \ldots, j_k) \end{bmatrix} = I_k \oplus A(\alpha \mid \gamma), \qquad (13)$$

which, by Theorem 2, has determinant equal to $\det(A(\alpha \mid \gamma))$. Thus from (9) and (11) we have

6.2 Two Classical Determinant Theorems

$$\det(A) = \sum_{\gamma \in Q_{k,n}} (-1)^{s(\alpha)-k(k+1)/2} \det(A[\,\alpha\,|\,\gamma\,]) \, (-1)^{s(\gamma)-k(k+1)/2} \det(A(\,\alpha\,|\,\gamma\,))$$

$$= \sum_{\gamma \in Q_{k,n}} (-1)^{s(\alpha)+s(\gamma)} \det(A[\,\alpha\,|\,\gamma\,]) \, \det(A(\,\alpha\,|\,\gamma\,)).$$

The latter formula follows because $k(k+1)$ is always even. ∎

Observe that if $k = 1$ and α is a single integer, then (6) collapses to

$$\det(A) = \sum_{\beta=1}^{n} (-1)^{\alpha+\beta} a_{\alpha\beta} \det(A(\,\alpha\,|\,\beta\,)). \qquad (14)$$

This is called the *expansion by row* α.

Since $\det(A) = \det(A^T)$ we can apply the Laplace Expansion Theorem to A^T and obtain an analogous theorem on expansion by columns. To do this we observe first that

$$(A^T[\,\alpha\,|\,\beta\,])_{ij} = (A^T)_{\alpha_i \beta_j} = A_{\beta_j, \alpha_i}.$$

Also, the (i, j) entry of $(A[\,\beta\,|\,\alpha\,])^T$ is the (j, i) entry of $A[\,\beta\,|\,\alpha\,]$, namely A_{β_j, α_i}. Hence

$$A^T[\,\alpha\,|\,\beta\,] = (A[\,\beta\,|\,\alpha\,])^T; \quad \det(A^T[\,\alpha\,|\,\beta\,]) = \det(A[\,\beta\,|\,\alpha\,]). \qquad (15)$$

Also, the (i, j) entry of $A^T(\,\alpha\,|\,\beta\,)$ is the α'_i, β'_j entry of A^T, where α' is the ordered complementary sequence to α in $\{1, ..., n\}$. Thus

$$(A^T(\,\alpha\,|\,\beta\,))_{i,j} = (A^T)_{\alpha'_i, \beta'_j} = A_{\beta'_j, \alpha'_i}.$$

The (i, j) entry of $(A(\,\beta\,|\,\alpha\,))^T$ is the (j, i) entry of $A(\,\beta\,|\,\alpha\,)$, i.e., $A_{\beta'_j, \alpha'_i}$. Hence

$$A^T(\,\alpha\,|\,\beta\,) = (A(\,\beta\,|\,\alpha\,))^T; \quad \det(A^T(\,\alpha\,|\,\beta\,)) = \det(A(\,\beta\,|\,\alpha\,)).$$

We can now apply the Laplace Expansion Theorem to A^T to obtain

$$\det(A) = \det(A^T) = \sum_{\gamma \in Q_{k,n}} (-1)^{s(\alpha)+s(\gamma)} \det(A^T[\,\alpha\,|\,\gamma\,]) \det(A^T(\,\alpha\,|\,\gamma\,))$$

$$= \sum_{\gamma \in Q_{k,n}} (-1)^{s(\alpha)+s(\gamma)} \det(A[\,\gamma\,|\,\alpha\,]) \det(A(\,\gamma\,|\,\alpha\,)). \tag{16}$$

An expansion by a column, which parallels the expansion (14) by a row, should be evident.

As an example of the use of (16) we compute $\det(A)$ where

$$A = \begin{bmatrix} 0 & 1 & 2 & 3 \\ 1 & 0 & -1 & 4 \\ 0 & 0 & 7 & 5 \\ -1 & 0 & 2 & 8 \end{bmatrix}.$$

It is appropriate to use the Laplace expansion on the first two columns. Thus let $\alpha = (1, 2)$, so that $(-1)^{s(\alpha)} = -1$. Clearly, if 3 appears in γ then $\det(A[\,\gamma\,|\,1,2\,]) = 0$. The only $\gamma \in Q_{2,4}$ that survive are $(1, 2), (1, 4), (2, 4)$. Thus (16) becomes

$$\det(A) = -[(-1)^3(-1)(46) + (-1)^5(1)(-33) + (-1)^6(0)(-11)] = -79.$$

Let A be $n \times n$. Define an $n \times n$ matrix B by

$$b_{ij} = (-1)^{i+j} \det(A(\,j\,|\,i\,)), \quad i, j = 1, \ldots, n. \tag{17}$$

Then B is called the *adjugate* of A and is denoted by $B = \text{adj } A$ or sometimes, using parentheses, by $B = \text{adj}(A)$.

6.2 Two Classical Determinant Theorems

Theorem 4.

If $A \in M_n(R)$ then

$$A(\text{adj } A) = (\text{adj } A)A = \det(A) I_n. \qquad (18)$$

If A is non-singular, then

$$A^{-1} = [\det(A)]^{-1} (\text{adj } A). \qquad (19)$$

Proof.

Let $B = \text{adj } A$ so that the (i, j) entry of AB is

$$\sum_{t=1}^{n} a_{it} b_{tj} = \sum_{t=1}^{n} a_{it}(-1)^{t+j} \det(A(j|t)).$$

If $i = j$ then this last summation becomes

$$\sum_{t=1}^{n} (-1)^{t+i} a_{it} \det(A(i|t))$$

which, by (14), is precisely the expansion of the determinant of A using row i. It remains to show that, for $i \ne j$, $(AB)_{ij} = 0$. Suppose then that $i \ne j$ and consider the matrix C which agrees with A except that the jth row of C is the ith row of A. It is clear that

$$A(j|t) = C(j|t), \quad t = 1, \ldots, n.$$

Also C has two rows the same (both row i and row j of C are $A_{(i)}$) and hence $\det(C) = 0$. Thus expanding C by its jth row we have

$$0 = \det(C) = \sum_{t=1}^{n} c_{jt}(-1)^{j+t} \det(C(j|t))$$

$$= \sum_{t=1}^{n} a_{it}(-1)^{j+t} \det(A(j|t))$$

$$= \sum_{t=1}^{n} a_{it} b_{tj}$$

$$= (AB)_{ij}.$$

Hence $AB = \det(A)I_n$. A similar argument will show that $BA = \det(A)I_n$. If $\det(A) \neq 0$ then, by §6.1 Theorem 3, A has an inverse and since the inverse is unique,

$$A^{-1} = [\det(A)]^{-1} (\text{adj } A). \qquad \blacksquare$$

Theorem 5 (Cramer's Rule).

Let A be an $n \times n$ nonsingular matrix and suppose that x and b are $n \times 1$ matrices. If

$$Ax = b,$$

i.e.,

$$\sum_{t=1}^{n} a_{it} x_t = b_i, \quad i = 1, \ldots, n,$$

then

$$x_t = (\det(A))^{-1} \det(A^{(1)}, \ldots, A^{(t-1)}, b, A^{(t+1)}, \ldots, A^{(n)}), \quad t = 1, \ldots, n. \qquad (20)$$

Proof.

Since A is nonsingular, $\det(A) \neq 0$ and $A^{-1} = [\det(A)]^{-1} (\text{adj } A)$. Thus

6.2 Two Classical Determinant Theorems

$$x = A^{-1}b = (\det(A))^{-1}(\text{adj } A)b.$$

The tth entry in $x = [x_1, \ldots, x_n]^T$ is given by

$$x_t = (\det(A))^{-1} (\text{adj } A)_{(t)} b$$

$$= (\det(A))^{-1} \sum_{k=1}^{n} (\text{adj } A)_{tk} b_k$$

$$= (\det(A))^{-1} \sum_{k=1}^{n} (-1)^{t+k} \det(A(k \mid t)) b_k.$$

The summation in the last expression is just the Laplace expansion by column t of

$$\det(A^{(1)}, \ldots, A^{(t-1)}, b, A^{(t+1)}, \ldots, A^{(n)})$$

and the result follows. ∎

We observe here that Theorem 5 can be used to compute particular columns of A^{-1}. For, we know that if we write

$$Ax = e_j = [0, \ldots, 0, 1, 0, \ldots, 0]^T$$

then $x = A^{-1} e_j = (A^{-1})^{(j)}$. But (20) tells us that

$$x_t = (\det(A))^{-1} \det(A^{(1)}, \ldots, A^{(t-1)}, e_j, A^{(t+1)}, \ldots, A^{(n)})$$

$$= (-1)^{j+t} (\det(A))^{-1} \det(A(j \mid t)), \quad t = 1, \ldots, n.$$

In other words, the jth column of A^{-1} is given by

$$(A^{-1})^{(j)} = (\det(A))^{-1} [(-1)^{j+1} \det(A(j \mid 1)), \ldots, (-1)^{j+n} \det(A(j \mid n))]^T. \quad (21)$$

For example, if

$$A = \begin{bmatrix} 1 & 1 & 0 \\ 0 & 1 & -1 \\ 1 & 0 & 3 \end{bmatrix}$$

then the first column of A^{-1} is

$$(\det(A))^{-1} \left((-1)^2 \det \begin{bmatrix} 1 & -1 \\ 0 & 3 \end{bmatrix}, \ (-1)^3 \det \begin{bmatrix} 0 & -1 \\ 1 & 3 \end{bmatrix}, \ (-1)^4 \det \begin{bmatrix} 0 & 1 \\ 1 & 0 \end{bmatrix} \right)^T$$

$$= \frac{1}{2} [\, 3 \ -1 \ -1 \,]^T.$$

The determinant can also be used to evaluate the rank of a matrix.

Theorem 6.

If A is m × n and $\rho(A) = k$, then there exists a k-square submatrix of A whose determinant is nonzero. Moreover, if k < min(m, n) then every (k + t)-square submatrix of A is singular, k < k + t ≤ min(m, n). Thus $\rho(A)$ is the dimension of the largest nonzero subdeterminant of A.

Proof.

Let P and Q be nonsingular such that

$$PAQ = \begin{bmatrix} I_k & 0 \\ 0 & 0 \end{bmatrix} = H$$

is in canonical form. Then, by the Cauchy-Binet theorem,

$$1 = \det(H[\, 1, \ldots, k \mid 1, \ldots, k \,])$$
$$= \det((PAQ)[\, 1, \ldots, k \mid 1, \ldots, k \,])$$
$$= \sum_{\omega, \gamma \in Q_{k,n}} \det(P[\, 1, \ldots, k \mid \omega \,]) \det(A[\, \omega \mid \gamma \,]) \det(Q[\, \gamma \mid 1, \ldots, k \,]).$$

Hence some k-square subdeterminant of A must be nonzero. Since $A = P^{-1}HQ^{-1}$, a similar application of the Cauchy-Binet Theorem shows that every $(k + t)$-square subdeterminant of A is 0. ∎

6.2 Exercises

1. Prove: if A and B are m × n then $\rho(A + B) \leq \rho(A) + \rho(B)$.

   Hint: Let $\rho(A) = r$ and $\rho(B) = s$. If $s = 1$, use §4.3 Exercises, #24, to assume B is in canonical form, i.e., $\rho(A + B) = \rho(P(A + B)Q) = \rho(PAQ + PBQ)$. Apply §6.2 Theorems 3 and 6 to see that every $(r + 2)$-square subdeterminant of $A + B$ is 0 and hence that $\rho(A + B) \leq r + 1$. Then use §4.3 Exercises, #31, and induction on s to complete the argument.

2. Let A be an n × n matrix with integer entries. Prove that A^{-1} has integer entries iff $\det(A) = \pm 1$.

3. Calculate the determinants of the following matrices using the Laplace Expansion Theorem.

   (a) $\begin{bmatrix} 1 & 0 & 2 & 3 \\ 0 & 0 & 5 & 8 \\ 0 & 0 & -1 & 7 \\ 0 & 0 & 6 & 2 \end{bmatrix}$

   (b) $\begin{bmatrix} 1 & 0 & -1 & 0 & 0 \\ 0 & 0 & 1 & -1 & 0 \\ 0 & 0 & 0 & 0 & 1 \\ 1 & 0 & 0 & 0 & 1 \\ 2 & 1 & 3 & -1 & 2 \end{bmatrix}$

   (c) $\begin{bmatrix} 1 & 2 & 3 \\ 0 & 1 & 2 \\ 0 & 1 & -1 \end{bmatrix}$

   (d) $\begin{bmatrix} -1 & 0 & 1 \\ 2 & 1 & 2 \\ 3 & -1 & 5 \end{bmatrix}$

4. Let $A \in M_{m,n}(\mathbb{R})$. Prove: $\det(AA^T) \geq 0$.

5. Prove: If $A \in M_n(\mathbb{C})$ then

$$\det\left(\begin{bmatrix} 0 & A \\ A^* & 0 \end{bmatrix}\right) = (-1)^n |\det(A)|^2.$$

6. Let

$$M = \begin{bmatrix} A & B \\ C & D \end{bmatrix}$$

where A is p-square and D is q-square and A is nonsingular. Prove:

$$\det(M) = \det(A)\det(D - CA^{-1}B)$$

by showing that

$$M = \begin{bmatrix} I_p & 0 \\ L & R \end{bmatrix}\begin{bmatrix} A & B \\ 0 & I_q \end{bmatrix}, \quad L = CA^{-1}, \quad R = D - CA^{-1}B.$$

7. Let

$$A = \begin{bmatrix} B & y \\ x & \alpha \end{bmatrix}$$

where B is an $(n-1)$-square nonsingular matrix, x is $1 \times (n-1)$, y is $(n-1) \times 1$ and α is a number. Assume that $B = L_1 R_1$ where $\det(L_1) = 1$, $\det(R_1) \neq 0$, L_1 is lower triangular and R_1 is upper triangular. Show that if $\det(A) \neq 0$ then $A = LR$ where L and R are n-square and satisfy the same conditions as L_1 and R_1, namely $\det(L) = 1$, $\det(R) \neq 0$, L is lower triangular, and R is upper triangular.

Hint: Write

$$L = \begin{bmatrix} L_1 & 0 \\ u & 1 \end{bmatrix}, \quad R = \begin{bmatrix} R_1 & v \\ 0 & q \end{bmatrix}$$

and then determine u, $1 \times (n-1)$, and v, $(n-1) \times 1$, and the scalar q. By block multiplication, the equation $LR = A$ becomes $L_1 R_1 = B$, $L_1 v = y$, $uR_1 = x$, $uv + q = \alpha$. Then define $v = L_1^{-1} y$, $u = xR_1^{-1}$. Finally, define $q = \alpha - uv$.

6.2 Exercises

8. Let A be an n-square matrix. Assume $\det(A[\,1, ..., k \mid 1, ..., k\,]) \neq 0$, $k = 1, ..., n$.

   (a) Prove: $A = LR$ where L is n-square lower triangular, R is n-square upper triangular, $\det(L) = 1$, and $\det(R) \neq 0$.

   (b) Note that in (a) we can assume that all the main diagonal entries of L are 1, e.g., if

   $$A = \begin{bmatrix} 3 & 0 \\ 5 & 2 \end{bmatrix} \begin{bmatrix} 8 & 2 \\ 0 & -1 \end{bmatrix}$$

   then

   $$A = \begin{bmatrix} 1 & 0 \\ 5/3 & 1 \end{bmatrix} \begin{bmatrix} 3 & 0 \\ 0 & 2 \end{bmatrix} \begin{bmatrix} 8 & 2 \\ 0 & -1 \end{bmatrix}$$

   $$= \begin{bmatrix} 1 & 0 \\ 5/3 & 1 \end{bmatrix} \begin{bmatrix} 24 & 6 \\ 0 & -2 \end{bmatrix}.$$

   Show that if it is assumed that all main diagonal entries of L are 1, then L and R are uniquely determined.

Hint:

(a) Since $a_{11} \neq 0$, $a_{11} = 1 \cdot a_{11}$ and thus a 1×1 matrix can be so expressed. The matrix $A(\,n \mid n\,)$ satisfies the induction hypothesis, i.e.,

$$\det(A(\,n \mid n\,)[\,1 ... k \mid 1 ... k\,]) = \det(A[\,1 ... k \mid 1 ... k\,]) \neq 0.$$

Hence by #7, if $A(\,n \mid n\,)$ is bordered in any way, so that the resulting matrix is nonsingular, then it too can be factored as required.

(b) Assume that $A = LR = TU$ in which L and T are lower triangular, R and U are upper triangular. Moreover, assume that all the main diagonal entries of L and T are 1. Then

$$T^{-1}L = UR^{-1},$$

so that $T^{-1}L$ must be upper triangular. But T^{-1} is lower triangular (why?) and moreover has 1's along the main diagonal (why?). Thus $T^{-1}L$ has 1's along its main diagonal (why?) and is upper and lower triangular (i.e., UR^{-1} is upper triangular). Thus $T^{-1}L = I_n$, $L = T$ and finally $U = R$. The decomposition $A = LR$ is sometimes called the *LR factorization* of A (see #17 below). In MatLab it is called the *LU factorization*.

9. Let $A \in M_n(\mathbb{C})$ be hermitian. If for all $0 \neq x \in M_{n,1}(\mathbb{C})$ the inequality $x^*Ax \geq 0$, then A is *nonnegative hermitian* (n.n.h.), sometimes referred to as *positive semi-definite hermitian*. If $x^*Ax > 0$ for all $x \neq 0$, then A is *positive definite hermitian* (p.d.h.). Prove: if

$$A = \begin{bmatrix} H & u \\ u^* & 0 \end{bmatrix}$$

is n.n.h., $u \in M_{n-1,1}(\mathbb{C})$, then $u = 0$. If A is n.n.h. (p.d.h.) we frequently write $A \geq 0$ ($A > 0$).

Hint: Let $x = \begin{bmatrix} z \\ r \end{bmatrix}$ where $z \in M_{n-1,1}(\mathbb{C})$ and $r \in \mathbb{R}$. Then by block multiplication, $x^*Ax = z^*Hz + 2\text{Re}(r\, z^*u)$. First note that $z^*Hz \geq 0$ for any z since for $r = 0$ this is x^*Ax, and A is n.n.h. Now, if $u \neq 0$ there is a z such that $z^*u = 1$. Then choose r sufficiently negative so that $x^*Ax < 0$, a contradiction.

10. Prove: If A is n.n.h. then there exists U, upper triangular, such that $A = UU^*$. Also prove that there exists a lower triangular matrix L such that $A = LL^*$. This result is called the *Cholesky factorization theorem*.

Hint: If A is 1×1 then $A = [\alpha^2]$ so that we can choose $L = [\alpha]$. Assume the result is correct for (n - 1)-square matrices. Let A be n-square n.n.h. If $a_{nn} = 0$ then A has the form indicated in the statement of #9 and an induction applied to H easily completes the proof. Since $e_n^*Ae_n = a_{nn}$, it follows that if $a_{nn} \neq 0$ then $a_{nn} > 0$. (Recall that e_n is the $n \times 1$ matrix whose only non-zero entry is a 1 in

6.2 Exercises

the (n, 1) position.) Write A in the form

$$A = \begin{bmatrix} H & \alpha u \\ \alpha u^* & \alpha^2 \end{bmatrix}$$

where $\alpha > 0$. Now let

$$\Delta = H - uu^*,$$

an $(n-1)$-square hermitian matrix. If $y \in M_{n-1,1}(\mathbb{C})$ then define $x \in M_{n,1}(\mathbb{C})$ by

$$x = \begin{bmatrix} y \\ \beta \end{bmatrix}$$

where $\beta = -\alpha^{-1}u^*y$. Then we compute that

$$x^*Ax = \begin{bmatrix} y^* & \bar{\beta} \end{bmatrix} \begin{bmatrix} H & \alpha u \\ \alpha u^* & \alpha^2 \end{bmatrix} \begin{bmatrix} y \\ \beta \end{bmatrix}$$

$$= \begin{bmatrix} y^* & \bar{\beta} \end{bmatrix} \begin{bmatrix} Hy + \alpha\beta u \\ \alpha u^*y + \alpha^2\beta \end{bmatrix}$$

$$= y^*Hy + \alpha\beta y^*u + \alpha\bar{\beta}u^*y + \alpha^2\bar{\beta}\beta$$

$$= y^*Hy - y^*uu^*y - y^*uu^*y + \alpha^2(-\alpha^{-1}y^*u)(-\alpha^{-1}u^*y)$$

$$= y^*Hy - y^*(uu^*)y - y^*(uu^*)y + y^*(uu^*)y$$

$$= y^*(H - uu^*)y.$$

Since $x^*Ax \geq 0$ and y is arbitrary it follows that $\Delta = H - uu^* \geq 0$. Next, note that

$$\begin{bmatrix} I_{n-1} & u \\ 0 & \alpha \end{bmatrix} \begin{bmatrix} \Delta & 0 \\ 0 & 1 \end{bmatrix} \begin{bmatrix} I_{n-1} & 0 \\ u^* & \alpha \end{bmatrix} = \begin{bmatrix} \Delta & u \\ 0 & \alpha \end{bmatrix} \begin{bmatrix} I_{n-1} & 0 \\ u^* & \alpha \end{bmatrix}$$

$$= \begin{bmatrix} \Delta + uu^* & \alpha u \\ \alpha u^* & \alpha^2 \end{bmatrix}$$

$$= \begin{bmatrix} H & \alpha u \\ \alpha u^* & \alpha^2 \end{bmatrix}$$

$$= A.$$

Applying the induction to Δ we can write

$$\Delta = U_1 U_1^*$$

where U_1 is upper triangular. Then from the above computation

$$A = V(U_1 \oplus 1)(U_1^* \oplus 1)V^*$$

where

$$V = \begin{bmatrix} I_{n-1} & u \\ 0 & \alpha \end{bmatrix}.$$

But $U = V(U_1 \oplus 1)$ is upper triangular and hence $A = UU^*$. The same kind of inductive argument can be used to show that A can be expressed as $A = LL^*$, L lower triangular.

11. Prove: if A is n.n.h. then $\det(A) \geq 0$.

12. If A and B are n-square n.n.h., prove that $\det(A + B) \geq \det(A) + \det(B)$.

    Hint: By #10 write $A = P^*P$ and $B = Q^*Q$. Then

    $$A + B = [P^* \quad Q^*] \begin{bmatrix} P \\ Q \end{bmatrix} = Z^*Z$$

where $Z = \begin{bmatrix} P \\ Q \end{bmatrix}$ is $2n \times n$. By the Cauchy-Binet Theorem,

$$\det(A+B) = \sum_{\omega \in Q_{n,2n}} \det(Z^*[1\ldots n \mid \omega]) \det(Z[\omega \mid 1\ldots n])$$

$$= \sum_{\omega \in Q_{n,2n}} \overline{\det(Z^T[1\ldots n \mid \omega])} \det(Z[\omega \mid 1\ldots n])$$

$$= \sum_{\omega \in Q_{n,2n}} |\det(Z[\omega \mid 1\ldots n])|^2$$

$$= |\det(P)|^2 + |\det(Q)|^2 + \text{nonnegative terms}$$

$$= \det(A) + \det(B) + \text{nonnegative terms}.$$

13. Assume that A is n.n.h. and that $A = LL^*$ as in #10. If

$$A = \begin{bmatrix} A_{11} & A_{12} \\ A_{21} & A_{22} \end{bmatrix}$$

where A_{11} and A_{22} are square, then partition L conformally and find formulas for A_{11} and A_{22} in terms of the blocks in L.

Hint: $A_{11} = L_{11}L_{11}^*$, $A_{22} = L_{22}L_{22}^* + L_{21}L_{21}^*$.

14. In the notation of #13, prove that $\det(A) \leq \det(A_{11}) \det(A_{22})$.

Hint: From the Hint for #13,

$$\det(A) = \det(LL^*) = \det(L)\det(L^*)$$

$$= |\det(L)|^2$$

$$= |\det(L_{11})|^2 |\det(L_{22})|^2$$

$$= \det(L_{11}L_{11}^*)\det(L_{22}L_{22}^*)$$

$$= \det(A_{11})\det(L_{22}L_{22}^*).$$

But

$$A_{22} = L_{22}L_{22}^* + L_{21}L_{21}^*$$

and both summands are n.n.h., i.e., any matrix of the form XX^* is n.n.h. (why?). Apply #12.

15. Prove: if A is n-square n.n.h., then

$$\det(A) \leq \prod_{i=1}^{n} a_{ii}.$$

This result is called the *Hadamard determinant inequality*.

16. Let $A \in M_n(R)$ and let α and β be fixed sequences in $Q_{k,n}$, $1 \leq k \leq n$. Prove that

$$\sum_{\gamma \in Q_{k,n}} (-1)^{s(\alpha) + s(\gamma)} \det(A[\alpha | \gamma]) \det(A(\beta | \gamma)) = \delta_{\alpha\beta} \det(A).$$

Also prove that

$$\sum_{\gamma \in Q_{k,n}} (-1)^{s(\alpha) + s(\gamma)} \det(A[\gamma | \alpha]) \det(A(\gamma | \beta)) = \delta_{\alpha\beta} \det(A).$$

17. Let A be an n-square nonsingular matrix. Prove that there exists an n-square permutation matrix P such that $B = PA$ satisfies

6.2 Exercises

$$\prod_{k=1}^{n} \det(B[\,1\ 2\ldots k\mid 1\ 2\ldots k\,]) \ne 0.$$

Hint: Consider the Laplace expansion of A by column n:

$$\det(A) = \sum_{i=1}^{n} (-1)^{i+n} a_{in} \det(A(\,i\mid n\,)).$$

Obviously, since $\det(A) \ne 0$, there exists an i_0 such that

$$a_{i_0 n}\det(A(\,i_0\mid n\,)) \ne 0.$$

Now let Q be a permutation matrix such that the first $n-1$ rows of $C = QA$ are rows $A_{(i)}$ of A, $i \ne i_0$, and row n of C is row i_0 of A. Then, clearly,

$$C[\,1\ldots n{-}1\mid 1\ldots n{-}1\,] = A(\,i_0\mid n\,)$$

and

$$c_{nn} = a_{i_0 n}.$$

Let $\Delta = C[1\ldots n{-}1\mid 1\ldots n{-}1\,]$. Since $A(\,i_0\mid n\,)$ is nonsingular, use induction on n to obtain an $(n-1)$-square permutation matrix S such

$$\prod_{k=1}^{n-1} \det((S\Delta)[1\ldots k\mid 1\ldots k\,]) \ne 0, \quad k = 1,\ldots, n{-}1.$$

Let $B = (S \oplus 1)QA = (S \oplus 1)\,C$. Since

$$B[\,1\ldots k\mid 1\ldots k\,] = (S\Delta)[\,1\ldots k\mid 1\ldots k\,], \quad k = 1,\ldots, n-1,$$

and $\det(B) = \det((S \oplus 1)QA) = \pm\det(A)$, the proof is complete upon noting that $P = (S \oplus 1)Q$ is a permutation matrix.

18. Let A be an n-square nonsingular matrix. Prove that there exists an n-square permutation matrix P such that $PA = LR$ where L is n-square lower triangular, R is n-square upper triangular, $\det(L) = 1$, and $\det(R) \ne 0$.

Hint: Use #17 to obtain the required permutation matrix P such that

$$\prod_{k=1}^{n} \det((PA)[1\cdots k \mid 1 \cdots k]) \neq 0.$$

Apply #8.

19. Let A be a square matrix. Prove:

$$\det(I_n \otimes A) = (\det(A))^n$$

and

$$\det(A \otimes I_n) = (\det(A))^n.$$

Hint: Write $I_n \otimes A$ as $A \oplus (I_{n-1} \otimes A)$. Then by Laplace expansion, $\det(I_n \otimes A) = \det(A) \det(I_{n-1} \otimes A)$. Use induction.

To prove that $\det(A \otimes I_n) = \det(A)^n$ first note that if A is singular we know that $A \otimes I_n$ is singular and both sides are 0. Otherwise, A is a product of elementary matrices $A = E_1 \cdots E_p$. Then

$$A \otimes I_n = (E_1 \otimes I_n) \cdots (E_p \otimes I_n)$$

so that

$$\det(A \otimes I_n) = \det(E_1 \otimes I_n) \cdots \det(E_p \otimes I_n).$$

If the result holds for an elementary matrix, then
$$\det(A \otimes I_n) = \det(E_1)^n \cdots \det(E_p)^n$$

$$= (\det(E_1 \cdots E_p))^n$$

$$= (\det(A))^n.$$

6.2 Exercises

Thus we need only confirm the result for elementary matrices. Moreover, since any row interchange can be accomplished by a sequence of adjacent interchanges, it follows that type I elementary matrices E_j can be assumed to be adjacent row interchanges. For simplicity, take E to correspond to an interchange of rows 1 and 2. Then

$$E \otimes I_n = \begin{bmatrix} 0 & I_n & 0 & \cdots & 0 \\ I_n & 0 & 0 & \cdots & 0 \\ 0 & 0 & I_n & \cdots & 0 \\ \vdots & & & \ddots & \vdots \\ 0 & 0 & \cdots & 0 & I_n \end{bmatrix}.$$

By #5,

$$\det(E \otimes I_n) = \det\left(\begin{bmatrix} 0 & I_n \\ I_n & 0 \end{bmatrix}\right) = (-1)^n = (\det(E))^n.$$

If E is of type II then $E \otimes I_n$ is lower triangular with 1's along the main diagonal. Hence

$$\det(E \otimes I_n) = 1 = 1^n = (\det(E))^n.$$

If E is type III then

$$E \otimes I_n = \begin{bmatrix} I_n & 0 & & \cdots & & 0 \\ & I_n & & \cdots & & 0 \\ \vdots & & cI_n & & & \\ & & & & \ddots & \\ 0 & & \cdots & & 0 & I_n \end{bmatrix}$$

and

$$\det(E \otimes I_n) = c^n = (\det(E))^n.$$

20. Prove: if A and B are m-square and n-square, respectively, then

$$\det(A \otimes B) = (\det(A))^n (\det(B))^m.$$

Hint: Use $A \otimes B = (I_m \otimes B)(A \otimes I_n)$ and #19.

6.2 MatLab

1. Write a MatLab function called cbin that

    - accepts $A \in M_{p,q}(\mathbb{C})$, $B \in M_{q,r}(\mathbb{C})$, $a \in Q_{m,p}$, $b \in Q_{m,r}$
    - returns $\det((AB)[a \mid b])$ as the value of cbin(A, B, a, b) using Theorem 1, the Cauchy-Binet Theorem.

    Save cbin in determ.

    Hint:

    ```
    function s = cbin(A, B, a, b);
    %CBIN uses Cauchy-Binet to compute the
    %determinant of the submatrix of AB lying
    %in rows a and columns b.
    %A is pxq, B is qxr, a is in Qmp and b
    %is in Qmr.
    [p q]=size(A);
    m=length(a);
    C=cmb(m,q);
    cnr=size(C);
    add=0;
    for k=1:cnr(1)
            add=add + det(A (a,C(k,:)))*det(B(C(k,:),b));
    end
    s=add;
    ```

2. Use cbin in #1 to compute det(AB) where A=rand(3,5), B=rand(5,3). Compare the value obtained with det(A*B). What is the value of det(B*A)? Why?

6.2 MatLab

3. Write a MatLab function called cmpl that accepts a positive integer n and a sequence $a \in Q_{k,n}$ ($k < n$) and returns the complementary sequence to a, $a' \in Q_{n-k,n}$. Save cmpl in determ.

   Hint:
   > function b=cmpl(a,n)
   > %CMPL accepts the positive integer n > 1
   > %and the sequence a in $Q_{k,n}$, k < n, and returns
   > %the complementary sequence to a in $Q_{n-k,n}$.
   > c=1:n;
   > c(a)=[ ];
   > b=c;

4. Write a MatLab function called lapl that

   - accepts $A \in M_n(\mathbb{C})$ and $a \in Q_{k,n}$
   - returns det(A) as the value of lapl(A,a) using Theorem 3, the Laplace Expansion Theorem by rows a.

   Save lapl in determ.

   Hint:
   > function f=lapl(A,a)
   > %LAPL computes det(A) using Laplace
   > %expansion by rows a in Qkn.
   > n=length(A);
   > r=length(a);
   > C=cmb(r,n);
   > cnr=size(C);
   > add=0;
   > for k=1:cnr(1)
   >   w=C(k,:);
   >   add=add+((-1)^sum(w))*det(A(a,w))*det(A(cmpl(a,n), cmpl(w,n)));
   > end
   > f=((-1)^(sum(a)))*add;

5. Write a MatLab script called posidet that

- calls for the input of two positive integers m and n, m < n
- generates 20 random $A \in M_{m,n}(\mathbb{C})$ and tabulates $\det(AA^*)$ and $\det(A^*A)$.

On the basis of the output it appears that $\det(AA^*) > 0$ and $\det(A^*A) = 0$ for any such A. Prove this is always the case.

Hint: From §4.3 Exercises, #35, we know that $\rho(A^*A) = \rho(A) \leq \min\{m, n\} = m$. But A^*A is n-square and $n > m$. Hence by Theorem 6, $\det(A^*A) = 0$. For a random $A \in M_{m,n}(\mathbb{C})$ it is nearly inevitable that $\rho(A) = m$ (why?) so that $\rho(AA^*) = m$. But then $\det(AA^*) \neq 0$ follows from §6.1 Theorem 3. If $x \in M_{m,1}(\mathbb{C})$ then $x^*AA^*x = (A^*x)^*(A^*x) \geq 0$, so $AA^* \geq 0$. Hence by §6.2 Exercises, #11, it follows that $\det(AA^*) \geq 0$.

```
%POSIDET calls for two positive integers m and n,
%m < n. The program then generates 20 random
%m x n complex matrices and tabulates det(A*A') and det(A'*A).
m=input('Enter m: ');
n=input('Enter n, n > m: ');
D=[ ];
for k=1:20
    A=(2*rand(m,n)-1) + i*2*(rand(m,n)-1);
    D=[D; det(A*A') det(A'*A)];
end
disp('     det(AA")                    det(A"A)         ');
disp('===========================================');
disp(D);
```

6. In §6.2 Exercises, #18 we saw that if A is nonsingular, then a permutation matrix P exists such that PA = LU (we use "U" rather than "R" from now on in order to conform with the choice made in MatLab). Moreover, as we saw in §6.2 Exercises, #8, if the main diagonal elements of L are 1 then L and U are uniquely determined. Read the Help entries for **lu**, **det**, and **inv** and note that there are two versions of the **lu** function:

6.2 MatLab

$$[L, U] = lu(A)$$

and

$$[L, U, P] = lu(A).$$

In the first version, instead of $PA = LU$, the function absorbs P^T into L so that $A = (P^TL)U$ and the "permuted" P^TL is actually stored in L. The second version delivers a legitimate lower triangular L, an upper triangular U and a permutation matrix P for which $PA = LU$. The LU decomposition is computed using the same elementary operations used in computing the Hermite normal form of A (see §4.2 MatLab, #5, #6). As the MatLab documentation points out, if $A = LU$ then $det(A) = det(L)det(U) = \prod_{i=1}^{n} L_{ii} \prod_{i=1}^{n} U_{ii}$ is simple to compute. Also, the system of linear equations $Ax = b$ is computed in MatLab by solving the two triangular systems $Ly = b$ and $Ux = y$. In fact, the command $x = A\backslash b$ does precisely this. Moreover A^{-1} is computed in MatLab as $A^{-1} = (LU)^{-1} = U^{-1}L^{-1} = inv(U) * inv(L)$, i.e., the inverse of a triangular matrix is relatively easier to compute.

For the following A use MatLab to compute lu(A) using both versions of the function. Confirm directly for $[L, U] = lu(A)$ that the output satisfies $A = L*U$. Also confirm for $[L, U, P] = lu(A)$, that $L*U = P*A$.

(a) $A = \begin{bmatrix} 3 & 1 & -2 & -1 \\ 1 & 5 & -4 & -1 \\ 3 & 1 & 2 & 3 \\ 2 & -2 & 2 & 3 \end{bmatrix}$

(b) $A = \begin{bmatrix} 3 & 2 & 5 & 4 \\ 2 & 3 & 6 & 8 \\ 1 & -6 & -9 & -20 \\ 4 & 1 & 4 & 1 \end{bmatrix}$

(c) $A = \begin{bmatrix} 4 & 2 & 4 & 1 \\ 30 & 20 & 45 & 12 \\ 20 & 15 & 36 & 10 \\ 35 & 28 & 70 & 20 \end{bmatrix}$

(d) $A = \begin{bmatrix} 5 & 5 & 5 & 5 & 3 \\ 30 & 40 & 45 & 48 & 30 \\ 70 & 105 & 125 & 140 & 90 \\ 70 & 112 & 140 & 160 & 105 \\ 42 & 70 & 90 & 105 & 70 \end{bmatrix}$

(e) $A = \begin{bmatrix} 0 & 1 \\ 1 & 0 \end{bmatrix}$

7. Is it possible to write

$$A = \begin{bmatrix} 0 & 1 \\ 1 & 0 \end{bmatrix}$$

as

$$A = LU$$

where L is lower triangular and U is upper triangular?

Hint: Suppose $A = \begin{bmatrix} s & 0 \\ r & t \end{bmatrix} \begin{bmatrix} u & w \\ 0 & v \end{bmatrix}$. Then su = 0 and sw = 1. Clearly $s \neq 0$. Hence u = 0 because su = 0. Thus the second factor on the right is singular, whereas A is non-singular.

8. Review §6.2 Exercises, #15. Write a script named hadamard that

- generates 20 random positive integers $n \leq 5$ and 20 corresponding n-square positive semidefinite hermitian matrices A
- prints out a table of values listing det(A) and $\prod_{i=1}^{n} a_{ii}$.

Save hadamard in determ.

Hint:
```
%HADAMARD
D=[ ];
for k=1:20
    n=floor(5*rand+1);
    X=(2*rand(n) - 1) + i*(2*rand(n)-1);
    A = X' * X;
    D = [D; det(A)  dprd(A, 1:n)];
end
disp('    det(A)       a(1,1)*...*a(n,n)   ');
disp('===============================');
disp(real(D));
```

9. Review §6.2 Exercises, #10. Read the Help entry for **chol**. Write a function named chl that accepts a nonnegative hermitian (n.n.h.) matrix A and returns a lower triangular matrix L for which $A = LL^*$. Save chl in determ.

Hint:
```
function L = chl(A)
%CHL accepts a n.n.h. matrix
%A and returns a l.t. matrix
%L for which A = LL*.
L = (chol(A.')).'
```

10. MatLab graphics allow for interesting depictions of 3-dimensional plots. Read the Help entries for the following items: **mesh; meshdom; title; label; text; grid; rot90**. Briefly, as the MatLab documentation indicates, if $A \in M_{m,n}(\mathbb{R})$ then mesh(A) produces a 3-dimensional mesh plot in which a_{ij} is plotted on the z-axis above (i, j) in the x,y-plane. Type each of the following at the Command prompt and observe the outputs in the Graph window.

(a)   mesh(eye(20)); title('Graph of eye(20)')

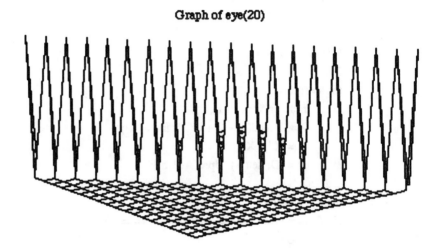

Fig. 1

(b)   mesh(rot90(eye(20)); title('Graph of rot90 (eye(20))')

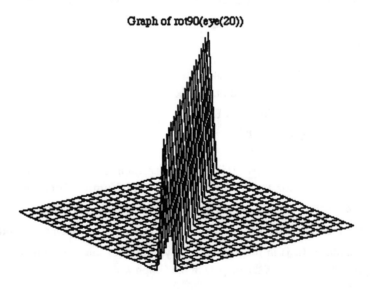

Fig. 2

(c)   X = 2*rand(20) - 1; A = X' * X; mesh(A); title('Graph of a n.n.h. A');

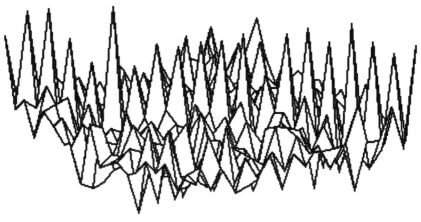

Graph of a n.n.h. A

Fig. 3

(d)   Repeat the execution of the plot in (c) several times. What appears to be the relationship between the main diagonal entries and the off-diagonal entries in a n.n.h. matrix?

Hint: The plots seem to indicate that the main diagonal entries are larger than the off-diagonal entries. Is this true for any n.n.h. matrix A? The answer to this question is 'no': consider

$$A = \begin{bmatrix} 6 & 2 \\ 2 & 1 \end{bmatrix}.$$

11. A general 3-dimensional plot of a function z = f(x,y) can also be produced in MatLab. The **meshdom** command is used for this purpose. If xx and yy are vectors then

$$[X, Y] = \text{meshdom}(xx, yy)$$

creates the cartesian product of xx and yy and makes it accessible for plotting purposes in the matrices X and Y. An example will make this clear. Enter the following script in an Edit window and then select Save and Go.

```
xx = -1:.05:1;
yy = -1:.05:1;
[X, Y] = meshdom(xx, yy);
Z=sqrt(1 - (X.^2 + Y.^2));
mesh(Z);
```

The result is a mesh plot of the upper half of a sphere of radius 1. In general, the xx and yy vectors must have their components in ascending order.

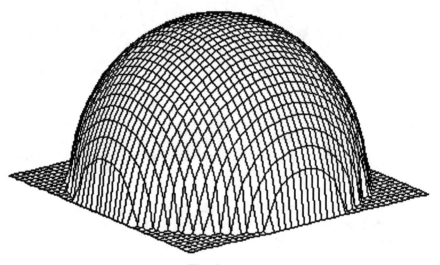

Fig. 4

To be more precise about how MatLab makes a 3-dimensional perspective plot of a function $z = f(x, y)$ defined over a rectangle R in the x,y-plane, here is what happens.

- Vectors xx and yy are defined that impose subdivisions on the x and y axes, respectively. Thus, xx and yy define an approximation to R, the rectangular domain of f.

- The command

$$[X, Y] = \text{meshdom}(xx, yy)$$

defines two matrices X and Y. Suppose xx is of length n and yy is of length m. Then each row of the matrix X is the vector xx, and the number of rows in X is the length of yy. Thus, X is m × n and the i, j entry is xx(j). Similarly, Y is a matrix each of whose columns is yy and the number of columns in Y is the length of xx. Hence Y is m × n and the i, j entry is yy(i).

- Then an m × n matrix Z is computed from the formula for f. For example, in

$$Z = \text{sqrt}(1-(X.^2 + Y.^2)),$$

the i, j entry is sqrt(1-(xx(j)^2 + yy(i)^2)).

- mesh(Z) then plots Z, replacing complex entries in Z by their real parts.

12. Show how to change the script in #11 so as to determine which direction in the plot corresponds to xx and which corresponds to yy.

    Hint: Change the size of the subdivisions, first in the xx direction, then in the yy direction and observe the effect on the plot corresponding to Fig. 4.

13. Plot the function z = sin(x) sin(y). Determine an appropriate rectangular region so that all the values that the function assumes are taken on, as (x,y) varies over the region.

    Hint:
    ```
    xx=-2*pi:.5:2*pi;
    yy=-2*pi:.1:2*pi;
    [X, Y]=meshdom(xx,yy);
    Z=sin(X).*sin(Y);
    mesh(Z);
    ```

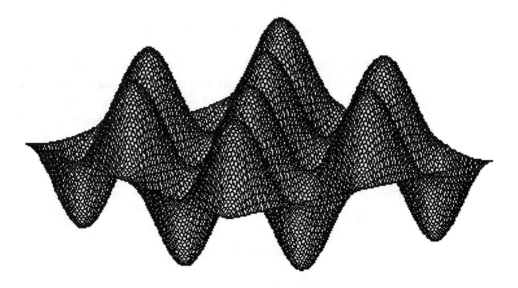

Fig. 5

6.2 Glossary

A > 0	316
A ≥ 0	316
adj	308
adjugate	308
Cauchy-Binet Theorem	300
chol	329
Cholesky factorization theorem	316
Cramer's Rule	310
det	326
expansion by column	307
expansion by row	307
grid	329
hadamard determinant inequality	320
inv	326
label	329

Laplace Expansion Theorem	303
LR factorization	316
LU factorization	316
lu	326
mesh	329
meshdom	329
n.n.h.	316
nonnegative hermitian	316
p.d.h.	316
positive definite hermitian	316
rot90	329
text	329
title	329

6.3 Compound Matrices

In this section we define a special type of matrix whose entries are subdeterminants of a given matrix. These so-called "compound" matrices have proven to be important and useful tools in applied and theoretical matrix theory.

Let $A \in M_{m,n}(R)$ and let k be a fixed integer, $1 \le k \le \min\{m, n\}$. For each $\alpha \in Q_{k,m}$ and $\beta \in Q_{k,n}$ let

$$c_{\alpha, \beta} = \det(A[\alpha \mid \beta]). \tag{1}$$

Construct an $\binom{m}{k} \times \binom{n}{k}$ matrix whose rows are ordered lexicographically according to the sequences $\alpha \in Q_{k,m}$, whose columns are ordered lexicographically according to the sequences $\beta \in Q_{k,n}$, and whose entry in row α column β is the subdeterminant (1). This matrix is called the kth *compound* of A and is denoted by

$$C_k(A). \tag{2}$$

Note that $C_k(A)$ is homogeneous of degree k in A. This means that for any scalar $r \in R$

$$C_k(rA) = r^k C_k(A).$$

For example, if

$$A = \begin{bmatrix} a_{11} & a_{12} & a_{13} \\ a_{21} & a_{22} & a_{23} \\ a_{31} & a_{32} & a_{33} \end{bmatrix}$$

and k = 2 then

$$C_2(A) = \begin{bmatrix} \det(A[\,1,2\,|\,1,2\,]) & \det(A[\,1,2\,|\,1,3\,]) & \det(A[\,1,2\,|\,2,3\,]) \\ \det(A[\,1,3\,|\,1,2\,]) & \det(A[\,1,3\,|\,1,3\,]) & \det(A[\,1,3\,|\,2,3\,]) \\ \det(A[\,2,3\,|\,1,2\,]) & \det(A[\,2,3\,|\,1,3\,]) & \det(A[\,2,3\,|\,2,3\,]) \end{bmatrix}.$$

There are a number of elementary results concerning compound matrices that are easy to prove.

Theorem 1.

(a) If $A \in M_{m,n}(R)$, $B \in M_{n,p}(R)$ and k is an integer satisfying $1 \leq k \leq \min\{m, n, p\}$ then

$$C_k(AB) = C_k(A)\, C_k(B). \tag{3}$$

(b) Let $A \in M_{m,n}(R)$. If r is a positive integer for which

$$C_k(A) \neq 0, \quad k = 1, \cdots, r, \tag{4}$$

and

$$C_{r+1}(A) = 0 \tag{5}$$

then $r = \rho(A)$.

(c) If $N = \binom{n}{k}$ then

$$C_k(I_n) = I_N. \tag{6}$$

6.3 Compound Matrices

(d) If $A \in M_n(R)$ and A is nonsingular then $C_k(A)$ is non-singular and

$$C_k(A)^{-1} = C_k(A^{-1}). \tag{7}$$

Proof.

(a) The formula (3) is simply another formulation of the Cauchy-Binet theorem (§6.2 Theorem 1). To see this, let $\alpha \in Q_{k,m}$ and $\beta \in Q_{k,p}$. Then from the defining formula (1) we have

$$\begin{aligned}
C_k(AB)_{\alpha,\beta} &= \det((AB)[\,\alpha\,|\,\beta\,]) \\
&= \sum_{\omega \in Q_{k,n}} \det(A[\,\alpha\,|\,\omega\,]) \det(B[\,\omega\,|\,\beta\,]) \\
&= \sum_{\omega \in Q_{k,n}} C_k(A)_{\alpha,\omega} C_k(B)_{\omega,\beta} \\
&= (C_k(A) C_k(B))_{\alpha,\beta}.
\end{aligned} \tag{8}$$

The second equality in (8) is an application of §6.2 Theorem 1, the third equality follows from the definition of the compound matrix, and the last equality is the definition of the product of the two matrices $C_k(A)$ and $C_k(B)$.

(b) The statement (4) implies that for $k = 1, \ldots, r$, not all k-square subdeterminants of A are 0. The statement (5) implies all $(r + 1)$-square subdeterminants of A are 0. Apply §6.2 Theorem 6.

(c) Let α and β be sequences in $Q_{k,n}$. Then from (1)

$$C_k(I_n)_{\alpha,\beta} = \det(I_n[\,\alpha\,|\,\beta\,]). \tag{9}$$

Clearly, (9) is 0 if $\alpha \neq \beta$ and 1 if $\alpha = \beta$. Hence (6) follows.

(d) Since A is nonsingular, there exists a unique $B \in M_n(R)$ such that $AB = BA = I_n$, i.e., $B = A^{-1}$. Then from (3) and (6) we have

$$I_N = C_k(I_n)$$

$$= C_k(AB)$$

$$= C_k(A)C_k(B),$$

and similarly,

$$I_N = C_k(B)C_k(A). \tag{10}$$

Thus $C_k(B)$ is the inverse of $C_k(A)$:

$$C_k(A^{-1}) = C_k(A)^{-1}. \qquad \blacksquare$$

For matrices in $M_{m,n}(R)$, where R is either $\mathbb{R}$ or $\mathbb{C}$, there are a number of interesting special results about the compound.

Theorem 2.

(a) If $A \in M_{m,n}(\mathbb{C})$ then

$$C_k(A)^* = C_k(A^*). \tag{11}$$

(b) If $A \in M_n(\mathbb{C})$, $1 \le k \le n$, and A is hermitian, then $C_k(A)$ is hermitian.

(c) If $A \in M_n(\mathbb{C})$, $1 \le k \le n$, and A is unitary, then $C_k(A)$ is unitary.

(d) If $A \in M_n(\mathbb{C})$, $1 \le k \le n$, and A is normal, then $C_k(A)$ is normal.

Proof.

(a) Let $\alpha \in Q_{k,m}$ and $\beta \in Q_{k,n}$. From §6.2 formula (15), we have

6.3 Compound Matrices

$$A^*[\alpha|\beta] = \overline{A}^T[\alpha|\beta]$$

$$= (\overline{A[\beta|\alpha]})^T$$

$$= \overline{A[\beta|\alpha]}^T$$

$$= A[\beta|\alpha]^*. \qquad (12)$$

Hence

$$C_k(A^*)_{\alpha,\beta} = \det(A^*[\alpha|\beta])$$

$$= \det((A[\beta|\alpha])^*)$$

$$= \overline{\det(A[\beta|\alpha])}$$

$$= \overline{C_k(A)_{\beta,\alpha}}$$

$$= (C_k(A)^*)_{\alpha,\beta}$$

and (11) is proved.

(b) Recall that for A to be hermitian means that $A^* = A$. But then from (11),

$$C_k(A)^* = C_k(A^*) = C_k(A)$$

and $C_k(A)$ is hermitian.

(c) If A is unitary then $AA^* = A^*A = I_n$. Then from (3), (6), and (11) we have

$$I_N = C_k(I_n) \qquad \text{(from (6))}$$

$$= C_k(AA^*)$$

$$= C_k(A)C_k(A^*) \qquad \text{(from (3))}$$

$$= C_k(A)C_k(A)^*. \qquad \text{(from (11))}$$

Similarly,

$$C_k(A)^* C_k(A) = I_N$$

and $C_k(A)$ is unitary.

(d) This follows immediately from (3) and (11). ∎

There is an associated matrix that is closely related to the compound. Let $A \in M_n(R)$ and assume that k is a fixed integer, $1 \le k \le n$. For each α and β in $Q_{k,n}$ define the number

$$d_{\alpha\beta} = (-1)^{s(\alpha)+s(\beta)} \det(A(\alpha|\beta)). \qquad (13)$$

Recall that $s(\alpha)$ is the sum of the integers in the sequence α. Construct an $\binom{n}{k}$-square matrix whose rows are ordered lexicographically according to the sequences $\alpha \in Q_{k,n}$, whose columns are ordered lexicographically according to the sequences $\beta \in Q_{k,n}$, and whose entry in row α and column β is the number defined in (13). This matrix is called the $(n-k)^{\text{th}}$ *supplementary compound* of A and is denoted by

$$C^{n-k}(A). \qquad (14)$$

The reason for the notation (14) is that the definition of the entries in (13) immediately implies that $C^{n-k}(A)$ is *homogeneous of degree* $n-k$ in A, i.e., if $r \in R$ then

$$C^{n-k}(rA) = r^{n-k} C^{n-k}(A).$$

For example, if $n = 3$, $k = 2$,

6.3 Compound Matrices

$$A = \begin{bmatrix} a_{11} & a_{12} & a_{13} \\ a_{21} & a_{22} & a_{23} \\ a_{31} & a_{32} & a_{33} \end{bmatrix}$$

then

$$C^1(A) = \begin{bmatrix} \pm\det(A(1\,2|1\,2)) & \pm\det(A(1\,2|1\,3)) & \pm\det(A(1\,2|2\,3)) \\ \pm\det(A(1\,3|1\,2)) & \pm\det(A(1\,3|1\,3)) & \pm\det(A(1\,3|2\,3)) \\ \pm\det(A(2\,3|1\,2)) & \pm\det(A(2\,3|1\,3)) & \pm\det(A(2\,3|2\,3)) \end{bmatrix}$$

where the sign preceding $\det(A(\alpha|\beta))$ is $(-1)^{s(\alpha)+s(\beta)}$. For example,

$$C^1(A) = \begin{bmatrix} a_{33} & -a_{32} & a_{31} \\ -a_{23} & a_{22} & -a_{21} \\ a_{13} & -a_{12} & a_{11} \end{bmatrix}.$$

Since $C^{n-k}(A)$ and $C_{n-k}(A)$ are both defined in terms of $(n-k)$-square subdeterminants of A, it is reasonable to expect that the two matrices are closely related. The precise connection is easy to understand but it is somewhat subtle to establish. To investigate the connection, first note that if the $\binom{n}{k}$ sequences $\alpha \in Q_{k,n}$ are ordered lexicographically, then the complementary sequences $\alpha' \in Q_{n-k,n}$ are ordered in precisely the reverse order. For example, in $Q_{2,4}$ we have

α	α'
(1 2)	(3 4)
(1 3)	(2 4)
(1 4)	(2 3)
(2 3)	(1 4)
(2 4)	(1 3)
(3 4)	(1 2)

If N is a positive integer, define an N-square permutation matrix P_N whose (i, j) entry is $\delta_{i,N-j+1}$. For example, if N = 3 then

$$P_3 = \begin{bmatrix} 0 & 0 & 1 \\ 0 & 1 & 0 \\ 1 & 0 & 0 \end{bmatrix}.$$

Note that if A is any matrix with rows $A_{(1)}, \ldots, A_{(N)}$ then the rows of $P_N A$ are, in succession, $A_{(N)}, A_{(N-1)}, \ldots, A_{(2)}, A_{(1)}$. Similarly, if the columns of A are $A^{(1)}, \ldots, A^{(N)}$ then the columns of AP_N are $A^{(N)}, A^{(N-1)}, \ldots, A^{(2)}, A^{(1)}$. In other words, pre-multiplying A by P_N reverses the order in which the rows occur, and post-multiplying A by P_N reverses the order in which the columns occur. It should also be obvious that P_N is a symmetric matrix, i.e.,

$$P_N^T = P_N. \tag{15}$$

Finally, if $N = \binom{n}{k}$, define an N-square diagonal matrix Δ by

$$\Delta_N = \mathrm{diag}(\ldots, (-1)^{s(\alpha)}, \ldots). \tag{16}$$

In other words, Δ_N is an N-square diagonal matrix whose rows are indexed by $Q_{k,n}$ and whose entry in row α and column α is $(-1)^{s(\alpha)}$, $\alpha \in Q_{k,n}$.

With these preliminary remarks we can easily establish the connection between the compound and the supplementary compound.

Theorem 3.

Let $A \in M_n(R)$ and let k be a fixed integer, $1 \leq k \leq n$. Set $N = \binom{n}{k} = \binom{n}{n-k}$ and define D_N to be the N-square matrix

$$D_N = \Delta_N P_N. \tag{17}$$

Then

$$D_N^T C^{n-k}(A) D_N = C_{n-k}(A). \tag{18}$$

6.3 Compound Matrices

Proof.

First compute that

$$D_N^T C^{n-k}(A) D_N = (\Delta_N P_N)^T C^{n-k}(A) \Delta_N P_N$$

$$= P_N \Delta_N^T C^{n-k}(A) \Delta_N P_N$$

$$= P_N \Delta_N C^{n-k}(A) \Delta_N P_N. \quad (19)$$

The product $\Delta_N C^{n-k}(A) \Delta_N$ alters the α, β entry of $C^{n-k}(A)$ by multiplying it by

$$(-1)^{s(\alpha)+s(\beta)}$$

where α and β are in $Q_{k,n}$. However

$$C^{n-k}(A)_{\alpha,\beta} = (-1)^{s(\alpha)+s(\beta)} \det(A(\alpha \mid \beta)).$$

Thus

$$(\Delta_N C^{n-k}(A) \Delta_N)_{\alpha,\beta} = \det(A(\alpha \mid \beta))$$

$$= \det(A[\alpha' \mid \beta']), \alpha, \beta \in Q_{k,n}, \quad (20)$$

where α' and β' are the sequences in $Q_{n-k,n}$ complementary to α and β respectively. As we saw earlier, the matrix

$$P_N(\Delta_N C^{n-k}(A) \Delta_N) P_N \quad (21)$$

is obtained by reversing the order in which the rows and columns of $\Delta_N C^{n-k}(A) \Delta_N$ occur. But to reverse the lexicographic order of the sequences α', $\alpha \in Q_{k,n}$, produces the sequences $\alpha \in Q_{n-k,n}$ in lexicographic order. Similarly, reversing the lexicographic order of the sequences β', $\beta \in Q_{k,n}$, produces the sequences $\beta \in Q_{n-k,n}$ in lexicographic order. It follows that if ω and γ are in $Q_{n-k,n}$ then

$$(P_N \Delta_N C^{n-k}(A) \Delta_N P_N)_{\omega,\gamma} = \det(A[\ \omega\ |\ \gamma\ ]). \tag{22}$$

In other words,

$$(D_N^T C^{n-k}(A) D_N)_{\omega,\gamma} = C_{n-k}(A)_{\omega,\gamma}$$

and (18) is established. ∎

A number of interesting results about the supplementary compound follow directly from Theorem 3.

Theorem 4.

Let A and B be matrices in $M_n(R)$ and let k be a fixed integer, $1 \le k \le n$. Set $N = \binom{n}{k}$.
Then

$$C^{n-k}(AB) = C^{n-k}(A)C^{n-k}(B). \tag{23}$$

Moreover,

$$C_k(A^T)C^{n-k}(A) = \det(A)I_N \tag{24}$$

and

$$C_k(\text{adj}(A)) = (\det(A))^{k-1} C^{n-k}(A^T). \tag{25}$$

If A is nonsingular and α and β are in $Q_{k,n}$ then

$$\det(A^{-1}[\ \alpha\ |\ \beta\ ]) = (-1)^{s(\alpha)+s(\beta)} \det(A^T(\ \alpha\ |\ \beta))/\det(A). \tag{26}$$

The formula (26) is called the *Jacobi determinant theorem*.

6.3 Compound Matrices

Proof.

From (18) and (3) we have

$$C^{n-k}(AB) = D_N C_{n-k}(AB) D_N^T$$

$$= D_N C_{n-k}(A) C_{n-k}(B) D_N^T$$

$$= D_N C_{n-k}(A) D_N^T D_N C_{n-k}(B) D_N^T$$

$$= C^{n-k}(A) C^{n-k}(B). \qquad (27)$$

The first and third equalities above follow from

$$D_N D_N^T = I_N \qquad (28)$$

which is easily proved (see §6.3 Exercises, #13).

To confirm (24), simply examine the α, β entry of the product on the left:

$$(C_k(A^T) C^{n-k}(A))_{\alpha,\beta} = \sum_{\omega \in Q_{k,n}} C_k(A^T)_{\alpha,\omega} C^{n-k}(A)_{\omega,\beta}$$

$$= \sum_{\omega \in Q_{k,n}} \det(A^T[\,\alpha \mid \omega\,])(-1)^{s(\omega)+s(\beta)} \det(A(\,\omega \mid \beta\,))$$

$$= \sum_{\omega \in Q_{k,n}} \det(A[\,\omega \mid \alpha\,]^T)(-1)^{s(\omega)+s(\beta)} \det(A(\,\omega \mid \beta\,))$$

$$= \sum_{\omega \in Q_{k,n}} (-1)^{s(\omega)+s(\beta)} \det(A(\,\omega \mid \beta\,)) \det(A[\,\omega \mid \alpha\,]). \qquad (29)$$

The third equality in (29) follows from §6.2, formula (15). By §6.2 Exercises, #16 it follows that

$$(C_k(A^T)C^{n-k}(A))_{\alpha,\beta} = \delta_{\alpha,\beta}\det(A)$$

and (24) is proved.

To prove (25) assume first that A is nonsingular. Then by §6.2 Theorem 4,

$$A \, \text{adj}(A) = \det(A)I_n. \tag{30}$$

Next replace A by A^T in (24) to obtain

$$C_k(A)C^{n-k}(A^T) = \det(A^T)I_N$$

$$= \det(A)I_N. \tag{31}$$

Then from (30) and (3),

$$C_k(A)C_k(\text{adj}(A)) = C_k(\det(A)I_n)$$

$$= \det(A)^k I_N. \tag{32}$$

Since $C_k(A)$ is nonsingular (see (7)), (31) and (32) imply that

$$C_k(\text{adj}(A)) = \det(A)^k C_k(A)^{-1}$$

$$= \det(A)^k C^{n-k}(A^T)/\det(A)$$

$$= \det(A)^{k-1} C^{n-k}(A^T).$$

Thus (25) holds if A is nonsingular. Assume then that A is singular and consider first the case $k = 1$ of (25), which is to be interpreted as meaning

$$C_1(\text{adj}(A)) = C^{n-1}(A^T).$$

But $C_1(X) = X$ for any matrix X, and $C^{n-1}(A^T) = \text{adj}(A)$ (see §6.3 Exercises, #9). If $k > 1$ then the right side of (25) is 0. The left side is $C_k(\text{adj}(A))$. But (see §6.3

Exercises, #10) ρ(adj(A)) is either 1 or 0. In either case every k-square subdeterminant of adj(A) is 0 and hence $C_k(adj(A)) = 0$. This establishes (25) in case A is singular.

Finally, to prove (26), observe that

$$\det(A^{-1}[\alpha|\beta]) = C_k(A^{-1})_{\alpha,\beta}$$

$$= C_k\left(\frac{adj(A)}{\det(A)}\right)_{\alpha,\beta}$$

$$= \frac{1}{\det(A)^k} C_k(adj(A))_{\alpha,\beta}$$

$$= \frac{\det(A)^{k-1}}{\det(A)^k} C^{n-k}(A^T)_{\alpha,\beta} \qquad \text{(from (25))}$$

$$= (-1)^{s(\alpha)+s(\beta)} \det(A^T(\alpha|\beta))/\det(A). \qquad \blacksquare$$

6.3 Exercises

1. Compute $C_2(A)$ for the matrices A in §6.2 Exercises, #3 (a), (c), (d) and compute $C_3(A)$ for the matrix A in §6.2 Exercises, #3 (c).

2. Prove: $C^k(rA) = r^k C^k(A)$ where r is any scalar.

3. Prove: if A is n × n upper (lower) triangular then $C_k(A)$ is upper (lower) triangular.

    Hint: If $\alpha, \beta \in Q_{k,n}$ then $A[\alpha|\beta] = [a_{\alpha(i),\beta(j)}]$. Suppose that $\alpha > \beta$: $\alpha(1) = \beta(1), \cdots, \alpha(p) = \beta(p), \alpha(p+1) > \beta(p+1)$. Then it is easy to check that $A[\alpha|\beta]$ has the form

$$A[\,\alpha\,|\,\beta\,] = \left[\begin{array}{c|c} X & * \\ \hline 0 & * \end{array}\right]$$

where the upper left block X is $(p + 1)$-square and has a zero last row. Laplace expansion using columns $1, \ldots, p + 1$ shows that $\det(A[\,\alpha\,|\,\beta\,]) = 0$. Hence $C_k(A)_{\alpha,\beta} = 0$ when $\alpha > \beta$ and we conclude that $C_k(A)$ is upper triangular.

4. Prove: if A is $n \times n$ is upper (lower) triangular then

$$C_k(A)_{\alpha,\alpha} = \prod_{j=1}^{k} a_{\alpha(j),\alpha(j)}, \quad \alpha \in Q_{k,m}.$$

Hint: In the same notation as the Hint for #3, we see that

$$A[\,\alpha\,|\,\alpha\,] = \begin{bmatrix} a_{\alpha(1),\alpha(1)} & & * \\ & \ddots & \\ 0 & & a_{\alpha(k),\beta(k)} \end{bmatrix}$$

so that

$$\det(A[\,\alpha\,|\,\alpha\,]) = \prod_{j=1}^{k} a_{\alpha(j),\alpha(j)}.$$

5. In how many sequences $\alpha \in Q_{k,n}$ does a given integer i appear?

    Hint: If i appears in α then there are $k - 1$ remaining integers in α chosen from $1, \ldots, i - 1, i + 1, \ldots, n$, i.e., there are $\binom{n-1}{k-1}$ such sequences.

6. Prove: If A is $n \times n$ upper (lower) triangular then $\det(C_k(A)) = \det(A)^N$ where $N = \binom{n-1}{k-1}$.

    Hint: By #3 and #4,

$$\det(C_k(A)) = \prod_{\alpha \in Q_{k,n}} \prod_{j=1}^{k} a_{\alpha(j),\alpha(j)}.$$

By #5, any particular a_{tt} appears in precisely $\binom{n-1}{k-1}$ products of the form

$$\prod_{j=1}^{k} a_{\alpha(j),\alpha(j)}.$$

Thus

$$\det(C_k(A)) = \left(\prod_{t=1}^{n} a_{tt}\right)^{\binom{n-1}{k-1}}$$

$$= \det(A)^{\binom{n-1}{k-1}}.$$

7. Prove: if A is n × n upper (lower) triangular then $\det(C_k(A)) = \det(A)^N$ where $N = \binom{n-1}{k-1}$. This result is called the *Sylvester-Franke theorem*.

Hint: By §6.1 Exercises, #20, $A = P^{-1}TP$ where T is upper triangular. Then

$$C_k(A) = C_k(P)^{-1} C_k(T) C_k(P)$$

and hence

$$\det(C_k(A)) = \det(C_k(T))$$
$$= (\det(T))^{\binom{n-1}{k-1}}$$
$$= (\det(A))^{\binom{n-1}{k-1}}.$$

(If R does not contain all the eigenvalues of A then inquire these calculations to be done in a larger R that does contain the eigenvalues of A.)

8. Use Theorem 3 to compute $C^2(A)$ for the matrix in §6.2 Exercises, #3(a).

9. Let $A \in M_n(R)$. Prove that $\text{adj}(A) = C^{n-1}(A^T)$.

Hint: $C^{n-1}(A^T)_{\alpha,\beta} = (-1)^{\alpha+\beta}\det(A^T(\alpha|\beta))$

$\qquad = (-1)^{\alpha+\beta}\det(A(\beta|\alpha)^T)$

$\qquad = (-1)^{\alpha+\beta}\det(A(\beta|\alpha))$

$\qquad = \text{adj}(A)_{\alpha,\beta}.$

10. Let $A \in M_n(R)$. Prove that if A is nonsingular then $\text{adj}(A)$ is nonsingular. Also prove that if $\rho(A) = n - 1$ then $\rho(\text{adj}(A)) = 1$, and if $\rho(A) < n - 1$ then $\text{adj}(A) = 0$.

Hint: The first assertion follows from §6.2 Theorem 4. Next, assume that $\rho(A) = r \leq n - 1$. Let P and Q be nonsingular matrices for which $A = PBQ$ and B is in canonical form. Then from #9 and (23)

$$\text{adj}(A) = C^{n-1}(A^T)$$

$$= C^{n-1}(Q^T B^T P^T)$$

$$= C^{n-1}(Q^T) C^{n-1}(B^T) C^{n-1}(P^T).$$

Since $C^{n-1}(Q^T)$ and $C^{n-1}(P^T)$ are nonsingular, it follows that $\text{adj}(A)$ has the same rank as $C^{n-1}(B^T) = C^{n-1}(B)$ (i.e., $B = I_r \oplus 0_{n-r}$). By definition,

$$C^{n-1}(B)_{\alpha,\beta} = (-1)^{\alpha+\beta}\det(B(\alpha|\beta)).$$

From the form of B it is clear that $C^{n-1}(B) = 0$ unless $r = n - 1$ in which case $C^{n-1}(B) = 1 \oplus 0_{n-1}$ and $\rho(C^{n-1}(B)) = 1$.

11. Let $A \in M_n(R)$. Prove that $\text{adj}(AB) = \text{adj}(B)\text{adj}(A)$.

Hint: From #9 and (23)

$$\text{adj}(AB) = C^{n-1}((AB)^T) = C^{n-1}(B^T)C^{n-1}(A^T) = \text{adj}(B)\text{adj}(A).$$

12. Let E be the $n \times k$ matrix, $n \geq k$, whose columns are the first k columns of the identity matrix I_n. Show that $C_k(E)$ is the $N \times 1$ matrix ($N = \binom{n}{k}$) whose first entry is 1 and whose remaining entries are 0.

    Hint: The entries of $C_k(E)$ are the subdeterminants $\det(E[\,\alpha\,|\,1\,\ldots\,k\,])$, $\alpha \in Q_{k,n}$. If $\alpha = (1\,\ldots\,k)$ then $\det(E[\,\alpha\,|\,1\,\ldots\,k\,]) = 1$; otherwise, this determinant has value 0.

13. Prove that $D_N D_N^T = I_N$, as in formula (28).

    Hint: Recall (formula (17)) that $D_N = \Delta_N P_N$ where P_N is a permutation matrix. Then

    $$D_N D_N^T = \Delta_N P_N P_N^T \Delta_N^T$$
    $$= \Delta_N I_N \Delta_N^T$$
    $$= \Delta_N \Delta_N^T$$
    $$= I_N.$$

    The last equality follows from the definition in the formula (16).

6.3 MatLab

1. Review the MatLab function cmpnd(r, A) in §6.1 MatLab, #15. Also review the Help entry for inv. Write a MatLab script called compinv that confirms formula (7) for $k = 2$ and 20 random $A \in M_5(\mathbb{R})$. Read the Help entry for **norm** to

incorporate a test for equality in your program: Use **norm(X, 'fro')**. Save compinv in determ.

Hint:

```
%COMPINV
D=[ ];
for m=1:20
        A=(2*rand(5)-1);
        CA=cmpnd(2,inv(A));
        CB=inv(cmpnd(2,A));
        D=[D;norm(CA-CB,'fro')];
end
disp(D);
```

2. Write a MatLab script called comph, similar to the script in #1, that confirms Theorem 2(b) for k = 2. Save comph in determ.

Hint:

```
%COMPH
D=[ ];
for m=1:20
        A=(2*rand(5) - 1) + i*(2*rand(5)-1);
        H=(A+A' )/2;
        CH=cmpnd(2,H);
        D=[D; norm(CH' - CH, 'fro')];
end
disp(D);
```

3. If A is an n-square nonsingular complex matrix, then the MatLab command U=orth(A) returns a related unitary matrix U. Write a MatLab script called compu, similar to the script in #1, that confirms Theorem 2(c) for k = 2. Save compu in determ.

Hint:

```
%COMPU
D=[ ];
for m=1:20
```

```
            X=(2*rand(5)-1)+i*(2*rand(5)-1);
            U=orth(X);
            CU=cmpnd(2,U);
            D=[D;norm(CU*CU'- eye(CU),'fro')];
        end
        disp(D);
```

4. As we will learn subsequently (see §8.1 Exercises, #8), any normal matrix has the form U^*DU where D is a diagonal matrix and U is a unitary matrix. Using **orth** in #3, write a MatLab function called nrand that accepts a positive integer n, and returns a random n-square normal matrix $A = U^*DU$ such that the elements of D are at most 1 in modulus. Save nrand in determ.

   Hint:
   ```
            function N=nrand(n);
            %NRAND accepts a positive integer n and
            %returns an n-square random normal matrix.
            A=(2*rand(n)-1)+i*(2*rand(n)-1);
            U=orth(A);
            theta=(2*pi*rand(1,n));
            r=rand(1,n);
            D=diag(r.*exp(i*theta));
            N=U' * D * U;
   ```

5. Use the function nrand in #4 to write a MatLab script named compn, similar to the script in #1, that confirms Theorem 2(d) for k = 2. Save compn in determ.

   Hint:
   ```
            %COMPN
            D=[ ];
            for m=1:20
                    A=nrand(5);
                    CA=cmpnd(2,A);
                    D=[D; norm(CA*CA' - CA'*CA, 'fro')];
            end
            disp(D);
   ```

6. Review the function cmpl in §6.2 MatLab, #3. Also review the function cmpnd in §6.1 MatLab, #14. Then write a function named scmpnd that accepts an n–square matrix A and an integer r, $1 \leq r \leq n$, and returns the (n-r)th supplementary compound of A, $C^{n-r}(A)$. Save scmpnd in determ.

   Hint:
   ```
   function SC = scmpnd(r, A)
   %SCMPND(r,A) returns the (n-r)th supplementary
   %compound of the n-square matrix A.
   n=length(A);
   C=cmb(r,n);
   cnr=size(C);
   for s=1:cnr(1)
           for t=1:cnr(1)
                   a=C(s,:);
                   b=C(t,:);
                   sn=(-1)^(sum(a)+sum(b));
                   ca=cmpl(a,n);
                   cb=cmpl(b,n);
                   B(s,t)=sn*det(A(ca,cb));
           end
   end
   SC=B;
   ```

7. Use scmpnd to confirm, for various 3×3 matrices $A = [a_{ij}]$, that

$$C^1(A) = \begin{bmatrix} a_{33} & -a_{32} & a_{31} \\ -a_{23} & a_{22} & -a_{21} \\ a_{13} & -a_{12} & a_{11} \end{bmatrix}.$$

   Note that scmpnd(r,A) returns the $(n-r)^{th}$ supplementary compound of A. Thus to compute $C^1(A)$ use scmpnd(2,A).

8. Review §6.3 Exercises, #9 to the effect that

$$\mathrm{adj}(A) = C^{n-1}(A^T).$$

   Confirm this formula by using scmpnd(1,A) in a script called scmpinv, similar to the script in #1.

6.3 MatLab

Hint:
```
%SCMPINV
D=[ ];
for m=1:20
    A=(2*rand(5) -1);
    adj=scmpnd(1,A' );
    D=[D; norm(A*adj- det(A)*eye(A), 'fro')];
end
disp(D);
```

9. Write a MatLab script called jacobi that

- generates a random integer k, $1 \le k \le 4$
- generates 20 random 5-square real matrices A
- generates 20 random pairs of sequences a and b in $Q_{k,\,n}$
- confirms the Jacobi determinant theorem in formula (26).

Review the functions rcmb in §5.2 MatLab, #9(c), and cmpl in §6.2 MatLab, #3. Save jacobi in determ.

Hint:
```
%JACOBI
D=[ ];
for m=1:20
    k=floor(4*rand+1);
    A=2*rand(5)-1;
    a=rcmb(k,5);
    b=rcmb(k,5);
    ca=cmpl(a,5);
    cb=cmpl(b,5);
    AI=inv(A);
    AT=A';
    LHS=det(AI(a,b));
    RHS=((-1)^(sum(a)+sum(b)))*det(AT(ca,cb))/det(A);
    D=[D;abs(LHS-RHS)];
end
disp(D)
```

10. At the Command prompt enter various A and k to confirm the formula

$$(C_k(A^T)C^{n-k}(A))_{\alpha,\beta} = \delta_{\alpha,\beta}\det(A) .$$

Hint: Use cmpnd(k,A') and scmpnd(k,A).

11. Write a MatLab script called sylfrank that confirms the Sylvester-Franke Theorem appearing in §6.3 Exercises, #7. Read the Help entry for **gamma** and review the function cmpnd in §6.1 MatLab, #14. Save sylfrank in determ.

Hint:
```
%SYLFRANK
D=[ ];
for m=1:20
        n=floor(5*rand+1);
        k=floor((n-1)*rand+1);      % 1 ≤ k < n
        A=2*rand(n)-1;
        N=gamma(n)/(gamma(k)*gamma(n-k+1));
        D=[ D; abs(det(cmpnd(k,A)) - det(A)^N)];
end
disp(D)
```

6.3 Glossary

$C_k(A)$	335
compound matrix	335
gamma	356
homogeneous of degree k	335
Jacobi determinant theorem	344
norm(X, 'fro')	352
supplementary compound matrix	340
Sylvester-Franke theorem	349

Chapter 7

Eigenvalues

Topics
- *characteristic polynomial*
- *eigenvalues*
- *Gram-Schmidt process*
- *QR Theorem*
- *Cauchy-Schwarz inequality*
- *singular values*

7.1 The Characteristic Polynomial

In this section the *characteristic polynomial* of a matrix is defined and its coefficients are related to the entries of the matrix. The roots of the characteristic polynomial are called *characteristic roots* or *eigenvalues* of the matrix. In §7.2 the basic facts concerning the algebraic and geometric multiplicities of the eigenvalues of a matrix are discussed. It is proven in this section that an arbitrary nonsingular matrix can be factored into the product of a lower-triangular matrix and a unitary matrix. The procedure for doing this is usually called the *Gram-Schmidt process*.

Let $A \in M_n(R)$, where R is a set of numbers contained in $\mathbb{C}$. In §6.1 Exercises, #18, a number $r \in R$ is defined to be an *eigenvalue* or a *characteristic root* of A if r satisfies the polynomial equation

$$\det(rI_n - A) = 0. \tag{1}$$

For example, if

$$A = \begin{bmatrix} 0 & 1 \\ 1 & 0 \end{bmatrix},$$

then

$$\det(\lambda I_2 - A) = \det\left(\begin{bmatrix} \lambda & -1 \\ -1 & \lambda \end{bmatrix}\right)$$

$$= \lambda^2 - 1.$$

Thus the eigenvalues, or characteristic roots, of A are the roots of the polynomial $\lambda^2 - 1$, namely $r = \pm 1$. As another example, let

$$A = \begin{bmatrix} 0 & 1 \\ -1 & 0 \end{bmatrix} \in M_2(\mathbb{R}).$$

Then

$$\det(\lambda I_2 - A) = \det\left(\begin{bmatrix} \lambda & -1 \\ 1 & \lambda \end{bmatrix}\right)$$

$$= \lambda^2 + 1.$$

The eigenvalues of A are $r = \pm i$. In this case, although A has entries in $\mathbb{R}$, the eigenvalues are numbers in a larger set, namely $\mathbb{C}$.

It is a famous result, known as the *Fundamental Theorem of Algebra*, that any polynomial with complex number coefficients $e_1, ..., e_n$,

$$p(\lambda) = \lambda^n + e_1 \lambda^{n-1} + \cdots + e_{n-1}\lambda + e_n, \tag{2}$$

has precisely n complex roots $r_1, ..., r_n$, when each root is counted with its proper multiplicity. For example,

7.1 The Characteristic Polynomial

$$p(\lambda) = (\lambda - 1)^2 (\lambda^2 + \lambda + 1)^3$$

has 8 roots: 1 (multiplicity 2); $-\frac{1}{2} + \frac{i\sqrt{3}}{2}$ (multiplicity 3); $-\frac{1}{2} - \frac{i\sqrt{3}}{2}$ (multiplicity 3).

The polynomial in λ, defined as

$$p(\lambda) = \det(\lambda I_n - A), \qquad (3)$$

is called the *characteristic polynomial* of A. Observe that the term $\prod_{i=1}^{n} (\lambda - a_{ii})$ in the expansion of $\det(\lambda I_n - A)$ is the only term in which λ appears n times as a factor. Thus the characteristic polynomial has the form (2) with leading coefficient equal to 1, i.e., it is *monic*. Note that

$$p(0) = e_n = \det(-A) = (-1)^n \det(A)$$

so that the constant term in the characteristic polynomial is $(-1)^n \det(A)$.

If A and B are n-square matrices over R (i.e., having entries in R) and $B = S^{-1}AS$ for some nonsingular S over R, then A and B are said to be *similar* over R. If A and B are similar over R, then they have the same characteristic polynomial:

$$\det(\lambda I_n - B) = \det(\lambda I_n - S^{-1}AS)$$

$$= \det(S^{-1}(\lambda I_n - A)S)$$

$$= \det(S^{-1}) \det(\lambda I_n - A) \det(S)$$

$$= \det(S^{-1}S) \det(\lambda I_n - A)$$

$$= \det(\lambda I_n - A).$$

In order to investigate the characteristic polynomial $\det(\lambda I_n - A)$ it is convenient to have a formula for the determinant of a sum of two matrices. This result is not widely known but it has a surprising number of applications.

Theorem 1.

Let A and B be n-square matrices. Then

$$\det(A + B) = \sum_{r=0}^{n} \sum_{\alpha,\beta \in Q_{r,n}} (-1)^{s(\alpha) + s(\beta)} \det(A[\alpha | \beta]) \det(B(\alpha | \beta)). \quad (4)$$

(Notation: when $r = 0$, $\det(A[\alpha | \beta]) = 1$ and $\det(B(\alpha | \beta)) = \det(B)$; when $r = n$, $\det(A[\alpha | \beta]) = \det(A)$ and $\det(B(\alpha | \beta)) = 1$; $s(\alpha)$ is the sum of the integers in the sequence α.)

Proof.

The proof of the formula (4) is based on the following general identity: for a product of binomials $x_1 + y_1, \ldots, x_n + y_n$,

$$\prod_{i=1}^{n} (x_i + y_i) = \sum_{r=0}^{n} \sum_{\alpha \in Q_{r,n}} \prod_{i=1}^{r} x_{\alpha(i)} \prod_{j=r+1}^{n} y_{\alpha'(j)}. \quad (5)$$

In (5), $\alpha' = (\alpha'(r + 1) \ldots \alpha'(n))$ is the increasing sequence in $\{1, \ldots, n\}$ complementary to α. Formula (5) is, in fact, a generalization of the usual binomial theorem. It is readily confirmed as follows. In forming a term in the expansion of the product on the left side of (5), either an x_i or a y_i is chosen from each of the binomials $(x_i + y_i)$. Then the n choices are multiplied together. The final expansion is obtained by summing over all possible choices. The choices can be organized as follows. First select r, $0 \leq r \leq n$, to designate the number of binomials $(x_i + y_i)$ from which the x_i are selected. Once r is fixed, select an $\alpha \in Q_{r,n}$ to designate precisely which r binomials $(x_i + y_i)$ are to contribute an x_i factor, i.e., these are the binomials $(x_{\alpha(1)} + y_{\alpha(1)}), \ldots, (x_{\alpha(r)} + y_{\alpha(r)})$. The y_i terms are chosen from the remaining binomials other than those numbered α, i.e., the y_i terms are chosen from the n - r binomials indexed by α'. Thus, once r is fixed and $\alpha \in Q_{r,n}$ is chosen, a term in the expansion of

$$\prod_{i=1}^{n} (x_i + y_i)$$

7.1 The Characteristic Polynomial

is determined:

$$\prod_{i=1}^{r} x_{\alpha(i)} \prod_{j=r+1}^{n} y_{\alpha'(j)}.$$

Then, to account for all such terms, the sum over $\alpha \in Q_{r,n}$ is computed, followed by the sum over all $r = 0, \ldots, n$. Using (5) it is straightforward to compute

$$\det(A + B) = \sum_{\sigma \in S_n} \varepsilon(\sigma) \prod_{i=1}^{n} (a_{i\sigma(i)} + b_{i\sigma(i)})$$

$$= \sum_{\sigma \in S_n} \varepsilon(\sigma) \sum_{r=0}^{n} \sum_{\alpha \in Q_{r,n}} \prod_{i=1}^{r} a_{\alpha(i), \sigma\alpha(i)} \prod_{j=r+1}^{n} b_{\alpha'(j), \sigma\alpha'(j)}$$

$$= \sum_{r=0}^{n} \sum_{\alpha \in Q_{r,n}} \sum_{\sigma \in S_n} \varepsilon(\sigma) \prod_{i=1}^{r} a_{\alpha(i), \sigma\alpha(i)} \prod_{j=r+1}^{n} b_{\alpha'(j), \sigma\alpha'(j)}. \tag{6}$$

Observe that the innermost summation over S_n in (6) is precisely $\det(X_\alpha)$ where X_α is the n-square matrix defined schematically by

$$X_\alpha = \begin{bmatrix} \vdots \\ B_{(k)} \\ \vdots \\ A_{(\alpha(1))} \\ \vdots \\ A_{(\alpha(r))} \\ \vdots \end{bmatrix} \begin{array}{l} \text{row } k, \ k \notin \alpha \\ \\ \text{row } \alpha(1) \\ \\ \text{row } \alpha(r) \end{array} \tag{7}$$

i.e., rows $\alpha(1), \ldots, \alpha(r)$ of X_α are rows $A_{(\alpha(1))}, \ldots, A_{(\alpha(r))}$ of A and for $k \notin \alpha$, row k of X_α is $B_{(k)}$. Notation: in case $r = 0$ define $X_\alpha = B$. Thus (6) becomes

$$\det(A + B) = \sum_{r=0}^{n} \sum_{\alpha \in Q_{r,n}} \det(X_\alpha). \tag{8}$$

We now use the Laplace expansion theorem in order to expand $\det(X_\alpha)$ by rows numbered α (when $1 \leq r \leq n$):

$$\det(X_\alpha) = (-1)^{s(\alpha)} \sum_{\beta \in Q_{r,n}} (-1)^{s(\beta)} \det(X_\alpha[\alpha|\beta]) \det(X_\alpha(\alpha|\beta)). \qquad (9)$$

But $X_\alpha[\alpha|\beta] = A[\alpha|\beta]$, $X_\alpha(\alpha|\beta) = X_\alpha[\alpha'|\beta'] = B[\alpha'|\beta'] = B(\alpha|\beta)$. Hence, from (8) and (9),

$$\det(A + B) = \sum_{r=0}^{n} \sum_{\alpha, \beta \in Q_{r,n}} (-1)^{s(\alpha) + s(\beta)} \det(A[\alpha|\beta]) \det(B(\alpha|\beta)). \blacksquare$$

Observe that if $A = D = \mathrm{diag}(d_1, \ldots, d_n)$ is a diagonal matrix, then $\det(D[\alpha|\beta]) = 0$ unless $\alpha = \beta$, and if $\alpha = \beta$ then

$$\det(D[\alpha|\alpha]) = d_{\alpha(1)} \cdots d_{\alpha(r)}.$$

Thus (4) becomes

$$\det(D + B) = \sum_{r=0}^{n} \sum_{\alpha \in Q_{r,n}} d_{\alpha(1)} \cdots d_{\alpha(r)} \det(B(\alpha|\alpha)). \qquad (10)$$

From (10) we can compute the coefficients e_r appearing in (2) in the characteristic polynomial of a matrix.

Theorem 2.

Let A be an n-square matrix with characteristic polynomial $p(\lambda)$ as indicated in (2). Then

$$e_k = (-1)^k \sum_{\alpha \in Q_{k,n}} \det(A[\alpha|\alpha]), \quad k = 1, \ldots, n. \qquad (11)$$

In other words, if

$$p(\lambda) = \det(\lambda I_n - A) = \lambda^n + e_1 \lambda^{n-1} + e_2 \lambda^{n-2} + \cdots + e_k \lambda^{n-k} + \cdots + e_n$$

7.1 The Characteristic Polynomial

then the coefficient of λ^{n-k} is $(-1)^k$ times the sum of all k-square principal subdeterminants of A.

Proof.

Replace D by λI_n and B by $-A$ in (10):

$$p(\lambda) = \det(\lambda I_n - A)$$

$$= \sum_{r=0}^{n} \sum_{\alpha \in Q_{r,n}} \lambda^r \det(-A(\alpha|\alpha))$$

$$= \sum_{r=0}^{n} \lambda^r \sum_{\alpha \in Q_{r,n}} (-1)^{n-r} \det(A(\alpha|\alpha)). \qquad (12)$$

It is obvious from (12) that the coefficient e_k of λ^{n-k} in $p(\lambda)$ is the right side of (11). ∎

Note that

$$e_1 = -\sum_{j=1}^{n} a_{jj} .$$

The expression $\sum_{j=1}^{n} a_{jj}$ is called the *trace* of A and is denoted mathematically by

$$\text{tr}(A) . \qquad (13)$$

Also note that if $\lambda_1, \ldots, \lambda_n$ are the eigenvalues of A, then from (10)

$$p(\lambda) = \prod_{i=1}^{n} (\lambda - \lambda_i)$$

$$= \det(\lambda I_n - \text{diag}(\lambda_1, \ldots, \lambda_n))$$

$$= \sum_{k=0}^{n} \lambda^{n-k} (-1)^k \sum_{\alpha \in Q_{k,n}} \lambda_{\alpha(1)} \cdots \lambda_{\alpha(k)} . \qquad (14)$$

The expression appearing in (14), namely,

$$\sum_{\alpha \in Q_{k,n}} \lambda_{\alpha(1)} \cdots \lambda_{\alpha(k)},$$

is called the kth *elementary symmetric function* of the numbers $\lambda_1, \ldots, \lambda_n$, and is denoted by

$$E_k(\lambda_1, \ldots, \lambda_n).$$

Matching coefficients in (11) and (14) we have

Theorem 3.

If A is n-square with eigenvalues $\lambda_1, \ldots, \lambda_n$, then

$$E_k(\lambda_1, \ldots, \lambda_n) = \sum_{\sigma \in Q_{k,n}} \det(A[\alpha \mid \alpha]), \quad k = 1, \ldots, n. \tag{15}$$

In particular,

$$\lambda_1 + \cdots + \lambda_n = \operatorname{tr}(A) \quad (k = 1),$$

$$\lambda_1 \cdots \lambda_n = \det(A) \quad (k = n).$$

7.1 Exercises

1. Use Theorem 2 to compute the characteristic polynomial of each of the following matrices:

   (a) $\begin{bmatrix} 1 & 2 \\ 0 & 3 \end{bmatrix}$
   (b) $\begin{bmatrix} 0 & 0 \\ 0 & 0 \end{bmatrix}$

(c) $\begin{bmatrix} 1 & 0 & 1 \\ 0 & 0 & 1 \\ 1 & 0 & 0 \end{bmatrix}$

(d) $\begin{bmatrix} 1 & 1 & 1 \\ 1 & 1 & 1 \\ 1 & 1 & 1 \end{bmatrix}$

(e) $\begin{bmatrix} 0 & 1 & 0 \\ 0 & 0 & 1 \\ 1 & 0 & 0 \end{bmatrix}$

(f) $\begin{bmatrix} 1 & -1 & 0 \\ 0 & 1 & -1 \\ 1 & -1 & 0 \end{bmatrix}$

(g) $\begin{bmatrix} a_{11} & a_{12} \\ a_{21} & a_{22} \end{bmatrix}$

(h) I_n

(i) 0_n

(j) $J_n = [1/n]$ $(n \times n)$.

2. Prove: if $A \in M_n(\mathbb{C})$, then A is nonsingular iff every eigenvalue of A is non-zero.

   Hint: From §6.1 Theorem 3, A is nonsingular iff $\det(A) \neq 0$. But $\det(A) = \lambda_1 \cdots \lambda_n$ from Theorem 3.

3. Find the sum of the squares of the eigenvalues of each matrix in #1.

   Hint: We compute that for any n-square A with eigenvalues $\lambda_1, \ldots, \lambda_n$,

   $$\sum_{i=1}^n \lambda_i^2 = \left(\sum_{i=1}^n \lambda_i\right)^2 - 2E_2(\lambda_1, \ldots, \lambda_n)$$

   $$= \operatorname{tr}(A)^2 - 2 \sum_{\alpha \in Q_{2,n}} \det(A[\alpha \mid \alpha]).$$

   Thus to compute $\sum_{i=1}^n \lambda_i^2$ we need only compute the square of the coefficient of λ^{n-1} and subtract twice the coefficient of λ^{n-2}.

   (a) $16 - 6 = 10$
   (b) 0
   (c) $1 + 2 = 3$

(d) 9
(e) 0
(f) 4
(g) $(a_{11} + a_{22})^2 - 2(a_{11}a_{22} - a_{12}a_{21}) = a_{11}^2 + a_{22}^2 + 2a_{12}a_{21}$
(h) $n^2 - 2\binom{n}{2} = n^2 - 2\frac{n(n-1)}{2} = n$
(i) 0
(j) 1

4. Prove: if $A \in M_{m,n}(\mathbb{R})$ then
$$\text{tr}(AA^T) = \sum_{i=1, j=1}^{m,n} a_{ij}^2 .$$

5. Prove: if $A \in M_{m,n}(\mathbb{C})$, then $A = 0_{m,n}$ iff $\text{tr}(AA^*) = 0$.

6. Prove: $\text{tr}(AB) = \text{tr}(BA)$ if both matrix products are defined.

7. Prove: if $A = S^{-1}BS$ then $\text{tr}(A) = \text{tr}(B)$.

8. Exhibit an example showing that the converse of #7 is false.

9. Prove: if A is m × m and B is n × n then
$$\text{tr}(A \otimes B) = \text{tr}(A)\text{tr}(B) .$$

10. Prove: if $A_1, \ldots, A_p$ are square matrices then
$$\text{tr}(A_1 \otimes \cdots \otimes A_p) = \text{tr}(A_1) \cdots \text{tr}(A_p) .$$

11. Let A be n × n and assume that $1 \leq k \leq n$. Prove:
$$\text{tr}(C_k(A)) = \sum_{\alpha \in Q_{k,n}} \det(A[\alpha \mid \alpha]) .$$

12. Prove: if A and B are square matrices then $\text{tr}(A \oplus B) = \text{tr}(A) + \text{tr}(B)$.

7.1 Exercises

13. Let A be an n-square matrix with eigenvalues $\lambda_1, ..., \lambda_n$. Prove: if k is a positive integer, then the eigenvalues of A^k are $\lambda_1^k, ..., \lambda_n^k$.

    Hint: From §6.1 Exercises, #20, there exists a nonsingular P such that

    $$P^{-1}AP = \begin{bmatrix} \lambda_1 & & * \\ & \ddots & \\ 0 & & \lambda_n \end{bmatrix}.$$

    Now

    $$(P^{-1}AP)^k = \begin{bmatrix} \lambda_1^k & & * \\ & \ddots & \\ 0 & & \lambda_n^k \end{bmatrix}.$$

    But

    $$(P^{-1}AP)^k = P^{-1}A^kP$$

    so

    $$\det(\lambda I_n - A^k) = \det(\lambda I_n - P^{-1}A^kP)$$

    $$= \prod_{i=1}^{n} (\lambda - \lambda_i^k).$$

14. Let $\lambda_1, ..., \lambda_n$ be the eigenvalues of A and let $E_k(A)$ denote the kth elementary symmetric function of the eigenvalues of A, i.e., $E_k(A) = E_k(\lambda_1, ..., \lambda_n)$. Prove: $E_k(S^{-1}AS) = E_k(A)$ for any nonsingular S.

    Hint: From Theorems 2 and 3, $E_k(A) = (-1)^k e_k$ (see formula (11)) and the characteristic polynomials of A and $S^{-1}AS$ are the same.

15. Prove: If A and B are n × n matrices then $\text{tr}(A + B) = \text{tr}(A) + \text{tr}(B)$, and $\text{tr}(cA) = c\,\text{tr}(A)$ for any scalar c.

16. Let U_m be the m-square matrix with 1's on the first *superdiagonal*, i.e., in positions (1, 2), (2, 3), ..., (m - 1, m), and 0's elsewhere. (if m = 1, then $U_m = [0]$.) The matrix U_m is called the m-square *auxiliary unit matrix*. If $g(\lambda) = (\lambda - \alpha)^m$, then $H(g(\lambda))$ is defined to be the matrix $\alpha I_m + U_m$ and is called the *hypercompanion matrix of the polynomial* $g(\lambda)$. Prove:

$$\det(\lambda I_m - H(g(\lambda))) = g(\lambda).$$

Hint: Note that $H(g(\lambda)) = \alpha I_m + U_m$ is upper triangular and

$$\lambda I_m - H(g(\lambda)) = (\lambda - \alpha)I_m - U_m.$$

Thus $\det(\alpha I_m - H(g(\lambda))) = (\lambda - \alpha)^m$.

17. Let $f(\lambda) = \lambda^m + c_{m-1}\lambda^{m-1} + c_{m-2}\lambda^{m-2} + \cdots + c_1\lambda + c_0$. Define the *companion matrix of the polynomial* $f(\lambda)$, denoted by $C(f(\lambda))$, to be the m-square matrix satisfying the following specifications.

In the first m - 1 columns the only nonzero entries are the 1's in positions (2, 1), (3, 2), ..., (m, m - 1), while the last column contains the negatives of the coefficients of $f(\lambda)$ starting at the top with the constant term and going down to the coefficient of λ^{m-1} (if m = 1 and $f(\lambda) = \lambda + c_0$ then $C(f(\lambda)) = [-c_0]$). Prove:

$$\det(\lambda I_m - C(f(\lambda))) = f(\lambda).$$

Hint: Use Laplace expansion on the first row of $X = \lambda I_m - C(f(\lambda))$ and note that $X(1 \mid 1) = \lambda I_m - C(g(\lambda))$ where $g(\lambda) = \lambda^{m-1} + c_{m-1}\lambda^{m-2} + \cdots + c_2\lambda + c_1$. Use induction on m to conclude that $\det(X(1 \mid 1)) = g(\lambda)$. Also $\det(X(1 \mid m)) = (-1)^{m-1}$. Hence

$$\det(X) = \lambda \det(X(1 \mid 1)) + (-1)^{m+1}c_0\det(X(1 \mid m))$$

$$= \lambda g(\lambda) + (-1)^{m+1}(-1)^{m-1}c_0$$

$$= f(\lambda).$$

18. Prove: if $A \in M_n(\mathbb{R})$ then the complex eigenvalues of A occur in complex conjugate pairs.

19. Let $\lambda_1, \lambda_2, \lambda_3$ be the roots of the polynomial $\lambda^3 - \lambda^2 - 1$. Construct a matrix $A \in M_3(\mathbb{R})$ such that the eigenvalues of A are $\gamma_1 = \lambda_2 + \lambda_3$, $\gamma_2 = \lambda_1 + \lambda_3$, $\gamma_3 = \lambda_1 + \lambda_2$.

20. Let $A = C(f(\lambda))$ where $f(\lambda) = \lambda^3 - 2\lambda^2 + \lambda + 1$. Find a matrix $B \in M_3(\mathbb{R})$ such that the eigenvalues of B are $\gamma_1 = \lambda_1\lambda_2$, $\gamma_2 = \lambda_1\lambda_3$, $\gamma_3 = \lambda_2\lambda_3$ where $\lambda_1, \lambda_2, \lambda_3$ are the eigenvalues of A.

7.1 MatLab

1. Read the Help entries for **roots**, **roots1**, and **poly**. The following are the important features of these commands.

    - If A is n-square then

    $$\text{poly}(A) = [\, c_1 \ c_2 \ c_3 \ \ldots \ c_n \ c_{n+1} \,]$$

    in which

    $$\det(\lambda I_n - A) = c_1\lambda^n + c_2\lambda^{n-1} + \ldots + c_n\lambda + c_{n+1} \tag{1}$$

    ($c_1 = 1$).

    - If $r = [\, r_1 \ r_2 \ \ldots \ r_n \,]^T$ then

    $$\text{poly}(r) = [\, c_1 \ c_2 \ c_3 \ \ldots \ c_n \ c_{n+1} \,] \tag{2}$$

    where

$$\prod_{t=1}^{n} (\lambda - r_t) = c_1 \lambda^n + c_2 \lambda^{n-1} + \cdots + c_n \lambda + c_{n+1}. \quad (3)$$

- If

$$c = [\, c_1 \; c_2 \; c_3 \; \cdots \; c_n \; c_{n+1} \,]$$

then

$$\text{roots}(c) = [\, r_1 \; r_2 \; \cdots \; r_n \,]^T \quad (4)$$

where $r_1, r_2, \ldots, r_n$ are the roots of

$$c_1 \lambda^n + c_2 \lambda^{n-1} + \cdots + c_n \lambda + c_{n+1}. \quad (5)$$

- roots1 is a more accurate version of roots.

(a) Use poly to obtain the characteristic polynomials of the n-square matrix $J_n = [\frac{1}{n}]$ for n = 2, 3, 4, 5.

(b) Based on the results in (a) state and prove a result describing the eigenvalues of $A = I_n + J_n$.

(c) Use poly to print out the binomial coefficients

$$\binom{10}{0}, \binom{10}{1}, \ldots, \binom{10}{10}.$$

Hint: (b) The eigenvalues of A are 2, and 1 (n - 1 times). For, any eigenvalue of $I_n + J_n$ must be of the form 1 + (an eigenvalue of J_n), i.e., if $J_n x = rx$ then $(I_n + J_n)x = x + rx = (1 + r)x$. The characteristic polynomial of J_n is $\lambda^n - \lambda^{n-1}$ (i.e., see Theorem 2 and note that all subdeterminants of J_n of size 2 or higher are 0). Thus the eigenvalues of J_n are 1, and 0 (n - 1 times).

(c) poly(-ones(10, 1)).

2. Use poly and roots to compute the eigenvalues of each of the following matrices:

(a) $A = \begin{bmatrix} 5 & 3 & 6 \\ 2 & 6 & 6 \\ 2 & 3 & 9 \end{bmatrix}$ (b) ones(20)

(c) $A = \begin{bmatrix} 0 & 0 & 0 & 1 \\ 0 & 0 & 1 & 0 \\ 0 & -1 & 0 & 0 \\ -1 & 0 & 0 & 0 \end{bmatrix}$

(d) orth((2*rand(5)-1)+i*(2*rand(5)-1))

What property do you expect the numbers in (d) to have?

Hint: The numbers in (d) have modulus 1 because any unitary matrix U has eigenvalues of modulus 1 (i.e., $Ux = \lambda x$, $(Ux)^*(Ux) = (\lambda x)^*(\lambda x)$, $x^*U^*Ux = \overline{\lambda}\lambda x^*x$, $x^*x = |\lambda|^2 x^*x$, $|\lambda|^2 = 1$).

3. Write a MatLab function named eigsq that accepts an n-square matrix A and returns the sum of the squares of the eigenvalues of A. Do not use roots, roots1, or eig in eigsq. Create a library called eigen and save eigsq in eigen.

Hint:
function s = eigsq(A)
%EIGSQ accepts an n-square
%matrix A and returns
%the sum of the squares of
%the eigenvalues of A.
c = poly(A);
s = c(2)^2 - 2*c(3)

4. Verify by direct computation that

$$s_3 = E_1^3 - 3E_1E_2 + 3E_3$$

where $s_3 = \lambda_1^3 + \lambda_2^3 + \cdots + \lambda_n^3$ and E_k is the kth elementary symmetric function of the λ's.

5. Write a MatLab function named eigcube that accepts an n-square matrix A and returns the sum of the cubes of the eigenvalues of A. Do not use roots, roots1, or eig in eigcube. Save eigcube in eigen.

    Hint:
    ```
    function s = eigcube(A)
    %EIGCUBE accepts an n-square
    %matrix A and returns the
    %sum of the cubes of
    %the eigenvalues of A.
    c = poly(A);
    s = - (c(2)^3) + 3*(c(2)*c(3)) - 3*c(4)
    ```

6. Review the function psdet(r, A) in §6.1 MatLab, #12. Use psdet to write a function named esf that

    - accepts an n-square matrix A

    - returns the vector poly(A) = [c_1 c_2 c_3 ... c_n c_{n+1}] where

    $$\det(\lambda I_n - A) = c_1\lambda^n + c_2\lambda^{n-1} + \cdots + c_n\lambda + c_{n+1}.$$

    Do not use poly itself in constructing esf. Save esf in eigen.

    Hint: According to Theorem 2, $c_1 = 1$, $c_2 = $ -psdet(1, A), $c_3 = $ psdet(2, A), ..., $c_r = (-1)^{r-1}$ psdet(r-1, A), ..., $c_{n+1} = (-1)^n$ psdet(n, A).

    ```
    function e = esf(A)
    %ESF accepts an n-square
    %matrix A and returns the
    %vector poly(A).
    n = length(A);
    ```

7.1 MatLab

```
        D = [ 1 ];
        for r = 1:n
                D = [ D ((-1)^r)*psdet(r, A) ];
        end
        e = D
```

7. Read the help entry for **compan**. The MatLab documentation describes $C = \text{compan}(p)$ as the companion matrix of the polynomial

$$p = p_1\lambda^n + p_2\lambda^{n-1} + \cdots + p_n\lambda + p_{n+1},$$

i.e.,

$$C = \begin{bmatrix} -p_2/p_1 & -p_3/p_1 & \cdots & & -p_{n+1}/p_1 \\ 1 & 0 & 0 & \cdots & 0 \\ 0 & 1 & 0 & \cdots & 0 \\ \vdots & & \ddots & & \vdots \\ 0 & \cdots & 0 & 1 & 0 \end{bmatrix}. \qquad (6)$$

Note that this definition is somewhat different from the definition of the companion matrix given in §7.1 Exercises, #17.

(a) At the command prompt compute poly(compan(p)) for several p of the form

$$p = \lambda^5 + p_2\lambda^4 + p_3\lambda^3 + p_4\lambda^2 + p_5\lambda + p_6$$

in which $p_2, \ldots, p_6$ are random integers in the range $[-10, 10]$. Compare the output with the input in each case.

(b) Prove that in general the characteristic polynomial of the matrix C in (6) is the polynomial

$$\lambda^n + \frac{p_2}{p_1}\lambda^{n-1} + \frac{p_3}{p_1}\lambda^{n-2} + \cdots + \frac{p_n}{p_1}\lambda + \frac{p_{n+1}}{p_1}.$$

Hint: (a)   p = [1 -10+floor(21*rand(1,5))]; poly(compan(p))

(b)   The case n = 4 is typical:

$$\lambda I_4 - C = \begin{bmatrix} \lambda + \frac{p_2}{p_1} & \frac{p_3}{p_1} & \frac{p_4}{p_1} & \frac{p_5}{p_1} \\ -1 & \lambda & 0 & 0 \\ 0 & -1 & \lambda & 0 \\ 0 & 0 & -1 & \lambda \end{bmatrix}.$$

Use Laplace expansion on the last column to obtain

$$\det(\lambda I_4 - C) = -\frac{p_5}{p_1} \det\begin{bmatrix} -1 & \lambda & 0 \\ 0 & -1 & \lambda \\ 0 & 0 & -1 \end{bmatrix} + \lambda \det\begin{bmatrix} \lambda + \frac{p_2}{p_1} & \frac{p_3}{p_1} & \frac{p_4}{p_1} \\ -1 & \lambda & 0 \\ 0 & -1 & \lambda \end{bmatrix}. \quad (7)$$

The first determinant on the right is trivial to compute. The second determinant is simply the case n = 3 and can be evaluated by induction. The general situation is identical.

8. Read the Help entry for **eig**. For our purposes, the principal aspects of the eig command are:

- if A is n-square then

$$\text{eig}(A) = [\lambda_1 \ \lambda_2 \ \cdots \ \lambda_n]$$

where the λ_i are the eigenvalues of A, i = 1, ..., n.

- the command

$$[X \ D] = \text{eig}(A)$$

returns a diagonal matrix

7.1 MatLab

$$D = \text{diag}(\lambda_1, \ldots, \lambda_n)$$

and a matrix X whose columns are eigenvectors of A corresponding to $\lambda_1, \ldots, \lambda_n$, i.e.,

$$AX^{(t)} = \lambda_t X^{(t)}, \quad t = 1, \ldots, n.$$

Use eig in the Command window to determine the eigenvalues and corresponding eigenvectors of each of the following matrices:

(a) $\begin{bmatrix} -4 & -3 & -2 \\ 2 & -1 & -2 \\ 3 & -3 & -1 \end{bmatrix}$
(b) $\begin{bmatrix} 1 & 1 & -1 \\ -1 & 3 & -1 \\ -1 & 1 & 1 \end{bmatrix}$

(c) $\begin{bmatrix} 3 & 2 & -6 \\ 0 & 2 & -1 \\ 1 & 1 & -2 \end{bmatrix}$
(d) $\begin{bmatrix} 0 & 1 & 0 & 0 \\ 0 & 0 & 1 & 0 \\ 0 & 0 & 0 & 1 \\ 0 & 0 & 0 & 0 \end{bmatrix}$

Note that for the matrix A in (d), [X D] = eig(A) produces

$$X = \begin{bmatrix} 1 & -1 & 1 & -1 \\ 0 & 0 & 0 & 0 \\ 0 & 0 & 0 & 0 \\ 0 & 0 & 0 & 0 \end{bmatrix} \quad D = \begin{bmatrix} 0 & 0 & 0 & 0 \\ 0 & 0 & 0 & 0 \\ 0 & 0 & 0 & 0 \\ 0 & 0 & 0 & 0 \end{bmatrix}.$$

Thus AX = XD. Is A similar to D?

9. Use eig and compan in the Command window to determine the roots of each of the following polynomials. Compare the results with the roots as produced by the roots command.

(a) $\lambda^3 - 9\lambda - 9$

Hint: p = [1 0 -9 -9];
eig(compan(p))
roots(p)

The roots are 3.4115, -2.2267, and -1.1848 (to the indicated number of decimal places).

(b) $\lambda^5 + \lambda^4 - 7\lambda^3 - 22\lambda^2 + \lambda + 1$

(c) $\lambda^4 - 6\lambda + 2$

10. Use MatLab to show that the matrices

$$A = \begin{bmatrix} 1 & 0 & 0 \\ 0 & 2 & 0 \\ 0 & 0 & 3 \end{bmatrix}, \qquad B = \begin{bmatrix} 1 & 1 & 1 \\ 0 & 2 & 1 \\ 0 & 0 & 3 \end{bmatrix}.$$

are similar. Find an X such that $X^{-1}BX = A$.

Hint: Try [X D] = eig(B). The resulting D is A and X has the form

$$X = \begin{bmatrix} 1 & \alpha & \beta \\ 0 & \alpha & \beta \\ 0 & 0 & \beta \end{bmatrix}.$$

Confirm that BX = XA so that any choices of α and β for which $\alpha\beta \neq 0$ will exhibit the similarity of A and B.

7.1 Glossary

auxiliary unit matrix 368
C(f(λ)) 368
characteristic polynomial 359
characteristic roots 357

compan	373
companion matrix	368
eig	374
eigenvalues	357
E_k	364
elementary symmetric function	364
Fundamental Theorem of Algebra	358
Gram-Schmidt process	357
$H(g(\lambda))$	368
hypercompanion matrix	368
poly	369
roots	369
roots1	369
similar	359
superdiagonal	368
tr	363
trace	363
U_m	368

7.2 The Gram-Schmidt Process

In §6.1 Exercises, #18, we saw that a number r is an eigenvalue of the n-square matrix A if and only if there exists a column vector $x \neq 0$ satisfying

$$Ax = rx. \qquad (1)$$

The vector x is called an *eigenvector* or *characteristic vector* of A *corresponding to* r. Let k be the largest integer for which there exists an n × k matrix X of rank k such that

$$AX = rX. \qquad (2)$$

The integer k is called the *geometric multiplicity* of the eigenvalue r. The *algebraic multiplicity* of r is defined to be the largest integer m such that $(\lambda - r)^m$ exactly divides the characteristic polynomial of A.

As an example, suppose

$$A = \begin{bmatrix} 0 & 1 \\ 0 & 0 \end{bmatrix}. \tag{3}$$

Then the algebraic multiplicity of the eigenvalue 0 is m = 2. To compute the geometric multiplicity of 0, consider the equation (1) with r = 0 and $x = [x_1 \ x_2]^T$. The equation (1) becomes

$$x_2 = 0.$$

Thus $[x_1 \ 0]^T = x_1[1 \ 0]^T$, $x_1 \neq 0$, is the form of any eigenvector of A corresponding to 0. Note that an n × k matrix X of rank k can satisfy (2) iff each column of X is an eigenvector corresponding to r. For the matrix A in (3) it follows that the geometric multiplicity of the eigenvalue 0 is k = 1, because any column of an X satisfying (2) must be a multiple of $[1 \ 0]^T$ and hence $\rho(X) = 1$.

In order to analyze eigenvalues of various classes of matrices, we require the following important result. The constructive process described in the proof of this theorem is known as the *Gram-Schmidt process*.

Theorem 1.

Let X be a nonsingular matrix in $M_n(\mathbb{C})$. Then there exists a lower triangular matrix L and a unitary matrix U, both in $M_n(\mathbb{C})$, such that

$$LX = U. \tag{4}$$

Proof.

The matrix X is nonsingular so that no row of X is zero and hence (why?)

$$X_{(i)} X_{(i)}^* > 0, \ i = 1, \ldots, n.$$

7.2 The Gram-Schmidt Process

Let L_1 be the lower triangular matrix

$$L_1 = \begin{bmatrix} 1 & 0 \\ t_{21} & 1 \end{bmatrix} \oplus I_{n-2}$$

where t_{21} will be specified below, and form the product

$$L_1 X = \begin{bmatrix} X_{(1)} \\ t_{21}X_{(1)} + X_{(2)} \\ X_{(3)} \\ \vdots \\ X_{(n)} \end{bmatrix} = \begin{bmatrix} S_{(1)} \\ S_{(2)} \\ X_{(3)} \\ \vdots \\ X_{(n)} \end{bmatrix}. \qquad (5)$$

Choose t_{21} so that

$$\begin{aligned} 0 &= S_{(2)} S_{(1)}^* \\ &= (t_{21} X_{(1)} + X_{(2)}) X_{(1)}^* \\ &= t_{21} X_{(1)} X_{(1)}^* + X_{(2)} X_{(1)}^*, \end{aligned}$$

i.e., choose t_{21} to be

$$t_{21} = -X_{(2)} X_{(1)}^* / X_{(1)} X_{(1)}^*.$$

Note that $S_{(2)} \neq 0$, since otherwise $L_1 X$ would be singular, whereas both L_1 and X are non-singular. Next, define the lower triangular matrix

$$L_2 = \begin{bmatrix} 1 & 0 & 0 \\ 0 & 1 & 0 \\ t_{31} & t_{32} & 1 \end{bmatrix} \oplus I_{n-3}, \qquad (6)$$

where t_{31} and t_{32} will be specified below. Then from (5) we have

$$L_2 L_1 X = L_2 \begin{bmatrix} S_{(1)} \\ S_{(2)} \\ X_{(3)} \\ \vdots \\ X_{(n)} \end{bmatrix}$$

$$= \begin{bmatrix} S_{(1)} \\ S_{(2)} \\ t_{31} S_{(1)} + t_{32} S_{(2)} + X_{(3)} \\ X_{(4)} \\ \vdots \\ X_{(n)} \end{bmatrix}$$

$$= \begin{bmatrix} S_{(1)} \\ S_{(2)} \\ S_{(3)} \\ X_{(4)} \\ \vdots \\ X_{(n)} \end{bmatrix},$$

where $S_{(3)} = t_{31} S_{(1)} + t_{32} S_{(2)} + X_{(3)}$. Choose t_{31} so that

$$0 = S_{(3)} S_{(1)}^*$$

$$= t_{31} S_{(1)} S_{(1)}^* + t_{32} S_{(2)} S_{(1)}^* + X_{(3)} S_{(1)}^*$$

$$= t_{31} S_{(1)} S_{(1)}^* + X_{(3)} S_{(1)}^*$$

since $S_{(2)} S_{(1)}^* = 0$. Thus define t_{31} by the formula

$$t_{31} = -X_{(3)} S_{(1)}^* / S_{(1)} S_{(1)}^* .$$

7.2 The Gram-Schmidt Process

Similarly, choose

$$t_{32} = -X_{(3)}S_{(2)}^* / S_{(2)}S_{(2)}^*$$

so as to ensure that $S_{(3)}S_{(2)}^* = 0$. Observe that $S_{(3)} \neq 0$ since L_2, L_1, and X are all non-singular. We continue in this way, finally obtaining

$$L_{n-1}L_{n-2}\cdots L_1 X = \begin{bmatrix} S_{(1)} \\ S_{(2)} \\ \vdots \\ S_{(n)} \end{bmatrix} = S$$

where

$$S_{(j)}S_{(i)}^* = 0, \quad 1 \leq i < j \leq n.$$

Note that by taking the conjugate transpose we also have

$$S_{(i)}S_{(j)}^* = 0, \quad 1 \leq i < j \leq n.$$

Moreover,

$$S_{(i)}S_{(i)}^* > 0, \quad i = 1, \ldots, n.$$

Now let

$$D = \mathrm{diag}((S_{(1)}S_{(1)}^*)^{1/2}, \ldots, (S_{(n)}S_{(n)}^*)^{1/2}),$$

and observe that

$$((D^{-1}S)(D^{-1}S)^*)_{ij} = \frac{S_{(i)}}{(S_{(i)}S_{(i)}^*)^{1/2}} \frac{S_{(j)}^*}{(S_{(j)}S_{(j)}^*)^{1/2}} = \delta_{ij}.$$

Hence $U = D^{-1}S$ is unitary, and clearly $L = D^{-1}L_{n-1}\cdots L_1$ is lower triangular. ∎

Note that in the proof of Theorem 1, if X is real, then the process for determining L produces a real matrix and moreover U is then a real orthogonal matrix. It is easy to see that if L is nonsingular and lower triangular then so is L^{-1}. Hence, if W is unitary and lower triangular then, in fact, W must be a diagonal matrix, i.e., $W^* = W^{-1}$. Thus suppose

$$LX = U,$$

$$TX = V$$

are two factorizations, as in (4), that satisfy the conclusions of Theorem 1. Then

$$L^{-1}U = X = T^{-1}V,$$

and

$$TL^{-1} = VU^{-1}. \qquad (7)$$

The matrix TL^{-1} is lower triangular so that the unitary matrix $W = VU^{-1}$ must be diagonal, $W = \text{diag}(\theta_1, \ldots, \theta_n)$ where $|\theta_i| = 1$ for $1 \le i \le n$ (see §7.2 Exercises, #2). Since

$$V = WU,$$

we conclude that the rows of U are determined to within multiples of absolute value 1. Similarly, $TL^{-1} = W$, so $T = WL$ implies that the rows of T are also determined to within multiples of absolute value 1.

Let A be an $r \times n$ matrix of rank r ($r \le n$):

$$A = \begin{bmatrix} A_{(1)} \\ \vdots \\ A_{(r)} \end{bmatrix}.$$

7.2 The Gram-Schmidt Process

From §6.2 Theorem 6 there is a nonsingular $r \times r$ submatrix of A, say $A[1, \ldots, r \mid \omega]$, $\omega \in Q_{r,n}$. Let $\omega' \in Q_{n-r,n}$ be the sequence complementary to ω and define the $1 \times n$ row vectors

$$A_{(r+t)} = [0, \ldots, 0, 1, 0, \ldots, 0], \quad t = 1, \ldots, n-r,$$

where the 1 appears in position $\omega'(t)$. Then the n-square matrix

$$X = \begin{bmatrix} A_{(1)} \\ A_{(2)} \\ \vdots \\ A_{(n)} \end{bmatrix}$$

is obviously nonsingular because $\det(X) = \pm\det(A[1, \ldots, r \mid \omega])$ by the Laplace expansion theorem. Let

$$LX = U$$

as in (4) and observe that this equation implies

$$\begin{bmatrix} 1 & & & 0 \\ t_{21} & 1 & & \\ \vdots & & \ddots & \\ t_{r1} & \cdots & t_{r,r-1} & 1 \end{bmatrix} A = \begin{bmatrix} U_{(1)} \\ \vdots \\ U_{(r)} \end{bmatrix}. \tag{8}$$

Suppose that the rows of A satisfy $A_{(i)}A_{(j)}^* = \delta_{ij}$, $i, j = 1, \ldots, r$. Then $A_{(1)} = U_{(1)}$,

$$t_{21}A_{(1)} + A_{(2)} = U_{(2)},$$

and

$$t_{21}A_{(1)}A_{(1)}^* + A_{(2)}A_{(1)}^* = U_{(2)}A_{(1)}^*$$

$$= U_{(2)}U_{(1)}^*$$

$$= 0.$$

Thus $t_{21} = 0$ and $A_{(2)} = U_{(2)}$. Similarly, we prove in succession that

$$t_{31} = t_{32} = 0, \quad A_{(3)} = U_{(3)},$$

$$t_{41} = t_{42} = t_{43} = 0, \quad A_{(4)} = U_{(4)},$$

$$\vdots$$

$$t_{r1} = \cdots = t_{r, r-1} = 0, \quad A_{(r)} = U_{(r)}.$$

Since U is unitary we have proved the following interesting fact: if the first r rows of $A \in M_{r, n}(\mathbb{C})$ satisfy $A_{(i)} A_{(j)}^* = \delta_{ij}$, $i, j = 1, \ldots, r$, then A can be completed to a unitary matrix by adjoining n - r rows. Moreover, the procedure can be carried out by the Gram-Schmidt process described in the proof of Theorem 1.

7.2 Exercises

1. Prove: if L_1 and L_2 are n-square lower (upper) triangular matrices, then $L_1 L_2$ is a lower (upper) triangular matrix.

2. Prove: if U is an n-square unitary matrix and U is lower (upper) triangular, then U must be a diagonal matrix.

   Hint: Assume U is upper triangular. The eigenvalues of any triangular matrix are on the main diagonal. Since U is unitary, the eigenvalues $\lambda_1, \ldots, \lambda_n$ have modulus 1. (From $Ux = \lambda x$ it follows that $(Ux)^*(Ux) = |\lambda|^2 x^* x$, or $x^* U^* U x = |\lambda|^2 x^* x$. But $U^* U = I_n$.) Thus

   $$\sum_{i=1}^{n} |U_{ii}|^2 = \sum_{i=1}^{n} |\lambda_i|^2 = n.$$

   On the other hand,

7.2 Exercises

$$tr(U^*U) = tr(I_n) = n,$$

and

$$tr(U^*U) = \sum_{i=1,j=1}^{n} |u_{ij}|^2$$

$$= n + \sum_{1 \leq i < j \leq n} |u_{ij}|^2.$$

Hence

$$\sum_{1 \leq i < j \leq n} |u_{ij}|^2 = 0$$

and U is diagonal.

3. Let

$$A = \begin{bmatrix} A_{11} & 0 \\ A_{21} & A_{22} \end{bmatrix}$$

be an n-square complex normal matrix partitioned so that A_{11} square. Prove: $A_{21} = 0$.

Hint: We compute that

$$AA^* = \begin{bmatrix} A_{11}A_{11}^* & A_{11}A_{21}^* \\ A_{21}A_{11}^* & A_{21}A_{21}^* + A_{22}A_{22}^* \end{bmatrix}$$

and

$$A^*A = \begin{bmatrix} A_{11}^*A_{11} + A_{21}^*A_{21} & A_{21}^*A_{22} \\ A_{22}^*A_{21} & A_{22}^*A_{22} \end{bmatrix}.$$

Matching the upper left blocks we have $A_{11}^*A_{11} + A_{21}^*A_{21} = A_{11}A_{11}^*$. Take traces to conclude $A_{21} = 0$.

4. Let A be an n × n matrix and let r be an eigenvalue of A of algebraic multiplicity m and geometric multiplicity k. Prove: $k \leq m$.

Hint: By definition, there exists an n × k matrix X of rank k such that $AX = rX$ (see(2)). Complete X to a nonsingular matrix S by adjoining n - k columns. Then for $1 \leq j \leq k$,

$$(S^{-1}AS)^{(j)} = S^{-1}(AS)^{(j)}$$
$$= S^{-1}AS^{(j)}$$
$$= S^{-1}AX^{(j)}$$
$$= S^{-1}(rX^{(j)})$$
$$= S^{-1}X^{(j)}r$$
$$= (S^{-1}S^{(j)})r$$
$$= (S^{-1}S)^{(j)}r$$
$$= I_n^{(j)}r.$$

In other words,

$$S^{-1}AS = \begin{bmatrix} rI_k & Z \\ 0 & Y \end{bmatrix}.$$

Expanding: $\det(\lambda I_n - A) = \det(\lambda I_n - S^{-1}AS) = (\lambda - r)^k \det(\lambda I_{n-k} - Y)$. Hence $(\lambda - r)^k$ divides the characteristic polynomial of A so that $k \leq m$.

5. Find matrices L and U in formula (4) for each of the following X.

(a) $\begin{bmatrix} 1 & 1 \\ 1 & 0 \end{bmatrix}$
(b) $\begin{bmatrix} 1 & 1 \\ 1 & 2 \end{bmatrix}$

(c) $\begin{bmatrix} 1 & -1 \\ -1 & 0 \end{bmatrix}$
(d) $\begin{bmatrix} 1 & 0 & 1 & 1 \\ 1 & 1 & 0 & 1 \\ 1 & 1 & 1 & 0 \\ 0 & 0 & 0 & 1 \end{bmatrix}$

6. Let X be an n-square nonsingular complex matrix. Prove: there exist an upper triangular matrix R and a unitary matrix Q such that X = QR. This is called the *QR factorization of* X.

Hint: Apply Theorem 1 to X^T to obtain L_1 and U_1, lower triangular and unitary respectively, such that

$$L_1 X^T = U_1.$$

Then

$$X^T = L_1^{-1} U_1,$$

$$X = U_1^T (L_1^{-1})^T.$$

The matrix $Q = U_1^T$ is unitary and the matrix $R = (L_1^{-1})^T$ is upper triangular.

7. In #6, show that the matrices Q and R are determined uniquely up to scalar multiples of the columns, in which the scalars can be arbitrary complex numbers of modulus 1.

Hint: Apply the uniqueness discussion immediately following the proof of Theorem 1 to $X^T = R^T Q^T$.

8. Complete each of the following matrices to a nonsingular matrix by adjoining appropriate rows or columns:

(a) $\begin{bmatrix} 1 & 0 & 1 \\ 0 & 0 & 1 \\ \cdot & \cdot & \cdot \end{bmatrix}$ (b) $\begin{bmatrix} 1 & 1 & 1 \\ 2 & 2 & 1 \\ \cdot & \cdot & \cdot \end{bmatrix}$

(c) $\begin{bmatrix} 1 & 1 & 3 & 7 \\ 0 & 1 & -8 & 6 \\ \cdot & \cdot & \cdot & \cdot \\ \cdot & \cdot & \cdot & \cdot \end{bmatrix}$ (d) $\begin{bmatrix} 1 & \cdot & \cdot & \cdot \\ 2 & \cdot & \cdot & \cdot \\ 3 & \cdot & \cdot & \cdot \\ 4 & \cdot & \cdot & \cdot \end{bmatrix}$

(e) $\begin{bmatrix} 1 & 1 & 1 & 1 \\ \cdot & \cdot & \cdot & \cdot \\ \cdot & \cdot & \cdot & \cdot \\ 2 & 2 & 2 & 0 \end{bmatrix}$

9. Complete each of the following matrices to a unitary (or orthogonal) matrix by adjoining appropriate rows or columns:

(a) $\begin{bmatrix} \frac{1}{\sqrt{2}} & 0 & \frac{1}{\sqrt{2}} \\ \frac{1}{\sqrt{2}} & 0 & -\frac{1}{\sqrt{2}} \\ \cdot & \cdot & \cdot \end{bmatrix}$ (b) $\begin{bmatrix} \frac{1}{\sqrt{3}} & \frac{1}{\sqrt{3}} & \frac{1}{\sqrt{3}} \\ \frac{1}{\sqrt{2}} & 0 & -\frac{1}{\sqrt{2}} \\ \cdot & \cdot & \cdot \end{bmatrix}$

(c) $\begin{bmatrix} \frac{1}{2} & \frac{1}{2} & \cdot & \cdot \\ \frac{1}{2} & -\frac{1}{2} & \cdot & \cdot \\ \frac{1}{2} & -\frac{1}{2} & \cdot & \cdot \\ \frac{1}{2} & \frac{1}{2} & \cdot & \cdot \end{bmatrix}$

10. Let $0 \neq [a_{11}\ a_{12}\ \ldots\ a_{1n}] \in M_{1,n}(\mathbb{C})$. Prove: there exists a matrix A in $M_n(\mathbb{C})$ such that $A_{(1)} = [a_{11}\ a_{12}\ \ldots\ a_{1n}]$ and AA^* is diagonal.

Hint: Complete $A_{(1)}$ as in the discussion concerning (8).

7.2 Exercises

11. Let $D = \text{diag}(d_1, ..., d_n)$ and $\Delta = \text{diag}(\delta_1, ..., \delta_n)$. Assume that $d_1 \geq \cdots \geq d_n \geq 0$ and $\delta_1 \geq \cdots \geq \delta_n \geq 0$. Also assume that U_1, V_1, U_2, and V_2 are n-square unitary matrices that satisfy $U_1 D V_1 = U_2 \Delta V_2$. Prove: $D = \Delta$.

    Hint: Write

    $$DV_1 V_2^* = U_1^* U_2 \Delta,$$

    $$DV = U\Delta,$$

    where

    $$V_1 V_2^* = V,$$

    $$U_1^* U_2 = U.$$

    Then

    $$DVV^*D = U\Delta^2 U^*,$$

    $$D^2 = U\Delta^2 U^*.$$

    Thus

    $$\det(\lambda I_n - D^2) = \det(\lambda I_n - \Delta^2),$$

    $$\prod_{i=1}^{n}(\lambda - d_i^2) = \prod_{i=1}^{n}(\lambda - \delta_i^2).$$

    Thus the eigenvalues of D^2 and Δ^2 must be the same with the same multiplicities, and since they are nonnegative, the ordering completes the argument.

12. Let x and y be nonzero matrices in $M_{1,n}(\mathbb{C})$. Show that $p(\lambda) = (\lambda x + y)(\lambda x + y)^*$ is a real quadratic polynomial in the real scalar λ and that $p(\lambda) \geq 0$ for all real λ.

    Hint: The function $p(\lambda)$ is a sum of absolute values squared and hence $p(\lambda) \geq 0$. Also
    $$p(\lambda) = \lambda^2 xx^* + \lambda(xy^* + yx^*) + yy^*,$$
    and $(xy^* + yx^*)^* = xy^* + yx^*$ is real.

13. In #12 compute the discriminant of the quadratic polynomial $p(\lambda)$.

    Hint: The discriminant is $(2\operatorname{Re} xy^*)^2 - 4xx^*yy^* = 4((\operatorname{Re} xy^*)^2 - xx^*yy^*)$.

14. Show that if x and y are in $M_{1,n}(\mathbb{C})$ then $|xy^*|^2 \leq (xx^*)(yy^*)$ with equality iff x or y is 0, or both are nonzero and $x = \alpha y$, $\alpha \in \mathbb{C}$. This is called the *Cauchy-Schwarz Inequality*.

    Hint: If x or y is 0 then both sides of the inequality are 0. Thus assume neither is 0. From #12, $p(\lambda) \geq 0$ so that from #13, $(\operatorname{Re} xy^*)^2 - xx^*yy^* \leq 0$, $(\operatorname{Re} xy^*)^2 \leq (xx^*)(yy^*)$. Now write $|xy^*| = e^{i\theta}xy^*$ and replace x in the preceding inequality by $e^{i\theta}x$. Then $\operatorname{Re} e^{i\theta}xy^* = \operatorname{Re} |xy^*| = |xy^*|$, while $(e^{i\theta}x)(e^{i\theta}x)^* = xx^*$. Thus $|xy^*|^2 \leq (xx^*)(yy^*)$. If equality holds, then consider the pair $e^{i\theta}x$ and y:
    $$(e^{i\theta}x)y^* = e^{i\theta}(xy^*) = |xy^*|,$$
    $$\operatorname{Re}(e^{i\theta}x)y^* = |xy^*|$$
    so that
    $$(\operatorname{Re}(e^{i\theta}x)y^*)^2 = |xy^*|^2 = (xx^*)(yy^*) = (e^{i\theta}x(e^{i\theta}x)^*)(yy^*).$$

    Thus the discriminant of $p(\lambda)$ (for $e^{i\theta}x$ and y) is 0 so that $p(\lambda)$ has a real root λ_0:

7.2 Exercises

$$0 = p(\lambda_0) = (\lambda_0 e^{i\theta}x + y)(\lambda_0 e^{i\theta}x + y)^*.$$

But then $\lambda_0 e^{i\theta}x + y = 0$ and we are done.

15. Prove: if $x_1, \ldots, x_n, y_1, \ldots, y_n$ are two sets of complex numbers then

$$\left|\sum_{i=1}^{n} x_i \overline{y_i}\right|^2 \le \sum_{i=1}^{n} |x_i|^2 \sum_{i=1}^{n} |y_i|^2.$$

Discuss the cases of equality in the inequality.

16. Prove: if A and B are in $M_{m,n}(\mathbb{C})$ then

$$|\mathrm{tr}(B^*A)|^2 \le \mathrm{tr}(B^*B)\,\mathrm{tr}(A^*A)$$

with equality iff one of A or B is 0 or neither is 0 and $A = \alpha B$, $\alpha \in \mathbb{C}$.

Hint: Compute

$$\mathrm{tr}(B^*A) = \sum_{i=1, j=1}^{m,n} \overline{b_{ij}}\, a_{ij}.$$

If A and B are regarded as in $M_{1, mn}(\mathbb{C})$, we can apply the Cauchy-Schwarz inequality to obtain the result.

17. Let $x_1, \ldots, x_m$ be in $M_{1,n}(\mathbb{C})$, $m \le n$. Prove: $\det([x_i x_j^*]) \ge 0$ with equality iff there exist $\alpha_1, \ldots, \alpha_m \in \mathbb{C}$, not all 0, such that $\alpha_1 x_1 + \cdots + \alpha_m x_m = 0$.

Hint: Let X be the $m \times n$ matrix whose rows are $x_1, \ldots, x_m$ in succession. Then

$$XX^* = [(x_i x_j^*)].$$

But

$$\det(XX^*) = \sum_{\omega \in Q_{m,n}} |\det(X[1, ..., m \mid \omega])|^2 \geq 0$$

(Cauchy-Binet). If equality holds then every m-square subdeterminant of X is 0, so that $\rho(X) < m$ (§6.2 Theorem 6). Hence $\rho(X^*) < m$, and thus by §4.3 Exercises, #33, there is a $y \neq 0$, $m \times 1$, such that $X^* y = 0$. This becomes

$$\sum_{j=1}^{m} x^{*(j)} y_j = 0$$

or

$$\sum_{j=1}^{m} x_j^* y_j = 0.$$

Take conjugate transposes and set $\alpha_j = \overline{y_j}$, $j = 1, ..., m$.

18. Let $x_1, ..., x_m$ be in $M_{1,n}(\mathbb{C})$, $m \leq n$. Let $\alpha_1, ..., \alpha_m$ be arbitrary complex numbers. Prove:

$$\sum_{s,t=1}^{m} \alpha_s \overline{\alpha_t} x_s x_t^* \geq 0$$

with equality iff $\alpha_1 x_1 + \cdots + \alpha_m x_m = 0$.

Hint: We compute that $\displaystyle\sum_{s,t=1}^{m} \alpha_s \overline{\alpha_t} x_s x_t^* = \sum_{s=1}^{m} \alpha_s x_s \left(\sum_{s=1}^{m} \alpha_s x_s \right)^* \geq 0$.

19. Let $A \in M_{m,n}(\mathbb{C})$. Prove: the eigenvalues of $A^* A$ are real nonnegative numbers.

Hint: Write $A^* A x = rx$, $0 \neq x$, $n \times 1$. Then $x^* A^* A x = r x^* x$. Since $x^* x > 0$ and $x^* A^* A x = (Ax)^* (Ax) \geq 0$, it follows that $r \geq 0$.

7.2 Exercises

20. Let $A \in M_{m,n}(\mathbb{C})$. The nonnegative square roots of the eigenvalues of A^*A (see #19) are called the *singular values* of A. Compute the singular values of each of the following matrices.

    (a) $\begin{bmatrix} 1 & 0 \\ 0 & 1 \end{bmatrix}$ (b) $\begin{bmatrix} 1 & 0 \end{bmatrix}$

    (c) $\begin{bmatrix} 1 \\ 0 \end{bmatrix}$ (d) $\begin{bmatrix} 1 & 0 \\ 0 & i \end{bmatrix}$

    (e) $\begin{bmatrix} 1 & 0 \\ 0 & 1 \\ -1 & 1 \\ 1 & -1 \end{bmatrix}$ (f) $\begin{bmatrix} 1 & 0 & -1 & 1 \\ 0 & 1 & 1 & -1 \end{bmatrix}$

    (g) A, where A is any n-square unitary matrix.

21. Let A be an m × n matrix. Prove: $\begin{bmatrix} I_m & A \\ 0 & I_n \end{bmatrix}^{-1} = \begin{bmatrix} I_m & -A \\ 0 & I_n \end{bmatrix}$. (See §6.1 Exercises, #16.)

22. Let A be an m × n matrix and B be an n × m matrix. Prove:

    $$\begin{bmatrix} I_m & -A \\ 0 & I_n \end{bmatrix} \begin{bmatrix} AB & 0 \\ B & 0 \end{bmatrix} \begin{bmatrix} I_m & A \\ 0 & I_n \end{bmatrix} = \begin{bmatrix} 0 & 0 \\ B & BA \end{bmatrix}.$$

23. Using the same notation as in #22, prove:

    $$\lambda^m \det(\lambda I_n - BA) = \lambda^n \det(\lambda I_m - AB).$$

24. Let $A \in M_{m,n}(\mathbb{C})$. Prove:

    $$\lambda^m \det(\lambda I_n - A^*A) = \lambda^n \det(\lambda I_m - AA^*).$$

25. Prove: if $A \in M_{m,n}(\mathbb{C})$, then A^*A and AA^* have the same positive eigenvalues, including multiplicities.

26. Prove: if $A \in M_{m,n}(\mathbb{C})$, then $A = 0$ iff all the singular values of A are 0.

    Hint: Notice that $\text{tr}(A^*A)$ is the sum of the squares of the singular values of A.

27. Let A be an n-square hermitian matrix. Prove: the eigenvalues of A are real.

28. Let A be an n-square unitary matrix. Prove: the eigenvalues of A all have absolute value 1.

29. Let A be an n-square hermitian matrix with eigenvalues $\lambda_1, \ldots, \lambda_n$, $|\lambda_1| \geq \ldots \geq |\lambda_n|$, and singular values $\alpha_1, \ldots, \alpha_n$, $\alpha_1 \geq \ldots \geq \alpha_n$. Prove: $\alpha_i = |\lambda_i|$, $i = 1, \ldots, n$.

    Hint: From $A^*A = A^2$ we conclude that the eigenvalues of A^*A are the squares of the eigenvalues of A (see §7.1 Exercises, #13).

30. Let $A \in M_n(\mathbb{C})$. Prove: the eigenvalues of $C_k(A)$ are the $\binom{n}{k}$ products

$$\prod_{j=1}^{k} \lambda_{\omega(j)}, \quad \omega \in Q_{k,n},$$

where $\lambda_1, \ldots, \lambda_n$ are the eigenvalues of A.

Hint: From §6.1 Exercises, #20, there exists a nonsingular matrix $P \in M_n(\mathbb{C})$ such that $P^{-1}AP = T$ is upper triangular. Then from §6.3 Theorem 1(a), (d) we have

$$C_k(T) = C_k(P^{-1}AP)$$
$$= C_k(P)^{-1}C_k(A)C_k(P).$$

Hence $C_k(A)$ and $C_k(T)$ have the same eigenvalues. The main diagonal entries of T are $\lambda_1, \ldots, \lambda_n$ so that by §6.3 Exercises, #3, #4, we conclude that $C_k(T)$ is upper triangular with main diagonal entries

$$\prod_{j=1}^{k} \lambda_{\omega(j)}, \quad \omega \in Q_{k,n}.$$

31. Let $A \in M_{m,n}(\mathbb{C})$. Prove: the singular values of $C_k(A)$, $1 \leq k \leq \min\{m, n\}$, are the products

$$\prod_{j=1}^{k} \alpha_{\omega(j)}, \quad \omega \in Q_{k,n},$$

where $\alpha_1, \ldots, \alpha_n$ are the singular values of A (i.e., the square roots of the eigenvalues of A^*A).

Hint: By §6.3 Theorem 2(a), $C_k(A)^* = C_k(A^*)$. Use §6.3 Theorem 1(a) to compute that

$$C_k(A)^* C_k(A) = C_k(A^*A).$$

Use #30.

32. Assume that $A \in M_n(\mathbb{C})$ has eigenvalues $\lambda_1, \ldots, \lambda_n$ and $B \in M_m(\mathbb{C})$ has eigenvalues $\mu_1, \ldots, \mu_m$. Prove: the eigenvalues of $A \otimes B$ are the nm numbers

$$\lambda_i \mu_j, \quad i = 1, \ldots, n, \quad j = 1, \ldots, m.$$

Hint: From §6.1 Exercises, #20, #21, there are nonsingular matrices P and Q such that

$$P^{-1}AP = \begin{bmatrix} \lambda_1 & & * \\ & \ddots & \\ 0 & & \lambda_n \end{bmatrix}, \quad Q^{-1}BQ = \begin{bmatrix} \mu_1 & & * \\ & \ddots & \\ 0 & & \mu_n \end{bmatrix}.$$

Then $(P \otimes Q)^{-1} (A \otimes B) (P \otimes Q) = (P^{-1}AP) \otimes (Q^{-1}BQ)$ is upper triangular with main diagonal entries $\lambda_i \mu_j$ (see §3.3 Exercises, #22).

33. Let $A_t \in M_{n_t}(\mathbb{C})$ with eigenvalues $\lambda_{t1}, ..., \lambda_{tn}$, $t = 1, ..., p$. Prove: the eigenvalues of $A_1 \otimes \cdots \otimes A_p$ are the products

$$\lambda_{1\alpha(1)} \lambda_{2\alpha(2)} \cdots \lambda_{p\alpha(p)}, \quad \alpha \in \Gamma(n_1, ..., n_p).$$

Recall that $\Gamma(n_1, ..., n_p)$ is the set of integer sequences α for which $1 \leq \alpha(i) \leq n_i$, $i = 1, ..., p$.

Hint: Use #32 and induction on p.

34. Let $\sigma \in S_n$ have a cycle structure in which there are k_t cycles of length t, $t = 1, ..., n$. (Some of the k_t will necessarily be 0 since $\sum_{t=1}^{n} t k_t = n$.) Prove that the characteristic polynomial of the permutation matrix $A(\sigma)$ is

$$(\lambda - 1)^{k_1} (\lambda^2 - 1)^{k_2} (\lambda^3 - 1)^{k_3} (\lambda^4 - 1)^{k_4} \cdots (\lambda^n - 1)^{k_n}.$$

Hint: From §5.2 Exercises, #3, there exists a permutation matrix $A(\theta)$ such that

$$A(\theta) A(\sigma) A(\theta)^T$$

is a direct sum of k_t t-square full cycle permutation matrices, $t = 1, ..., n$,

$$\sum_{t=1}^{n} t k_t = n.$$

The characteristic polynomial of a t-square full cycle permutation matrix is $(\lambda^t - 1)$.

35. Using the same notation as in #32, prove: the characteristic polynomial of

$$I_m \otimes A + B \otimes I_n$$

7.2 Exercises

is

$$\prod_{i=1}^{n}\prod_{j=1}^{m}(\lambda - (\lambda_i + \mu_j)).$$

36. Let $A \in M_n(\mathbb{C})$. Prove: the eigenvalues of $C^{n-k}(A)$ are the $\binom{n}{k}$ numbers

$$\prod_{j=1}^{n-k} \lambda_{\omega(j)}, \quad \omega \in Q_{n-k,n},$$

where $\lambda_1, \ldots, \lambda_n$ are the eigenvalues of A.

Hint: From formula (18) in §6.3 Theorem 3, we know that

$$D_N^T C^{n-k}(A) D_N = C_{n-k}(A).$$

The matrix D_N is $\Delta_N P_N$ where

$$\Delta_N = \text{diag}(\cdots, (-1)^{s(\alpha)}, \ldots)$$

and P_N is a permutation matrix (see §6.3 formulas (15) and (16)). Thus

$$D_N^T D_N = P_N^T \Delta_N^T \Delta_N P_N = I_N.$$

In other words, $D_N^T = D_N^{-1}$ so that $C^{n-k}(A)$ and $C_{n-k}(A)$ are similar matrices and hence have the same eigenvalues. (See §7.1, in particular the narrative preceding Theorem 1.) By #30, the eigenvalues of $C_{n-k}(A)$ are the $\binom{n}{k}$ numbers

$$\prod_{j=1}^{n-k} \lambda_{\omega(j)}, \quad \omega \in Q_{n-k,n}.$$

37. Assume that $\rho(A) = r$, i.e., $A \in M_{m,n}(R)$ has rank r. Show that the rank of $C_k(A)$ is

$$\rho(C_k(A)) = \begin{cases} \binom{r}{k} & \text{if } k \leq r \\ 0 & \text{if } k > r. \end{cases}$$

Hint: Let P and Q be m-square and n-square nonsingular matrices, respectively, for which PAQ = D is in canonical form (see §4.3 Theorem 1). From §6.3 Theorem 1(a), we conclude that

$$C_k(D) = C_k(P) \, C_k(A) \, C_k(Q)$$

and that $C_k(P)$ and $C_k(Q)$ are both nonsingular. Thus

$$\rho(C_k(D)) = \rho(C_k(A)).$$

The matrix D has r 1's in positions $(1, 1), \ldots, (r, r)$. If $k > r$ then every k-square subdeterminant of D is 0 so that $\rho(C_k(A)) = 0$. If $k \leq r$ then there are precisely $\binom{r}{k}$ choices of nonzero k-square subdeterminants that lie on the main diagonal of $C_k(D)$. All other entries of $C_k(D)$ are 0.

Exercises #38 - 43 have a common notation.

38. Let $A \in M_n(\mathbb{C})$. Assume that A is similar over $\mathbb{C}$ to a diagonal matrix $D = \text{diag}(\lambda_1, \lambda_2, \ldots, \lambda_n)$ and that $X^{-1}AX = D$. Prove: $\lambda_1, \lambda_2, \ldots, \lambda_n$ are the eigenvalues of A and the corresponding eigenvectors are the columns, $x_t = X^{(t)}$, $t = 1, \ldots, n$, of the matrix X.

Hint: As we saw immediately preceding §7.1 Theorem 1, the characteristic polynomials of A and D are the same and hence $\lambda_1, \ldots, \lambda_n$ are the eigenvalues of A. From $X^{-1}AX = D$ we have $AX = XD$. Thus $AX^{(t)} = \lambda_t X^{(t)}$, $t = 1, \ldots, n$.

39. Under the same circumstances as in #38, show that if $c_1, \ldots, c_n$ are numbers for which

$$\sum_{t=1}^{n} c_t x_t = 0$$

then $c_1 = c_2 = \cdots = c_n = 0$.

Hint: The equation $\sum_{t=1}^{n} c_t x_t = 0$ is equivalent to $Xc = 0$ where $c = [c_1, c_2, ..., c_n]^T$. Since X is nonsingular it follows that $c = 0$.

40. Let $A \in M_n(\mathbb{C})$ be a nonsingular matrix and let $\lambda_1, \lambda_2, ..., \lambda_n$ denote the eigenvalues of A. Assume that A is similar to D as in #38. Suppose that x_1 is an eigenvector of A corresponding to λ_1 and that $u_1 \in M_{n,1}(\mathbb{C})$ satisfies $u_1^T x_1 = \lambda_1$. Define a matrix $B \in M_n(\mathbb{C})$ by

$$B = A - x_1 u_1^T.$$

Show that x_1 is an eigenvector of B corresponding to 0 and that the remaining eigenvalues of B are $\lambda_2, ..., \lambda_n$ with corresponding eigenvectors

$$v_i = x_i - \alpha_i x_1, \quad \alpha_i = (u_1^T x_i / \lambda_i), \quad i = 2, ..., n.$$

The method of obtaining B from A is called the *Wielandt deflation*.

Hint: Compute that

$$\begin{aligned} Bx_1 &= (A - x_1 u_1^T) x_1 \\ &= Ax_1 - x_1(u_1^T x_1) \\ &= \lambda_1 x_1 - \lambda_1 x_1 \\ &= 0. \end{aligned}$$

As in #38, let $x_2, x_3, ..., x_n$ be the eigenvectors of A corresponding to $\lambda_2, ..., \lambda_n$ and define numbers α_i by

$$\alpha_i = (u_i^T x_i / \lambda_i), \quad i = 2, ..., n.$$

Set

$$v_i = x_i - \alpha_i x_1, \quad i = 2, ..., n$$

and using $Bx_1 = 0$, compute that

$$Bv_i = B(x_i - \alpha_i x_1) = Bx_i - \alpha_i Bx_1$$

$$= Bx_i = (A - x_1 u_1^T)x_i = Ax_i - x_1 u_1^T x_i$$

$$= \lambda_i x_i - x_1(u_1^T x_i)$$

$$= \lambda_i(x_i - (u_1^T x_i / \lambda_i) x_1)$$

$$= \lambda_i (x_i - \alpha_i x_1)$$

$$= \lambda_i v_i.$$

41. As in Exercise #39, show that if $c_1 x_1 + c_2 v_2 + \cdots + c_n v_n = 0$ then $c_1 = c_2 = \cdots = c_n = 0$.

Hint: We compute that

$$c_1 x_1 + c_2 v_2 + \cdots + c_n v_n = c_1 x_1 + \sum_{j=2}^{n} c_j(x_j - \alpha_j x_1)$$

$$= (c_1 - \sum_{j=2}^{n} c_j \alpha_j) x_1 + \sum_{j=2}^{n} c_j x_j.$$

Hence by #39, $c_2 = \cdots = c_n = 0$ and

$$c_1 - \sum_{j=2}^{n} c_j \alpha_j = 0.$$

Clearly this implies that $c_1 = c_2 = \cdots = c_n = 0$.

7.2 Exercises

42. Let A satisfy the conditions of #40. Show that u_1 may be chosen to be a multiple of some (transposed) row of A, say $A_{(r)}$.

    Hint: Let $x_1 = [x_{11}\ x_{21}\ \ldots\ x_{n1}]^T$, i.e., x_1 is the first column of the matrix X and

    $$X^{-1}AX = D = \text{diag}(\lambda_1, \lambda_2, \ldots, \lambda_n).$$

    Then $Ax_1 = \lambda_1 x_1$ so that for each row $A_{(i)}$ of A there is an equation of the form

    $$A_{(i)}x_1 = \lambda_1 x_{i1},$$

    Now $x_1 \neq 0$ so that some $x_{r1} \neq 0$. Then

    $$A_{(r)}x_1 = \lambda_1 x_{r1},$$

    $$\left(\frac{A_{(r)}}{x_{r1}}\right)x_1 = \lambda_1.$$

    Set $u_1 = \left(\dfrac{A_{(r)}}{x_{r1}}\right)$ as in #40.

43. If u_1 is chosen as in #42, show that the rth row of B in #40 is 0.

    Hint: We have

    $$B = A - x_1 u_1^T = A - x_1 \left(\frac{A_{(r)}}{x_{r1}}\right).$$

    Thus,

    $$B_{(r)} = A_{(r)} - x_{r1}\frac{A_{(r)}}{x_{r1}} = A_{(r)} - A_{(r)} = 0.$$

7.2 MatLab

1. Read the Help entry for **qr**. Also review Theorem 1 and §7.2 Exercises, #6. MatLab implements the Gram-Schmidt process with the qr command. According to the MatLab documentation the important facts about the command are:

    - [Q, R] = qr(X) produces Q, a unitary matrix, and R an upper-triangular matrix, such that X = QR

    - [Q, R, E] = qr(X) produces a permutation matrix E, an upper triangular matrix R, and a unitary Q such that XE = QR and the diagonal entries of R are in decreasing order.

    Use the qr command to solve each of the following problems.

    (a) Complete X to an orthogonal matrix

    $$X = \begin{bmatrix} 1/2 & 1/2 & \cdot & \cdot \\ 1/2 & -1/2 & \cdot & \cdot \\ 1/2 & -1/2 & \cdot & \cdot \\ 1/2 & 1/2 & \cdot & \cdot \end{bmatrix}$$

    (b) Complete X to an orthogonal matrix

    $$X = \begin{bmatrix} 1/\sqrt{3} & 1/\sqrt{3} & 1/\sqrt{3} \\ 1/\sqrt{2} & 0 & -1/\sqrt{2} \\ \cdot & \cdot & \cdot \end{bmatrix}$$

    Hint: (a) Note that the matrix X can be completed to a nonsingular matrix as follows:

    $$X = \begin{bmatrix} 1/2 & 1/2 & 0 & 0 \\ 1/2 & -1/2 & 0 & 0 \\ 1/2 & -1/2 & 1 & 0 \\ 1/2 & 1/2 & 0 & 1 \end{bmatrix} \qquad (1)$$

If $X = QR$ then $Q = XR^{-1}$. The matrix R^{-1} is also upper triangular so that if $R^{-1} = [r_{ij}]$ then matching columns we have

$$Q^{(1)} = X^{(1)}r_{11}, \tag{2}$$

$$Q^{(2)} = X^{(1)}r_{12} + X^{(2)}r_{22}, \tag{3}$$

$$Q^{(3)} = X^{(1)}r_{13} + X^{(2)}r_{23} + X^{(3)}r_{33},$$

$$Q^{(4)} = X^{(1)}r_{14} + X^{(2)}r_{24} + X^{(3)}r_{34} + X^{(4)}r_{44}.$$

Since $(Q^{(i)})^* Q^{(j)} = \delta_{ij}$ (why)?, from (1) and (2) we have

$$1 = Q^{(1)*}Q^{(1)} = (r_{11}X^{(1)})^* (r_{11}X^{(1)})$$

$$= |r_{11}|^2 X^{(1)*}X^{(1)}$$

$$= |r_{11}|^2.$$

Thus $Q^{(1)}$ is a unit modulus multiple of $X^{(1)}$, i.e., $Q^{(1)} = X^{(1)}r_{11}$. Next, from (2) and (3) we have

$$0 = X^{(1)*}Q^{(2)} = X^{(1)*}X^{(1)}r_{12} + X^{(1)*}X^{(2)}r_{22}$$

$$= r_{12}$$

so that $Q^{(2)}$ is a unit modulus multiple of $X^{(2)}$, i.e., $Q^{(2)} = X^{(2)}r_{22}$. In other words, except for possible unit modulus multiples in the first two columns, Q is the required completion of X. However, unit multiples have no effect on the unitary property so that we can simply augment X by making $X^{(3)} = Q^{(3)}$ and $X^{(4)} = Q^{(4)}$. Hence, the required MatLab commands are (with X the matrix in (1))

$$[Q, R] = qr(X);$$

$$Y = [\ X(:, [1\ 2])\quad Q(:, [3\ 4])\ ].$$

Actually, the qr command will sometimes take advantage of the fact that $X^{(1)}$ and $X^{(2)}$ are the first two columns of an orthogonal matrix, and use them in Q. Thus, the second command above is not always necessary.

2. Review the Help entry for **rref** and **find**. Write a MatLab function called comp that accepts an n × r matrix X of rank r, n > r, and returns a nonsingular matrix S such that $S[1, ..., n \mid 1, ..., r] = X$. Save comp in eigen.

Hint: First review §4.2 MatLab, #5. Let $Y = X^T$ and (in the notation of that exercise) obtain the indices $n_1 < \cdots < n_r$ in the Hermite normal form of Y. Let the complementary set to $\{n_1, ..., n_r\}$ in $\{1, ..., n\}$ be $c_1 < \cdots < c_{n-r}$. In columns $c_1, ..., c_{n-r}$ of Y adjoin (in order) the columns of the identity matrix I_{n-r} and in the remaining columns simply adjoin zeros. The transpose of the resulting matrix is the required S. In §6.2 MatLab, #3 the function cmpl is defined: it returns the complement of a sequence.

```
function S = comp(X)
%COMP accepts an n x r matrix
%X of rank r < n and returns an n x n
%non-singular S in which X appears
%in the first r columns.
[n, r] = size(X);
cc=[ ];                    %cc = [n1, ..., nr]
Y = X.';
H = rref(Y);
for row = 1:r
        v = find(H(row,:));
        cc = [cc v(1)];
end
c = cmpl(cc, n);           %c is complement to cc in 1, ..., n
Y((r+1):n, c) = eye(n - r); %adjoins eye(n - r) to Y
S = Y.'
```

3. Write a MatLab function called isom that accepts an n × r matrix X of rank r < n that satisfies $X^*X = I_r$, and returns an n-square unitary matrix U such that

7.2 MatLab

$U[1, ..., n \mid 1, ..., r] = X$. A matrix X that can be completed to a unitary matrix is sometimes called a *partial isometry*. Save isom in eigen.

Hint: The condition $X^*X = I_r$ simply means that X is a partial isometry.

```
function U=isom(X)
%isom accepts an n x r, n > r
%partial isometry X, and
%returns a unitary U
%with X in the first r
%columns of U
[n, r] = size(X);
S = comp(X);
[Q, R] = qr(S);
U = [X Q(:, (r+1):n)];
```

4. Use the function comp in #2 to work §7.2 Exercises, #8.

5. Use the function isom in #3 to work §7.2 Exercises, #9.

6. Use eig and sqrt to work §7.2 Exercises, #20(a) - (f).

7. Read the Help entries for **sort, find,** and **kron**. Then write a MatLab function called kronsv that accepts two matrices A and B, and a positive number tol, and returns in ascending order the positive singular values of $A \otimes B$ that exceed tol. Do not use the svd command in your function. Save kronsv in eigen.

Hint: Recall that $(A \otimes B)(A \otimes B)^* = AA^* \otimes BB^*$. Also, the eigenvalues of any kronecker product, $A \otimes B$, are all the products $\lambda_i \mu_j$ where the λ's are the eigenvalues of A and the μ's are the eigenvalues of B (see §7.2 Exercises, #32). Thus the singular values of $A \otimes B$ are all possible products $\alpha_i \beta_j$ where the α's are the singular values of A and the β's are the singular values of B.

```
function s = kronsv(A, B, tol)
%KRONSV accepts two matrices
%A and B and a positive tol.
%It returns in ascending order the
```

```
%singular values of the kronecker
%product of A and B that exceed tol.
K = kron(A*A', B*B');
ev=sqrt(eig(K));
I = find(ev > tol);
s = sort(ev(I));
```

8. Read the Help entries for **eig**, **polyval**, and **polyvalm**. The MatLab documentation states that:

    - if $p = [p_n \; p_{n-1} \; \cdots \; p_0]$ and s is a number then

    $$\text{polyval}(p, s) = p_n s^n + p_{n-1} s^{n-1} + \cdots + p_1 s + p_0$$

    - if A is an m-square matrix then

    $$\text{polyvalm}(p, A) = p_n A^n + p_{n-1} A^{n-1} + \cdots + p_1 A + p_0 I_m$$

    Write a MatLab function named polymat that

    - accepts a polynomial $p = p_n \lambda^n + p_{n-1} \lambda^{n-1} + \cdots + p_1 \lambda + p_0$

    - accepts an m-square complex matrix A

    - returns a row vector $[p(\lambda_1) \; p(\lambda_2) \; \cdots \; p(\lambda_m)]$ where $\lambda_1, \ldots, \lambda_m$ are the eigenvalues of A

    - evaluates the matrix $p(A) = p_n A^n + p_{n-1} A^{n-1} + \cdots + p_1 A + p_0 I_m$

    - returns a row vector $[\mu_1 \; \mu_2 \; \cdots \; \mu_m]$ where the μ's are the eigenvalues of $p(A)$

    Hint:
    ```
    function [pval, eigp] = polymat(p, A)
    %polymat accepts a polynomial p
    ```

```
%and a matrix A.  It returns the
%eigenvalues of p(A) and also the
%values of p on the eigenvalues of A.
mu=eig(polyvalm(p, A));
c = polyval(p, eig(A));
pval = sort(c).';
eigp = sort(mu).';
```

9. Run polymat for several random p and A. What appears to be the case? Prove your guess.

Hint: It appears that the eigenvalues of $p(A)$ are the numbers $p(\lambda_t)$, $t = 1, ..., m$, where $\lambda_1, ..., \lambda_m$ are the eigenvalues of A.

Use §7.1 Exercises, #13 to obtain a nonsingular P such that

$$P^{-1}AP = \begin{bmatrix} \lambda_1 & & * \\ & \ddots & \\ 0 & & \lambda_m \end{bmatrix}.$$

Then

$$(P^{-1}A^kP) = \begin{bmatrix} \lambda_1^k & & * \\ & \ddots & \\ 0 & & \lambda_m^k \end{bmatrix}$$

and hence if $p(\lambda) = c_n\lambda^n + c_{n-1}\lambda^{n-1} + \cdots + c_1\lambda + c_0$, it follows that

$$P^{-1}p(A)P = \begin{bmatrix} p(\lambda_1) & & * \\ & \ddots & \\ 0 & & p(\lambda_m) \end{bmatrix}.$$

7.2 Glossary

algebraic multiplicity 377
Cauchy-Schwarz Inequality 390
eig 406
find 404
geometric multiplicity 377
Gram-Schmidt process 378
kron 405
partial isometry 405
polyval 406
polyvalm 406
QR factorization 387
qr 402
rref 404
singular values 393
sort 405
Wielandt deflation 399

Chapter 8

Triangularization

Topics
- *Schur triangularization theorem*
- *normal matrices*
- *Cayley parameterization*
- *polar factorization*
- *singular value decomposition*
- *numerical radius*

8.1 The Triangular Form

In this chapter the important problem of reducing a complex matrix to triangular form is explained. The first section contains the proof and some generalizations of the Schur triangularization theorem. Theorem 3 shows that a set of pairwise commuting matrices can be simultaneously triangularized with the same unitary similarity. In the second section the problem of unitarily diagonalizing a set of pairwise commuting normal matrices is carefully analyzed. This section also contains the Cayley parameterization theorem and the polar factorization theorem. The final section is devoted to the important singular value decomposition (SVD) and to inequalities that relate eigenvalues, singular values and the numerical radius.

The single most important result concerning general square complex matrices is the *Schur triangularization theorem,* published by I. Schur in 1909.

Theorem 1 (Schur Triangularization Theorem).

Let A be an n-square complex matrix. Then there exists a unitary matrix U such that

$$U^*AU = T. \qquad (1)$$

The matrix T is upper triangular with the eigenvalues of A along the main diagonal in any prescribed order. In other words, if $\lambda_1, \lambda_2, ..., \lambda_n$ are the eigenvalues of A in some prescribed order, then there exists a unitary U such that

$$U^*AU = \begin{bmatrix} \lambda_1 & & & * \\ 0 & \lambda_2 & & \\ \vdots & & \ddots & \\ 0 & \cdots & 0 & \lambda_n \end{bmatrix}.$$

Proof.

Suppose λ_1 is an eigenvalue of A to be placed in the (1, 1) position of T. Let $x_1 \in M_{n,1}(\mathbb{C})$ be an eigenvector of A corresponding to λ_1, chosen so that $x_1^*x_1 = 1$. Let X be a unitary matrix with x_1 as the first column (see §7.2 formula (8) et seq.). Then

$$(X^*AX)^{(1)} = X^*Ax_1$$

$$= X^*x_1\lambda_1$$

$$= X^*X^{(1)}\lambda_1$$

$$= (X^*X)^{(1)}\lambda_1$$

$$= I_n^{(1)}\lambda_1.$$

Thus

8.1 The Triangular Form

$$X^*AX = \begin{bmatrix} \lambda_1 & * \\ 0 & \\ \vdots & B \\ 0 & \end{bmatrix}$$

where $B \in M_{n-1}(\mathbb{C})$. It is easy to relate the eigenvalues of A and B:

$$\det(\lambda I_n - A) = \det(X^*(\lambda I_n - A)X)$$

$$= \det(\lambda I_n - X^*AX)$$

$$= (\lambda - \lambda_1)\det(\lambda I_{n-1} - B).$$

Thus the remaining eigenvalues of A, namely $\lambda_2, \ldots, \lambda_n$, are the eigenvalues of B. Suppose $\lambda_2, \ldots, \lambda_n$ are to be placed, in that order, down the main diagonal of T. Then by induction on n choose Y, an (n - 1)-square unitary matrix, such that

$$Y^*BY = \begin{bmatrix} \lambda_2 & & * \\ & \ddots & \\ 0 & & \lambda_n \end{bmatrix}.$$

Set $U = X([1] \oplus Y)$ and observe that $U^*AU = T$. The matrix U is clearly unitary. ∎

Suppose that the eigenvalues of A are distinct and $T = [\,t_{ij}\,]$, $\Delta = [\Delta_{ij}]$ are two upper triangular matrices unitarily similar to A with the same ordering of $\lambda_1, \ldots, \lambda_n$ along the main diagonal:

$$U^*AU = T,$$

$$V^*AV = \Delta.$$

Then $UTU^* = V\Delta V^*$, $V^*UT = \Delta V^*U$. Set $W = V^*U$. Then $WT = \Delta W$ so that matching first columns of WT and ΔW we have:

$$[w_{11}, w_{21}, w_{31}, \ldots, w_{n1}]^T \lambda_1 = [\lambda_1, 0, \ldots, 0]^T w_{11} + [\Delta_{12}, \lambda_2, 0, \ldots, 0]^T w_{21}$$

$$+ \cdots + [\Delta_{1n}, \Delta_{2n}, \ldots, \Delta_{n-1,n}, \lambda_n]^T w_{n1}. \qquad (2)$$

Equating the (n,1) entries we obtain $w_{n1}\lambda_1 = w_{n1}\lambda_n$ so $w_{n1} = 0$ because $\lambda_1 \neq \lambda_n$. Continue by equating, in order, the (n - 1, 1), (n - 2, 1), ..., (2, 1) entries to conclude that $w_{21} = \ldots = w_{n1} = 0$. Hence $W^{(1)} = [w_{11}, 0, \ldots, 0]^T$ and, since W is unitary, it also follows that $W_{(1)} = [w_{11}, 0, \ldots, 0]$ (i.e., $|w_{11}| = 1$ so that $\sum_{j=2}^{n} |w_{1j}|^2 = 0$). Let $w_{11} = \theta_1, |\theta_1| = 1$. Then $W = [\theta_1] \oplus Z$ where Z is an (n - 1)-square unitary matrix. By block multiplication,

$$ZT(1 \mid 1) = \Delta(1 \mid 1)Z \qquad (3)$$

so that an inductive argument yields $Z = \text{diag}(\theta_2, \ldots, \theta_n), |\theta_2| = \cdots = |\theta_n| = 1$. This argument has established the following: if the eigenvalues of A are distinct and their order along the main diagonal in (1) is prescribed, then the unitary matrix U in (1) is determined uniquely to within post-multiplication by a diagonal unitary matrix, and the triangular matrix T is determined uniquely to within a unitary similarity by a diagonal unitary matrix.

We remark that Theorem 1 is not the same as the result in §6.1 Exercises, #20. In that problem the matrix effecting the similarity to triangular form is not necessarily unitary.

Suppose next that A has distinct eigenvalues $\mu_1, \ldots, \mu_p$ of algebraic multiplicities $m_1, \ldots, m_p$ respectively. Then A is similar (unitarily, if A is a complex matrix) to a partitioned triangular matrix:

8.1 The Triangular Form

$$T = [T_{ij}] = \begin{bmatrix} \begin{bmatrix} \mu_1 & & * \\ & \ddots & \\ 0 & & \mu_1 \end{bmatrix} & * & * & & * \\ 0 & \begin{bmatrix} \mu_2 & & * \\ & \ddots & \\ 0 & & \mu_2 \end{bmatrix} & * & & * \\ 0 & 0 & \ddots & & * \\ 0 & 0 & 0 & & \begin{bmatrix} \mu_p & & * \\ & \ddots & \\ 0 & & \mu_p \end{bmatrix} \end{bmatrix}. \qquad (4)$$

The (i, j) block in T is $m_i \times m_j$. Fix a particular pair of integers, $1 \leq s < t \leq p$, and let E be a matrix partitioned precisely as in (4) with identity blocks down the main diagonal, an $m_s \times m_t$ matrix X in the (s, t) block, and all other blocks equal to 0. The matrix E^{-1} is precisely the same as E except that X is replaced by -X. The effect of the similarity

$$E^{-1}TE \qquad (5)$$

is to replace T by a matrix identical to T except that block rows and columns s and t are affected and, in particular, the (s, t) block is replaced by

$$T_{st} - XT_{tt} + T_{ss}X. \qquad (6)$$

In fact, letting solid lines denote block rows and columns, only the blocks on the double lines are changed.

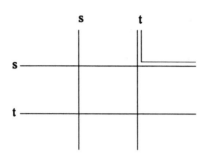

Fig. 1

For example, suppose $s = 2$, $t = 4$:

$$E^{-1}TE = \begin{bmatrix} I_{m_1} & 0 & 0 & 0 \\ 0 & I_{m_2} & 0 & -X \\ 0 & 0 & I_{m_3} & 0 \\ 0 & 0 & 0 & I_{m_4} \end{bmatrix} \begin{bmatrix} T_{11} & T_{12} & T_{13} & T_{14} \\ 0 & T_{22} & T_{23} & T_{24} \\ 0 & 0 & T_{33} & T_{34} \\ 0 & 0 & 0 & T_{44} \end{bmatrix} E$$

$$= \begin{bmatrix} T_{11} & T_{12} & T_{13} & T_{14} \\ 0 & T_{22} & T_{23} & T_{24} - XT_{44} \\ 0 & 0 & T_{33} & T_{34} \\ 0 & 0 & 0 & T_{44} \end{bmatrix} \begin{bmatrix} I_{m_1} & 0 & 0 & 0 \\ 0 & I_{m_2} & 0 & X \\ 0 & 0 & I_{m_3} & 0 \\ 0 & 0 & 0 & I_{m_4} \end{bmatrix}$$

$$= \begin{bmatrix} T_{11} & T_{12} & T_{13} & T_{12}X + T_{14} \\ 0 & T_{22} & T_{23} & T_{24} - XT_{44} + T_{22}X \\ 0 & 0 & T_{33} & T_{34} \\ 0 & 0 & 0 & T_{44} \end{bmatrix}. \quad (7)$$

Since $\mu_t \neq \mu_s$ it follows from §6.1 Exercises, #23 that X may be chosen so that (6) is 0. We can replace T by a succession of similar matrices, annihilating (i.e., making 0) off-diagonal blocks in succession in the following order: $(p - 1, p)$; $(p - 2, p - 1)$, $(p - 2, p)$; $(p - 3, p - 2)$, $(p - 3, p - 1)$, $(p - 3, p)$; ...; $(1, 2)$, $(1, 3)$, ..., $(1, p)$. For example, in (7) we annihilate, in order, blocks $(3, 4)$; $(2, 3)$, $(2, 4)$; $(1, 2)$, $(1, 3)$, $(1, 4)$. According to Fig. 1, the annihilation of a block at any stage does not affect the blocks already annihilated. We have proved the following important result.

8.1 The Triangular Form

Theorem 2.

Let A have distinct eigenvalues $\mu_1, \ldots, \mu_p$ of algebraic multiplicities $m_1, \ldots, m_p$ respectively. Then A is similar to a direct sum of upper triangular matrices of the form

$$\begin{bmatrix} \mu_s & & * \\ & \ddots & \\ 0 & & \mu_s \end{bmatrix}, \quad m_s \times m_s, \quad s = 1, \ldots, p. \tag{8}$$

Theorems 1 and 2 are concerned with reducing a single matrix to upper triangular form. If the matrix is in $M_n(\mathbb{C})$ then the matrix effecting the similarity can be chosen to be unitary. Under certain circumstances a set of matrices can all be brought to upper triangular form with the same matrix effecting the similarity.

Let $A_0 \in M_n(\mathbb{R})$ be a fixed matrix of rank r. The result in §6.2 Theorem 6 then implies that A_0 contains an r-square non-zero subdeterminant, and moreover every larger subdeterminant of A_0 must be 0. Suppose an r-square nonzero subdeterminant of A_0 lies in rows α and columns ω, $\alpha, \omega \in Q_{r,n}$. Then obviously the n × r matrix

$$B = A_0[1, \ldots, n \mid \omega] \tag{9}$$

has rank r. Next let $A \in M_n(\mathbb{R})$ be any matrix that commutes with A_0:

$$AA_0 = A_0A. \tag{10}$$

The n × r matrix AB is the n × r submatrix of AA_0 lying in columns ω, i.e.,

$$AB = (AA_0)[1, \ldots, n \mid \omega]. \tag{11}$$

Moreover,

$$(AB)^{(k)} = (AA_0)^{(\omega_k)} \quad \text{(from (11))}$$

$$= (A_0A)^{(\omega_k)} \quad \text{(from (10))}$$

$$= A_0 A^{(\omega_k)}$$

$$= \sum_{j=1}^{n} A_0^{(j)} a_{j\omega_k}, \quad k = 1, \ldots, r. \tag{12}$$

Consider next the $n \times (r + 1)$ matrix C obtained by appending column j of A_0 to the matrix B, where j is any integer not in ω, i.e.,

$$C = [B \mid A_0^{(j)}], \quad j \notin \omega. \tag{13}$$

Obviously, since $\rho(A_0) = \rho(B) = r$ it follows from §6.2 Theorem 6 that the largest non-vanishing subdeterminant of C is r-square (otherwise $\rho(A_0)$ would exceed r). But then, by §4.3 Exercises, #37, it follows that the equation

$$Bu = A_0^{(j)} \tag{14}$$

can be solved for the $r \times 1$ matrix u. The equation (14) can be viewed as follows:

$$A_0^{(j)} = \sum_{t=1}^{r} B^{(t)} u_t$$

$$= \sum_{t=1}^{r} A_0^{(\omega_t)} u_t. \tag{15}$$

Of course, u depends on $A_0^{(j)}$, $j \notin \omega$, but only columns ω of A_0 appear on the right in (15). If (15) is used to replace each of the columns $A_0^{(j)}$, $j \notin \omega$, that appear on the right in (12), the resulting expression will involve only columns $A_0^{(\omega_t)}$, $t = 1, \ldots, r$, of A_0:

$$(AB)^{(k)} = \sum_{t=1}^{r} A_0^{(\omega_t)} w_{tk}, \quad k = 1, \ldots, r, \tag{16}$$

for appropriate scalars w_{tk}, $t = 1, \ldots, r$, $k = 1, \ldots, r$. If we set

$$W = [w_{tk}] \in M_r(R)$$

8.1 The Triangular Form

then (16) has the simple formulation

$$AB = BW. \tag{17}$$

Assume now that A_0 is actually singular so that $r < n$. The $n \times r$ matrix B can be augmented with $n - r$ columns so that the resulting n-square matrix P is nonsingular: in fact, if $\alpha' \in Q_{n-r, n}$ is the sequence complementary to α then it is clear from the Laplace expansion theorem that if P is the n-square matrix

$$P = [\ B\ |\ e_{\alpha'_1}\ |\ e_{\alpha'_2}\ |\ \cdots\ |\ e_{\alpha'_{n-r}}\ ] \tag{18}$$

then

$$\det(P) = \pm \det(B[\alpha\ |\ 1, ..., r])$$
$$= \pm \det(A_0[\alpha\ |\ \omega])$$

and hence P is nonsingular (the vertical bars in (18) indicate column separators). It is important in what follows to observe that P depends only on the matrix A_0 and not on the matrix A. Let $x_t \in M_{n,1}(R)$ be defined by

$$x_t = P^{-1} A^{(\alpha'_t)}, \quad t = 1, ..., n-r. \tag{19}$$

We compute that

$$AP = [AB\ |\ Ae_{\alpha'_1}\ |\ Ae_{\alpha'_2}\ |\ \dots\ |\ Ae_{\alpha'_{n-r}}\ ] \qquad \text{(from (18))}$$

$$= [\ BW\ |\ A^{(\alpha'_1)}\ |\ A^{(\alpha'_2)}\ |\ \dots\ |\ A^{(\alpha'_{n-r})}\ ] \qquad \text{(from (17))}$$

$$= [BW\ |\ Px_1\ |\ Px_2\ |\ \dots\ |\ Px_{n-r}\ ]. \qquad \text{(from (19))}$$

But obviously, from the definition (18), B and P are identical in columns $1, ..., r$ so that since W is r-square,

$$BW = P[1, ..., n\ |\ 1, ..., r\ ]W. \tag{20}$$

Substituting (20) in the preceding formula for AP we have

$$AP = [P[1, ..., n \mid 1, ..., r]W \mid Px_1 \mid Px_2 \mid ... \mid Px_{n-r}]. \tag{21}$$

But since W is r-square and P is nonsingular, (21) has the following form:

$$P^{-1}AP = \begin{bmatrix} W & X \\ 0 & Z \end{bmatrix} \tag{22}$$

where the 0 block is $(n - r) \times r$ and Z is $(n - r)$-square.

The preceding discussion makes it possible to prove the following important result.

Theorem 3.

Let $F \subset M_n(\mathbb{C})$ be a set of n-square pairwise commuting matrices. Then there exists a vector v, $v \in M_{n,1}(\mathbb{C})$, such that v is a common eigenvector of every matrix $A \in F$.

Proof.

Since any multiple of I_n has every nonzero v as an eigenvector, we lose no generality in assuming that all multiples of I_n have been deleted from the set F. Let $H \in F$ and let μ be an eigenvalue of H. Then $A_0 = H - \mu I_n$ is singular, and since all multiples of I_n have been deleted from F, it follows that $n > \rho(A_0) = r > 0$. Moreover, any A that commutes with H must commute with A_0. Thus we are in a position to apply (22). Namely, there exists a nonsingular matrix P, depending only on $A_0 = H - \mu I_n$, so that if $A \in F$ then

$$P^{-1}AP = \begin{bmatrix} W_A & X \\ 0 & Z \end{bmatrix} \tag{23}$$

where W_A is r-square, $r < n$, and we have indicated the dependence of W on A with the subscript. We emphasize that the matrix P reduces every $A \in F$ to the same form on the right in (23) and that one matrix P works for all $A \in F$. Notice also that if A and K are

8.1 The Triangular Form

two matrices in F for which (23) holds then

$$(P^{-1}AP)(P^{-1}KP) = P^{-1}AKP$$

$$= P^{-1}KAP$$

$$= (P^{-1}KP)(P^{-1}AP). \qquad (24)$$

A simple block multiplication using (23) then shows that the matrices W_A and W_K also commute. Hence the set of r-square matrices, $\{W_A, A \in F\}$, are also pairwise commutative. Since $r < n$ we can use induction on n to conclude that there exists an $r \times 1$ vector z which is a common eigenvector of every W_A, $A \in F$. Define the $n \times 1$ vector v as

$$v = P[z_1, ..., z_r, 0, ..., 0]^T, \qquad (25)$$

and compute that

$$Av = P \begin{bmatrix} W_A & X \\ 0 & Z \end{bmatrix} P^{-1} v$$

$$= P \begin{bmatrix} W_A z \\ 0 \\ \vdots \\ 0 \end{bmatrix}$$

$$= \lambda_A P \begin{bmatrix} z_1 \\ \vdots \\ z_r \\ 0 \\ \vdots \\ 0 \end{bmatrix}$$

$$= \lambda_A v \qquad (26)$$

where λ_A is the eigenvalue of W_A corresponding to the eigenvector z. The vector v is the required common eigenvector of every $A \in F$. ∎

Theorem 4.

Let $F \subset M_n(\mathbb{C})$ be a set of n-square pairwise commuting matrices. Then there exists a unitary matrix P such that P^*AP is upper triangular for every $A \in F$.

Proof.

From Theorem 3 obtain $v \in M_{n,1}(\mathbb{C})$, a common eigenvector of every $A \in F$. The vector v may be normalized so that $v^*v = 1$. Define a unitary matrix Q whose first column is v. Then it is simple to check that

$$Q^*AQ = \begin{bmatrix} \lambda_A & * \\ 0 & \\ \vdots & Z_A \\ 0 & \end{bmatrix}$$

as in (23). Moreover, as in the proof of Theorem 3, the pairwise commutativity of the matrices $A \in F$ implies that the matrices Z_A are also pairwise commutative. By induction on n we can obtain an (n - 1)-square unitary matrix L such that L^*Z_AL is upper triangular for every Z_A. It is simple to confirm that if

$$P = Q([1] \oplus L)$$

then P is unitary and P^*AP is upper triangular for every $A \in F$. ∎

There are several observations that can be made about Theorems 3 and 4. First, if the matrices in F are normal then, in fact, there is a unitary P such that P^*AP is a diagonal matrix for every $A \in F$. The reason for this is quite simple: as we shall see in the proof of §8.2 Theorem 1, a normal triangular matrix must be diagonal.

8.1 The Triangular Form

Next, if $F \subset M_n(\mathbb{R})$ and if the eigenvalues of every $A \in F$ lie in $\mathbb{R}$ as well, then the matrix P in Theorem 4 can be chosen to be real orthogonal. The reason for this is the simple observation that the common eigenvector v of every $A \in F$ may be chosen to be real since, except for the computation of eigenvalues, the construction of v involves only rational operations.

Finally, in Theorem 4 it is possible to arrange for the eigenvalues of a particular A to appear in any prescribed order along the main diagonal in its triangular form. But, in general, this cannot be done for more than one $A \in F$.

We close this section by stating, without proof, a version of the Schur theorem (Theorem 1) for matrices in $M_n(\mathbb{R})$. Note first that if A has real entries then the nonreal eigenvalues appear in complex conjugate pairs. Thus, let the real eigenvalues of A in $M_n(\mathbb{R})$ be $r_1, \ldots, r_p$, and let the nonreal eigenvalues be $a_1 \pm ib_1, \ldots, a_m \pm ib_m$, $p + 2m = n$. For $t = 1, \ldots, m$ let D_t be the 2-square real matrix

$$D_t = \begin{bmatrix} a_t & b_t \\ -b_t & a_t \end{bmatrix}.$$

Note that D_t is a real matrix whose eigenvalues are $a_t \pm ib_t$. The following is a statement of the real version of the Schur Theorem. We use the preceding notation.

Theorem 5.

Let $A \in M_n(\mathbb{R})$. Then there exists a real orthogonal matrix U such that $U^T A U$ has the following form: the eigenvalues $r_1, \ldots, r_p$ appear in any desired order down the main diagonal; the 2 × 2 blocks D_t appear down the main diagonal in any desired order, $t = 1, \ldots, m$; all entries below the main diagonal outside of the blocks D_t are 0.

MatLab implements the real Schur form as we will see in §8.1 MatLab.

8.1 Exercises

1. Prove: if A is in $M_n(\mathbb{C})$ and has distinct eigenvalues then A is similar to a diagonal matrix. Can the matrix effecting the similarity always be chosen to be unitary?

   Hint: Apply Theorem 2 with $p = n$ to conclude that A is similar to a diagonal matrix. To see that this similarity can not always be achieved with a unitary matrix, consider the matrix

   $$A = \begin{bmatrix} 1 & 1 \\ 0 & 0 \end{bmatrix}.$$

   If A were unitarily similar to a diagonal matrix, the diagonal matrix would necessarily be

   $$D = \begin{bmatrix} 1 & 0 \\ 0 & 0 \end{bmatrix}.$$

   But then $2 = \text{tr}(A^*A) = \text{tr}(U^*D^*UU^*DU) = \text{tr}(D^*D) = 1$, a contradiction.

2. For each of the following matrices A, find a matrix S such that $S^{-1}AS$ is a diagonal matrix by using the method in the proof of Theorem 2.

   (a) $\begin{bmatrix} 1 & 1 \\ 0 & 2 \end{bmatrix}$

   (b) $\begin{bmatrix} 2 & 1 \\ 0 & 0 \end{bmatrix}$

   (c) $\begin{bmatrix} -1 & 1 \\ 0 & 1 \end{bmatrix}$

   (d) $\begin{bmatrix} 1 & 1 \\ 0 & 3 \end{bmatrix}$

   (e) $\begin{bmatrix} 1 & 0 & 1 \\ 0 & -1 & 1 \\ 0 & 0 & 0 \end{bmatrix}.$

8.1 Exercises

Hint: We use equation (6) to eliminate the off-diagonal elements. Because each matrix in this exercise has distinct eigenvalues, each matrix X in §8.1 formula (6) is 1×1 and so we can easily solve for X in terms of A:

$$X = \frac{a_{st}}{a_{tt} - a_{ss}}, \quad s < t.$$

This yields:

(a) $S^{-1}AS = \begin{bmatrix} 1 & -1 \\ 0 & 1 \end{bmatrix} \begin{bmatrix} 1 & 1 \\ 0 & 2 \end{bmatrix} \begin{bmatrix} 1 & 1 \\ 0 & 1 \end{bmatrix} = \begin{bmatrix} 1 & 0 \\ 0 & 2 \end{bmatrix}$

(b) $S^{-1}AS = \begin{bmatrix} 1 & 1 \\ 0 & 1 \end{bmatrix} \begin{bmatrix} 2 & 1 \\ 0 & 1 \end{bmatrix} \begin{bmatrix} 1 & -1 \\ 0 & 1 \end{bmatrix} = \begin{bmatrix} 2 & 0 \\ 0 & 1 \end{bmatrix}$

(c) $S^{-1}AS = \begin{bmatrix} 1 & -1/2 \\ 0 & 1 \end{bmatrix} \begin{bmatrix} -1 & 1 \\ 0 & 1 \end{bmatrix} \begin{bmatrix} 1 & 1/2 \\ 0 & 1 \end{bmatrix} = \begin{bmatrix} -1 & 0 \\ 0 & 1 \end{bmatrix}$

(d) $S^{-1}AS = \begin{bmatrix} 1 & -1/2 \\ 0 & 1 \end{bmatrix} \begin{bmatrix} 1 & 1 \\ 0 & 3 \end{bmatrix} \begin{bmatrix} 1 & 1/2 \\ 0 & 1 \end{bmatrix} = \begin{bmatrix} 1 & 0 \\ 0 & 1 \end{bmatrix}$

(e) $S^{-1}AS =$

$\begin{bmatrix} 1 & 0 & 1 \\ 0 & 1 & 0 \\ 0 & 0 & 1 \end{bmatrix} \begin{bmatrix} 1 & 0 & 0 \\ 0 & 1 & -1 \\ 0 & 0 & 1 \end{bmatrix} \begin{bmatrix} 1 & 0 & 1 \\ 0 & -1 & 1 \\ 0 & 0 & 0 \end{bmatrix} \begin{bmatrix} 1 & 0 & 0 \\ 0 & 1 & 1 \\ 0 & 0 & 1 \end{bmatrix} \begin{bmatrix} 1 & 0 & -1 \\ 0 & 1 & 0 \\ 0 & 0 & 1 \end{bmatrix} = \begin{bmatrix} 1 & 0 & 0 \\ 0 & -1 & 0 \\ 0 & 0 & 0 \end{bmatrix}$

3. Reduce the matrix

$$\begin{bmatrix} 2 & 1 & 0 & -1 \\ -1 & 1 & 1 & 1 \\ 0 & 1 & 1 & 0 \\ 1 & 1 & 0 & 0 \end{bmatrix}$$

to upper triangular form by means of a unitary similarity.

Hint: The required triangular form is

$$\begin{bmatrix} 2 & -1 & -\frac{1}{\sqrt{2}} & \frac{1}{\sqrt{2}} \\ 0 & 1 & -\frac{1}{\sqrt{2}} & \frac{3}{\sqrt{2}} \\ 0 & 0 & 0 & 1 \\ 0 & 0 & 0 & 1 \end{bmatrix}$$

4. Prove: if $A \in M_n(\mathbb{R})$ has real eigenvalues, then the matrix U in Theorem 1 can be chosen to be real orthogonal.

   Hint: Let λ_1 be an eigenvalue of A, as in the proof of Theorem 1. Since λ_1 is real and $A \in M_n(\mathbb{R})$, $Ax_1 = \lambda_1 x_1$ is solvable for real x_1, $x_1^T x_1 = 1$. Then x_1 can be completed to a real orthogonal matrix X with first column x_1, and $X^*AX = X^TAX$ has first column equal to $\lambda_1 e_1$, as in the proof of Theorem 1. (The vector e_1 is $[1, 0, ..., 0]^T$). Since $B = X^*AX [2, ..., n | 2, ..., n] \in M_{n-1}(\mathbb{R})$ and has real eigenvalues, induction can be easily applied to obtain the result.

5. Prove: if $A_1, ..., A_p$ are square complex matrices then there exist unitary matrices $U_1, ..., U_p$ such that

$$(U_1 \otimes \cdots \otimes U_p)^*(A_1 \otimes \cdots \otimes A_p)(U_1 \otimes \cdots \otimes U_p)$$

   is upper triangular.

   Hint: For each $k \in \{1, ..., p\}$, apply Schur's Theorem (Theorem 1) to obtain a unitary matrix U_k for which $U_k^*A_kU_k$ is upper triangular. Then

$$(U_1 \otimes \cdots \otimes U_p)^*(A_1 \otimes \cdots \otimes A_p)(U_1 \otimes \cdots \otimes U_p)$$

$$= (U_1^* \otimes \cdots \otimes U_p^*)(A_1 \otimes \cdots \otimes A_p)(U_1 \otimes \cdots \otimes U_p)$$

$$= U_1^*A_1U_1 \otimes \cdots \otimes U_p^* A_pU_p \, .$$

Since the Kronecker product of upper triangular matrices is again upper triangular (see §3.3 Exercises, #22), the result follows.

6. Assume $A \in M_n(\mathbb{C})$ and let $A = H + iK$, $H = (A + A^*)/2$, $K = (A - A^*)/2i$. Prove: H and K are hermitian and if $A = H_1 + iK_1$, H_1 and K_1 hermitian, then $H = H_1$, and $K = K_1$. (This should be a familiar result by now.)

7. Show that for the matrix A in #6, A is normal if and only if $HK = KH$.

8. Use Theorem 4 to prove that A is normal iff there exists a unitary matrix U such that U^*AU is diagonal.

   Hint: It is easy to see that if $A = U^*DU$, where D is a diagonal and U is unitary, then $AA^* = A^*A$, i.e., A is normal. Conversely, if A is normal then A and A^* commute and so Theorem 4 yields a (single) unitary matrix U for which both U^*A^*U and U^*AU are upper triangular. But $U^*A^*U = (U^*AU)^*$, and hence U^*AU is diagonal.

9. Let $A = \text{diag}(\lambda_1, ..., \lambda_n) \in M_n(\mathbb{C})$. Prove: A is unitary iff $|\lambda_t| = 1$, $t = 1, ..., n$; A is hermitian iff $\lambda_t \in \mathbb{R}$, $t = 1, ..., n$; A is skew-hermitian iff every λ_t is pure imaginary.

10. Prove: if $A = \text{diag}(\lambda_1, ..., \lambda_n)$ is hermitian then A is positive semi-definite (positive definite) (see §6.2 Exercises, #9) iff $\lambda_t \geq 0$ ($\lambda_t > 0$), $t = 1, ..., n$.

    Hint: Let $A \in M_n(\mathbb{C})$ be hermitian with $A = \text{diag}(\lambda_1, ..., \lambda_n)$. As in #9 above, we know that each λ_k is real. Let $x = [x_1, ..., x_n]^T \in M_{n,1}(\mathbb{C})$ be an arbitrary nonzero vector. Then

    $$x^*Ax = \sum_{t=1}^{n} \lambda_t |x_t|^2.$$

Clearly, every value of x^*Ax is nonnegative (positive) iff each λ_t is nonnegative (positive). Thus A is positive (semi-)definite hermitian iff each $\lambda_t > 0$ ($\lambda_t \geq 0$) for $t = 1, ..., n$.

8.1 MatLab

1. Read the Help entry for **schur** and **rsf2csf**. According to the MatLab documentation,

    - [U, T]=schur(A) returns an upper triangular matrix T and a unitary U such that $U^*AU = T$

    - T=schur(A) returns the upper triangular factor

    - if $A \in M_n(\mathbb{R})$ then [U, T] = schur(A) returns the real version described in Theorem 5

    - rsf2csf ("real schur form to complex schur form") converts the real version of schur(A), $A \in M_n(\mathbb{R})$, to the complex version

    Use schur and rsf2csf to find the unitary U and upper triangular T such that $U^*AU = T$:

    (a) $A = \begin{bmatrix} 2 & 1 & 0 & -1 \\ -1 & 1 & 1 & 1 \\ 0 & 1 & 1 & 0 \\ 1 & 1 & 0 & 0 \end{bmatrix}$

    (b) $A = \begin{bmatrix} 0 & 1 & 0 \\ -1 & 0 & 0 \\ 0 & 0 & 1 \end{bmatrix}$

(c) $A = \begin{bmatrix} 4 & -5 & 0 & 3 \\ 0 & 4 & -3 & -5 \\ 5 & -3 & 4 & 0 \\ 3 & 0 & 5 & 4 \end{bmatrix}$

(d) $A = \begin{bmatrix} 15 & 11 & 6 & -9 & -15 \\ 1 & 3 & 9 & -3 & -8 \\ 7 & 6 & 6 & -3 & -11 \\ 7 & 7 & 5 & -3 & -11 \\ 17 & 12 & 5 & -10 & -16 \end{bmatrix}$

Hint: (b) [U, T]=schur(A) returns

$$U = \begin{bmatrix} 1 & 0 & 0 \\ 0 & -1 & 0 \\ 0 & 0 & 1 \end{bmatrix}, \quad T = \begin{bmatrix} 0 & -1 & 0 \\ 1 & 0 & 0 \\ 0 & 0 & 1 \end{bmatrix}.$$

If this is followed by [U, T]=rsf2csf(U, T) the output is

$$U = \begin{bmatrix} .7071i & -.7071 & 0 \\ -.7071 & .7071i & 0 \\ 0 & 0 & 1 \end{bmatrix}, \quad T = \begin{bmatrix} 1.0000i & 0 & 0 \\ 0 & -1.0000i & 0 \\ 0 & 0 & 1 \end{bmatrix}.$$

(c) The eigenvalues are 12, 1 ± 5i, 2.

(d) The eigenvalues are 1.5 ± 3.5707i (twice), -1.

2. Let A and B be n-square and m-square complex matrices respectively, and let $K = A \otimes B$. If $U^*AU = T$ and $V^*BV = S$, both upper-triangular, then

$$(U \otimes V)^* K (U \otimes V) = T \otimes S$$

is upper triangular and $U \otimes V$ is unitary. Experiment with MatLab to determine whether the command

$$[W, L] = \text{schur}(\text{kron}(A, B))$$

returns $W = U \otimes V$ and $L = T \otimes S$. Is every mn-square unitary matrix of the form $U \otimes V$ in which U is m-square unitary and V is n-square unitary?

3. Devise a MatLab function called srand to generate a random real symmetric n–square matrix A. Save srand in a library called shur. Use srand to confirm that if [U, T]=schur(A) and A is real symmetric then T is diagonal.

   Hint:
   > function A = srand(n)
   > %SRAND returns a random real
   > %symmetric matrix of size n
   > %whose entries are between -1 and 1.
   > M = (2*rand(n) - 1);
   > T = triu(M, 1);
   > A = diag(diag(M)) + T + T.';

   Then at the Command prompt enter [U, T] = schur(srand(n)), where n is a postive integer.

4. Devise a MatLab function called hrand to generate a random complex hermitian n–square matrix A. Save hrand in the library called shur. Use hrand to confirm that if [U, T]=schur(A) and A is hermitian then T is diagonal.

   Hint:
   > function A = hrand(n)
   > %HRAND returns a random complex
   > %hermitian matrix of size n
   > %whose real and imaginary parts
   > %are between -1 and 1.
   > M=(2*rand(n)-1) + i*(2*rand(n)-1);
   > T=triu(M,1);
   > A=diag(diag(real(M))) + T + T';

   Then at the Command prompt enter [U, T]=schur(srand(n)) where n is a positive integer.

5. Devise a MatLab function called orand to generate a random real orthogonal n–square matrix A. Save orand in shur. Use orand to confirm that if

[U, T]=schur(A) is followed by [U, T]=rsf2csf(U,T) for a real orthogonal A then T is diagonal.

Hint:
> function A = orand(n)
> %ORAND returns a random real
> %orthogonal matrix of size n.
> M=(2*rand(n)-1);
> A = orth(M);

6. Devise a MatLab function called urand to generate a random complex unitary n–square matrix A. Save urand in shur. Use urand to confirm that if [U, T]=schur(A) is followed by [U, T]=rsf2csf(U,T) for a complex unitary A then T is diagonal.

Hint:
> function U = urand(n)
> %URAND returns a random complex
> %unitary matrix of size n
> %whose real and complex parts are between -1 and 1
> M=(2*rand(n)-1)+i*(2*rand(n)-1);
> th=2*pi*rand(1,n);
> D=diag(exp(i*th));
> U=D*orth(M);

7. Devise a MatLab function called skrand to generate a random real skew-symmetric n–square matrix A. Save skrand in shur. Use skrand to confirm that if [U, T]=schur(A) is followed by [U, T]=rsf2csf(U,T) for a real skew-symmetric A then T is diagonal.

Hint:
> function A = skrand(n)
> %SKRAND returns a random real
> %skew-symmetric matrix of size n
> %whose entries are between -1 and 1.
> M=(2*rand(n)-1);
> T=triu(M,1);
> A=T+(-1)*(T.');

8. Devise a MatLab function called skhrand to generate a random complex skew–hermitian n-square matrix A. Save skhrand in shur. Use skhrand to confirm that if [U, T]=schur(A) is followed by [U, T]=rsf2csf(U,T) for a complex skew–hermitian A then T is diagonal.

   Hint:

   function A = skhrand(n)
   %SKHRAND returns a random complex
   %skew-hermitian matrix of size n
   %whose real and imaginary parts
   %are between -1 and 1.
   M = (2*rand(n) - 1) + i*(2*rand(n) - 1);
   T = triu(M, 1);
   A = (T - T') + i*diag(diag(imag(M))));

9. Devise a MatLab function called normrand to generate a random complex normal n–square matrix A. Save normrand in shur. Use normrand to confirm that if [U, T]=schur(A) is followed by [U, T]=rsf2csf(U,T) for a complex normal A then T is diagonal.

   Hint:

   function A=normrand(n)
   %NORMRAND returns a random complex
   %normal matrix of size n
   %with real and imaginary parts
   %of eigenvalues between -1 and 1.
   U = urand(n);
   D = diag(diag((2*rand(n)-1) + i*(2*rand(n)-1)));
   A = U * D * U';

10. Let T be the upper-triangular matrix in the complex Schur form of the matrix A. The main diagonal entries of T are the eigenvalues of A, $\lambda_1, \lambda_2, ..., \lambda_n$. Use Help to review norm(A, 'fro'), diag, max, and triu. Then write a MatLab function called nrdg ("near diagonal") that

    • accepts an n-square matrix A and a positive number tol

- returns an upper-triangular matrix S, similar to A, for which f, the Frobenius norm of S - diag($\lambda_1, ..., \lambda_n$), is at most tol, together with the actual value of f

Save nrdg in shur.

Hint: Let $c_1, ..., c_n$ be positive numbers and let $C = \text{diag}(c_1, ..., c_n)$. Then the i, j entry of $C^{-1}TC$, $i \le j$, is

$$c_i^{-1} t_{ij} c_j .$$

Thus in the similarity $C^{-1}TC$ the main diagonal of T is left unaltered and the entry t_{ij} above the main diagonal is multiplied by c_j / c_i to obtain $c_j t_{ij} / c_i$. If $c_1, ..., c_n$ can be chosen so that

$$\sum_{i<j} | c_j t_{ij} / c_i |^2 \le \text{tol}^2 \qquad (1)$$

then the required matrix S can be taken to be $S = C^{-1}TC$. Note that

$$\sum_{i<j} | c_j t_{ij} / c_i |^2 \le t^2 \sum_{i<j} | c_j / c_i |^2 \qquad (2)$$

where $t = \max_{i<j} | t_{ij} |$. Let $\xi_c^2 = \sum_{i<j} | c_j / c_i |^2$. Then from (1) and (2), if we can choose $c_1, ..., c_n$ so that

$$t^2 \xi_c^2 \le \text{tol}^2 \qquad (3)$$

then $S = C^{-1}TC$ satisfies the requirements. Note that from (1), if $\frac{n(n-1)}{2} t^2 \le \text{tol}^2$ then we take $c_1 = \cdots = c_n = 1$ and $S = T$. On the other hand, if

$$\frac{n(n-1)}{2} t^2 > \text{tol}^2 \qquad (4)$$

then from (1) it will suffice to chose $c_1, \ldots, c_n$ such that

$$\left(\max_{i<j} (c_j/c_i)\right)^2 \frac{n(n-1)}{2} t^2 \leq \text{tol}^2,$$

or equivalently,

$$\max_{i<j} (c_j/c_i) \leq \frac{\text{tol}}{t}\sqrt{\frac{2}{n(n-1)}}. \qquad (5)$$

Let β be the number on the right hand side of (5). Then from (4), $\beta < 1$. Now let $c_1 = 1, c_2 = \beta, c_3 = \beta^2, \ldots, c_n = \beta^{n-1}$ so that

$$\frac{c_j}{c_i} = \frac{\beta^{j-1}}{\beta^{i-1}} = \beta^{j-i}.$$

Since β is less than 1, the largest of the c_j/c_i, $j > i$ is β. Then from (1)

$$\sum_{i<j} \left|\frac{c_j}{c_i} t_{ij}\right|^2 \leq \beta^2 t^2 \frac{n(n-1)}{2}$$

$$= \frac{\text{tol}^2}{t^2} \frac{2}{n(n-1)} t^2 \frac{n(n-1)}{2}$$

$$= \text{tol}^2.$$

The problem with forming $C^{-1} = \text{diag}(1, \beta^{-1}, \beta^{-2}, \ldots, \beta^{-(n-1)})$ is that $\beta^{-(n-1)}$ may cause an error because, in general, it is very large. Instead we can compute $C^{-1}TC$ by leaving the main diagonal of T unaltered and multiplying the kth superdiagonal of T by β^k, $k = 1, \ldots, n-1$.

```
function [S,f]=nrdg(A,tol)
%NRDG accepts an n-square matrix A
%and a positive number tol and
%returns an upper triangular matrix
%S similar to A for which the
```

```
            %frobenius norm of S-diag(diag(T))
            %is at most tol.
            n=length(A);
            [U,T]=schur(A);
            [U,T]=rsf2csf(U,T);
            t=max(max(abs(triu(T,1))));
            b=tol*sqrt(2/(n*(n-1)))/t;
            if b >= 1
                    S1=T;
            else
                    S1=diag(diag(T));
                    for k=1:(n-1)
                            v=(b^k)*diag(T,k);
                            S1=S1+diag(v,k);
                    end
            end
            S=S1;
            f=norm(triu(S1,1),'fro');
```

11. Review the Help entry for sort. Then write a MatLab function called sschur such that for a hermitian matrix A the function returns a pair [U, D] such that $U^*AU = D = \mathrm{diag}(d_1, ..., d_n)$, $d_1 \geq \cdots \geq d_n$. Save sschur in shur.

    Hint:
```
            function [U,D]=sschur(A)
            %SSCHUR returns a pair U,D such that
            %U'*A*U = D = diag(d1,...,dn) with
            %the d's in non-increasing order.
            n=length(A);
            [U,D]=schur(A);
            [order, I]=sort(real(diag(D)));
            I=I(n:-1:1);
            U=U(:,I);
            D=real(D(I,I));
```

8.1 Glossary

rsf2csf 426
Schur triangularization theorem 410
schur 426
sort 433

8.2 Normal Matrices

Using the results in §8.1 we can prove the principal result about normal matrices.

Theorem 1.

Let $A \in M_n(\mathbb{C})$. Then A is normal iff there exists a unitary matrix U such that

$$U^*AU = \text{diag}(\lambda_1, ..., \lambda_n) \qquad (1)$$

where $\lambda_1, ..., \lambda_n$ are the eigenvalues of A in any prescribed order.

Proof.

Assume A is normal. By §8.1 Theorem 1, there exists a unitary matrix U such that $U^*AU = T$ is upper triangular. Clearly T is normal:

$$TT^* = U^*AUU^*A^*U = U^*AA^*U = U^*A^*AU$$
$$= U^*A^*UU^*AU = (U^*AU)^*(U^*AU) = T^*T.$$

The (1, 1) entry of TT^* is

$$|t_{11}|^2 + \sum_{j=2}^{n} |t_{1j}|^2$$

8.2 Normal Matrices

whereas the (1, 1) entry of T^*T is $|t_{11}|^2$. It follows that $t_{1j} = 0$, $j = 2, \ldots, n$. By successively equating the (2, 2), (3, 3), ..., (n, n) entries of TT^* and T^*T, a similar argument shows that T is diagonal.

Conversely, if U^*AU is a diagonal matrix for some unitary matrix U, it follows immediately that A is normal. ∎

The equation (1) is equivalent to the n equations

$$AU^{(t)} = \lambda_t U^{(t)}, \quad t = 1, \ldots, n. \tag{2}$$

Moreover,

$$U^{(s)*}U^{(t)} = \delta_{st}, \quad s, t = 1, \ldots, n.$$

A set of vectors $U^{(1)}, \ldots, U^{(n)}$ in $M_{n,1}(\mathbb{C})$ which satisfy this last set of conditions is called *orthonormal* (usually abbreviated o.n.). Thus Theorem 1 states that A is normal iff A possesses an o.n. set of eigenvectors.

Theorem 2.

Let F be a set of pairwise commuting normal matrices in $M_n(\mathbb{C})$. Then there exists a unitary matrix U such that U^*AU is diagonal for every $A \in F$.

Proof.

From §8.1 Theorem 4, there exists a unitary matrix U such that U^*AU is upper triangular for every $A \in F$. Since every U^*AU is also normal it follows as in the proof of Theorem 1 that every U^*AU is diagonal. ∎

Theorem 3.

Let A be a normal matrix. Then

(a) A is hermitian iff every eigenvalue of A is real;

(b) A is unitary iff every eigenvalue of A has absolute value 1;
(c) A is skew-hermitian iff every eigenvalue of A is pure imaginary;
(d) if A is hermitian, it is positive definite (positive semi-definite) iff every eigenvalue of A is positive (non-negative).

Proof.

To prove (a), suppose $\text{diag}(\lambda_1, ..., \lambda_n)$ is the diagonal form of the matrix A. The condition $A^* = A$ is then equivalent to $\overline{\lambda_j} = \lambda_j$, $j = 1, ..., n$. Thus A is hermitian iff all the eigenvalues of A are real. The proofs of (b) and (c) are similarly completed. To prove (d), recall that A is positive definite iff $x^*Ax > 0$ for every non-zero $x \in M_{n,1}(\mathbb{C})$. If $A = U^*DU$, $D = \text{diag}(\lambda_1, ..., \lambda_n)$ then

$$x^*Ax = x^*U^*DUx$$
$$= (Ux)^*DUx. \qquad (3)$$

Let $y = Ux$ and note that since U is non-singular, y runs over all non-zero vectors as x does. From (3),

$$x^*Ax = y^*Dy$$
$$= \sum_{j=1}^{n} \lambda_j |y_j|^2 \qquad (4)$$

where $y = [y_1, ..., y_n]^T$. Since the λ_j can be arranged in any order, we can assume that $\lambda_1, ..., \lambda_p$ are positive, $\lambda_{p+1} = \cdots = \lambda_{p+k} = 0$, and $\lambda_{p+k+1}, ..., \lambda_n$ are negative. Of course, p could be 0 or n. Then from (4),

$$x^*Ax = \sum_{j=1}^{p} \lambda_j |y_j|^2 + \sum_{j=p+k+1}^{n} \lambda_j |y_j|^2. \qquad (5)$$

If $p < n$ it is obvious that a nonzero y can be chosen so that the right side of (5) is non-positive, e.g., set $y_n = 1$, $y_1 = \cdots = y_{n-1} = 0$. Thus $x^*Ax > 0$ iff $p = n$. Similarly if $\lambda_n < 0$ the same choice of y makes (5) strictly negative. Thus if A is positive definite, or positive semi-definite, the eigenvalues $\lambda_1, ..., \lambda_n$ must be positive or non-negative respectively. The converse is equally obvious from the equation (5). ∎

8.2 Normal Matrices

Observe that Theorems 1 and 3 imply that any hermitian, skew-hermitian, or unitary matrix is unitarily similar to a diagonal matrix.

The next theorem is interesting in that it actually shows how unitary and hermitian matrices are related. It is the matrix counterpart of the result about complex numbers to the effect that if $u \neq -1$ and $|u| = 1$ then u can be written as

$$u = (1 + ia) / (1 - ia)$$

where a is an appropriate real number.

Theorem 4 (Cayley Parameterization).

Let $U \in M_n(\mathbb{C})$ be unitary and assume -1 is not an eigenvalue of U. Then there exists a unique hermitian matrix A such that

$$U = (I_n + iA)(I_n - iA)^{-1}. \tag{6}$$

Conversely, if $A \in M_n(\mathbb{C})$ is hermitian then the matrix U defined by (6) is unitary and -1 is not an eigenvalue of U.

Proof.

If $\beta \in \mathbb{C}, |\beta| = 1$, and $\beta \neq -1$ then the reader will easily verify that there exists a unique real number α such that

$$\beta = (1 + i\alpha)(1 - i\alpha)^{-1}.$$

Let V be a unitary matrix for which

$$V^*UV = \text{diag}(\beta_1, \ldots, \beta_n) = D,$$

$|\beta_k| = 1$, $k = 1, \ldots, n$, $\beta_k \neq -1$. Write $\beta_k = (1 + i\alpha_k)(1 - i\alpha_k)^{-1}$, where α_k is real. Obviously if $\Delta = \text{diag}(\alpha_1, \ldots, \alpha_n)$ then

$$D = (I_n + i\Delta)(I_n - i\Delta)^{-1}. \tag{7}$$

We can then compute that

$$U = VDV^* = V(I_n + i\Delta)V^*V(I_n - i\Delta)^{-1}V^*$$
$$= (I_n + iV\Delta V^*)(I_n - iV\Delta V^*)^{-1}.$$

Let $A = V\Delta V^*$ and note that by Theorem 3(a), A is hermitian. To see that A is uniquely determined by the equation (6), simply solve for A in (6):

$$A = i(I_n + U)^{-1}(I_n - U). \tag{8}$$

To prove the converse, if U has the form (6) then by diagonalizing A we see immediately that the eigenvalues of U are of the form $(1 + i\alpha)(1 - i\alpha)^{-1}$, α real. Such numbers have absolute value 1 and cannot equal -1. Thus U is unitarily similar to a diagonal matrix with eigenvalues of modulus 1. By Theorem 3(b), U is unitary. ∎

If A is a hermitian matrix then $A > 0$ ($A \geq 0$) is a notation we have used to denote the fact that A is positive definite (positive semi-definite). In general, if A and B are hermitian matrices then $A > B$ means that $A - B > 0$ and similarly $A \geq B$ means that $A - B \geq 0$. If $A \in M_n(\mathbb{C})$ is an arbitrary complex matrix and $p(\lambda)$ is a polynomial with complex number coefficients, then recall that $p(A)$ is the matrix that results by replacing λ by A; e.g., if

$$p(\lambda) = \lambda^3 + 2\lambda^2 - 3\lambda + 2$$

then

$$p(A) = A^3 + 2A^2 - 3A + 2I_n.$$

Theorem 5.

If $A \geq 0$ and k is a positive integer, then there exists $B \geq 0$ such that $B^k = A$. Also, there exists a scalar polynomial $p(\lambda)$ with real coefficients such that

$$B = p(A).$$

8.2 Normal Matrices

Proof.

Let $U^*AU = \text{diag}(\lambda_1, \ldots, \lambda_n) = D$, $\lambda_t \geq 0$, $t = 1, \ldots, n$, and define

$$B = U \, \text{diag}(\lambda_1^{1/k}, \ldots, \lambda_n^{1/k}) \, U^*.$$

Then $B^k = A$. Let $p(\lambda)$ be a polynomial for which $p(\lambda_t) = \lambda_t^{1/k}$, $t = 1, \ldots, n$. The fact that such a polynomial exists appears in §8.2 Exercises, #3. Moreover, that exercise produces a real polynomial because $\lambda_t \geq 0$ and $\lambda_t^{1/k} \geq 0$ are real numbers, $t = 1, \ldots, n$. Then

$$p(A) = p(UDU^*) = Up(D)U^* = U \, \text{diag}(\lambda_1^{1/k}, \ldots, \lambda_n^{1/k})U^* = B. \qquad (9)$$

The second equality in (9) follows from §8.2 Exercises, #2. ■

Theorem 5 can be improved somewhat to show that there is only one $B \geq 0$ for which $B^k = A$. To confirm this, first note that if C is normal and $C^k = A$, then

$$B = p(A) = p(C^k) = q(C)$$

where $q(\lambda)$ is an appropriate scalar polynomial. Thus $BC = CB$, and by Theorem 2 there is a unitary V such that V^*BV and V^*CV are both diagonal. Suppose that $C \geq 0$. Then V^*CV has non-negative main diagonal elements. Note that

$$(V^*CV)^k = V^*C^kV = V^*AV = V^*B^kV = (V^*BV)^k.$$

In other words, the two diagonal matrices V^*CV and V^*BV have the same k^{th} powers and have non-negative main diagonal entries. It follows that $V^*CV = V^*BV$ and hence $C = B$. Thus we can conclude that there is only one $B \geq 0$ such that $B^k = A$. The matrix B is denoted by

$$B = A^{1/k}. \qquad (10)$$

The following result is the generalization to matrices of the fact that any nonzero complex number z can be written as

$$z = re^{i\theta}$$

where $r = |z| > 0$ and θ is real.

Theorem 6 (Polar factorization).

If $A \in M_n(\mathbb{C})$ is non-singular, then there exists a unique unitary U and a unique hermitian $H > 0$ such that $A = UH$.

Proof.

The matrix A^*A is positive definite. Define $H = (A^*A)^{1/2}$ and then set $U = AH^{-1}$. Note that

$$UU^* = AH^{-1}H^{-1}A^* = AH^{-2}A^* = A(H^2)^{-1}A^* = A(A^*A)^{-1}A^* = AA^{-1}A^{*-1}A^* = I_n.$$

Thus U is unitary. Also, if $A = VK$, V unitary, $K > 0$, then $A^*A = K^2$, so $K = (A^*A)^{1/2}$ is uniquely determined. But then $V = AK^{-1}$ is uniquely determined. ∎

8.2 Exercises

1. Prove: if $H > 0$ and K is hermitian, then the roots of the polynomial $\det(\lambda H - K)$ are real.

   Hint: Since $H > 0$ there exists $H^{1/2} > 0$ with $(H^{1/2})^2 = H$. But then

   $$\det(\lambda H - K) = \det(\lambda (H^{1/2})^2 - K)$$
   $$= \det(H^{1/2}) \det(\lambda I - H^{-1/2}KH^{-1/2}) \det(H^{1/2}).$$

   It follows that the roots of $\det(\lambda H - K)$ are the eigenvalues of $H^{-1/2}KH^{-1/2}$, a hermitian matrix. Thus the roots are all real.

8.2 Exercises

2. Prove: if $p(\lambda) = a_n \lambda^n + a_{n-1} \lambda^{n-1} + \cdots + a_0$ is a scalar polynomial, A is an m–square matrix, and S is a non-singular m-square matrix, then $p(S^{-1}AS) = S^{-1} p(A) S$.

3. Let $b_1, \ldots, b_k$ be complex numbers and $a_1, \ldots, a_k$ be distinct complex numbers. Define a polynomial $p(\lambda)$ by

$$p(\lambda) = \sum_{j=1}^{k} \frac{\prod_{t \neq j} (\lambda - a_t)}{\prod_{t \neq j} (a_j - a_t)} b_j .$$

Show that $p(a_j) = b_j$, $j = 1, \ldots, k$. The polynomial $p(\lambda)$ is called a *Lagrange interpolation polynomial*.

4. Let U_k, $k = 1, 2, 3, \ldots$, be an infinite sequence of unitary matrices. Prove: there is a subsequence U_{k_t}, $t = 1, 2, 3, \ldots$, that converges to a unitary matrix.

5. Prove: if $A = UH$, U unitary, and $H \geq 0$, then $\operatorname{tr}(A^*A) = \operatorname{tr}(H^2)$.

6. Prove: if $A \in M_n(\mathbb{C})$ then $A = UH$ where U is unitary and $H \geq 0$.

   Hint: If A is nonsingular then use Theorem 6. If A is singular, define a sequence of matrices $A_k = A + \frac{1}{k} I_n$. There exists k_0 such that for $k \geq k_0$, A_k is nonsingular (why?). Use Theorem 6 to write $A_k = U_k H_k$, U_k unitary, $H_k > 0$, $k \geq k_0$. Clearly, the entries in A_k are bounded. Also, $H_k = U_k^* A_k$ and since U_k is unitary the entries in H_k are bounded. Thus we can choose subsequences $\{H_{k_t}\}$ and $\{U_{k_t}\}$ of $\{H_k\}$ and $\{U_k\}$, respectively, which converge. Hence $\lim_{t \to \infty} H_{k_t} = H$, $\lim_{t \to \infty} U_{k_t} = U$ and obviously $A = \lim_{t \to \infty} A_{k_t} = UH$, U unitary, $H \geq 0$.

7. Let $A \in M_{m,n}(\mathbb{C})$, $m \leq n$. Prove: $A = UH$ where U is $m \times n$, $UU^* = I_m$, and H is n-square, $H \geq 0$.

Hint: Let B denote the n-square matrix given by

$$B = \begin{bmatrix} A \\ 0_{n-m,n} \end{bmatrix}$$

(i.e., $B[1...m \mid 1...n] = A$ and $B[m+1 ... n \mid 1 ... n] = 0_{n-m,n}$). By #6 above, we may write $B = WH$ where W is an n-square unitary matrix and H is an n-square positive semi-definite hermitian matrix. Partition W conformally with B as

$$W = \begin{bmatrix} U \\ \hline V \end{bmatrix}.$$

Then block multiplication yields $A = UH$ and $UU^* = I_m$.

8. Prove: if H is hermitian and $h_{11} < 0$ then H has a negative eigenvalue.

   Hint: By §8.1 Exercises, #10, if the eigenvalues of H were all non-negative then $x^*Hx \geq 0$ for any x. Take $x = [1\ 0\ ...\ 0]^T$.

9. Prove: if H is hermitian and $h_{11} > 0$ then H has a positive eigenvalue.

10. Let H be hermitian. Prove: $H \geq 0\ (> 0)$ iff all principal subdeterminants of H are non-negative (positive).

    Hint: If $H \geq 0$ write $K = H^{1/2} \geq 0$, $H = K^2 = KK^*$. By Cauchy-Binet,

    $$\det(H[\omega \mid \omega])$$
    $$= \sum_{\gamma \in Q_{k,n}} \det(K[\omega \mid \gamma]) \det(K^*[\gamma \mid \omega]) = \sum_{\gamma \in Q_{k,n}} |\det(K[\omega \mid \gamma])|^2 \geq 0.$$

    If $H > 0$, then $K > 0$ so K is nonsingular. Hence $\rho(K[\omega \mid 1. ..., n]) = k$ for any $\omega \in Q_{k,n}$ and thus

$$H[\omega \mid \omega] = K[\omega \mid 1, ..., n] \, K[\omega \mid 1, ...,n]^*$$

has rank k (see §4.3 Exercises, #34). It follows that $\det(H[\omega \mid \omega]) > 0$.

Conversely, if every $\det(H[\omega \mid \omega]) > 0$, then $E_k(\lambda_1, ..., \lambda_n) > 0$, $k = 1, ..., n$. Write

$$\prod_{t=1}^{n} (\lambda + \lambda_t) = \lambda^n + \sum_{k=1}^{n} E_k(\lambda_1, ..., \lambda_n) \lambda^{n-k},$$

where $\lambda_1 \geq ... \geq \lambda_n$ are the eigenvalues of H. If $\lambda_n < 0$, set $\lambda = -\lambda_n$ so that

$$(-\lambda_n)^n + \sum_{k=1}^{n} E_k(\lambda_1, ..., \lambda_n)(-\lambda_n)^{n-k} = 0,$$

impossible. Thus every eigenvalue of H is nonnegative. If every principal subdeterminant is positive then, in particular, $\lambda_1 \cdots \lambda_n = E_n(\lambda_1, ..., \lambda_n) = \det(H)$ is positive and the nonnegative hermitian matrix H is non-singular, and hence positive-definite.

11. Find the square root of

$$H = \begin{bmatrix} 8 & -2 & -2 \\ -2 & 5 & -4 \\ -2 & -4 & 5 \end{bmatrix}.$$

Hint: The solution is $H^{1/2} = \frac{1}{3} H$.

12. Let

$$A = \begin{bmatrix} 1 & 1 & 1 \\ 1 & -1 & 1 \\ 0 & 0 & 0 \end{bmatrix}.$$

Find unitary U, V, $H \geq 0$ and $K \geq 0$ for which $A = UH = KV$.

13. Prove: if $H \geq 0$ and $h_{11} = 0$ then $H_{(1)} = 0$.

    Hint: From #10, $0 \leq \det(H[1, k \mid 1, k]) = h_{11}h_{kk} - |h_{1k}|^2$. Thus $h_{1k} = 0$, $k = 2, \ldots, n$.

14. Let $A \in M_n(\mathbb{R})$ be symmetric. Prove: there exists a real orthogonal U such that $U^T A U = \text{diag}(\lambda_1, \ldots, \lambda_n)$.

    Hint: From §8.1 Exercises, #4 there exists a real orthogonal U such that $U^T A U$ is upper triangular. But since A is symmetric, $U^T A U$ is symmetric, and hence diagonal.

15. Assume that $H > 0$ and that K is hermitian, both n-square. Prove: there exists a nonsingular P such that $P^* H P = I_n$ and $P^* K P = \text{diag}(\gamma_1, \ldots, \gamma_n)$, where $\gamma_1, \ldots, \gamma_n$ are the eigenvalues of $H^{-1} K$.

    Hint: First let $Q = H^{-1/2}$ so $Q^* H Q = I_n$, $Q^* K Q$ hermitian. Next, let V be unitary, $V^*(Q^* K Q)V = \text{diag}(\gamma_1, \ldots, \gamma_n)$ and define $P = QV$. Clearly $P^* H P = I_n$, $P^* K P = \text{diag}(\gamma_1, \ldots, \gamma_n)$ where the γ_i are the eigenvalues of $Q^* K Q = H^{-1/2} K H^{-1/2}$. But the latter matrix has the same eigenvalues as $H^{-1} K$ (why?).

8.2 MatLab

1. As we saw in §8.2 Exercises, #3, a polynomial can be defined for which $p(a_t) = b_t$, $t = 1, \ldots, k$, where $a_1, \ldots, a_k$ and $b_1, \ldots, b_k$ are given and $a_1, \ldots, a_k$ are distinct. MatLab has available a function named **polyfit** such that the command

    $$c = \text{polyfit}(a, b, n)$$

    returns a row vector

    $$[c_n \; c_{n-1} \; \cdots \; c_1 \; c_0].$$

The polynomial

$$p(\lambda) = c_n \lambda^n + c_{n-1} \lambda^{n-1} + \cdots + c_1 \lambda + c_0$$

"fits" b_t to a_t, $t = 1, \ldots, k$, in the least squares sense. That is, the system of linear equations

$$p(a_t) = a_t^n c_n + a_t^{n-1} c_{n-1} + \cdots + a_t c_1 + c_0 = b_t, \quad t = 1, \ldots, k,$$

is solved in the least squares sense for the coefficients $c_n, c_{n-1}, \ldots, c_0$ (see §8.3 Exercises, #5 and §4.3 MatLab, #10).

Use polyfit to solve the following problems.

(a) find the equation of a straight line through the points (0, 1) and (2, 3)

(b) find the equation of a parabola that goes through the points (-1, 2), (0, 1), (2, 4)

(c) generate two random 1 × 3 complex matrices, z and w, and obtain a polynomial p of degree two for which $p(z_k) = w_k$, $k = 1, 2, 3$. Check your answer by using polyval.

2. Write a MatLab function called nroot that accepts a normal matrix A and a positive integer k and returns a normal matrix B for which $B^k = A$. Save nroot in shur.

Hint:
```
function B=nroot(A, k)
%NROOT accepts a normal matrix A
%and a positive integer k
%and returns the kth root of A.
if norm(A'*A - A*A', 2) > 1e-10
      error('input not normal');
else
      [U,T]=schur(A);
```

```
        [U,T]=rsf2csf(U,T);
        D=diag(diag(T)).^(1/k);
        B=U*D*U';
end
```

3. Write a MatLab script named nroottest1 that

   - accepts a positive integer k

   - generates 20 random unitary matrices A of dimension at most 5

   - computes $B = A^{1/k}$ using the function nroot in #2

   - prints out the 2-norm of $A - B^k$

   Save nroottest1 in shur.

   Hint:
   ```
   %NROOTTEST1 accepts a positive integer k
   %computes A^(1/k) for 20 random unitary A
   %and prints out the 2-norm of A - B^k.
   k=input('Enter a positive integer k: ');
   D=[ ];
   for j=1:20
           n=floor(5*rand+1);
           A=urand(n);
           B=nroot(A,k);
           d=norm(A-B^k);
           D=[D d];
   end
   D'
   ```

4. Use the function nroot in #2 to find the square root of the matrix H in §8.2 Exercises, #11.

5. Write a MatLab function called polynorm that

8.2 MatLab

- accepts an n-square complex normal A

- returns a polynomial p such that $p(A) = (A + A^*)/2$.

Save polynorm in shur.

Hint:

```
function p = polynorm(A)
%POLYNORM accepts a normal nxn A
%and returns a polynomial p
%such that p(A) = (A + A')/2.
n=length(A);
[U,T]=schur(A);
[U,T]=rsf2csf(U,T);
ev=diag(T);
re=real(ev);
p=polyfit(ev,re,n);
```

6. Write a MatLab script named nroottest2 that

   - calls for the input of a positive integer k

   - generates 20 random n-square complex normal matrices A, $1 \le n \le 5$.

   - computes B, the kth root of each such A

   - prints out the Frobenius norm of the differences A - B^k for each A

Save nroottest2 in shur.

Hint:

```
%NROOTTEST2 calls for a postive integer k,
%generates 20 n-square random complex normal A,
%1 ≤ n ≤ 5, and prints out the Frobenius norm
%of A - B^k where B is the kth root of A.
k = input('Enter a positive integer k: ');
D=[ ];
```

```
            for j=1:20
                n=floor(5*rand+1);
                A=normrand(n);
                B=nroot(A,k);
                d=norm(A-B^k, 'fro');
                D=[D d];
            end
            D'
```

7. Write a MatLab function called polar1 that

   - accepts a nonsingular matrix A

   - returns a unitary U and a positive definite H such that A = UH

   Save polar1 in shur.

   Hint:
   ```
   function [U,H]=polar1(A)
   %POLAR1 accepts a non-singular A
   %and returns U unitary and
   %H > 0 such that A=UH
   H=(A'*A)^(1/2);
   U=A*inv(H);
   ```

8. Write a MatLab function called polar2 that

   - accepts an m × n matrix A of rank m, m < n

   - returns two matrices U and H where U is m × n, $UU^* = I_m$, H is n-square, H > 0, and A = UH.

   Save polar2 in shur.

   Hint: First, complete A to a nonsingular n-square matrix B by adjoining P:

$$B = \begin{bmatrix} A \\ P \end{bmatrix}.$$

(See §7.2 MatLab, #2). Then polar factorize B:

$$B = WH \qquad (1)$$

where W is unitary and H > 0. In partitioned form (1) becomes

$$\begin{bmatrix} A \\ \hline P \end{bmatrix} = \begin{bmatrix} U \\ \hline V \end{bmatrix} H \qquad (2)$$

where U is m × n. Performing the indicated block multiplication yields

$$A = UH$$

and since $WW^* = I_n$ it follows that $UU^* = I_m$.

Hint:
```
function [U, H] = polar2(A)
%POLAR2 accepts an m × n matrix A
%of rank m, m < n, and returns
%two matrices U and H where
%U is m × n, UU* = eye(m), H is n-square,
%H > 0 and A = UH.
[m, n]=size(A);
B=(comp(A.')).';
[W, K]=polar1(B);
U=W(1:m,:);
H=K;
```

9. Write a MatLab function called polyherm that

- accepts an n-square hermitian $A \geq 0$ and a postive integer k

- returns a polynomial p such that $p(A) = A^{1/k}$

Save polyherm in shur.

Hint:
```
function p=polyherm(A, k)
%POLYHERM accepts a
%hermitian n x n A ≥ 0
%and a positive integer k and
%returns a polynomial p such that
%p(A) = A^(1/k).
n=length(A);
[U, T]=schur(A);
[U,T] = rsf2csf(U, T);
ev=diag(T);
b=ev.^(1/k);
p = polyfit(ev, b, n);
```

10. Write a MatLab function called movieunit that

   - calls for the input of an n-square unitary matrix A

   - continuously (i.e., small discrete steps) changes A into the identity matrix, graphing each matrix in the sequence in the Graph window. Save movieunit in shur.

Hint: The basis of the program is this: write

$$U^*AU = \text{diag}(e^{i\varphi_1}, e^{i\varphi_2}, \ldots, e^{i\varphi_n}).$$

Then define a family of matrices A_t by

$$A_t = U\text{diag}(e^{it\varphi_1}, \ldots, e^{it\varphi_n})U^*, \quad 0 \le t \le 1.$$

Note that $A_1 = A$ and $A_0 = I_n$. Thus as t steadily decreases from 1 to 0, A_t changes continuously from A to I_n. The idea of the script is to graph A_t as t takes on closely spaced discrete values from 1 down to 0.

Hint:
```
%MOVIEUNIT calls for the input
%of a unitary matrix A. The program then
%continuously deforms A to the identity matrix
%with small changes, graphing each change.
A = input('Enter a unitary matrix: ');
[U, T]=schur(A);
[U, T]=rsf2csf(U, T);
d = diag(T);
for t = 1:-.01:0
    d = d.^t;
    At = U*diag(d)*U ';
    clg;
    mesh(At);
end
```

8.2 Glossary

$A^{1/k}$	439
Cayley parameterization	437
Lagrange interpolation polynomial	441
least squares	445
o.n.	435
orthonormal	435
polar factorization	440
polyfit	444

8.3 Singular Values

If $A \in M_{m,n}(\mathbb{C})$ then $AA^* \geq 0$ so that $(AA^*)^{1/2}$ is a uniquely determined positive semi–definite hermitian matrix in $M_m(\mathbb{C})$. The eigenvalues of $(AA^*)^{1/2}$ are called the *singular values* of A. Similarly, $(A^*A)^{1/2} \geq 0$ is a well defined positive semi-definite hermitian matrix in $M_n(\mathbb{C})$. As we saw in §4.3 Exercises, #35, A^*A and AA^* have the same rank (as A). Moreover, the positive eigenvalues of A^*A and AA^* are the same (§7.2 Exercises, #25). For the sake of definiteness we shall assume that $m \leq n$, and we shall call the m eigenvalues of $(AA^*)^{1/2}$ the singular values of A. With this convention, if $\rho(A) < m$ then A will certainly have singular values equal to 0. It is also convenient to notationally order the singular values of A as follows:

$$\alpha_1(A) \geq \alpha_2(A) \geq \cdots \geq \alpha_m(A) \geq 0. \tag{1}$$

Since the rank of a hermitian matrix is the number of its nonzero eigenvalues, it follows that there are r positive numbers in (1), where $r = \rho(A)$. Note that if $U \in M_m(\mathbb{C})$ is unitary and $V \in M_n(\mathbb{C})$ is unitary then

$$(UAV)(UAV)^* = UAA^*U^*$$

so that the singular values of UAV must be the same as the singular values of A. If $A \in M_n(\mathbb{C})$ is normal than there is a unitary U such that

$$U^*AU = \mathrm{diag}(\lambda_1, ..., \lambda_n).$$

Since the singular values of U^*AU obviously are the numbers $|\lambda_1|, ..., |\lambda_n|$, we can conclude that the singular values of a normal matrix are the absolute values of its eigenvalues, i.e.,

$$\alpha_j(A) = |\lambda_j(A)|, \quad j = 1, ..., n, \tag{2}$$

where the notation is so chosen that $\lambda_j(A), j = 1, ..., n$, are the eigenvalues of A arranged in decreasing order of their absolute values:

$$|\lambda_1(A)| \geq \cdots \geq |\lambda_n(A)|. \tag{3}$$

There is an important elementary fact that is useful in analyzing the relationships that hold between eigenvalues and singular values.

Theorem 1.

Let $H \in M_n(\mathbb{C})$ be hermitian with largest and least eigenvalues r_1 and r_n respectively. If $u \in M_{n,1}(\mathbb{C})$ and $u^*u = 1$ then

$$r_n \le u^*Hu \le r_1 . \qquad (4)$$

The upper equality can hold in (4) iff

$$Hu = r_1 u , \qquad (5)$$

i.e., u is an eigenvector of H corresponding to r_1. The lower inequality in (4) can hold iff u is an eigenvector of H corresponding to r_n.

Proof.

Let r_j, $j = 1, \ldots, n$, be the eigenvalues of H and suppose that

$$r_1 = \cdots = r_p > r_{p+1} \ge \cdots \ge r_n . \qquad (6)$$

Let V be a unitary matrix chosen so that

$$V^*HV = \text{diag}(r_1, r_2, \ldots, r_n) . \qquad (7)$$

If we let D denote the diagonal matrix on the right in (7) and set $v = V^*u = [v_1, \ldots, v_n]^T$, then $v^*v = 1$ and

$$\begin{aligned} u^*Hu &= u^*VDV^*u \\ &= v^*Dv \\ &= r_1 \sum_{i=1}^{p} |v_i|^2 + \sum_{i=p+1}^{n} r_i |v_i|^2 . \end{aligned} \qquad (8)$$

It is easy to check that the right side of (8) is at most r_1 and that it is equal to r_1 iff $v_i = 0$, $i = p+1, \ldots, n$, (see §8.3 Exercises, #2). But from (7) we know that the columns $V^{(1)}, \ldots, V^{(p)}$ are eigenvectors of H corresponding respectively to $r_1 = r_2 = \cdots = r_p$. Thus, if $v_{p+1} = \cdots = v_n = 0$ then

$$u = Vv$$

$$= V^{(1)}v_1 + \cdots + V^{(p)}v_p$$

and it follows that

$$Hu = HV^{(1)}v_1 + \cdots + HV^{(p)}v_p$$

$$= r_1(V^{(1)}v_1 + \cdots + V^{(p)}v_p)$$

$$= r_1 u.$$

The converse, namely that if (5) holds then $u^*Hu = r_1$, follows immediately from $u^*u = 1$. The assertions concerning the lower inequality in (4) are proved in exactly the same way. ∎

An interesting consequence of Theorem 1 is the following result that relates the eigenvalue of A of maximum modulus, $\lambda_1(A)$, to the largest singular value of A, $\alpha_1(A)$.

Theorem 2.

Let $A \in M_n(\mathbb{C})$ be a nonzero matrix. Then

$$|\lambda_1(A)| \leq \alpha_1(A) \tag{9}$$

with equality iff there is a $u \in M_{n,1}(\mathbb{C})$, $u^*u = 1$, such that

$$Au = \lambda_1(A)u \tag{10}$$

and

$$A^*Au = \alpha_1(A)^2 u. \tag{11}$$

8.3 Singular Values

Proof.

Write $H = (A^*A)^{1/2}$ and use §8.2 Exercises, #6 to obtain a unitary matrix U such that

$$A = UH. \qquad (12)$$

Choose u to satisfy (10), $u^*u = 1$, and compute directly that

$$|\lambda_1(A)| = |u^*Au|$$

$$= |u^*UHu|$$

$$= |v^*Hu|$$

where $v = U^*u$. Clearly $v^*v = 1$ so that

$$|\lambda_1(A)|^2 = |v^*Hu|^2$$

$$\leq (v^*v)(Hu)^*Hu$$

$$= u^*H^2u \qquad \text{(from } v^*v = 1\text{)}$$

$$\leq \alpha_1(A)^2. \qquad (13)$$

The first inequality in (13) is the Cauchy-Schwarz inequality (see §7.2 Exercises, #14) and the second inequality is an application of Theorem 1 to the hermitian matrix $H^2 = A^*A$. If equality holds in (9), and hence in (13), then Theorem 1 implies that

$$H^2u = \alpha_1(A)^2u,$$

i.e.,

$$A^*Au = \alpha_1(A)^2u.$$

Conversely, if (10) and (11) hold then, since $u^*u = 1$, we observe that

$$|\lambda_1(A)|^2 = (Au)^*(Au) \qquad \text{(from (10))}$$

$$= u^*A^*Au$$

$$= \alpha_1(A)^2. \qquad \text{(from (11))} \blacksquare$$

In studying rank we developed a canonical form for general m × n matrices under pre- and post- multiplication by nonsingular matrices. For general matrices $A \in M_{m,n}(\mathbb{C})$ there is an analogous result known as the *singular value decomposition* theorem (abbreviated SVD). The SVD theorem is useful and important in the numerical analysis of linear systems. Indeed, MatLab uses the singular value decomposition of A in computing the Moore-Penrose inverse, the rank, and the 2-norm of A.

Theorem 3. (SVD)

Let $A \in M_{m,n}(\mathbb{C})$, m ≤ n, with singular values

$$\alpha_1 \geq \alpha_2 \geq \cdots \geq \alpha_m. \qquad (14)$$

Then there exist unitary matrices $U \in M_m(\mathbb{C})$ and $V \in M_n(\mathbb{C})$ such that

$$UAV = D \qquad (15)$$

where $D \in M_{m,n}(\mathbb{C})$ and

$$D = \begin{bmatrix} \alpha_1 & & & & \\ & \alpha_2 & & 0 & \\ & & \ddots & & \\ & 0 & & & \\ & & & & \alpha_m \end{bmatrix},$$

i.e., $D_{ii} = \alpha_i$, i = 1, ..., m, and all other entries of D are 0.

8.3 Singular Values

Proof.

We have denoted the singular values of A, i.e., the eigenvalues of $(AA^*)^{1/2}$, by α_i rather than $\alpha_i(A)$ in order to simplify the notation.

Let $r = \rho(A) = \rho(A^*A) = \rho(AA^*)$. Then from (14), $\alpha_1, \ldots \alpha_r$ are positive and $\alpha_{r+1} = \cdots = \alpha_m = 0$. Moreover, since AA^* and A^*A have the same eigenvalues except possibly for zero eigenvalues (see §7.2 Exercises, #25), we can regard $\alpha_1^2, \ldots, \alpha_m^2$ as eigenvalues of the n-square hermitian matrix A^*A. Now let X be an n-square unitary matrix for which

$$X^*A^*AX = \text{diag}(\alpha_1^2, \alpha_2^2, \ldots, \alpha_r^2, 0, \ldots, 0). \tag{16}$$

Next, let

$$Y^{(t)} = (1/\alpha_t)AX^{(t)}, \quad t = 1, \ldots, r, \tag{17}$$

be vectors in $M_{m,1}(\mathbb{C})$. Then for any s and t chosen from $1, \ldots, r$ we have

$$Y^{(s)*}Y^{(t)} = \frac{1}{\alpha_s\alpha_t}(AX^{(s)})^*AX^{(t)}$$

$$= \frac{1}{\alpha_s\alpha_t}X^{(s)*}A^*AX^{(t)}$$

$$= \frac{1}{\alpha_s\alpha_t}(X^*A^*AX)_{st}$$

$$= \frac{\alpha_t^2}{\alpha_s\alpha_t}\delta_{st}. \qquad \text{(from (16))} \tag{18}$$

In other words, $Y^{(1)}, \ldots, Y^{(r)}$ are orthonormal column vectors in $M_{m,1}(\mathbb{C})$. As we saw in the narrative immediately preceding §7.2 Exercises, $Y^{(1)}, \ldots, Y^{(r)}$ can be completed with $Y^{(r+1)}, \ldots, Y^{(m)}$ so that the matrix $Y \in M_m(\mathbb{C})$ whose columns are $Y^{(1)}, \ldots, Y^{(m)}$ is unitary.

Observe that (16) also implies that for $t = r + 1, \ldots, n$,

$$0 = (X^*A^*AX)_{tt}$$
$$= (AX)^{(t)*} (AX)^{(t)}$$

and hence

$$AX^{(t)} = 0, \quad t = r + 1, \ldots, n. \tag{19}$$

If we combine (17) and (18), keeping in mind that $\alpha_{r+1} = \cdots = \alpha_m = 0$ and $m \leq n$, we have

$$AX^{(t)} = Y^{(t)}\alpha_t, \quad t = 1, \ldots, m,$$

or

$$AX = YD.$$

Simply take $U = Y^*$ and $V = X$ and the proof of (15) is complete. ∎

In the narrative for §4.3 (see formula (24)) we deferred a proof of the SVD theorem until now. That promise has been made good with Theorem 3. Thus the arguments in Chapter 4 concerning the existence of the Moore-Penrose inverse are justified. The SVD theorem is useful for solving least squares problems, but we shall relegate this topic to some of the exercises for the present section.

The maximum singular value of an arbitrary matrix $A \in M_{m,n}(\mathbb{C})$ has a number of important and interesting properties. In fact, the MatLab command norm(A) produces $\alpha_1(A)$. The number $\alpha_1(A)$ goes under several names: "hilbert norm of A"; "2-norm of A". Most properties of $\alpha_1(A)$ are relatively easy to confirm. For example:

$$\alpha_1(A)^2 = \max_{x^*x=1} x^*A^*Ax; \tag{20}$$

if $B \in M_{n,p}(\mathbb{C})$ then

8.3 Singular Values

$$\alpha_1(AB) \leq \alpha_1(A)\,\alpha_1(B); \tag{21}$$

if $\alpha_1(A) \leq 1$ then

$$I_m - AA^* \geq 0 \tag{22}$$

and

$$I_n - A^*A \geq 0; \tag{23}$$

if A and B are both in $M_{m,n}(\mathbb{C})$ then

$$\alpha_1(A + B) \leq \alpha_1(A) + \alpha_1(B); \tag{24}$$

if $B \in M_n(\mathbb{C})$ and $\alpha_1(B) \leq 1$ then

$$B(I_n - B^*B)^{1/2} = (I_n - BB^*)^{1/2}B; \tag{25}$$

if $B \in M_n(\mathbb{C})$ and $\alpha_1(B) \leq 1$ then the 2n-square matrix

$$U = \begin{bmatrix} B & (I_n - BB^*)^{1/2} \\ (I_n - B^*B)^{1/2} & -B^* \end{bmatrix} \tag{26}$$

is unitary; the matrix U is called the *unitary dilation* of B. Thus, if $\alpha_1(B) \leq 1$, then B can be embedded as an n-square principal submatrix in a 2n-square unitary matrix. This fact is useful in obtaining information about B itself.

The equality (20) is simply Theorem 1 applied to the hermitian matrix A^*A. If v is any complex n-tuple in $M_{n,1}(\mathbb{C})$ then (20) obviously implies that

$$v^*A^*Av \leq \alpha_1(A)^2 v^*v. \tag{27}$$

To prove (21), use (20) as follows:

$$\alpha_1(AB)^2 = \max_{x^*x=1} x^*(AB)^*(AB)x$$

$$= \max_{x^*x=1} x^*B^*A^*ABx$$

$$= \max_{x^*x=1} (Bx)^*A^*A(Bx)$$

$$\leq \alpha_1(A)^2 \max_{x^*x=1} (Bx)^*(Bx) \qquad \text{(from (27))}$$

$$= \alpha_1(A)^2 \max_{x^*x=1} x^*B^*Bx$$

$$\leq \alpha_1(A)^2 \alpha_1(B)^2 .$$

The eigenvalues of the hermitian matrix (22) are

$$1 - \alpha_i(A)^2$$

and hence are nonnegative, and (22) follows. The results (23) and (24) are equally simple and will be left to the reader.

To confirm (25) we first use §8.2 Theorem 6 (or §8.2 Exercises, #6) to write $B = UH$ in polar form. Then, since $H^2 = B^*B$, we have

$$B(I_n - B^*B)^{1/2} = UH(I_n - H^2)^{1/2} . \qquad (28)$$

On the other hand,

$$(I_n - BB^*)^{1/2}B = (I_n - UH^2U^*)^{1/2}UH$$

$$= U(I_n - H^2)^{1/2}U^*UH$$

$$= U(I_n - H^2)^{1/2}H$$

$$= UH(I_n - H^2)^{1/2}$$

$$= B(I_n - B^*B)^{1/2} . \qquad (29)$$

8.3 Singular Values

The second equality in (29) follows from the fact that $(I_n - UH^2U^*)^{1/2}$ is a polynomial in $I_n - UH^2U^* = U(I_n - H^2)U^*$ (see §8.2 Theorem 5) and hence §8.2 Exercises, #2 can be applied. The fourth equality in (29) also follows from §8.2 Theorem 5 and the fact that H commutes with any polynomial in H. The proof that the 2n-square matrix U in (26) is unitary is simply a matter of block multiplication in computing UU^*, and using (25).

We close this chapter with a sequence of results concerning the *numerical radius* of a matrix $A \in M_n(\mathbb{C})$. The numerical radius is a relatively new concept in matrix theory. In fact, it has been studied extensively in recent years by a large group of talented research workers. The numerical radius is related to the maximum singular value, but it has a somewhat different definition. Let $A \in M_n(\mathbb{C})$. Then the numerical radius of A is defined as

$$w(A) = \max_{x^*x = 1} |x^*Ax|. \tag{30}$$

The reason for choosing the letter "w" to denote the numerical radius is not all that mysterious. The original (and best) work done in this general area is due to O. Toeplitz and F. Hausdorff over 70 years ago. The set of complex numbers x^*Ax, for $x^*x = 1$, was given the name "Wertvorrat" (field of values) by those authors - thus the "w". The next theorem gives us some important information about w(A): namely, it is an upper bound for the absolute values of the eigenvalues of A.

Theorem 4.

If $A \in M_n(\mathbb{C})$ then

$$|\lambda_1(A)| \leq w(A) \leq \alpha_1(A) \leq 2w(A). \tag{31}$$

Proof.

To prove that $|\lambda_1(A)| \leq w(A)$, simply choose u so that $u^*u = 1$ and $Au = \lambda_1(A)u$. The inequality $w(A) \leq \alpha_1(A)$ is not difficult:

$$w(A) = \max_{x^*x = 1} |x^*Ax|$$

$$\leq \max_{x^*x = 1} (x^*A^*Ax)^{1/2}$$

$$= \alpha_1(A). \tag{32}$$

The inequality in (32) is the Cauchy-Schwarz inequality (see §7.2 Exercises, #14) and the second equality is an application of Theorem 1.

Now write $A = H + iK$ where H and K are hermitian. Then

$$x^*Ax = x^*Hx + ix^*Kx$$

so that

$$|x^*Ax| \geq |x^*Hx|,$$

and hence

$$w(A) \geq \max_{x^*x = 1} |x^*Hx|.$$

Similarly

$$w(A) \geq \max_{x^*x = 1} |x^*Kx|$$

and it follows that

$$2w(A) \geq w(H) + w(K). \tag{33}$$

Since H and K are both hermitian it follows easily from Theorem 1 that $w(H) = |\lambda_1(H)|$ and $w(K) = |\lambda_1(K)|$. But, the hermitian property of H and K implies that

$$|\lambda_1(H)| = \alpha_1(H) \tag{34}$$

and

8.3 Singular Values

$$|\lambda_1(K)| = \alpha_1(K). \tag{35}$$

Both (34) and (35) are immediate consequences of the fact that any hermitian matrix is unitarily similar to a diagonal matrix, and unitary similarity does not affect either the eigenvalues or the singular values. Finally, we combine (24), (33), (34) and (35) to conclude that

$$\alpha_1(A) = \alpha_1(H + iK)$$

$$\leq \alpha_1(H) + \alpha_1(K)$$

$$= |\lambda_1(H)| + |\lambda_1(K)|$$

$$= w(H) + w(K)$$

$$\leq 2w(A). \qquad \blacksquare$$

In general, the numerical radius of a matrix is a difficult number to compute. However, if A is a direct sum, the problem can be simplified. Let

$$A = \sum_{t=1}^{k} \oplus A_t$$

where $A_t \in M_{n_t}(\mathbb{C})$, $t = 1, \ldots, k$. Then

$$w(A) = \max_t w(A_t). \tag{36}$$

To confirm (36), first let $u_t \in M_{n_t,1}(\mathbb{C})$, $u_t^* u_t = 1$. Set $n = \sum_{t=1}^{k} n_t$ and define $x \in M_{n,1}(\mathbb{C})$ to have 0 entries except in positions

$$n_1 + \cdots + n_{t-1} + 1, \ldots, n_1 + \cdots + n_{t-1} + n_t \tag{37}$$

where the n_t-tuple u_t appears. Obviously $x^*x = 1$, and by a simple block multiplication

$$x^*Ax = u_t^* A_t u_t$$

and hence

$$w(A_t) \le w(A), \quad t = 1, \ldots, k. \tag{38}$$

On the other hand, if $x \in M_{n,1}(\mathbb{C})$ and $x^*x = 1$, let u_t denote the n_t-tuple that occurs in positions (37) of x. Once again, by block multiplication we compute that

$$|x^*Ax| = \left| \sum_{t=1}^{k} u_t^* A_t u_t \right|$$

$$\le \sum_{t=1}^{k} |u_t^* A_t u_t|$$

$$\le \sum_{t=1}^{k} w(A_t) u_t^* u_t$$

$$\le \max_t w(A_t) \sum_{t=1}^{k} u_t^* u_t$$

$$= \max_t w(A_t) x^* x$$

$$= \max_t w(A_t). \tag{39}$$

The inequalities (38) and (39) combine to produce (36).

It is instructive to attempt to compute $w(A)$ for matrices of modest dimension. As an example, consider

$$A = \begin{bmatrix} 0 & 1 \\ 0 & 0 \end{bmatrix}. \tag{40}$$

Then set $x = [x_1 \ x_2]^T$, $x^*x = |x_1|^2 + |x_2|^2 = 1$, and compute directly that

$$x^*Ax = \bar{x}_1 x_2. \tag{41}$$

Write $x_1 = re^{i\theta}$ and $x_2 = \sqrt{1-r^2}\, e^{i\varphi}$ where $0 \le r \le 1$ and θ and φ are the arguments of x_1 and x_2 respectively, $0 \le \theta < 2\pi$, $0 \le \varphi < 2\pi$. Then (41) becomes

$$x^*Ax = r\sqrt{1-r^2}\, e^{i(\varphi-\theta)}$$

so that

$$|x^*Ax| = r\sqrt{1-r^2}.$$

It is a simple calculus problem to maximize $r\sqrt{1-r^2}$ for $0 \le r \le 1$. The maximum value is 1/2 and hence

$$w(A) = \frac{1}{2}.$$

Despite the preceding elementary calculation, it is surprisingly difficult to compute $w(A)$, even for low dimensional matrices. Notice that for the matrix (40)

$$\alpha_1(A) = 1$$

so that the equality $\alpha_1(A) = 2w(A)$ in (31) can certainly hold.

There is an analogue of (24) for the numerical radius:

$$w(A+B) \le w(A) + w(B), \tag{42}$$

which follows immediately from the definition of w. However, there is no counterpart to (21) for w. As an example, let

$$A = \begin{bmatrix} 0 & 0 \\ 1 & 0 \end{bmatrix}, \quad B = \begin{bmatrix} 0 & 1 \\ 0 & 0 \end{bmatrix}.$$

Then $w(A) = w(B) = 1/2$ and $w(AB) = 1$. Thus $w(AB) > w(A)w(B)$. Even if A and B commute it cannot be concluded that

$$w(AB) \le w(A)w(B),$$

however an example is somewhat more difficult to construct. Let

$$A = \begin{bmatrix} 0 & 1 & 0 & 0 \\ 0 & 0 & 1 & 0 \\ 0 & 0 & 0 & 1 \\ 0 & 0 & 0 & 0 \end{bmatrix}.$$

Then

$$A^2 = \begin{bmatrix} 0 & 0 & 1 & 0 \\ 0 & 0 & 0 & 1 \\ 0 & 0 & 0 & 0 \\ 0 & 0 & 0 & 0 \end{bmatrix}$$

and

$$A^3 = \begin{bmatrix} 0 & 0 & 0 & 1 \\ 0 & 0 & 0 & 0 \\ 0 & 0 & 0 & 0 \\ 0 & 0 & 0 & 0 \end{bmatrix}.$$

If $x = [x_1\ x_2\ x_3\ x_4]^T$, $x^*x = 1$, and $x_t = r_t e^{i\varphi_t}$ is the polar form of x_t, $t = 1, \ldots, 4$, it is easy to compute that

$$x^*A^2 x = r_1 r_3 e^{i(\varphi_3 - \varphi_1)} + r_2 r_4 e^{i(\varphi_4 - \varphi_2)}.$$

Obviously the largest value of $|x^*A^2 x|$ has the form

$$r_1 r_3 + r_2 r_4 \tag{43}$$

in which

8.3 Singular Values

$$\sum_{i=1}^{4} r_i^2 = x^*x = 1. \tag{44}$$

It is another easy calculus problem to confirm that the largest value of (43), subject to the condition (44), is 1/2. Thus $w(A^2) = 1/2$. The computation of $w(A^3)$ is identical to what we did for the matrix (40) so that $w(A^3) = 1/2$. For A itself we compute (using the preceding notation) that

$$x^*Ax = \bar{x}_1 x_2 + \bar{x}_2 x_3 + \bar{x}_3 x_4$$

so that

$$|x^*Ax| \leq r_1 r_2 + r_2 r_3 + r_3 r_4$$

$$\leq \frac{r_1^2 + r_2^2}{2} + \frac{r_2^2 + r_3^2}{2} + \frac{r_3^2 + r_4^2}{2}$$

$$= \frac{r_1^2 + 2r_2^2 + 2r_3^2 + r_4^2}{2}$$

$$\leq \frac{2(r_1^2 + r_2^2 + r_3^2 + r_4^2)}{2}$$

$$= r_1^2 + r_2^2 + r_3^2 + r_4^2$$

$$= x^*x$$

$$= 1.$$

But actually, this inequality must be strict. For, at the third inequality we added in r_1^2 and r_4^2 to the numerator. That would make the inequality strict unless $r_1 = r_4 = 0$. But if $r_1 = r_4 = 0$ then

$$|x^*Ax| \leq r_2 r_3$$

with $r_2^2 + r_3^2 = 1$. As we computed for the matrix (40), the largest such value of $r_2 r_3$ is

1/2. Thus in any event, $w(A) < 1$. It follows that

$$w(A)w(A^2) < w(A^2) = \frac{1}{2}$$

and

$$w(A^3) = \frac{1}{2},$$

and hence that

$$w(A^3) > w(A)w(A^2). \qquad (45)$$

We see then from (45) that commutativity of A and B is not enough to conclude that

$$w(AB) \leq w(A)w(B). \qquad (46)$$

Despite the example constructed in (45) it is not difficult to show that if A commutes with a normal matrix N then

$$w(AN) \leq w(A)w(N).$$

Before proving this we assemble a few more elementary facts about the numerical radius in the following result.

Theorem 5.

Let $A \in M_n(\mathbb{C})$.

(i)　If A is normal then

$$w(A) = |\lambda_1(A)|. \qquad (47)$$

(ii)　If C is a principal submatrix of A then

$$w(C) \leq w(A). \qquad (48)$$

8.3 Singular Values

(iii) The equalities

$$w(A) = w(A^T) = w(A^*) \tag{49}$$

and

$$w(zA) = |z|w(A) \tag{50}$$

hold for any $z \in \mathbb{C}$.

Proof.

(i) Let $U^*AU = \text{diag}(\lambda_1(A), ..., \lambda_n(A))$. Obviously, since $(Ux)^*Ux = x^*U^*Ux = x^*x$ for any $x \in M_{n,1}(\mathbb{C})$ it follows that

$$w(U^*AU) = \max_{x^*x = 1} |x^*U^*AUx|$$

$$= \max_{x^*x = 1} |(Ux)^*A(Ux)|$$

$$= \max_{y^*y = 1} |y^*Ay|$$

$$= w(A).$$

In other words, the numerical radius of A and U^*AU are the same for any $A \in M_n(\mathbb{C})$ and any unitary $U \in M_n(\mathbb{C})$. Thus

$$w(A) = w(\text{diag}(\lambda_1(A), ..., \lambda_n(A)))$$

$$= \max_t w([\lambda_t(A)])$$

$$= |\lambda_1(A)|.$$

(ii) Suppose that $C = A[\,\omega\,|\,\omega\,]$, $\omega \in Q_{p,\,n}$. If $u \in M_{p,\,1}(\mathbb{C})$ and $u^*u = 1$, define $x \in M_{n,\,1}(\mathbb{C})$ to be the n-tuple for which $x_{\omega_t} = u_t$, $t = 1, \ldots, p$, and $x_j = 0$ if $j \notin \omega$. Then obviously $x^*x = 1$ and by block multiplication

$$|u^*Cu| = |x^*Ax|$$

$$\leq w(A).$$

Since u was arbitrary the inequality (48) follows.

(iii) The proofs of (49) and (50) are very easy and are left to the reader. ∎

In view of the example (45), the following theorem is interesting.

Theorem 6.

If A and N are in $M_n(\mathbb{C})$, N is normal, and $AN = NA$, then

$$w(AN) \leq w(A)w(N). \tag{51}$$

Proof.

Let the distinct eigenvalues of N be $\mu_1, \ldots, \mu_k$ of algebraic multiplicities $n_1, \ldots, n_k$, respectively. Let U be a unitary matrix that brings N to diagonal form:

$$U^*NU = \sum_{t=1}^{k} \oplus \mu_t I_{n_t}. \tag{52}$$

Since A and N commute, so do U^*AU and U^*NU. It is an easy exercise in block multiplication to show that any matrix that commutes with (52) must have the same block diagonal structure as (52):

$$U^*AU = \sum_{t=1}^{k} \oplus A_t \tag{53}$$

where A_t is an n_t - square matrix (see §8.3 Exercises, #10). Thus (52) and (53) combine

8.3 Singular Values

to produce

$$U^*ANU = U^*AUU^*NU$$

$$= \sum_{t=1}^{k} \oplus A_t \sum_{t=1}^{k} \oplus \mu_t I_{n_t}$$

$$= \sum_{t=1}^{k} \oplus \mu_t A_t .$$

Thus

$$w(AN) = w(U^*ANU)$$

$$= w\left(\sum_{t=1}^{k} \oplus \mu_t A_t \right)$$

$$= \max_t w(\mu_t A_t) \qquad \text{(from (36))}$$

$$= \max_t |\mu_t| w(A_t) \qquad \text{(from (50))}$$

$$\leq |\lambda_1(N)| \max_t w(A_t)$$

$$\leq w(N)\, w(A) . \qquad \text{(from (47))} \blacksquare$$

Two matrices A and B in $M_n(\mathbb{C})$ are said to *double commute* if $AB = BA$ and $AB^* = B^*A$. The reader should be able to see that if A and B double commute, then so do B and A (i.e., compute $(AB^*)^* = (B^*A)^*$). Also, double commutativity is invariant under the same unitary similarity on A and B (i.e., $(U^*AU)(U^*BU) = U^*ABU = U^*BAU = (U^*BU)(U^*AU)$). Also, if B happens to be a normal matrix N then (52) and (53) show that $AN = NA$ automatically implies that $AN^* = N^*A$. Thus commutativity and double commutativity are the same if one of the matrices is normal. Of course,

$$A = \begin{bmatrix} 0 & 1 \\ 0 & 0 \end{bmatrix}$$

commutes with itself but it certainly does not commute with A^*.

As a final result in this chapter we can obtain a somewhat weaker inequality than (46) if A and B double commute.

Theorem 7.

Assume that A and B are in $M_n(\mathbb{C})$. If A and B double commute it follows that

$$w(AB) \leq \min \{ w(A)\alpha_1(B), w(B)\alpha_1(A) \}.$$

Proof.

Let

$$B_1 = \frac{1}{\alpha_1(B)} B$$

so that $\alpha_1(B_1) = 1$. As in (26), construct the 2n-square unitary dilation of B_1:

$$U = \begin{bmatrix} B_1 & (I_n - B_1 B_1^*)^{1/2} \\ (I_n - B_1^* B_1)^{1/2} & -B_1^* \end{bmatrix}.$$

We know that $(I_n - B_1 B_1^*)^{1/2}$ and $(I_n - B_1^* B_1)^{1/2}$ are scalar polynomials in $B_1 B_1^*$ and $B_1^* B_1$ respectively. From the double commutativity of A and B (and hence of A and B_1) it follows that

$$AB_1 B_1^* = B_1 AB_1^*$$

$$= B_1 B_1^* A.$$

8.3 Singular Values

Thus A commutes with $B_1 B_1^*$ and similarly, A commutes with $B_1^* B_1$. It follows that A commutes with both $(I_n - B_1 B_1^*)^{1/2}$ and $(I_n - B_1^* B_1)^{1/2}$. Another way of saying the same thing is: $A \oplus A$ commutes with U. Since U is unitary, and hence normal, we can apply Theorem 6 to conclude that

$$w((A \oplus A)U) \leq w(A \oplus A)w(U). \tag{54}$$

But the eigenvalues of U have modulus 1 and Theorem 5(i) then implies that $w(U) = 1$. Also, by (36)

$$w(A \oplus A) = w(A).$$

Hence (54) becomes

$$w((A \oplus A)U) \leq w(A). \tag{55}$$

But by block multiplication, $AB_1 = \dfrac{1}{\alpha_1(B)} AB$ is a principal submatrix of $(A \oplus A)U$ and hence (48) and (55) combine to produce

$$w(\dfrac{1}{\alpha_1(B)} AB) \leq w(A),$$

$$w(AB) \leq w(A)\alpha_1(B). \tag{56}$$

The matrices B^* and A^* also double commute (this follows immediately from the definition) and hence (56) implies that

$$w(B^* A^*) \leq w(B^*)\alpha_1(A^*).$$

But $\alpha_1(A^*) = \alpha_1(A)$ and $B^* A^* = (AB)^*$. Apply (49) to conclude that

$$w(AB) \leq w(B)\alpha_1(A)$$

and the proof is complete. ∎

8.3 Exercises

1. Prove: if $A \in M_n(\mathbb{C})$ then

$$\sum_{j=1}^{n} |\lambda_j(A)|^2 \leq \sum_{j=1}^{n} \alpha_j(A)^2$$

with equality holding iff A is normal.

Hint: Both singular values and eigenvalues are invariant under unitary similarity. Thus we can assume that A is upper triangular with $\lambda_1(A), \ldots, \lambda_n(A)$ along the main diagonal. It is simple to compute that

$$\text{tr}(AA^*) = \sum_{i,j=1}^{n} |a_{ij}|^2$$

$$= \sum_{j=1}^{n} |\lambda_j(A)|^2 + \sum_{1 \leq i < j \leq n} |a_{ij}|^2.$$

But

$$\text{tr}(AA^*) = \sum_{j=1}^{n} \alpha_j(A)^2$$

and the desired inequality follows. If equality holds then

$$\sum_{1 \leq i < j \leq n} |a_{ij}|^2 = 0$$

and the upper triangular form of A must be diagonal.

2. Let $r_1 \geq r_2 \geq \cdots \geq r_n$ be real numbers and let $\sigma_1, \ldots, \sigma_n$ be nonnegative numbers that add to 1. Prove:

8.3 Exercises

$$r_n \le \sum_{t=1}^{n} \sigma_t r_t \le r_1.$$

Moreover, the upper inequality can be equality iff the σ_t which are coefficients of those r_t for which $r_t < r_1$ must, in fact, be 0. A similar statement holds with respect to the lower inequality.

Hint: Let $r_1 = \cdots = r_p > r_{p+1} \ge \cdots \ge r_n$. Then

$$\sum_{t=1}^{n} \sigma_t r_t = \sigma r_1 + \sum_{t=p+1}^{n} \sigma_t r_t$$

where $\sigma = \sigma_1 + \cdots + \sigma_p$. Then, since all of $r_{p+1}, \ldots, r_n$ are strictly less than r_1,

$$\sum_{t=1}^{n} \sigma_t r_t \le \sigma r_1 + r_1 \sum_{t=p+1}^{n} \sigma_t$$

$$= r_1.$$

Moreover, equality can hold iff $\sigma_{p+1} = \cdots = \sigma_n = 0$.

3. Show that if $A \in M_{m,n}(\mathbb{C})$ and $m \ge n$ then there exist unitary matrices $U \in M_m(\mathbb{C})$ and $V \in M_n(\mathbb{C})$ such that

$$UAV = \begin{bmatrix} \alpha_1 & & 0 \\ & \ddots & \\ & & \alpha_n \\ 0 & & \end{bmatrix}$$

where $\alpha_1, \ldots, \alpha_n$ are singular values of A.

Hint: Apply Theorem 3 to $A^* \in M_{n,m}(\mathbb{C})$, $n \le m$, to obtain unitary matrices $R \in M_n(\mathbb{C})$ and $S \in M_m(\mathbb{C})$ for which

$$RA^*S = D,$$

$$D = \begin{bmatrix} \alpha_1 & & 0 & \\ & \ddots & & 0 \\ 0 & & \alpha_n & \end{bmatrix}.$$

Now since $n \leq m$ there can be at most n nonzero singular values of A, i.e., $\rho(A) \leq \min\{m, n\}$. Simply compute the conjugate transpose of both sides of the latter equation and set $U = S^*$ and $V = R^*$.

4. In the notation of §4.3, formula (22), if $A = PDQ$ is the SVD of $A \in M_{m,n}(\mathbb{C})$ then the unique Moore-Penrose inverse of A is $A^+ = Q^*\Delta P^*$ (see §4.3 formula (24); the location of the 0 blocks may be slightly altered depending on the relative sizes of m and n). The matrix Δ is $n \times m$ with $\alpha_1^{-1}, \ldots, \alpha_r^{-1}$ appearing in positions $(1, 1), \ldots, (r, r)$ in Δ, 0's elsewhere. Hence $r = \rho(A)$.

(a) Prove: if A^*A is nonsingular then $A^+ = (A^*A)^{-1}A^*$;

(b) Prove: if A is nonsingular then $A^+ = A^{-1}$.

Hint: (a) $A = PDQ$, so $A^*A = Q^*D^*P^*PDQ = Q^*D^*I_m DQ = Q^*D^*DQ$. Now D^*D is $n \times n$ with $\alpha_1^2, \ldots, \alpha_r^2$ leading down the main diagonal, 0 elsewhere. If A^*A is nonsingular then $r = n$ and $(A^*A)^{-1} = Q^*\text{diag}(\alpha_1^{-2}, \ldots, \alpha_n^{-2})Q$. Hence

$$(A^*A)^{-1}A^* = Q^*\text{diag}(\alpha_1^{-2}, \ldots, \alpha_n^{-2})\, QQ^*D^*P^*$$

$$= Q^*\text{diag}(\alpha_1^{-2} \ldots \alpha_n^{-2}) \begin{bmatrix} \alpha_1 & & 0 \\ & \ddots & \\ 0 & & \alpha_n \\ & 0 & \end{bmatrix} P^*$$

$$= Q^*\Delta P^*$$

$$= A^+.$$

(b) If A is nonsingular then $n = m = r$ and by (a), $A^+ = (A^*A)^{-1}A^* = A^{-1}A^{*-1}A^* = A^{-1}$.

5. If $u \in M_{n,1}(\mathbb{C})$ then recall that the nonnegative number $(u^*u)^{1/2}$ is called the 2-norm of u. The 2-norm of u is frequently written as $\| u \|$ or $\| u \|_2$. Let $A \in M_{m,n}(\mathbb{C})$ and assume that $r = \rho(A) \leq n$.

(a) Let $b \in M_{m,1}(\mathbb{C})$. Show that $x = A^+b$ minimizes $\| Ax - b \|$. The minimum is taken over all $x \in M_{n,1}(\mathbb{C})$.

(b) Show that if $z \in M_{n,1}(\mathbb{C})$ and z also minimizes $\| Ax - b \|$, $z \neq A^+b$, then $\| z \| > \| A^+b \|$.

Thus, if $\rho(A) \leq n$, then $x = A^+b$ minimizes $\| Ax - b \|$ and among all such minimizers it has least 2-norm.

Hint: Write the singular value decomposition of A as $A = PDQ$

$$D = \begin{bmatrix} \alpha_1 & & & 0 \\ & \ddots & & \\ & & \alpha_r & \\ \hline 0 & & & 0 \end{bmatrix} \in M_{m,n}(\mathbb{C}).$$

Since $\| u \| = \| Uu \|$ for any unitary U (why?), we compute that

$$\| Ax - b \|^2 = \| P^*(Ax - b) \|^2$$
$$= \| DQx - P^*b \|^2.$$

Write

$$Qx = \begin{bmatrix} y \\ w \end{bmatrix} \in M_{n,1}(\mathbb{C})$$

where $y \in M_{r,1}(\mathbb{C})$, and also write

$$P^*b = \begin{bmatrix} c \\ d \end{bmatrix} \in M_{m,1}(\mathbb{C})$$

where $c \in M_{r,1}(\mathbb{C})$. Then, from the form of D,

$$\| DQx - P^*b \|^2 = \left\| D \begin{bmatrix} y \\ w \end{bmatrix} - \begin{bmatrix} c \\ d \end{bmatrix} \right\|^2$$

$$= \sum_{i=1}^{r} |\alpha_i y_i - c_i|^2 + \| d \|^2 .$$

Clearly, this is minimized by choosing $y_i = c_i / \alpha_i$, $i = 1, \ldots, r$. Since the value of w does not affect the value of $\| Ax - b \|^2$ we can take $w = 0$. Thus

$$x = Q^* \begin{bmatrix} c_1 / \alpha_1 \\ \vdots \\ c_r / \alpha_r \\ 0 \\ \vdots \\ 0 \end{bmatrix}$$

$$= Q^* \begin{bmatrix} \alpha_1^{-1} & & & & 0 \\ & \ddots & & & \\ & & \alpha_r^{-1} & & \\ \hline & 0 & & & 0 \end{bmatrix} \begin{bmatrix} c \\ 0 \end{bmatrix}$$

$$= Q^* \Delta P^* b$$

$$= A^+ b.$$

8.3 Exercises

(i.e., the d disappears in the product $\Delta \begin{bmatrix} c \\ d \end{bmatrix}$). This proves (a). Also, $\| x \|^2 = \| Qx \|^2 = \| y \|^2 + \| w \|^2$, so that any other minimizer will have 2-norm greater than $x = A^+b$ unless $w = 0$. Since y is completely determined, i.e., $y_i = c_i / \alpha_i$, $i = 1, \ldots, r$, $x = A^+b$ is the unique minimizer with least 2-norm. This establishes (b).

6. (a) If α is a complex number, what is $w(A)$ for

$$A = \begin{bmatrix} 0 & \alpha \\ 0 & 0 \end{bmatrix} ?$$

   (b) Exhibit an example of a matrix A for which $w(A) \leq 1$ and yet $I_n - A^*A$ is not positive semi-definite.

   Hint: (a) $w(A) = |\alpha|/2$.

   (b) Try $A = \begin{bmatrix} 0 & 2 \\ 0 & 0 \end{bmatrix}$.

7. Find the unitary dilation of

$$B = \begin{bmatrix} 0 & 1 \\ 0 & 0 \end{bmatrix}.$$

8. Show that if U is unitary then $w(U^*AU) = w(A)$.

9. Show that the precise value of $w(A)$ for the matrix

$$A = \begin{bmatrix} 0 & 1 & 0 & 0 \\ 0 & 0 & 1 & 0 \\ 0 & 0 & 0 & 1 \\ 0 & 0 & 0 & 0 \end{bmatrix}$$

is

$$w(A) = \frac{(3 + 5^{1/2})^{1/2}}{2^{3/2}}$$

which, to 3 significant decimal places, is 0.809.

Hint: Setting $|x_i| = r_i$, $i = 1, \ldots, 4$, it is clear that the maximum absolute value of x^*Ax is the maximum of the function $f(r) = r_1 r_2 + r_2 r_3 + r_3 r_4$. Note that $f(r) = r^T S r / 2$ where S is the 4-square symmetric matrix.

$$S = \begin{bmatrix} 0 & 1 & 0 & 0 \\ 1 & 0 & 1 & 0 \\ 0 & 1 & 0 & 1 \\ 0 & 0 & 1 & 0 \end{bmatrix}.$$

Hence the largest value of $r^T S r$ for $r^T r = 1$ is the largest eigenvalue of S. The characteristic polynomial of S is easily computed to be $\lambda^4 - 3\lambda^2 + 1$. If we solve this polynomial, then the indicated value of $w(A)$ results.

10. Let $N = \sum_{t=1}^{k} \oplus \mu_t I_{n_t}$, where the μ_t are distinct, $t = 1, \ldots, k$. Assume that A commutes with N. Prove that

$$N = \sum_{t=1}^{k} \oplus A_t$$

where A_t is an n_t-square matrix, $t = 1, \ldots, k$.

Hint: Write $A = [A_{st}]$ where A_{st} is an $n_s \times n_t$ block in A, conformable with the block structure of N. Then

$$NA = [\mu_s A_{st}], \quad AN = [A_{st} \mu_t]$$

so that

$$0 = NA - AN = \left[(\mu_s - \mu_t) A_{st}\right].$$

Since $\mu_s - \mu_t \neq 0$ for $s \neq t$, it follows that $A_{st} = 0$ for $s \neq t$.

11. Let n and q be positive integers. Prove that there exists an nq-square permutation matrix Q such that for any $x \in M_{n,1}(R)$ and any $y \in M_{q,1}(R)$,

$$Q(x \otimes y) = y \otimes x.$$

If $n = q$ then Q is symmetric as well.

Hint: In §3.3 formula (17) we saw that

$$e_i^n \otimes e_j^q = e_{(i-1)q+j}^{nq}, \quad i = 1, \ldots, n, \; j = 1, \ldots, q.$$

Moreover, distinct pairs i and j produce distinct $e_i^n \otimes e_j^q$ and every e_t^{nq}, $t = 1, \ldots, nq$, appears precisely once as $e_i^n \otimes e_j^q$ for appropriate i and j. Thus the nq-tuples e_t^{nq} can be indexed in order by simply finding the i and j for which $t = (i-1)q + j$. Similarly,

$$e_j^q \otimes e_i^n = e_{(j-1)n+i}^{nq}, \quad j = 1, \ldots, q, \; i = 1, \ldots, n,$$

are the e_t^{nq}, $t = 1, \ldots, nq$, in some order. Simply define an nq-square permutation matrix Q that satisfies

$$Q e_{(i-1)q+j}^{nq} = e_{(j-1)n+i}^{nq}, \quad i = 1, \ldots, n, \; j = 1, \ldots, q.$$

Then

$$Q(e_i^n \otimes e_j^q) = e_j^q \otimes e_i^n, \quad i = 1, \ldots, n, \; j = 1, \ldots, q.$$

If

$$x = \sum_{i=1}^n x_i e_i^n,$$

and

$$y = \sum_{j=1}^{q} y_j e_j^q$$

then it is clear that

$$Q(x \otimes y) = Q(\sum_{i=1}^{n} x_i e_i^n \otimes \sum_{j=1}^{q} y_j e_j^q)$$

$$= Q(\sum_{i,j} x_i y_j e_i^n \otimes e_j^q)$$

$$= \sum_{i,j} x_i y_j Q(e_i^n \otimes e_j^q)$$

$$= \sum_{i,j} x_i y_j e_j^q \otimes e_i^n$$

$$= \sum_{j} y_j e_j^q \otimes \sum_{i} x_i e_i^n$$

$$= y \otimes x.$$

If $n = q$ then $Q^2(x \otimes y) = Q(y \otimes x) = x \otimes y$ so that $Q^2 = I_{n^2}$ and hence $Q^T = Q$.

12. Let m, n, p, q be positive integers. Prove that there exists an mp-square permutation matrix P and an nq-square permutation matrix Q such that for any $A \in M_{m,n}(R)$ and $B \in M_{p,q}(R)$

$$P(A \otimes B)Q = B \otimes A.$$

If $m = q$ and $n = p$ then $P = Q$. If $m = n = p = q$ then $P = Q = Q^T$.

Hint: Choose the nq-square permutation matrix Q such that $x \otimes y = Q(y \otimes x)$ for any $x \in M_{n,1}(R)$ and $y \in M_{q,1}(R)$. Similarly, choose the mp-square permutation matrix P such that $P(u \otimes v) = v \otimes u$ for any $u \in M_{m,1}(\mathbb{C})$ and $v \in M_{p,1}(\mathbb{C})$. Then

$$(P(A \otimes B)Q)y \otimes x = P(A \otimes B) x \otimes y$$

$$= P(Ax \otimes By)$$

$$= By \otimes Ax$$

$$= (B \otimes A)y \otimes x.$$

Since x and y are arbitrary it follows that $P(A \otimes B)Q = B \otimes A$. In the solution to #11 the nq-square permutation matrix Q is defined by

$$e_i^n \otimes e_j^q = Q(e_j^q \otimes e_i^n)$$

or

$$Q e_{(j-1)n+i}^{nq} = e_{(i-1)q+j}^{nq}, \quad i = 1, \ldots, n, \quad j = 1, \ldots, q.$$

Similarly, P is defined by

$$P(e_i^m \otimes e_j^p) = e_j^p \otimes e_i^m,$$

or

$$P e_{(i-1)p+j}^{mp} = e_{(j-1)m+i}^{mp}, \quad i = 1, \ldots, m, \quad j = 1, \ldots, p.$$

Thus if $m = q$ and $n = p$ then these defining equations for P and Q are identical. If $m = n = p = q$ then Q is symmetric from #11.

13. Let $A \in M_m(\mathbb{C})$ and $B \in M_n(\mathbb{C})$. Show that $A \otimes I_n$ and $I_m \otimes B$ double commute.

14. Prove: if $A \in M_n(\mathbb{C})$ and $B \in M_n(\mathbb{C})$ then

$$w(A \otimes B) \leq \min \{w(A)\alpha_1(B), w(B)\alpha_1(A)\}.$$

Hint: Since $A \otimes I_n$ and $I_n \otimes B$ double commute it follows from Theorem 7 that

$$w(A \otimes B) = w((A \otimes I_n)(I_n \otimes B))$$

$$\leq \min \{w(A \otimes I_n)\alpha_1(I_n \otimes B), w(I_n \otimes B)\alpha_1(A \otimes I_n)\}.$$

Now, $\alpha_1(I_n \otimes B) = \alpha_1(B)$ and $\alpha_1(A \otimes I_n) = \alpha_1(A)$. Also, $w(I_n \otimes B) = w(B)$ and $w(I_n \otimes A) = w(A)$ by (36). However, by #12, $A \otimes I_n = Q^T(I_n \otimes A)Q$ for an appropriate permutation matrix Q so that $w(A \otimes I_n) = w(I_n \otimes A) = w(A)$.

15. Let $A \in M_n(\mathbb{C})$ and $B \in M_n(\mathbb{C})$ and let U be unitary. Prove that

$$|\text{tr}(U^*AUB)| \leq n\tau$$

where $\tau = \min\{w(A)\alpha_1(B), w(B)\alpha_1(A)\}$.

Hint: First note that if X and Y are any two n-square matrices then $\text{tr}(X^*Y) = c(X)^*c(Y)$ where $c(X)$ is the column form of X. We also know that $c(AXB) = (B^T \otimes A)c(X)$ (see §3.3 formula (10)). Thus for any X, $\text{tr}(X^*AXB) = c(X)^*(B^T \otimes A)c(X)$. It follows from the definition of $w(B^T \otimes A)$ that $|\text{tr}(X^*AXB)| \leq w(B^T \otimes A)c(X)^*c(X) = w(B^T \otimes A)\text{tr}(X^*X)$. Since $w(B^T) = w(B)$ we can apply #14 to conclude that $|\text{tr}(X^*AXB)| \leq \tau \text{tr}(X^*X)$. If X = U is unitary then $\text{tr}(X^*X) = n$.

8.3 MatLab

1. Theorem 3, the SVD theorem, is implemented in MatLab with the function **svd**. Let $A \in M_{m,n}(\mathbb{C})$. Then the command

8.3 MatLab

$$[U, D, V] = svd(A) \quad (1)$$

produces an m-square unitary U, an n-square unitary V, and an m × n matrix D with positive entries $\alpha_1, ..., \alpha_r$ in positions (t, t), t = 1, ..., r = ρ(A), 0's elsewhere, such that

$$A = UDV^*.$$

The positive singular values are arranged in descending order:

$$\alpha_1 \geq ... \geq \alpha_r > 0.$$

The command

$$u = svd(A) \quad (2)$$

returns the vector of singular values of A.

(a) At the Command prompt compare u(1) in (2) with norm(A, 2) for random complex n-square matrices A. Also use eig to compute $|\lambda_1(A)|$, the modulus of the eigenvalue of maximum modulus, and confirm Theorem 4 to the effect that $|\lambda_1(A)| \leq \alpha_1(A)$.

Hint: To simplify entering several random A first define a function named arand by

function A=arand(m, n)
%ARAND(m, n) generates a random
%m × n complex matrix.
A = (2*rand(m,n)-1) + i*(2*rand(m,n) - 1)

and save arand in shur. Then (for example) enter the following commands:

A=arand(5,5); alfa1=norm(A,2)
u = svd(A); u(1)
ev=eig(A); lamda1 = max(abs(ev))

(b)   Use normrand to confirm that

$$|\lambda_1(A)| = \alpha_1(A) = \text{norm}(A, 2)$$

for any normal A.

(c)   Use [U, D, V] =svd(A) to write a function called psinv such that psinv(A) is the Moore-Penrose inverse of A. Compare psinv(A) with pinv(A) for several random m × n matrices A. Save psinv in shur.

Hint:
```
function X=psinv(A)
%PSINV accepts an mxn
%matrix A and returns
%the Moore-Penrose inverse
%of A.
[m, n]=size(A);
[U,D,V]=svd(A);
r=rank(A);
D1=inv(D(1:r,1:r));
Y=[D1 zeros(r,m-r); zeros(n-r,m)];
X=V*Y*U';
```

2. Write a MatLab function called nrunit that

- accepts an arbitrary n-square complex matrix A

- returns an n-square unitary R for which

$$\text{norm}(A - R, \text{'fro'}) \le \text{norm}(A - W, \text{'fro'})$$

for any other n-square unitary matrix W, i.e., nrunit(A) is the unitary matrix closest to A, measured in terms of the Frobenius norm.

Save nrunit in shur.

Hint: Let W be any n-square unitary matrix and let $A = UDV^*$ be the SVD of

A. For notational convenience, let $\|X\|_F = (tr(X^*X))^{1/2}$ denote the Frobenius norm of any X (remember that tr is the trace). Then

$$\|A - W\|_F^2 = tr((A - W)^*(A - W))$$

$$= tr(A^*A) + tr(W^*W) - 2\operatorname{Re} tr(A^*W)$$

$$= \|A\|_F^2 + n - 2\operatorname{Re}(tr(VDU^*W))$$

$$= \|A\|_F^2 + n - 2\operatorname{Re}(tr(DU^*WV)).$$

Now $Q = U^*WV$ is a unitary matrix so that each main diagonal entry has modulus at most 1, say $Q_{tt} = r_t e^{i\varphi_t}$, $r_t \leq 1$, $t = 1, \ldots, n$. Now, if $k = \rho(A)$ then

$$tr(DQ) = D_{11} r_1 e^{i\varphi_1} + \cdots + D_{kk} r_k e^{i\varphi_k}$$

and obviously the largest value the $\operatorname{Re}(tr(DQ))$ can take on is $D_{11} + \cdots + D_{kk} = tr(D)$. In fact, choose W so that $Q = U^*WV = I_n$, i.e., $W = UV^*$. Then

$$tr(DQ) = tr(D).$$

Thus,

$$\|A - W\|_F \geq \|A - UV^*\|_F.$$

Hint:
```
function R = nrunit(A)
%NRUNIT(A) returns the
%unitary matrix closest to
%A measured in terms of the
%Frobenius norm.
[U, D, V] = svd(A);
R = U*V ';
```

3. Write a MatLab script named clsunit that confirms the definition of nrunit by

   - generating 20 random complex n-square matrices A, and 20 n-square random unitary matrices W, $1 \leq n \leq 5$

   - printing out a table consisting of the values $\| A - \text{nrunit}(A) \|_F$ and $\| A - W \|_F$

   Hint:
   ```
   %CLSUNIT compares the Frobenius
   %norms of A-nrunit(A) and A-W
   %where nrunit(A) is the closest
   %unitary matrix to A in
   %Frobenius norm.
   T=[ ];
   for k=1:20
           n=floor(5*rand+1);
           W=urand(n);
           A=arand(n,n);
           R=nrunit(A);
           clu=norm(A-R,'fro');
           dw=norm(A-W,'fro');
           T=[T; clu dw];
   end
   disp(' ');
   disp(T)
   ```

4. Experiment with svd(A) when A is a row vector or a column vector. What is the SVD of a row or column vector?

   Hint: If A is m × 1 then the command (1) returns an m-square unitary U, a 1–square unitary V, and an m × 1 D such that $A = UDV^*$. Hence

   $$A = U\alpha_1(A)e_1 z$$

   $$= \alpha_1(A)zu$$

where u is the first column of U, z is a complex number of modulus 1, and $\alpha_1(A) = \left(\sum_{i=1}^{m} |a_{i1}|^2 \right)^{1/2}$, the 2-norm of A. In other words,

$$A = \alpha_1(A)v$$

where v is an m × 1 column vector that satisfies $v^*v = 1$, and $\alpha_1(A)$ is the 2-norm of A.

5. According to the MatLab documentation, any singular value of $A \in M_{m,n}(\mathbb{C})$ for which

$$\alpha_t(A) < \max([m, n]) * \alpha_1(A) * \text{eps} \tag{3}$$

is counted as 0 in computing svd(A). The number eps is the smallest floating point number for which $1 + \text{eps} > 1$. It is about $2.220446 \cdot 10^{-16}$. The rank function, rank(A), counts the number of singular values of A that are no less than the number on the right in (3). The command rank(A, tol) counts the number of singular values of A exceeding tol. Let $A = \text{hilb}(10) = \left[\frac{1}{i+j-1} \right] \in M_{10}(\mathbb{R})$. Use MatLab to evaluate svd(A) as well as each of the following numbers. Explain the outputs.

(a) rank(A, 1e - 4)   (b) rank(A, 1e - 7)

(c) rank(A, 1e - 11)

Hint: (a) The output is 5 because 5 singular values of A exceed 10^{-4}.

(b) The output is 7 because precisely 7 of the singular values of A exceed 10^{-7}. The remaining 3 singular values are less than 10^{-7} and hence do not get counted in rank(A, tol).

(c) The output is 9.

6. Write a MatLab function called factr that

- accepts an m × n complex matrix A

- returns an m × r matrix B and an r × n matrix C such that A = BC where $r = \rho(A) = \rho(B) = \rho(C)$.

Save factr in shur.

Hint: Let [U, D, V] = svd(A) so that $A = UDV^*$, U m-square unitary and V n-square unitary. Let $d_t = \alpha_t^{1/2}$, t = 1, ..., r, where $r = \rho(A)$ and $\alpha_1 \geq ... \geq \alpha_r > 0$ are the singular values of A. Then check that

$$A = U \begin{bmatrix} d_1^2 & & & 0 \\ & \ddots & & \\ & & d_r^2 & \\ 0 & & & 0 \end{bmatrix} V^*$$

$$= U \begin{bmatrix} d_1 & & & 0 \\ & \ddots & & \\ & & d_r & \\ 0 & & & 0 \end{bmatrix} \begin{bmatrix} d_1 & & & 0 \\ & \ddots & & \\ & & d_r & \\ 0 & & & 0 \end{bmatrix} V^* \quad (4)$$

$$= [\, U^{(1)}d_1 \ \cdots \ U^{(r)}d_r \,] \begin{bmatrix} d_1 V^*_{(1)} \\ \vdots \\ d_r V^*_{(r)} \end{bmatrix}. \quad (5)$$

In (4), the second factor on the right is m × n and the third factor is n × n. The first factor on the right in (5) is m × r and the second factor is r × n.

(The reader should take m = 3, n = 4, r = 2 to confirm that the product in (4) is correct). Since U and V are both unitary the ranks of both factors on the right in (5) are equal to r.

8.3 MatLab

```
function [B,C]=factr(A)
%FACTR accepts an mxn complex
%matrix A of rank r and returns
%B, m x r, C, r x n, both of rank r, for
%which A=BC.
[U,D,V]=svd(A);
r=rank(A);
E=D.^(1/2);
B1=U*E;
B=B1(:,1:r);
C2=E*V';
C=C2(1:r,:);
```

7. Write a MatLab function called rankr such that rankr(m, n, r) returns a random m × n complex matrix of rank r, (r ≤ min{m, n}). Save rankr in shur.

   Hint:
```
function A=rankr(m,n,r)
%RANKR accepts positive integers
%m,n,r, r ≤ min([m, n]),
%and returns a random mxn
%complex A of rank r.
mu=min([m, n]);
A1=arand(m,n);
k=rank(A1);
while k~=mu                %in the unlikely event
     A1 = arand(m,n)       %k ~= mu, replace A1.
       k=rank(A1);
end;
if r < k
     [U,D,V]=svd(A1);
     D((r+1):m,(r+1):n)=zeros(m-r,n-r);
     A=U*D*V';
else
     A=A1;
end
```

8. Evaluate [B, C] = factr(A) for several matrices A = rankr(m, n, r). Confirm that A = BC.

9. As we saw in §8.2 Theorem 5, the kth root of a hermitian matrix A can be computed as a scalar polynomial with real coefficients in the matrix A itself. Also polyvalm(p, A) returns p(A) where p(λ) is a scalar polynomial. (See §7.2 MatLab, #8.) Read the Help entries for **expm, logm, sqrtm,** and **funm**. Then compute the following matrices:

   (a) expm(zeros(2))

   (b) expm(eye(2))

   (c) X = expm([0 1; -1 0]); X '*X

   (d) X = expm(i * hrand(5)); X '*X

   (e) A = arand(3,3); X=logm(A); expm(X)

   (f) X = sqrtm([0 1; 0 0]), X^2

   (g) X = funm(i*hrand(3), 'exp'); X'*X

   Hint: (f) If $\begin{bmatrix} a & b \\ c & d \end{bmatrix}^2 = \begin{bmatrix} 0 & 1 \\ 0 & 0 \end{bmatrix}$ then

   $a^2 + bc = 0$

   $d^2 + bc = 0$

   $ac + dc = 0$

   $ab + bd = 1$

   These equations are inconsistent.

10. It is possible to encode strings and execute them with the MatLab command **eval**. Read the Help entry for eval and then try the following experiment:

8.3 MatLab

```
t = 'x + y ';
x = 2;
y = 3;
s = eval(t)
```

Write a MatLab function named anyf that

- accepts an m × n complex matrix A

- accepts a correct MatLab function f encoded as a string, e.g., f = 'exp'

- uses eval to return the matrix whose i, j entry is f(A(i, j))

In general, a command such as X(i, j) = eval(['exp', 'A(i, j)']) will evaluate X(i, j) as

$$X(i, j) = \exp(A(i, j))$$

Save anyf in shur. Then define a function named cube that cubes a scalar variable and set f = 'cube'. Compare anyf(A, f) with A.^3.

Hint:
```
function X=anyf(A,f)
%ANYF accepts an mxn
%complex matrix A and
%a function f encoded
%in a string.  anyf(A,f)
%returns the mxn matrix X,
%X(i,j)=f(A(i,j)).
[m, n]=size(A);
for i=1:m
    for j=1:n
        X(i,j)=eval([f, '(A(i,j))']);
    end
end
```

8.3 Glossary

2-norm of A	458
$\alpha_1(A)$	458
double commute	471
eval	492
expm	492
field of values	461
Frobenius norm	487
funm	492
hilb	489
hilbert norm	458
logm	492
Moore-Penrose inverse	458
numerical radius	461
singular value decomposition	456
singular values	452
sqrtm	492
SVD	456
svd	484
unitary dilation	459
$w(A)$	461

Chapter 9

Congruence

Topics
- *forms*
- *congruence*
- *inertia*
- *convex set*
- *elliptical range theorem*
- *numerical range*
- *Toeplitz-Hausdorff Theorem*

9.1 Forms

In this section we present the standard reduction theory for symmetric and hermitian forms and matrices. Various conditions for positive definiteness in terms of principal subdeterminants are derived and the essential properties of inertia are proved. In §9.2 the basic geometry of the set of values assumed by a hermitian form in the 2-dimensional case is analyzed. In §9.3 the 2-dimensional geometric analysis of §9.2 is extended to arbitrary complex hermitian forms. The Hausdorff-Toeplitz theorem is proved and two algorithms for depicting the numerical range are described.

A homogeneous polynomial in variables $x_1, ..., x_n$ is called a *form* or *quantic* in the variables $x_1, ..., x_n$. Recall that *homogeneous* means that every term has the same degree. Thus,

$$x_1^2 x_2 x_3 + x_1^4 + 3x_2^3 x_3$$

is homogeneous of degree 4 in the variables x_1, x_2, x_3, while

$$x_1 + x_2^2 x_3$$

is not homogeneous.

A *bilinear form* in variables $x_1, ..., x_m, y_1, ..., y_n$ is a polynomial of the form

$$\sum_{i=1, j=1}^{m,n} a_{ij} x_i y_j . \qquad (1)$$

If we write $x = [x_1, ..., x_m]^T$, $y = [y_1, ..., y_n]^T$ then (1) can be written as

$$\beta(x,y) = x^T A y \qquad (2)$$

where the m × n matrix $A = [a_{ij}]$ is called the *matrix of the form*. If in (2) the substitutions $x = Pu$ and $y = Qv$ are made, in which P and Q are nonsingular, $u = [u_1, ..., u_m]^T$ and $v = [v_1, ..., v_n]^T$, then

$$x^T A y = u^T P^T A Q v.$$

Thus the bilinear form in variables x and y is changed into the bilinear form $\gamma(u, v) = u^T B v$ in variables $u = [u_1, ..., u_m]^T$ and $v = [v_1, ..., v_n]^T$. The matrices A and B are related by the equation $B = P^T A Q$. Note that since P and Q are nonsingular, any value taken on by β is taken on by γ and conversely. Thus, if we think of β and γ as functions,

$$\beta, \gamma: M_{m,1}(R) \times M_{n,1}(R) \to R,$$

then the ranges of β and γ are the same. If we choose P and Q so that

$$P^T A Q = \begin{bmatrix} I_r & 0_{r,n-r} \\ 0_{m-r,r} & 0_{m-r,n-r} \end{bmatrix} = B, \qquad (3)$$

9.1 Forms

the canonical form of A, then $\gamma(u, v)$ has the particularly simple form

$$u_1v_1 + \cdots + u_rv_r. \tag{4}$$

The integer r is called the *rank* of β. The nonsingularity of P and Q implies that $r = \rho(A)$.

Suppose that A is square (m = n) and y = x. Then (2) becomes

$$q(x) = x^T A x. \tag{5}$$

The function q(x) is called a *quadratic form* in the variables $x_1, \ldots, x_n$. Note that $q(x)^T = (x^T A x)^T = x^T A^T x$. But q(x) is a scalar, so that obviously $q(x)^T = q(x)$. Hence

$$2q(x) = x^T A x + x^T A^T x$$

or

$$q(x) = \frac{x^T(A + A^T)x}{2}. \tag{6}$$

In view of (6) there is no loss of generality if we assume that A is a symmetric matrix in studying the quadratic form (5). The rank of A is called the *rank* of q(x) and det(A) is called the *discriminant* of q(x). Observe that if x = Pu, P nonsingular, then

$$x^T A x = u^T P^T A P u \tag{7}$$

so that q(x) is changed into the quadratic form in the variables u:

$$f(u) = u^T B u,$$

$$B = P^T A P. \tag{8}$$

If $A \in M_n(\mathbb{R})$ is symmetric then we know that there exists a real orthogonal matrix P such that $B = P^T A P = \text{diag}(\lambda_1, \ldots, \lambda_n)$: the numbers $\lambda_1, \ldots, \lambda_n$ are the eigenvalues of A (see §8.2 Exercises, #14). Hence q(x) may be changed into the quadratic form

$$f(u) = u^T B u$$

$$= \sum_{k=1}^{n} \lambda_k u_k^2 \qquad (9)$$

by making the change of variables $x = Pu$.

Matrices A and B in $M_n(R)$ related as in (8) (with P non-singular in $M_n(R)$) are said to be *congruent* over R. Note that $B^T = (P^T A P)^T = P^T A^T P = P^T A P$ so that symmetry is preserved by a congruence.

The problem of reducing $q(x)$ to a sum of multiples of squares as in (9) depends on knowing the eigenvalues of A, which in general are difficult to compute. Moreover, if A is complex and symmetric we may not be able to obtain a complex orthogonal matrix P ($P^T = P^{-1}$) for which $P^T A P$ is diagonal. However, the reduction of $q(x) = x^T A x$ to a sum of multiples of squares can be done constructively as we shall see in Theorem 1 below. First, however, it is helpful to examine precisely the effect on A of a congruence of the form $P^T A P$ when P is an elementary matrix. Suppose that P is an elementary matrix of type I. Then $P = A(\tau)$ where τ is a transposition say $\tau = (i\ j)$, $i < j$. It follows that

$$P^T A P = A(\tau) A A(\tau). \qquad (10)$$

According to §5.1 Exercises, #9, the (s, t) entry of $A(\tau) A A(\tau)$ is

$$a_{\tau(s), \tau(t)}. \qquad (11)$$

If $\{s, t\}$ and $\{i, j\}$ are disjoint then (11) is simply a_{st}, i.e., the entries of A outside of rows and columns i, j are unaltered. It is not difficult to check from (11) that $P^T A P$ is obtained from A by first interchanging columns i and j in A and then in the resulting matrix interchanging rows i and j. For example, if $n = 3$ and $\tau = (2\ 3)$ then

9.1 Forms

$$P^TAP = \begin{bmatrix} a_{\tau(1),\tau(1)} & a_{\tau(1),\tau(2)} & a_{\tau(1),\tau(3)} \\ a_{\tau(2),\tau(1)} & a_{\tau(2),\tau(2)} & a_{\tau(2),\tau(3)} \\ a_{\tau(3),\tau(1)} & a_{\tau(3),\tau(2)} & a_{\tau(3),\tau(3)} \end{bmatrix}$$

$$= \begin{bmatrix} a_{11} & a_{13} & a_{12} \\ a_{31} & a_{33} & a_{32} \\ a_{21} & a_{23} & a_{22} \end{bmatrix}. \quad (12)$$

Notice that, in general, the main diagonal of P^TAP is $a_{\tau(1),\tau(1)}, \ldots, a_{\tau(n),\tau(n)}$.

Next let P be a type II elementary matrix. Then

$$P = I_n + cE_{ij} \quad (13)$$

where E_{ij} has a 1 in position (i, j), 0 elsewhere. Thus

$$P^TAP = (I_n + cE_{ji})A(I_n + cE_{ij}) \quad (14)$$

is obtained from A as follows: add c times column i to column j of A; in the resulting matrix, add c times row i to row j. For example, if n = 3, i = 1, j = 3 then

$$P^TAP = P^T(AP)$$

$$= P^T \begin{bmatrix} a_{11} & a_{12} & a_{13} + ca_{11} \\ a_{21} & a_{22} & a_{23} + ca_{21} \\ a_{31} & a_{32} & a_{33} + ca_{31} \end{bmatrix}$$

$$= \begin{bmatrix} a_{11} & a_{12} & a_{13} + ca_{11} \\ a_{21} & a_{22} & a_{23} + ca_{21} \\ a_{31} + ca_{11} & a_{32} + ca_{12} & a_{33} + ca_{13} + ca_{31} + c^2a_{11} \end{bmatrix} \quad (15)$$

Thus, only rows and columns i and j are effected.

Finally, it is clear that if $P = E_{c(i)}$ then P^TAP results from A by multiplying column i of A by c and then in the resulting matrix multiplying row i by c. The effect on A is to multiply row and column i by c with the result that c^2 multiplies a_{ii}. For example, if n = 3, i = 2 then

$$P^T A P = \begin{bmatrix} a_{11} & ca_{12} & a_{13} \\ ca_{21} & c^2 a_{22} & ca_{23} \\ a_{31} & ca_{32} & a_{33} \end{bmatrix}. \tag{16}$$

Several simple facts follow from these observations. If $A \neq 0$ is assumed to be symmetric, then A is congruent to a matrix with a nonzero main diagonal entry. For, either some $a_{ii} \neq 0$, in which case there is no problem, or every main diagonal entry is zero. From (15), the (j, j) entry in (14) is

$$c a_{ij} + c a_{ji} = 2 c a_{ij}. \tag{17}$$

Simply choose i and j so that $a_{ij} \neq 0$. Also observe from (15) that if $a_{11} \neq 0$ then each of $a_{21}, a_{31}, \ldots, a_{n1}$ can be reduced to 0 by a succession of congruences of the type (14).

From (12) it should be equally clear that if A possesses a non-zero main diagonal entry then A is congruent to a matrix whose (1, 1) entry is non-zero.

We can put these observations together to obtain the following theorem.

Theorem 1.

If R is a set in which division is possible (e.g., $\mathbb{Q}, \mathbb{R}, \mathbb{C}$) and $A \in M_n(R)$ is a symmetric matrix, then A is congruent over R to a diagonal matrix.

Proof.

If $A = 0_n$ there is nothing to prove. If $A \neq 0_n$ then the above remarks show that A is congruent over R to a matrix with a nonzero (1, 1) entry. By type II congruences the entries in positions (2, 1), ..., (n, 1) can be reduced to 0, and since symmetry is preserved, we conclude that A is congruent to a matrix of the form

$$b_{11} \oplus C \tag{18}$$

where C is a symmetric matrix in $M_{n-1}(R)$. By induction on n there is a nonsingular

$Q \in M_{n-1}(R)$ such that

$$Q^T C Q$$

is a diagonal matrix. Since

$$(1 \oplus Q)^T (b_{11} \oplus C) (1 \oplus Q) = b_{11} \oplus Q^T C Q,$$

the proof is complete. ∎

From (16) we note that if A is a diagonal matrix then A is congruent to a matrix in which each main diagonal entry can be modified by multiplying by the square of a nonzero element in R. We have the following result.

Theorem 2.

If $A \in M_n(\mathbb{R})$ and A is symmetric, then A is congruent over $\mathbb{R}$ to a unique diagonal matrix of the form

$$P^T A P = I_p \oplus (-I_{r-p}) \oplus 0_{n-r} \qquad (19)$$

where $r = \rho(A)$. The integer p is the number of positive eigenvalues of A.

Proof.

The fact that A is congruent to a matrix of the form (19) is clear from Theorem 1, from the remark preceding the statement of the present theorem, and from the fact that rank is preserved by a congruence. To prove uniqueness, suppose that A is also congruent over $\mathbb{R}$ to the matrix

$$Q^T A Q = I_q \oplus (-I_{r-q}) \oplus 0_{n-r}. \qquad (20)$$

Consider the quadratic form

$$q(x) = x^T A x. \qquad (21)$$

If we set $u = P^{-1}x$ then $q(x)$ becomes

$$u^T P^T A P u = \sum_{i=1}^{p} u_i^2 - \sum_{i=p+1}^{r} u_i^2. \tag{22}$$

Similarly, if we set $v = Q^{-1}x$ then $q(x)$ becomes

$$v^T Q^T A Q v = \sum_{i=1}^{q} v_i^2 - \sum_{i=q+1}^{r} v_i^2. \tag{23}$$

Suppose that $p > q$ and consider the following system of homogeneous linear equations for $x = [\, x_1, \ldots, x_n \,]^T$:

$$\begin{aligned} P^{-1}[\, p+1, \ldots, n \,|\, 1, \ldots, n \,] x &= 0, \\ Q^{-1}[\, 1, \ldots, q \,|\, 1, \ldots, n \,] x &= 0. \end{aligned} \tag{24}$$

In the homogeneous system (24) there are $n - p + q = n - (p - q) < n$ equations in n unknowns. Hence there exists a nonzero solution x^0 (§4.3 Exercises, #36). Set

$$u^0 = P^{-1} x^0 \tag{25}$$

and

$$v^0 = Q^{-1} x^0. \tag{26}$$

From (24), $u^0_{p+1} = \cdots = u^0_n = 0$ and $v^0_1 = \cdots = v^0_q = 0$. From (22),

$$q(x^0) = q(Pu^0)$$

$$= u^{0T} P^T A P u^0$$

$$= \sum_{i=1}^{p} (u_i^0)^2 - \sum_{i=p+1}^{r} (u_i^0)^2$$

$$= \sum_{i=1}^{p} (u_i^0)^2.$$

9.1 Forms

Since $u^0 \neq 0$ it follows that

$$q(x^0) > 0. \tag{27}$$

A similar computation using (23) shows that

$$q(x^0) < 0 \tag{28}$$

which is incompatible with (27). Thus $p = q$ and the form (19) is unique.

To prove that p is the number of positive eigenvalues of A, first bring A to diagonal form with a real orthogonal P:

$$P^T A P = P^{-1} A P$$

$$= \text{diag}(\lambda_1, \cdots, \lambda_p, \lambda_{p+1}, \cdots, \lambda_r, 0, \cdots, 0) \tag{29}$$

in which $\lambda_i > 0$, $i = 1, \ldots, p$, $\lambda_i < 0$, $i = p + 1, \ldots, r$. Since each diagonal entry can be modified by multiplying by a square, the result follows immediately. ∎

If $A \in M_n(\mathbb{C})$ then a theory that parallels the real case is available for hermitian matrices. Rather than the quadratic form the appropriate concept is the *hermitian form*

$$q(x) = x^* A x, \tag{30}$$

in which it is assumed that $x \in M_{n,1}(\mathbb{C})$, i.e., $x_1, \ldots, x_n$ are complex variables. Assume then that A is hermitian, $A^* = A$, and note that

$$\overline{q(x)} = (x^* A x)^*$$

$$= x^* A^* x$$

$$= x^* A x$$

$$= q(x).$$

Thus, a hermitian form is always real. The converse of this statement is also true. That is, if A is any complex matrix and q(x) is real for all $x \in M_{n,1}(\mathbb{C})$ then A is hermitian. To see this, write A = H + iK, where H and K are hermitian. If q(x) is real for all $x \in M_{n,1}(\mathbb{C})$ it follows that

$$(Hx, x) + i(Kx, x)$$

is real and hence

$$(Kx, x) = 0 \tag{31}$$

for all $x \in M_{n,1}(\mathbb{C})$. The equation (31) immediately implies that every eigenvalue of K is 0 and hence that K itself is the 0 matrix. Thus A = H is hermitian.

If x = Pu, $P \in M_n(\mathbb{C})$ nonsingular, then (30) is changed into the hermitian form

$$f(u) = u^*P^*APu. \tag{32}$$

Note that P*AP is hermitian. Two matrices A and B in $M_n(\mathbb{C})$ for which

$$P^*AP = B,$$

P nonsingular in $M_n(\mathbb{C})$, are said to be *hermitian congruent* or, sometimes, *conjunctive*. Thus the problem of constructively reducing A to a diagonal matrix by means of a hermitian congruence is reduced to examining P*AP when P is an elementary matrix in $M_n(\mathbb{C})$.

For a type I hermitian congruence the general situation is precisely as before - see equation (12). For a type II hermitian congruence the equations corresponding to (14) and (15) are

$$P^*AP = (I_n + \bar{c}E_{ji}) A (I_n + cE_{ij}) \tag{33}$$

and

$$P^*AP = \begin{bmatrix} a_{11} & a_{12} & a_{13} + ca_{11} \\ a_{21} & a_{22} & a_{23} + ca_{21} \\ a_{31} + \bar{c}a_{11} & a_{32} + \bar{c}a_{12} & a_{33} + \bar{c}a_{13} + ca_{31} + |c|^2 a_{11} \end{bmatrix}. \quad (34)$$

The entire argument in Theorem 1 hinged on assuming that a nonzero symmetric A is congruent to a matrix with a non-zero entry in position (1, 1). If some main diagonal entry of A is non-zero simply use a type I hermitian congruence. If every main diagonal entry is 0 and $a_{ij} \neq 0$ then the hermitian property of A implies

$$a_{ij} = \bar{a}_{ji}. \quad (35)$$

Then

$$\bar{c}a_{ij} + ca_{ji} = \bar{c}a_{ij} + c\bar{a}_{ij}$$
$$= 2\operatorname{Re} \bar{c}a_{ij}. \quad (36)$$

It is always possible to choose c so that (36) is not 0 (see (34)).

The hermitian version of Theorem 2 is almost identical to Theorem 2 in both its statement and proof.

Theorem 3.

If $A \in M_n(\mathbb{C})$ and A is hermitian, then A is hermitian congruent over $\mathbb{C}$ to a unique diagonal matrix of the form

$$P^*AP = I_p \oplus (-I_{r-p}) \oplus 0_{n-r} \quad (37)$$

where $r = \rho(A)$. The integer p is the number of positive eigenvalues of A.

It should also be clear from Theorem 3 that if $P^*AP = \operatorname{diag}(d_1, \ldots, d_n)$ then the number of positive d_i is p, the number of negative d_i is $r - p$, and the number of zero d_i is $n - r$.

There is an interesting connection between the principal subdeterminants of a hermitian matrix $A \in M_n(\mathbb{C})$ and its conjunctive reduction to a diagonal form. It is convenient in the statement and proof of the following theorem to have a notation for congruence and hermitian congruence. Thus

$$A \equiv B \tag{38}$$

will denote that A and B are congruent over R. If A and B are hermitian then

$$A \stackrel{c}{\equiv} B \tag{39}$$

will denote that A and B are hermitian congruent. Observe that $A \equiv B$ means that $P^T A P = B$, whereas $A \stackrel{c}{\equiv} B$ means that $P^* A P = B$ for an appropriate non-singular P. In the case of hermitian congruence, $P \in M_n(\mathbb{C})$.

Theorem 4.

Assume that $A \in M_n(\mathbb{C})$ is hermitian. Let

$$\Delta_k = \det([\,1, \ldots, k \mid 1, \ldots, k\,]), \quad k = 1, \ldots, n, \tag{40}$$

and set $\Delta_0 = 1$. If

$$\Delta_1 \Delta_2 \cdots \Delta_n \neq 0 \tag{41}$$

then

$$A \stackrel{c}{\equiv} \mathrm{diag}\left(\frac{\Delta_0}{\Delta_1}, \frac{\Delta_1}{\Delta_2}, \ldots, \frac{\Delta_{n-1}}{\Delta_n}\right). \tag{42}$$

Proof.

The proof is by induction on n. For $n = 1$,

$$[a_{11}] \stackrel{c}{\equiv} \left[\frac{1}{a_{11}}\right] [a_{11}] \left[\frac{1}{a_{11}}\right] = \left[\frac{1}{a_{11}}\right].$$

We remark that every Δ_k is a real number because it is the determinant of a hermitian

9.1 Forms

matrix. Thus assume that (42) is valid for hermitian matrices of order n - 1. Since $\Delta_1 \cdots \Delta_{n-1} \ne 0$ we can use induction to conclude that

$$A \stackrel{c}{\equiv} \begin{bmatrix} D & w^* \\ w & \alpha \end{bmatrix} \qquad (43)$$

where

$$D = \operatorname{diag}\left(\frac{\Delta_0}{\Delta_1}, \frac{\Delta_1}{\Delta_2}, \ldots, \frac{\Delta_{n-2}}{\Delta_{n-1}}\right),$$

$w \in M_{1,n-1}(\mathbb{C})$, and $\alpha \in \mathbb{R}$. By elementary hermitian congruences of type II using rows and columns 1, ..., n - 1 in succession, the vector w can be reduced to 0 so that

$$A = P^*(D \oplus [\,\beta\,])P. \qquad (44)$$

Now

$$\Delta_n = \det(A)$$

$$= |\det(P)|^2 \, \beta \prod_{i=0}^{n-2} \frac{\Delta_i}{\Delta_{i+1}}$$

$$= |\det(P)|^2 \, \beta \, \frac{\Delta_0}{\Delta_{n-1}} \ .$$

Since $\Delta_0 = 1$, it follows that

$$\beta = \frac{\Delta_{n-1}\Delta_n}{|\det(P)|^2} \ . \qquad (45)$$

Now multiply column n and row n of $D \oplus [\,\beta\,]$ by $|\det(P)|/\Delta_n$. Since Δ_n is real this is a hermitian congruence and hence $D \oplus [\,\beta\,]$ is hermitian congruent to the matrix on the right in (42). Then (44) completes the induction. ∎

As an immediate consequence of Theorem 4 we have (using the same notation):

Theorem 5.

A necessary and sufficient condition that A be positive definite hermitian is that the sequence of determinants Δ_k be positive, $k = 1, \ldots, n$.

Proof.

Assume that $\Delta_k > 0$, $k = 1, \ldots, n$. Then from Theorem 4

$$A \stackrel{c}{\equiv} \operatorname{diag}(d_1, \ldots, d_n) \tag{46}$$

in which $d_i > 0$, $i = 1, \ldots, n$. Multiplying row i and column i by $d_i^{-1/2}$ is a hermitian congruence and hence

$$\operatorname{diag}(d_1, \ldots, d_n) \stackrel{c}{\equiv} I_n,$$

so that $A \stackrel{c}{\equiv} I_n$. Theorem 3 then implies that every eigenvalue of A is positive so that $A > 0$.

Conversely, if $A > 0$ then obviously every $A_k = A[\,1, \ldots, k \mid 1, \ldots, k\,] > 0$ (look at the hermitian form when the last $n - k$ variables are 0) and hence $\Delta_k = \det(A_k) > 0$. ∎

As a final topic in this section we discuss the important concept of the *inertia* of a complex matrix. If $A \in M_n(\mathbb{C})$ then the inertia of A is a triple of nonnegative integers:

$$\operatorname{In}(A) = (\pi(A), \nu(A), \partial(A)); \tag{47}$$

the integers $\pi(A)$, $\nu(A)$, $\partial(A)$ are, respectively, the number of eigenvalues of A, counted with their algebraic multiplicities, lying in the open right half-plane (real part positive), in the open left half-plane (real part negative), and on the imaginary axis (real part zero). For example, the inertia of the matrix

$$A = \operatorname{diag}(1 + i, 1 + i, -1, -1, -1 + i, 2, 3i, 3i, 0)$$

is

$$\text{In}(A) = (3, 3, 3)$$

because $1 + i$ (counted twice) and 2 lie in the right half-plane, -1 (counted twice) and $-1 + i$ lie in the negative half-plane, and $3i$ (counted twice) and 0 lie on the imaginary axis.

If A is hermitian then every eigenvalue is real. If

$$A \stackrel{c}{=} I_p \oplus (-I_{r-p}) \oplus 0_{n-r} \tag{48}$$

then from Theorem 3, p is the number of positive eigenvalues of A and hence $\pi(A) = p$. If A is replaced by -A in (48), r - p is seen to be the number of positive eigenvalues of -A so that $\nu(A) = r - p$. Since $\rho(A) = r$ is the number of nonzero eigenvalues of A (remember that A is hermitian), it follows that $\partial(A) = n - r$. Hence

$$\text{In}(A) = (\pi(A), \nu(A), \partial(A))$$
$$= (p, \rho(A) - p, n - \rho(A)). \tag{49}$$

It is important to emphasize that (49) is valid only for hermitian matrices, e.g., for

$$A = \begin{bmatrix} 0 & 1 \\ 0 & 0 \end{bmatrix},$$

$\pi(A) = 0$, $\nu(A) = 0$, and $\partial(A) = 2$, so that

$$\text{In}(A) = (0, 0, 2). \tag{50}$$

However, $p = 0$, $\rho(A) - p = 1$, $n - \rho(A) = 1$ and (49) and (50) do not agree.

There is only one hermitian matrix of the type (37) in each equivalence class under hermitian congruence (see §9.1 Exercises, #13). Thus, for hermitian matrices, In(A) is a complete set of invariants with respect to hermitian congruence. This means that two hermitian matrices satisfy: $A \stackrel{c}{=} B$ iff In(A) = In(B). This observation about hermitian matrices has a famous name: it is called *Sylvester's law of inertia*.

Inertia is an important concept in applied linear algebra principally because the behavior of systems of linear differential equations (i.e., stability, periodicity, etc.) depends on the distribution in the complex plane of the eigenvalues of the coefficient matrix.

The definition of hermitian congruence, $A = P^*BP$, requires that P be nonsingular. Thus, it is interesting to investigate the extent to which the nonsingularity of P is actually required.

Theorem 6.

Assume that A and B in $M_n(\mathbb{C})$ are hermitian and have the same rank r. If

$$A = P^*BP \tag{51}$$

for some $P \in M_n(\mathbb{C})$ then

$$\text{In}(A) = \text{In}(B). \tag{52}$$

Proof.

Let $p = \pi(A)$, $q = \pi(B)$. Since $r = \rho(A) = \rho(B)$ it follows that

$$A \stackrel{c}{=} D_r \oplus 0_{n-r} \tag{53}$$

and

$$B \stackrel{c}{=} \Delta_r \oplus 0_{n-r} \tag{54}$$

where

$$D_r = I_p \oplus (-I_{r-p}),$$
$$\Delta_r = I_q \oplus (-I_{r-q}).$$

From (51), (53), and (54) it follows immediately that there exists a matrix $M \in M_n(\mathbb{C})$, not necessarily nonsingular, for which

$$D_r \oplus 0_{n-r} = M^*(\Delta_r \oplus 0_{n-r})M. \tag{55}$$

Partition M conformally for block multiplication in (55),

$$M = \begin{bmatrix} M_1 & M_2 \\ M_3 & M_4 \end{bmatrix},$$

and directly compute that

$$D_r = M_1^* \Delta_r M_1. \tag{56}$$

Since both D_r and Δ_r are nonsingular, (56) implies that M_1 is nonsingular and hence $D_r \stackrel{c}{=} \Delta_r$. But then $\text{In}(D_r) = \text{In}(\Delta_r)$. It is obvious that

$$\text{In}(D_r) = (p, r - p, 0),$$
$$\text{In}(\Delta_r) = (q, r - q, 0)$$

so that $p = q$. But then

$$\text{In}(A) = (p, r - p, n - r)$$
$$= (q, r - q, n - r)$$
$$= \text{In}(B). \qquad \blacksquare$$

9.1 Exercises

1. Write each of the following in the form $x^T A x$ where A is an n-square symmetric matrix:

    (a) $n = 2$: $x_1^2 + x_2^2$

    (b) $n = 2$: $-x_1^2$

    (c) $n = 2$: $x_1 x_2$

(d) $n = 2$: $x_1^2 + x_1 x_2 + x_2^2$

(e) $n = 2$: $-x_1 x_2 + x_2^2$

(f) $n = 3$: $x_1 x_2 + x_1 x_3 + x_2 x_3$

(g) $n = 3$: $x_1^2 - x_2^2 + x_2 x_3$

(h) $n = 3$: $x_1^2 + x_1 x_3 + x_2 x_3$

(i) $n = 4$: $x_1^2 + x_2^2 + 3x_3^2 - x_4^2 + x_1 x_2$

(j) $n = 4$: $x_1 x_2$

2. Let $A \in M_n(\mathbb{C})$. Prove that if $x^* A x = 0$ for all $x \in M_{n,1}(\mathbb{C})$, then $A = 0$.

3. Let $A \in M_n(\mathbb{C})$. Prove that if $x^* A x = x^* B x$ for all $x \in M_{n,1}(\mathbb{C})$, then $A = B$.

4. Let $A \in M_n(\mathbb{C})$. Prove that if $x^* A x \in \mathbb{R}$ for all $x \in M_{n,1}(\mathbb{C})$, then A is hermitian.

5. Let A be an n-square complex symmetric matrix. Prove that there exists a non–singular matrix $P \in M_n(\mathbb{C})$ such that $P^T A P = I_r \oplus 0_{n-r}$ where $r = \rho(A)$.

   Hint: The proof is the same as Theorem 2, but at the end we simply perform type III congruences to reduce the diagonal matrix to the form $I_r \oplus 0_{n-r}$.

6. Prove that if A and B are n-square complex symmetric matrices, and $x^T A x = x^T B x$ for all $x \in M_{n,1}(\mathbb{C})$, then $A = B$.

   Hint: By subtracting, it suffices to prove that $x^T A x = 0$ for all x implies $A = 0$. First set $x = [0, ..., 0, 1, 0, ..., 0]^T$, where the 1 occurs in position i, to see that a_{ii} is 0, $i = 1, ..., n$. Then take $x = [0, ..., 0, 1, 0, ..., 0, 1, 0, ..., 0]^T$, where the 1's occur in positions i and j, to obtain $x^T A x = 2a_{ij} = 0$.

9.1 Exercises

7. Let $A \in M_n(\mathbb{C})$ be a complex symmetric matrix. Let $q(x) = x^T A x$. Prove that $q(x)$ is a product of two complex linear polynomials in the variables $x = [x_1, \ldots, x_n]^T$ iff $\rho(A) \leq 2$.

   Hint: If $\rho(A) \leq 2$ then there exists a nonsingular P such that $P^T A P = I_r \oplus 0_{n-r}$, where $r = 1$ or 2. Let $x = Pu$ so that

   $$x^T A x = u^T P^T A P u = \begin{cases} u_1^2 + u_2^2, & \text{if } r = 2 \\ u_1^2, & \text{if } r = 1 \end{cases}.$$

   Now, $u_k = P_{(k)}^{-1} x$, $k = 1, 2$. If $r = 2$, $u_1^2 + u_2^2 = (u_1 + iu_2)(u_1 - iu_2)$ and this is a product of two linear polynomials in x. The other case, $r = 1$, is the same. Conversely, if $q(x) = (x^T a)(b^T x)$, $a, b \in M_{n,1}(\mathbb{C})$, then

   $$x^T A x = q(x) = x^T (ab^T) x = x^T \frac{(ab^T + ba^T)}{2} x.$$

   Apply #6 to conclude that $A = \frac{(ab^T + ba^T)}{2}$. But A is then a sum of two rank 1 matrices and hence has rank at most 2 (see §6.2 Exercises, #1).

8. Factor each of the following quadratic polynomials into a product of complex linear factors.

   (a) $x_1^2 + x_1 x_2 + x_2^2$

   (b) $2x_1^2 - 3x_2^2 - 2x_3^2 - 5x_1 x_2 + 3x_1 x_3 + 5x_2 x_3$

   (c) $x_1^2 + 2x_1 x_2 + 2x_1 x_3 + 2x_1 x_4 + x_2^2 + 2x_2 x_3 + 2x_2 x_4$

   Hint: Use #7.

9. Reduce each of the following real symmetric (complex hermitian) matrices to the form (19) by elementary congruence (conjunctive) operations. Find a non-singular matrix P effecting the congruence (conjunctivity).

(a) $\begin{bmatrix} 9 & 5 & 2 \\ 5 & 3 & 1 \\ 2 & 1 & 1 \end{bmatrix}$
(b) $\begin{bmatrix} 0 & 3 & 2 \\ 3 & 0 & 1 \\ 2 & 1 & 0 \end{bmatrix}$

(c) $\begin{bmatrix} 1 & 0 & 1 & 2 \\ 0 & 1 & 2 & 3 \\ 1 & 2 & 0 & 1 \\ 2 & 3 & 1 & 0 \end{bmatrix}$
(d) $\begin{bmatrix} 1 & 2+i \\ 2-i & -3 \end{bmatrix}$

10. Prove: the rank of a normal matrix $A \in M_n(\mathbb{C})$ is the size of the largest nonzero principal subdeterminant of A.

Hint: Let $A = U^*DU$, $D = \text{diag}(\lambda_1, \ldots, \lambda_r, 0, \ldots, 0)$, $r = \rho(A)$, U unitary, and use Cauchy-Binet to obtain $\det(A[\,\omega\,|\,\omega\,]) = 0$ if $\omega \in Q_{k,n}$, $k > r$. If $k = r$ then

$$\det(A[\,\omega\,|\,\omega\,]) = |\det(U[\,\omega\,|\,1,\ldots,r\,])|^2 \lambda_1 \cdots \lambda_r.$$

Now $C_r(U)$ is unitary so that

$$1 = \sum_{\alpha \in Q_{r,n}} |C_r(U)_{\alpha,(1,\ldots,r)}|^2 = \sum_{\alpha \in Q_{r,n}} |\det(U[\,\alpha\,|\,1,\ldots,r\,])|^2.$$

Hence there is an ω such that $\det(U[\,\omega\,|\,1,\ldots,r\,]) \neq 0$.

11. Calculate $\pi(A)$ for the matrix

$$\begin{bmatrix} 1 & 1 & 1 & 6 \\ 1 & 1 & -1 & 2 \\ 1 & -1 & 1 & -2 \\ 6 & 2 & -2 & 3 \end{bmatrix}.$$

Hint: It can be verified directly that A is congruent over $\mathbb{R}$ to $\text{diag}(1, 1, -1, -1)$ and hence $\pi(A) = 2$.

12. Let A be an n-square hermitian matrix. Prove: if $A \geq 0$ and $a_{11} = \cdots = a_{nn} = 1$ then $\det(A) \leq 1$.

    Hint: $\det(A) = \lambda_1 \cdots \lambda_n \leq \left(\dfrac{\lambda_1 + \cdots + \lambda_n}{n}\right)^n = 1$.

13. Prove that congruence is an equivalence relation in $M_n(\mathbb{R})$. Show that if the equivalence relation is restricted to symmetric matrices then each equivalence class contains exactly one matrix of the form $I_p \oplus (-I_{r-p}) \oplus 0_{n-r}$. State and prove the corresponding result for hermitian congruence.

14. Prove that if $A \in M_n(\mathbb{C})$ then $A \geq 0$ iff $A + \varepsilon I_n > 0$ for every $\varepsilon > 0$.

    Hint: Observe that $x^*(A + \varepsilon I_n)x = x^*Ax + \varepsilon x^*x$. Thus if $x \neq 0$ then $x^*x > 0$ so that $x^*(A + \varepsilon I_n)x > 0$. Conversely, suppose $A + \varepsilon I_n > 0$ for every $\varepsilon > 0$. Then $x^*Ax + \varepsilon x^*x > 0$ for any $x \neq 0$ and any $\varepsilon > 0$. It follows that x^*Ax cannot be negative and hence $x^*Ax \geq 0$. Thus $A \geq 0$.

15. Assume that $A \in M_n(\mathbb{C})$ is hermitian. Prove that if $A \geq 0$ then the sequence of leading principal subdeterminants defined in Theorem 4 satisfy $\Delta_k \geq 0$, $k = 1, \ldots, n$.

    Hint: If $x = [\, u \ 0\,]^T$ where $x \in M_{n,1}(\mathbb{C})$ is nonzero and $u \in M_{k,1}(\mathbb{C})$, then $x^*Ax = u^*A_k u$ where $A_k = A[\, 1, \ldots, k \mid 1, \ldots, k\,]$. Since $A \geq 0$ it follows that $A_k \geq 0$ and hence that $\Delta_k = \det(A_k) \geq 0$.

16. Prove that the converse of the statement in #15 is false. Namely, exhibit a hermitian A for which $\Delta_k \geq 0$, $k = 1, \ldots, n$, and yet it is not the case that $A \geq 0$.

    Hint: $A = \begin{bmatrix} 0 & 0 \\ 0 & -1 \end{bmatrix}$.

17. Assume that $A \in M_n(\mathbb{C})$ is hermitian. Prove that $A \geq 0$ iff every principal subdeterminant of A is nonnegative.

    Hint: In §7.1, formula (10) set $B = A_k$ and $D = \varepsilon I_k$. Then, in the notation used in the Hint for #15,

$$\det((A + \varepsilon I_n) [\, 1, \ldots, k \mid 1, \ldots, k \,]) = \det(A_k + \varepsilon I_k) = \det(\varepsilon I_k + A_k)$$

$$= \sum_{r=0}^{k} \varepsilon^{k-r} \sum_{\alpha \in Q_{r,k}} \det(A_k [\, \alpha \mid \alpha \,]).$$

By hypothesis, every $\det(A_k[\, \alpha \mid \alpha \,])$ is nonnegative and hence, setting

$$\Delta_k(\varepsilon) = \det((A + \varepsilon I_n) [\, 1, \ldots, k \mid 1, \ldots, k \,]),$$

we have

$$\Delta_k(\varepsilon) = \varepsilon^k + p_1 \varepsilon^{k-1} + \cdots + p_k,$$

in which every $p_t \geq 0$. Thus $\Delta_k(\varepsilon) \geq \varepsilon^k > 0$ for all $\varepsilon > 0$, $k = 1, \ldots, n$. Theorem 5 then implies that $A + \varepsilon I_n > 0$ for every $\varepsilon > 0$. Use #14 to conclude that $A \geq 0$.

18. Let $A = C \oplus M$ and $B = C \oplus N$ be two hermitian matrices in $M_n(\mathbb{C})$ in which C is k-square. Assume that $A \stackrel{c}{\equiv} B$. Prove that $M \stackrel{c}{\equiv} N$. In somewhat more general contexts this result is known as the *Witt cancellation theorem*.

Hint: Let

$$C \stackrel{c}{\equiv} I_{p_1} \oplus (-I_{r_1 - p_1}) \oplus 0_{k - r_1}.$$

Similarly, write

$$M \stackrel{c}{\equiv} I_{p_2} \oplus (-I_{r_2 - p_2}) \oplus 0_{n - k - r_2},$$

$$N \stackrel{c}{\equiv} I_{p_3} \oplus (-I_{r_3 - p_3}) \oplus 0_{n - k - r_3}.$$

Then

$$A \stackrel{c}{\equiv} I_{p_1 + p_2} \oplus -(I_{r_1 - p_1 + r_2 - p_2}) \oplus 0_{n - (r_1 + r_2)},$$

and

$$B \stackrel{c}{=} I_{p_1+p_3} \oplus -(I_{r_1-p_1} + r_3 P_3) \oplus 0_{n-(r_1+r_3)}.$$

Since $A \stackrel{c}{=} B$ it follows from Theorem 3 that

$$p_1 + p_2 = p_1 + p_3,$$

$$r_1 + r_2 - (p_1 + p_2) = r_1 + r_3 - (p_1 + p_3),$$

$$n - (r_1 + r_2) = n - (r_1 + r_3).$$

Hence $p_2 = p_3$ follows from the first of these equations. Using this, the second equation implies $r_2 = r_3$. But then the form (37) is the same for both M and N and it follows that $M \stackrel{c}{=} N$.

19. Let A and B be n-square hermitian and assume that $B > 0$. Prove: there exists a nonsingular $P \in M_n(\mathbb{C})$ such that P^*AP is diagonal and $P^*BP = I_n$.

Hint: Write $S = (B^{1/2})^{-1}$. Then $S^*BS = SBS = (B^{1/2})^{-1} (B^{1/2})^2 (B^{1/2})^{-1} = I_n$. Since S^*AS is hermitian, choose a unitary U such that $U^*(S^*AS)U$ is diagonal. Then set $P = SU$ and compute that P^*AP is diagonal and $P^*BP = I_n$.

9.1 MatLab

1. Review the Help entries for **sign** and **find**. If L is a real vector, describe the output of each of the following MatLab commands performed in the order given.

    (a)   s = sign(L)

    (b)   s == 1

    (c)   find(s == 1)

(d)   length(find(s ==1))

Hint: (a)   s = sign(L) is a vector formed by replacing each entry in L by 1, −1, or 0 according as that entry is positive, negative, or zero.

(b)   s == 1 is a vector formed from s by comparing each entry of s with 1 and replacing that entry with 1 or 0 according as the entry is 1 or not 1, e.g., if s = [1 −1 0 1] then s == 1 is the vector [1 0 0 1].

(c)   find(s == 1) is a vector containing the indices of the nonzero entries in s == 1. Thus find(s == 1) is the vector of indices t for which L(t) is positive.

(d)   length(find(s == 1)) is a count of the number of positive entries in L.

2. Write a sequence of MatLab commands that replace with 0 any entries of a real vector L that in absolute value are less than a given positive number p.

Hint:
```
zip = find(abs(L) < p);
L(zip) = zeros(zip);
```

3. Write a MatLab function named iner that

- accepts an n-square complex matrix X

- sets equal to 0 the real parts of those eigenvalues of X whose real parts are less in modulus than n*norm(X)*eps

- returns the triple In = [pos, neg, zer] = In(X) where pos, neg, zer are the number of eigenvalues of X with positive, negative and zero real parts respectively.

Create a library called congru and save iner in congru.

Hint:
```
function In=iner(X)
%INER accepts an n-square
```

```
              %complex matrix X and
              %returns In(X), the inertia
              %triple of X.
              n=length(X);
              L=real(eig(X));
              zip=find(abs(L) < n*norm(X)*eps);
              L(zip)=zeros(zip);
              s=sign(L);
              pos=length(find(s = = 1));
              neg=length(find(s = = -1));
              zer=length(find(s = = 0));
              In=[pos neg zer];
```

4. Write a MatLab function called inerh that uses the built-in function schur to returns the inertia triple of an n-square hermitian matrix. Save inerh in congru.

   Hint:
```
              function In=inerh(X)
              %INERH accepts an n-square
              %complex hermitian matrix X and
              %returns In(X), the inertia
              %triple of X.
              n=length(X);
              tol=norm(X-X','fro');
              if tol< (n*norm(X)*eps)
                  [U,T]=schur(X);
                  [U,T]=rsf2csf(U,T);
                  L=diag(real(T));
                  zip=find(abs(L) < n*norm(X)*eps);
                  L(zip)=zeros(zip);
                  s=sign(L);
                  pos=length(find(s = = 1));
                  neg=length(find(s = = -1));
                  zer=length(find(s = = 0));
                  In=[pos neg zer];
```

                else
                    disp('The matrix is not hermitian')
                    In=[0 0 0];
                end

5. Use hrand (see §8.1 MatLab #4), iner, and inerh to write a MatLab script called inerttest that

   • generates 20 random n-square complex hermitian matrices X, $1 \leq n \leq 5$

   • prints out a table consisting of the two inertia triples for X produced by iner and inerh.

   Save inerttest in congru.

   Hint:
   ```
   %INERTTEST produces a table
   %of inertia triples for 20
   %random n-square complex
   %hermitian matrices as
   %computed by INER and INERH.
   T1=[ ];
   for k=1:20
           n=floor(5*rand+1);
           X=hrand(n);
           In1=iner(X);
           In2=inerh(X);
           T2=[In1, In2];
           T1=[T1; T2];
   end
   disp('    iner         inerh  ');
   disp('===============================');
   disp(T1)
   ```

6. Explain why iner and inerh return the same inertia triples for hermitian matrices.

   Hint: The eigenvalues of any hermitian matrix are real.

9.1 MatLab

7. Devise a method for using MatLab to solve the factorization problems in §9.1 Exercises #8 (a), (b), (c).

Hint: First obtain the real symmetric matrix A such that the quadratic polynomial has the form $g(x) = x^T A x$ where x is the column vector $x = [x_1, x_2, ..., x_n]^T$. According to §9.1 Exercises #7, the factorization of $q(x)$ into linear factors is possible iff $r = \text{rank}(A) \leq 2$. Thus the next step is to obtain r. The commands

$$[U, T] = \text{schur}(A);$$
$$L = \text{diag}(T);$$

will return a real orthogonal U and a diagonal matrix T such that

$$A = U T U^T.$$

The diagonal entries of T are the eigenvalues of A and precisely r of these are non-zero. Next, observe that

$$q(x) = x^T A x = x^T U T U^T x = y^T T y, \quad (1)$$

where $y = U^T x$. First, take the case $r = 1$. Then the command

$$f = \text{find}(\text{abs}(L) > 1e-14) \quad (2)$$

should return the single index $p = f(1)$ for which $L(f(1))$ is not 0. From (1) we have

$$q(x) = L(p) y_p^2 . \quad (3)$$

Then $q(x)$ is the square of the linear form

$$\sqrt{L(p)}\, y_p = \sqrt{L(p)}\, U_{(p)}^T x.$$

Thus we can exhibit the coefficients of x in a row as

$$a = \text{sqrt}(L(p))*(U(:,p)).' \ . \qquad (4)$$

In case $r = 2$, the command (2) should return $[p, q] = f = [f(1), f(2)]$; $p = f(1)$ and $q = f(2)$ are the indices for which $L(p)$ and $L(q)$ are not 0. Then (1) becomes

$$q(x) = L(p)y_p^2 + L(q)y_q^2 \ . \qquad (5)$$

The form (5) factors into

$$q(x) = (\sqrt{L(p)}\ y_p + i\sqrt{L(q)}\ y_q)(\sqrt{L(p)}\ y_p - i\sqrt{L(q)}\ y_q)$$

$$= (\sqrt{L(p)}\ U_{(p)}^T x + i\sqrt{L(q)}\ U_{(q)}^T x)(\sqrt{L(p)}\ U_{(p)}^T x - i\sqrt{L(q)}\ U_{(q)}^T x) \ . \qquad (6)$$

The coefficients in the two linear factors in (6) are then given in MatLab by the two choices of sign in

$$a = \text{sqrt}(L(p)) * (U(:, p)).' \pm i * \text{sqrt}(L(q)) * (U(:, q)).' \ . \qquad (7)$$

Thus the script to be used is:

```
r=rank(A);
[U, T]=schur(A);
L=diag(T);
f=find(abs(L) > 1e-14);
```

Then if $r = 1$ continue with

```
p = f(1);
a = sqrt(L(p)) * (U(:, p)).'
```

If $r = 2$ continue with

```
p = f(1);  q = f(2);
a = sqrt(L(p)) * (U(:, p)).' + i*sqrt(L(q)) * (U(:, q)).'
b = sqrt(L(p)) * (U(:, p)) .' - i*sqrt(L(q)) * (U(:, q)).'
```

9.1 MatLab

The required linear factors are then

$$a(1)x_1 + \cdots + a(n)x_n$$

and

$$b(1)x_1 + \cdots + b(n)x_n .$$

8. Factor the quadratic polynomials in §9.1 Exercises, #8(a), (b), (c) using the method of #7.

Hint:

(a) A = [1 1/2; 1/2 1];
     r = rank(A);
     [U, T] = schur(A);
     L = diag(T);
     f = find(abs(L) > 1e-14);
     p = f(1);
     q = f(2);
     a = sqrt(L(p)) * (U(:, p)).' + i*sqrt(L(q)) * (U(:, q)).';
     b = sqrt(L(p)) * (U(:, p)).' - i*sqrt(L(q)) * (U(:, q)).';

a = 0.5000 + 0.8660i   -0.5000 + 0.8660i

b = 0.5000 - 0.8660i   -0.5000 - 0.8660i

(b) A = [2 -5/2 3/2; -5/2 -3 5/2; 3/2 5/2 -2]
     r = rank(A);
     [U, T] = schur(A);
     L = diag(T);
     f = find(abs(L) > 1e-14);
     p=f(1);
     p=f(2);
     a = sqrt(L(p)) * (U(:, p)).' + i*sqrt(L(q)) * (U(:, q)).';
     b = sqrt(L(p)) * (U(:, p)).' - i*sqrt(L(q)) * (U(:,q)).';

a = 2.4719   1.2359   -1.2359

b = 0.8091   -2.7273   1.6182

(c)   A = [1 1 1 1; 1 1 1 1; 1 1 0 0; 1 1 0 0];
      r = rank(A);
      [U, T] = schur(A);
      L = diag(T);
      f = find(abs(L) > 1e-14);
      p = f(1);
      q = f(2);
      a = sqrt(L(p)) * (U(:, p)).' + i*sqrt(L(q)) * (U(:, q)) .';
      b = sqrt(L(p)) * (U(:, p)).' - i*sqrt(L(q)) * (U(:, q)).';

```
a =    0 + 0.6687i    0 + 0.6687i    0 + 1.3375i    0 + 1.3375i
b =    0 - 1.4953i    0 - 1.4953i    0 - 0.0000i    0 + 0.0000i
```

9. Write a MatLab function called formdet that

  • accepts an n-square complex hermitian matrix A

  • computes the determinants Δ_k in Theorem 4, formula (40)

  • if formula (41) holds then the function returns the inertia of A

Save formdet in congru.

Hint:
```
function In=formdet(A)
%FORMDET accepts a complex
%hermitian A for which the
%leading principal subdeterminants
%are non-zero.  formdet returns
%the inertia triple of A in
%terms of these subdeterminants.
n=length(A);
del=1;
for k=1:n
    delk=det(A(1:k,1:k));
    del=[del delk];
end
```

```
            if prod(abs(del))<1e-14
                disp('matrix close to singular')
                In=zeros(1,3);
        else
                delta=del(1:n)./del(2:(n+1));
                In=inerh(diag(delta));
        end
```

10. Use formdet in #9 to evaluate the inertia triples for hermitian matrices of the form

$$U * \text{diag}(d) * U'$$

where U = urand(n).

(a)   n = 3, d = [1 -1 1]

(b)   n = 4, d = [1 1 -1 -1]

(c)   n = 5, d = [2 -3 4 7]

(d)   n = 6, d = [1 2 -1 -1 3 4]

Could you have predicted the outputs? Explain.

9.1 Glossary

bilinear form	496
congruent	498
conjunctive	504
discriminant of a quadratic form	497
find	517
form	495
hermitian congruent	504
hermitian form	503
homogeneous	495

In 508
inertia 508
matrix of a form 496
quantic 495
rank of a form 496
rank of a quadratic form 497
sign 517
Sylvester's law of inertia 509
Witt cancellation theorem 516

9.2 Geometry of Forms

For $A \in M_n(\mathbb{C})$ there is an important set of complex numbers associated with A called the *numerical range* or *field of values* of A. It is defined as the range of the hermitian form associated with A when the variables are restricted to satisfy $x^*x = 1$. The numerical range of A is usually denoted by W(A), so that

$$W(A) = \{ x^*Ax \mid x^*x = 1 \}. \tag{1}$$

(The notation in (1) is read as "the set of values x^*Ax obtained as $x \in M_{n,1}(\mathbb{C})$ varies over all vectors satisfying $x^*x = 1$"). In §8.3, formula (30), the *numerical radius* of A, w(A), was defined by

$$w(A) = \max_{x^*x = 1} | x^*Ax |. \tag{2}$$

Thus, geometrically, w(A) is the distance from the origin to a point in W(A) farthest from the origin:

$$w(A) = \max_{z \in W(A)} |z|. \tag{3}$$

9.2 Geometry of Forms

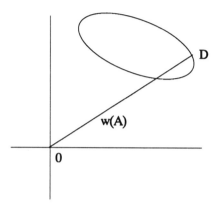

Fig. 1

In Fig. 1, the numerical range $W(A)$ is depicted as a laminar elliptical region and the line OD has length $w(A)$. If $C(z_0, r)$ denotes the circular region of radius r centered at z_0, then

$$W(A) \subset C(0, w(A)). \tag{4}$$

If $A = H + iK$, in which H and K are the unique hermitian components of A, then a rectangular containment region for $W(A)$ is easy to obtain in terms of the eigenvalues of H and K. From §8.3 Theorem 1, we know that $W(H) = [\, r_n, r_1\,]$ where $r_1 \geq \cdots \geq r_n$ are the eigenvalues of H, and similarly $W(K) = [\, s_n, s_1\,]$ where $s_1 \geq \cdots \geq s_n$ are the eigenvalues of K. Define the rectangular region $R(A)$ to be the region whose vertices are the four points (r_n, s_n), (r_n, s_1), (r_1, s_n), (r_1, s_1). Then

$$W(A) \subset R(A). \tag{5}$$

The containment (5), in various forms, has been known since 1900 and is due to I. Bendixon and A. Hirsch. There is nothing much to its proof:

$$x^*Ax = x^*Hx + ix^*Kx;$$

apply §8.3 Theorem 1 to x^*Hx and x^*Kx.

There are several other easily accessible results about $W(A)$ which we can list here. Before we do this, a few simple geometric facts about the complex plane should be observed. Let X and Y be any two sets in the complex plane and let c and d be any two complex numbers. Then $cX + dY$ is the set of all points $cx + dy$, $x \in X$, $y \in Y$. Also, $\overline{X}$ is the set of all $\overline{x}$ for $x \in X$, and XY is the set of all points xy, $x \in X$, $y \in Y$. Let $A \in M_n(\mathbb{C})$, let B be any principal submatrix of A and let $U \in M_n(\mathbb{C})$ be any unitary matrix. Also let $\sigma(A)$ denote the *spectrum* of A, i.e., $\sigma(A)$ is the set of all eigenvalues of A. We have

$$W(cA) = cW(A), \tag{6}$$

$$W(cI_n + A) = c + W(A), \tag{7}$$

$$W(A^*) = \overline{W(A)}, \tag{8}$$

$$W(B) \subset W(A), \tag{9}$$

$$W(U^*AU) = W(A), \tag{10}$$

$$W(A) \subset \mathbb{R} \text{ iff } A \text{ is hermitian}, \tag{11}$$

$$W(A) \subset i\mathbb{R} \text{ iff } A \text{ is skew-hermitian}, \tag{12}$$

$$W(A) = \{0\} \text{ iff } A = 0_n, \tag{13}$$

$$\sigma(A) \subset W(A), \tag{14}$$

$$W(A) = \{c\} \text{ iff } A = cI_n. \tag{15}$$

The short proofs of these facts appear in §9.2 Exercises.

The concept of *convexity* plays an important role in the study of the numerical range. A subset C of the complex plane is said to be convex if for every x and y in C and every θ, $0 \le \theta \le 1$, the set of all points of the form $z = (1 - \theta)x + \theta y$ is contained in C. It is an elementary exercise to confirm that

$$\{z \mid z = (1 - \theta)x + \theta y, 0 \le \theta \le 1\}$$

9.2 Geometry of Forms

consists precisely of those points that lie on the closed line segment in the plane joining x to y. We will let [x, y] denote this line segment in analogy with the notation for closed intervals on the real line. Thus C is convex iff for every x and y in C,

$$[x, y] \subset C.$$

Clearly, a single point is a convex set, as is any line segment, and the entire complex plane. Several other examples of convex sets appear in §9.2 Exercises.

In §9.2 Exercises, #12 it is confirmed that if $z = x + iy$, and x and y are real, then any line in the plane has an equation of the form

$$\text{Re}(az) = r, \tag{16}$$

where $r = 1$ if the line does not go through the origin and $r = 0$ if it does. In (16) it is assumed that a is fixed and nonzero, and z varies so as to satisfy (16).

The set of points z each of which satisfies

$$\text{Re}(az) < r, \tag{17}$$

is called the *negative open half-plane* defined by the line (16). Similarly, the set of all points z each of which satisfies

$$\text{Re}(az) > r, \tag{18}$$

is called the *positive open half-plane* defined by the line (16). If the strict inequality in (17) or (18) is relaxed to "≤" or "≥", the half-planes are called *closed*. Clearly, any line separates $\mathbb{C}$ into three disjoint convex sets: the line itself and each of the open half-planes (see §9.2 Exercises, #15).

Let $\lambda_1, \lambda_2, \ldots, \lambda_n$ be any set of points in the plane. If $\theta_1, \theta_2, \ldots, \theta_n$ are non-negative numbers with sum 1 then

$$z = \sum_{t=1}^{n} \theta_t \lambda_t$$

is called a *convex combination* of $\lambda_1, \lambda_2, \ldots, \lambda_n$. If X is any subset of $\mathbb{C}$ then the *convex hull* of X is the totality of convex combinations of finite subsets consisting of points in X. The convex hull of X is denoted by H(X).

Theorem 1.

If $X \subset \mathbb{C}$ then H(X) is a convex set.

Proof.

If $\{ x_1, \ldots, x_p \}$ and $\{ y_1, \ldots, y_q \}$ are two finite subsets of X, we can let $\{ \lambda_1, \ldots, \lambda_n \}$ denote their union. Moreover if

$$x = \sum_{i=1}^{p} \sigma_i x_i \in H(X)$$

and

$$y = \sum_{i=1}^{q} \tau_i y_i \in H(X)$$

are convex combinations, then we can rewrite x as

$$x = \sum_{t=1}^{n} \theta_t \lambda_t \in H(X), \qquad \theta_t \geq 0, \quad \sum_{t=1}^{n} \theta_t = 1,$$

and y as

$$y = \sum_{t=1}^{n} \mu_t \lambda_t \in H(X), \qquad \mu_t \geq 0, \quad \sum_{t=1}^{n} \mu_t = 1.$$

If $0 \leq \theta \leq 1$ then

$$(1 - \theta)x + \theta y = \sum_{t=1}^{n} [\,(1 - \theta)\theta_t + \theta \mu_t\,] \lambda_t$$

and it is simple to check that

9.2 Geometry of Forms

$$\sum_{t=1}^{n} [(1-\theta)\theta_t + \theta\mu_t] = 1.$$

Since $(1-\theta)\theta_t + \theta\mu_t \geq 0$, it follows that $(1-\theta)x + \theta y \in H(X)$. ∎

We are now in a position to completely determine the numerical range of any normal $A \in M_n(\mathbb{C})$.

Theorem 2.

Let $A \in M_n(\mathbb{C})$ be a normal matrix. Then

$$W(A) = H(\sigma(A)).$$

In other words, $W(A)$ is the convex hull of the set of eigenvalues of A.

Proof.

Let $\sigma(A) = \{\lambda_1, \ldots, \lambda_n\}$. Let U be a unitary matrix for which $U^*AU = \text{diag}(\lambda_1, \ldots, \lambda_n)$. Since U is nonsingular it is clear that any $x \in M_{n,1}(\mathbb{C})$ is of the form $x = Uu$ and moreover, $x^*x = u^*u$. Thus, if $x^*x = 1$ we compute that

$$x^*Ax = u^*U^*AUu$$
$$= u^*\text{diag}(\lambda_1, \ldots, \lambda_n)u$$
$$= \sum_{t=1}^{n} \lambda_t \bar{u}_t u_t$$
$$= \sum_{t=1}^{n} \lambda_t |u_t|^2.$$

Since

$$1 = u^*u$$

$$= \sum_{t=1}^{n} |u_t|^2,$$

it follows that $x^*Ax \in H(\sigma(A))$. Conversely, let

$$z = \sum_{t=1}^{n} \theta_t \lambda_t$$

be any point in $H(\sigma(A))$. Define $x = U[\theta_1^{1/2}, \theta_2^{1/2}, \ldots, \theta_n^{1/2}]^T$. Then

$$x^*x = \sum_{t=1}^{n} \theta_t$$

$$= 1,$$

and

$$x^*Ax = [\theta_1^{1/2}, \ldots, \theta_n^{1/2}] U^*AU [\theta_1^{1/2}, \ldots, \theta_n^{1/2}]^T$$

$$= [\theta_1^{1/2}, \ldots, \theta_n^{1/2}] \operatorname{diag}(\lambda_1, \ldots, \lambda_n) [\theta_1^{1/2}, \ldots, \theta_n^{1/2}]^T$$

$$= \sum_{t=1}^{n} \theta_t \lambda_t.$$

Thus $z = x^*Ax$ is a point in $W(A)$. ∎

In view of Theorem 2 it is important to understand the geometry of the convex hull of a set of points $\lambda_1, \ldots, \lambda_n$ in $\mathbb{C}$. Perhaps the easiest way to do this is to realize that the convex hull of any set $X \subset \mathbb{C}$ is the "smallest" convex set containing X as a subset.

Theorem 3.

If $X \subset \mathbb{C}$ then $H(X)$ is the intersection of all convex sets each of which contains X as a subset.

Proof.

Let Y be a convex set which contains X. Clearly any convex combination of points in X is a convex combination of points in Y, and by the convexity of Y, is in Y. Thus

$$H(X) \subset Y.$$

In other words, $H(X)$ is a subset of any convex set Y containing X. On the other hand, by Theorem 1, $H(X)$ is a convex set containing X. Thus $H(X)$ must be the intersection of all convex sets containing X. ■

It is geometrically evident from Theorem 3 that the smallest convex set containing $\lambda_1, \ldots, \lambda_n$ is the smallest convex polygonal region containing these points, and that the vertices of this polygonal region are selected from among $\lambda_1, \ldots, \lambda_n$. It is instructive to consider several examples. Let

$$A = \begin{bmatrix} 0 & 1 & 0 \\ 0 & 0 & 1 \\ 1 & 0 & 0 \end{bmatrix}. \tag{19}$$

The eigenvalues of A are easily computed to be 1, $e^{i2\pi/3}$, $e^{i4\pi/3}$. Thus W(A) is the polygonal region with these points as vertices:

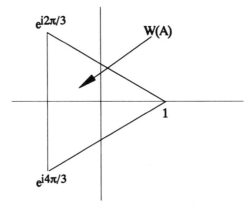

Fig. 2

If

$$A = \begin{bmatrix} 0 & 1 \\ 0 & 0 \end{bmatrix}$$

then

$$x^*Ax = \overline{x_1}x_2$$
$$= |x_1| e^{-i\omega_1} |x_2| e^{i\omega_2}$$
$$= |x_1||x_2| e^{i(\omega_2 - \omega_1)}.$$

Since $x^*x = 1$ it follows that if $|x_1| = s^{1/2}$, $s \geq 0$, then $|x_2| = \sqrt{1-s}$. Moreover, the angle $\omega_2 - \omega_1$ can be chosen arbitrarily while still maintaining the condition $x^*x = 1$. Thus

$$x^*Ax = \sqrt{s(1-s)}\, e^{i\theta}, \quad 0 \leq \theta \leq 2\pi. \tag{20}$$

For any fixed $s \in [0, 1]$, as θ varies over $[0, 2\pi]$, the points x^*Ax run over a circle of radius $\sqrt{s(1-s)}$ centered at the origin, i.e., x^*Ax describes the circle $C(0, \sqrt{s(1-s)}\,)$. Obviously, as s runs over $[0, 1]$ the union of these circles is simply the circular disk centered at 0 whose radius is the maximum value of $\sqrt{s(1-s)}$ for $s \in [0, 1]$. It is an easy calculus problem to confirm that this maximum value is $1/2$. Hence W(A) is the circular disk bounded by $C(0, 1/2)$.

9.2 Geometry of Forms

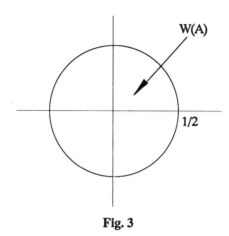

Fig. 3

Note that in contrast to Fig. 2, neither eigenvalue of A is on the boundary of W(A).

At this point we are about to state and prove an important result that determines the general form of W(A) when A is an arbitrary matrix in $M_2(\mathbb{C})$. The proof itself depends on a few elementary facts about *envelopes*. We digress somewhat to remind the reader about this topic from elementary calculus.

Suppose that

$$F(x, y, s) = 0 \qquad (21)$$

represents a one-parameter family of curves in the x,y-plane with s the parameter. Frequently there is a curve C in the plane with the properties:

(a) each curve of the family is tangent to C;

(b) at each point P of C there is a curve in the family which is tangent to C at P.

If such a curve C exists, it is called the *envelope* of the family (21).

As an example, consider the family of lines written in normal form:

$$F(x, y, s) = x \cos(s) + y \sin(s) - 1 = 0, \quad 0 \leq s \leq 2\pi. \tag{22}$$

From elementary geometry, each line in the family is distance 1 from the origin. Obviously, the unit circle is the envelope of the family.

Suppose that the envelope C of the family (21) exists. Then each point P of C corresponds to a value of the parameter s that determines the curve in the family to which C is tangent at P. Thus C itself may be regarded as given parametrically in terms of s, say,

$$x = f(s), \quad y = g(s). \tag{23}$$

Since $(f(s), g(s))$ is also a point on the curve in the family corresponding to s it follows from (21) that

$$F(f(s), g(s), s) = 0. \tag{24}$$

Assume that F has continuous first partial derivatives F_1, F_2, F_3 and that f and g have derivatives. Then from (24),

$$F_1 f' + F_2 g' + F_3 = 0. \tag{25}$$

The slope of the tangent line to the curve in the family (21), corresponding to the value s of the parameter, is obtained by differentiating with respect to x:

$$F_1 + F_2 \frac{dy}{dx} = 0,$$

or

$$\frac{dy}{dx} = -\frac{F_1}{F_2}. \tag{26}$$

On the other hand the slope of the curve C given by (23), for the value s, is

$$\frac{dy}{dx} = \frac{g'(s)}{f'(s)}. \tag{27}$$

9.2 Geometry of Forms

Combining (26) and (27),

$$F_1 f' + F_2 g' = 0,$$

and hence from (25),

$$F_3 = 0. \tag{28}$$

In other words, for each value of s,

$$F_3(f(s), g(s), s) = 0. \tag{29}$$

The two equations (21) and (29),

$$F(x, y, s) = 0$$
$$F_3(x, y, s) = 0, \tag{30}$$

can sometimes be solved for x and y in terms of s to yield the parametric equations of the envelope (23), or perhaps s can be eliminated between the two equations to obtain an implicit equation in x and y for C.

For example, consider the family of lines (22). The equations (30) become

$$x \cos(s) + y \sin(s) = 1,$$

$$-x \sin(s) + y \cos(s) = 0.$$

Multiply the first equation by sin(s), the second by cos(s) and add to obtain y = sin(s). Replace y by sin(s) in the second equation to obtain x = cos(s). Clearly

$$x = \cos(s), \; y = \sin(s)$$

are the parametric equations of the unit circle.

As another example for which we have some serious use, consider the family of circles

$$F(x, y, s) = (x - s)^2 + y^2 - \rho^2 s(1 - s) = 0, \quad 0 \leq s \leq 1, \tag{31}$$

in which $\rho > 0$. We compute directly that (30) becomes

$$2(x - s) + \rho^2(1 - 2s) = 0, \tag{32}$$

a linear equation in s. It is a routine, if tedious, calculation to solve for s in terms of x in (32) and substitute in (31) to obtain the implicit equation of the envelope:

$$\frac{(x - 1/2)^2}{(\rho^2 + 1)/4} + \frac{y^2}{\rho^2/4} = 1. \tag{33}$$

The curve (33) is an ellipse with center at (1/2, 0), semi-major axis of length $(\rho^2 + 1)^{1/2}/2$, and semi-minor axis of length $\rho/2$. The foci are at (0, 0) and (1, 0).

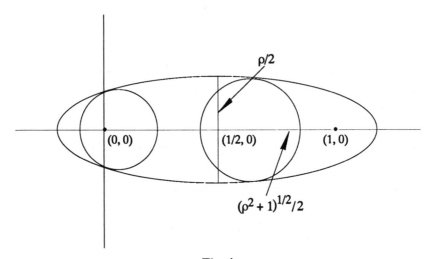

Fig. 4

Fig. 4 shows two typical circles in the family (31) and their points of tangency with the envelope.

The following theorem, known as the *elliptical range theorem*, completely describes the structure of W(A) for any $A \in M_2(\mathbb{C})$. It is the basis for proving the fact that for any

9.2 Geometry of Forms

$A \in M_n(\mathbb{C})$, $W(A)$ is always convex, the most important theorem about the numerical range.

Theorem 4 (Elliptical Range Theorem).

Let $A \in M_2(\mathbb{C})$ with eigenvalues λ and μ. Define the following numbers associated with A:

$$v = \left(\sum_{i,j=1}^{2} |a_{ij}|^2 \right)^{1/2} \tag{34}$$

and

$$\alpha = (v^2 - |\lambda|^2 - |\mu|^2)^{1/2}. \tag{35}$$

Then $W(A)$ is an elliptical region bounded by an ellipse (possibly degenerate) whose description is as follows:

$$\text{foci: } \lambda, \mu; \tag{36}$$

$$\text{semi-major axis: } \frac{(v^2 - 2\text{Re}(\lambda\bar{\mu}))^{1/2}}{2}; \tag{37}$$

$$\text{semi-minor axis: } \alpha/2. \tag{38}$$

Proof.

By §8.1 Theorem 1, the Schur triangularization theorem, there exists a unitary U such that

$$U^*AU = B = \begin{bmatrix} \lambda & \alpha \\ 0 & \mu \end{bmatrix} \tag{39}$$

in which we can assume $\alpha \geq 0$ (why?). From (10), $W(A) = W(B)$. Next set

$$C = B - \mu I_2 = \begin{bmatrix} \lambda - \mu & \alpha \\ 0 & 0 \end{bmatrix} \quad (40)$$

and note that

$$W(A) = W(B) = W(C) + \mu. \quad (41)$$

Assume first that $\lambda = \mu$ in (40). If $\alpha = 0$ then (40) implies that $C = 0$ and hence from (41), $W(A) = \mu$ is a single point. If $\alpha > 0$ then write

$$D = \alpha^{-1} C = \begin{bmatrix} 0 & 1 \\ 0 & 0 \end{bmatrix}$$

and by (6),

$$W(A) = W(C) + \mu$$
$$= W(\alpha D) + \mu$$
$$= \alpha W(D) + \mu. \quad (42)$$

However, from Fig. 3, $W(D)$ is a circular disk centered at the origin of radius 1/2. Hence from (42), $W(A)$ is a circular disk centered at μ of radius $\alpha/2$.

Next assume that $\lambda \neq \mu$ and from (40) write

$$C = (\lambda - \mu) \begin{bmatrix} 1 & \dfrac{\alpha}{(\lambda - \mu)} \\ 0 & 0 \end{bmatrix} = (\lambda - \mu) E \quad (43)$$

where

$$E = \begin{bmatrix} 1 & \beta \\ 0 & 0 \end{bmatrix}, \quad \beta = \frac{\alpha}{(\lambda - \mu)}. \quad (44)$$

In computing $W(E)$ we can (by a diagonal unitary similarity) assume that β is replaced by $|\beta|$. Then

9.2 Geometry of Forms

$$W(A) = W(C) + \mu$$
$$= W((\lambda - \mu)E) + \mu$$
$$= (\lambda - \mu) W(E) + \mu. \tag{45}$$

If $\alpha = 0$ then $\beta = 0$ and it is simple to compute that $W(E) = [\,0, 1\,]$. Hence from (45),

$$W(A) = [\,\mu, \lambda\,], \tag{46}$$

the line segment joining μ and λ.

If $\alpha > 0$ then we let $v = [\,\xi\ \ \eta\,]^T$ be an arbitrary complex column vector that satisfies $1 = v^*v = |\xi|^2 + |\eta|^2$. Let $s = |\xi|^2$ and compute directly from (44) that

$$v^*Ev = |\xi|^2 + |\beta|\overline{\xi}\eta$$
$$= |\xi|^2 + |\beta||\xi||\eta|e^{i\theta}$$
$$= s + \rho\sqrt{s(1-s)}\,e^{i\theta}, \tag{47}$$

in which $\rho = |\beta|$ and θ is any argument in $[\,0, 2\pi\,]$. For a fixed s, $0 \le s \le 1$, the right side of (47) describes a circle centered at s with radius $\rho\sqrt{s(1-s)}$, generated as θ runs over $[\,0, 2\pi\,]$. In rectangular coordinates the equation of $C(s, \rho\sqrt{s(1-s)}\,)$ is

$$(x - s)^2 + y^2 - \rho^2 s(1 - s) = 0, \tag{48}$$

and in terms of complex numbers the equation is

$$|z - s| = \rho\sqrt{s(1-s)}\,. \tag{49}$$

By §9.2 Exercises, #14, the union of all circles (49), as s runs over $[\,0, 1\,]$, is a convex set. Moreover, by (31) et seq., we saw that this family of circles has the ellipse (33) as its envelope. Thus W(E) is an elliptical region bounded by the ellipse (33). Since the foci of (33) are at (0, 0) and (1, 0), the elliptical region W(E) is described as the set of all z satisfying

$$|z| + |z - 1| \le (\rho^2 + 1)^{1/2}. \tag{50}$$

(Remember that for an ellipse the sum of the distances to the foci has a constant value equal to the length of the major axis.) Now

$$\rho = |\beta| = |\alpha/(\lambda - \mu)|$$

so that (50) becomes

$$|z| + |z - 1| \le (\alpha^2 + |\lambda - \mu|^2)^{1/2}/|\lambda - u|. \tag{51}$$

According to (45), $W(A)$ is obtained by multiplying each point $z \in W(E)$ by $(\lambda - \mu)$ and then translating the results by μ. In other words, $z \in W(E)$ iff $w = (\lambda - \mu)z + \mu \in W(A)$, i.e., the relationship between points w in $W(A)$ and z in $W(E)$ is

$$z = \frac{w - \mu}{\lambda - \mu}. \tag{52}$$

Making this substitution in (51) and multiplying through by $|\lambda - \mu|$ produces

$$|w - \mu| + |w - \lambda| \le (\alpha^2 + |\lambda - \mu|^2)^{1/2} \tag{53}$$

as describing all points w lying in the elliptical region $W(A)$.

Finally, from §9.2 Exercises, #19,

$$\alpha^2 + |\lambda - \mu|^2 = \alpha^2 + |\lambda|^2 + |\mu|^2 - 2\operatorname{Re}(\lambda\overline{\mu})$$

$$= v^2 - 2\operatorname{Re}(\lambda\overline{\mu}),$$

and we can now say, using (53): if $\lambda \ne \mu$ and $\alpha > 0$ then $W(A)$ is the elliptical region bounded by the ellipse with foci at λ and μ, semi-major axis equal in length to $(v^2 - 2\operatorname{Re}(\lambda\overline{\mu}))^{1/2}/2$ and semi-minor axis equal in length to $\alpha/2$.

We summarize these results. For the matrix

$$U^*AU = B = \begin{bmatrix} \lambda & \alpha \\ 0 & \mu \end{bmatrix}, \quad \alpha \ge 0,$$

9.2 Geometry of Forms

W(A) = W(B) has the following description.

$\lambda = \mu, \alpha = 0$:

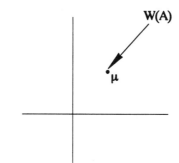

Fig. 5

$\lambda = \mu, \alpha > 0$:

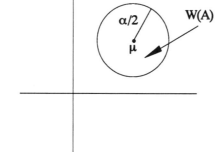

Fig. 6

$\lambda \neq \mu, \alpha = 0$:

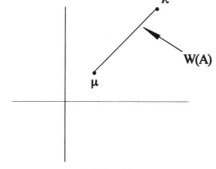

Fig. 7

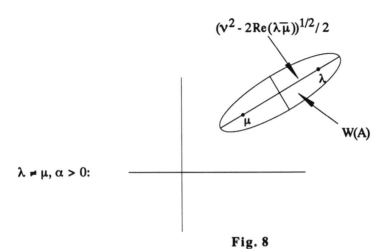

Fig. 8

The most general situation is depicted in Fig. 8. Note that if α becomes 0 in Fig. 8 then the semi-minor axis collapses and W(A) becomes the line segment joining λ and μ. As a check, note that $\nu^2 = |\lambda|^2 + |\mu|^2 + \alpha^2$ so that ν^2 becomes $|\lambda|^2 + |\mu|^2$ and $(\nu^2 - 2\text{Re}(\lambda\bar{\mu}))^{1/2}/2 = |\lambda - \mu|/2$, half the distance between λ and μ. Thus the $\alpha = 0$ case in Fig. 8 is depicted in Fig. 7.

If λ collapses to μ in Fig. 8, $\alpha > 0$, then Fig. 8 collapses to the circular disk, centered at $\mu = \lambda$, of radius $\alpha/2$. Thus the $\lambda = \mu$, $\alpha > 0$ case of Fig. 8 is depicted in Fig. 6.

If λ collapses to μ and α becomes 0 in Fig. 8 then Fig. 8 collapses to the single point μ. Thus the $\lambda = \mu$, $\alpha = 0$ case of Fig. 8 is Fig. 5.

Summarizing, the description of W(A) in the statement of the theorem specializes automatically to all degenerate cases. ∎

An ellipse is completely determined once its foci, λ and μ, are prescribed and the length, α, of its minor axis is given. According to Theorem 4, for the matrix

$$A = \begin{bmatrix} \lambda & \alpha \\ 0 & \mu \end{bmatrix}, \quad \alpha \geq 0, \tag{54}$$

9.2 Geometry of Forms

$W(A)$ is an elliptical region with foci λ and μ and minor axis of length α. Hence, any elliptical region is the numerical range of an appropriate A. Moreover, by (10), any matrix unitarily similar to A also has the same $W(A)$. Thus any ellipse is the envelope of a family of circles centered on the major axis. In fact, let the ellipse be the boundary of $W(A)$ where A is the matrix in (54). If $x = [\ \xi\ \ \eta\ ]^T$ and $x^*x = 1$ then, as we have computed before,

$$x^*Ax = |\xi|^2\lambda + |\eta|^2\mu + \alpha\bar{\xi}\eta$$

$$= s\lambda + (1-s)\mu + \alpha\sqrt{s(1-s)}\ e^{i\theta} \tag{55}$$

where $s = |\xi|^2$, $0 \le s \le 1$, and $0 \le \theta \le 2\pi$. For a fixed s, as θ varies over $[\ 0, 2\pi\ ]$, a circle of radius $\alpha\sqrt{s(1-s)}$ is generated with center at $s\lambda + (1-s)\mu$. Notice that the centers of the circles in the family (55) are in fact situated on the line joining μ and λ. Thus the boundary of $W(A)$ is precisely the envelope of the family of circles (55). Of course, if $\lambda = \mu$ then (55) is a circle with fixed center $\mu = \lambda$ and radius $\alpha\sqrt{s(1-s)}$. The numerical range $W(A)$ is then the circular disk of largest radius, namely $\alpha/2$. The other degenerate cases are similar.

The preceding remarks are helpful in geometrically determining the distance from the origin to the most distant point in an elliptical region $W(A)$ that depicts the numerical range of $A \in M_2(\mathbb{C})$. By definition, this distance is the numerical radius $w(A)$.

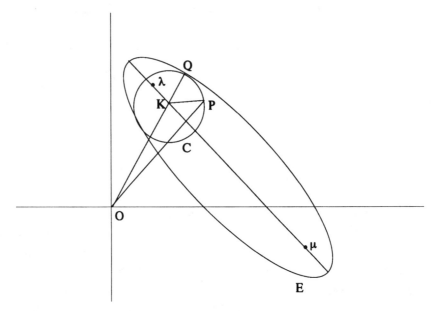

Fig. 9

Consider the ellipse E in Fig. 9. Since E is the envelope of a family of circles centered on the line joining μ and λ, let C denote a circle in the family with the property that the distance of some point on its circumference to the origin is the numerical radius. Join the origin O to the center K of the circle C. From K consider any radius KP to a point P on the circle. By the triangle inequality, OK + KP $\geq$ OP and the inequality is strict unless the line segment KP is a continuation of OK. In other words, the point on C most distant from the origin can be obtained by extending the line OK until it intersects C at the point labelled Q. Since OQ = OK + KQ = OK + KP > OP for any point P (P $\neq$ Q) on C, it follows that OQ is the numerical radius, i.e., there is a point Q on C whose distance from the origin is the numerical radius. We assert that Q is on the ellipse E. Otherwise, if Q were inside of E it would follow that points of E exist whose distance from the origin exceed the numerical radius. Thus we have the following algorithm for determining the distance from the origin to the most distant point on E.

- For each of the circles in the family whose envelope is E, construct the line segment from the origin to the center of the circle.

- Extend this line segment until it intersects the circle.

- Compute the distance from O to this point of intersection. Maximize this distance over all circles in the family.

In view of these remarks we can now state the following useful result for computing the numerical radius of a 2 × 2 matrix.

Theorem 5.

Let A be unitarily similar to

$$\begin{bmatrix} \lambda & \alpha \\ 0 & \mu \end{bmatrix}, \quad \alpha \geq 0. \tag{56}$$

For $s \in [0, 1]$ define the function

$$d(s) = |s\lambda + (1-s)\mu| + \alpha\sqrt{s(1-s)}. \tag{57}$$

Then

$$w(A) = \max d(s) \tag{58}$$

where the max in (58) is computed for $s \in [0, 1]$.

Proof.

If W(A) is a non-degenerate ellipse then the preceding geometric argument, together with (55), establishes the result. If $\alpha = 0$ then obviously w(A) is either $|\lambda|$ or $|\mu|$ and this is precisely the maximum of (57). If $\alpha > 0$ and $\lambda = \mu$ then W(A) is a circle centered at $\lambda = \mu$ with radius $\alpha/2$, and the maximum of (57) is $|\lambda| + \alpha/2$. Thus, in all cases, w(A) is the maximum of (57) for $0 \leq s \leq 1$. ∎

For matrices $A \in M_2(\mathbb{C})$ which are unitarily similar to a real matrix it is possible to give an explicit formula for w(A) in terms of the entries of A. In §9.2 Exercises, #20, w(A) is explicitly exhibited in terms of the entries in the upper triangular form of A. But this can be combined with §9.2 Exercises, #19, (and obvious formulas for the eigenvalues of A in terms of its entries) to produce the explicit formula in terms of A itself.

9.2 Exercises

(Note: Exercises #20 - #22 are more difficult than usual and can be omitted without effecting any subsequent material.)

1. Prove: $W(cA) = cW(A)$.

2. Prove: $W(cI_n + A) = c + W(A)$.

3. Prove: $W(A^*) = \overline{W(A)}$.

4. Prove: if B is a principal submatrix of A, then $W(B) \subset W(A)$.

5. Prove: $W(U^*AU) = W(A)$ for any unitary matrix U.

6. Prove: $W(A) \subset \mathbb{R}$ iff A is hermitian.

7. Prove: $W(A) \subset i\mathbb{R}$ iff A is skew-hermitian.

8. Prove: $W(A) = \{0\}$ iff $A = 0$.

9. Prove: $\sigma(A) \subset W(A)$.

10. Prove: $W(A) = \{c\}$ iff $A = cI_n$.

11. Recall that an ellipse is the set of all points in the plane the sum of whose distances from two fixed points (called the foci) is a constant. If λ and μ are the foci then the equation for the ellipse is usually written as

$$|z - \lambda| + |z - \mu| = 2a$$

where 2a represents the constant appearing in the definition.

9.2 Exercises

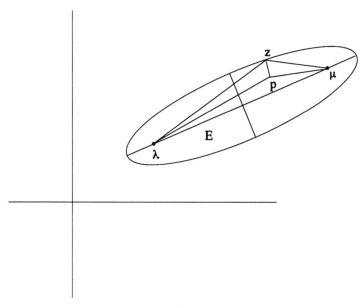

Fig. 1

It is clear that p is a point in the region E enclosed by the ellipse iff

$$|p - \lambda| + |p - \mu| \leq 2a$$

Prove: E is a convex set.

Hint: Let p and q be points in E. Then

$$|p - \lambda| + |p - \mu| \leq 2a$$

and

$$|q - \lambda| + |q - \mu| \leq 2a.$$

If $0 \leq \theta \leq 1$ and $m = (1 - \theta)p + \theta q$ then

$$|m - \lambda| + |m - \mu| = |(1 - \theta)p + \theta q - \lambda| + |(1 - \theta)p + \theta q - \mu|$$

$$= |(1 - \theta)p + \theta q - (1 - \theta)\lambda - \theta\lambda| + |(1 - \theta)p + \theta q - (1 - \theta)\mu - \theta\mu|$$

$$= |(1-\theta)(p-\lambda) + \theta(q-\lambda)| + |(1-\theta)(p-\mu) + \theta(q-\mu)|$$

$$\leq (1-\theta)|p-\lambda| + \theta|q-\lambda| + (1-\theta)|p-\mu| + \theta|q-\mu|$$

$$= (1-\theta)\{|p-\lambda| + |p-\mu|\} + \theta\{|q-\lambda| + |q-\mu|\}$$

$$\leq (1-\theta) \cdot 2a + \theta \cdot 2a$$

$$= 2a.$$

12. A line which does not pass through the origin has an equation of the form $Lx + my = 1$ ((x, y) are rectangular coordinates). Show that any line not through the origin has a complex equation of the form

$$\text{Re}(az) = 1, \quad z = x + iy.$$

Show that a line is a convex set in the plane. Also show that any line not through the origin and perpendicular to $\text{Re}(az) = 1$ has an equation of the form $\text{Re}((ia)z) = 1$.

Hint: Let $a = L - im$. Then if $z = x + iy$ it follows that $\text{Re}(az) = Lx + my$. If $0 \leq \theta \leq 1$, $\text{Re}(ap) = 1$, and $\text{Re}(aq) = 1$, then

$$\text{Re}(a[(1-\theta)p + \theta q]) = (1-\theta)\text{Re}(ap) + \theta\text{Re}(aq) = 1.$$

13. Use the diagram in #11 (Fig. 1) to prove that the length of the semi-major axis of the ellipse appearing in #11 is a, the center of the ellipse is the point $(\lambda + \mu)/2$, the distance from the center to a focus is $|\mu - \lambda|/2$, and the length of the semi-minor axis is $b = (4a^2 - |\mu - \lambda|^2)^{1/2}/2$.

14. Let s be a real variable in the interval $[0, 1]$ and let $p > 0$. Define C to be the set

$$C = \{z \mid |z - s| = p\sqrt{s(1-s)}, \ s \in [0, 1]\}.$$

Thus C is the union of all circles with center at $(s, 0)$ and radius $p\sqrt{s(1-s)}$, $0 \leq s \leq 1$. Prove that C is a convex set.

Hint: Let $f(s) = \sqrt{s(1-s)}$. It is an elementary calculus problem to compute that

$$f'(s) = \frac{1-2s}{2(s(1-s))^{1/2}}$$

and

$$f''(s) = \frac{-1}{4(s(1-s))^{3/2}}.$$

Thus $f''(s) < 0$ for $0 < s < 1$ and the graph of $f(s)$ is concave downward. In other words, if s_1 and s_2 are any two values in the open interval $[0, 1]$ and $0 \le \theta \le 1$, then

$$f((1-\theta)s_1 + \theta s_2) \ge (1-\theta) f(s_1) + \theta f(s_2). \tag{a}$$

Let C_s denote the circle $|z - s| = p\sqrt{s(1-s)} = pf(s)$ and suppose that $z_1 \in C_{s_1}$, $z_2 \in C_{s_2}$ and $0 \le \theta \le 1$. Consider the point $z_0 = (1-\theta)z_1 + \theta z_2$. Define $s_0 = (1-\theta)s_1 + \theta s_2$ and consider the function

$$g(s) = |z_0 - s| - pf(s).$$

Obviously, $g(0) = |z_0| \ge 0$. Now consider

$g(s_0) = |z_0 - s_0| - pf(s_0)$

$= \big|[(1-\theta)z_1 + \theta z_2] - [(1-\theta)s_1 + \theta s_2]\big| - pf((1-\theta)s_1 + \theta s_2)$

$\le |(1-\theta)(z_1 - s_1) + \theta(z_2 - s_2)| - p\{(1-\theta)f(s_1) + \theta f(s_2)\}$

$\le (1-\theta)|z_1 - s_1| + \theta|z_2 - s_2| - p\{(1-\theta)f(s_1) + \theta f(s_2)\}$

$= (1-\theta)[|z_1 - s_1| - pf(s_1)] + \theta[|z_2 - s_2| - pf(s_2)]$

$= 0.$

The first inequality in the above computation uses (a) above. The second is simply the triangle inequality. Thus $g(0) \geq 0$ and $g(s_0) \leq 0$. Obviously g is a continuous function of s and hence there is an s_3, $0 \leq s_3 \leq s_0 \leq 1$ such that $g(s_3)$ is 0. But from the definition of $g(s)$

$$0 = g(s_3) = |z_0 - s_3| - pf(s_3)$$

and hence $z_0 \in C_{s_3}$, i.e.,

$$(1 - \theta)z_1 + \theta z_2 \in C.$$

15. Prove: the open half-planes defined by (17) and (18) are convex.

16. Prove: if $A \in M_n(\mathbb{R})$ then $W(A) = \overline{W(A)}$. Geometrically this means that $W(A)$ is symmetrically situated in the plane with respect to the real axis, i.e., the portion of $W(A)$ in the upper half-plane is the mirror image of the portion in the lower half-plane.

17. Prove: if $A \in M_{m,n}(\mathbb{C})$ then $\text{tr}(A^*A) = \sum_{i=1}^{m}\sum_{j=1}^{n} |a_{ij}|^2$ (familiar by now).

18. Prove: if $A \in M_{m,n}(\mathbb{C})$ and U and V are unitary, $U \in M_m(\mathbb{C})$, $V \in M_n(\mathbb{C})$, and $B = UAV$ then $\text{tr}(B^*B) = \text{tr}(A^*A)$.

19. Let $A \in M_2(\mathbb{C})$ and assume that A is unitarily similar to the upper triangular matrix

$$\begin{bmatrix} \lambda & \alpha \\ 0 & \mu \end{bmatrix}.$$

Show that

$$|\lambda|^2 + |\mu|^2 + |\alpha|^2 = \sum_{i,j=1}^{2} |a_{ij}|^2.$$

Hint: Use #17.

9.2 Exercises

20. Let

$$A = \begin{bmatrix} \lambda & \alpha \\ 0 & \mu \end{bmatrix} \in M_2(\mathbb{C}), \quad \alpha \geq 0.$$

Show that if A is unitarily similar to a real matrix then w(A) can be determined as follows:

I. λ and μ are real. Then w(A) is the larger of the two numbers

$$\frac{|\lambda + \mu \pm \sqrt{(\lambda - \mu)^2 + \alpha^2}|}{2}.$$

II. λ and μ are complex conjugates: $\lambda = h + ik$, $\mu = h - ik$, $k \neq 0$. If $2k^2 \geq \alpha |h|$ then

$$w(A) = \frac{|\lambda|}{|k|} \frac{\sqrt{\alpha^2 + 4k^2}}{2}.$$

III. λ and μ are complex conjugates: $\lambda = h + ik$, $\mu = h - ik$, $k \neq 0$. If $2k^2 < \alpha |h|$ then

$$w(A) = |h| + \frac{\alpha}{2}.$$

Hint: First note that the cases in which $\alpha = 0$ are simple, for then A is diagonal and W(A) is the line segment or (point) connecting λ and μ, and w(A) is the distance from the origin to the farthest endpoint of this line. Consider case (I) when $\alpha = 0$. The formula in (I) then gives

$$w(A) = \max \frac{1}{2} \left| \lambda + \mu \pm \sqrt{(\lambda - \mu)^2 + \alpha^2} \right|$$

$$= \frac{1}{2} \max \left| \lambda + \mu \pm \sqrt{(\lambda - \mu)^2} \right|.$$

The larger of $|\lambda|$ and $|\mu|$ is always the value in the preceding formula.

Next consider case (II). In this case the eigenvalues are nonreal. When $\alpha = 0$, $2k^2 \geq \alpha |h| = 0$ holds trivially. The formula in (II) implies that

$$w(A) = \frac{|\lambda|}{|k|} \frac{\sqrt{\alpha^2 + 4k^2}}{2}$$

$$= \frac{|\lambda|}{|k|} \frac{2|k|}{2}$$

$$= |\lambda| = |\mu|.$$

Case (III) does not apply when $\alpha = 0$. This dispenses with all cases in which $\alpha = 0$. For the remainder of the argument assume $\alpha > 0$.

Since the characteristic polynomial of A has real coefficients, the eigenvalues of A are either both real or complex conjugates $\lambda = h + ik$, $\mu = h - ik$. The two situations for the non-degenerate case of W(A) are depicted in the following diagrams (see #16 above).

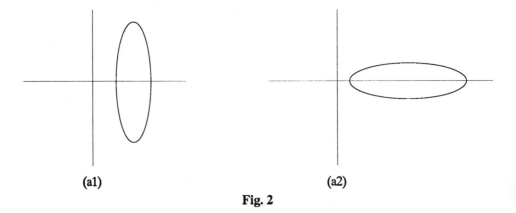

(a1)　　　　　　　　　　　　(a2)

Fig. 2

In case (I), in which both eigenvalues are real, W(A) is situated as in diagram (a2). It is obvious that w(A) is the distance from the origin to the most distant end of the major axis. It is simple to confirm from the statement of Theorem 4 that in this case w(A) is the larger of the two numbers

$$\frac{1}{2}\left|\lambda + \mu \pm \sqrt{(\lambda - \mu)^2 + \alpha^2}\right|,$$

which can also be expressed as

$$\frac{1}{2}\left| \operatorname{tr}(A) \pm \sqrt{\operatorname{tr}(A^*A) - 2\det(A)} \right|.$$

This concludes case (I).

In cases (II) and (III), in which the eigenvalues are nonreal complex conjugates, $W(A)$ is situated as in the diagram (a1) of Fig. 2. Rewrite $d(s)$ in (57) as

$$d(s) = \sqrt{h^2 + k^2(2s - 1)^2} + \alpha\sqrt{s(1 - s)}.$$

We easily compute that

$$d'(s) = (2s - 1)\left\{ \frac{2k^2}{\sqrt{h^2 + k^2(2s - 1)^2}} - \frac{\alpha}{2\sqrt{s(1 - s)}} \right\}. \qquad (b)$$

To minimize the notation somewhat, set $\rho_1 = h^2 + k^2(2s - 1)^2$, $\rho_2 = s(1 - s)$. Then (b) becomes

$$d'(s) = (2s - 1)(2k^2\rho_1^{-1/2} - \alpha\rho_2^{-1/2}/2),$$

and the maximum value of d on $[0, 1]$ is $d(0)$, $d(1)$, $d(1/2)$ or possibly the value of d at that $s \in [0, 1]$, (if any), for which

$$2k^2\rho_1^{-1/2} = \alpha\rho_2^{-1/2}/2. \qquad (c)$$

From (c) we obtain

$$\alpha^2\rho_1 = 16k^4\rho_2 \qquad (d)$$

so that

$$\sqrt{\rho_1} = \frac{4k^2}{\alpha} \sqrt{\rho_2} . \tag{e}$$

Now

$$d(s) = \left(\frac{4k^2}{\alpha} + \alpha\right) \sqrt{\rho_2} . \tag{f}$$

To determine ρ_2, rewrite equation (d) as

$$\alpha^2 [h^2 + k^2(2s - 1)^2] = 16k^4 \rho_2. \tag{g}$$

Notice that

$$(2s - 1)^2 = 4s^2 - 4s + 1$$

$$= 4s(s - 1) + 1$$

$$= -4\rho_2 + 1.$$

Making this substitution in (g) we have

$$\alpha^2 [h^2 + k^2(1 - 4\rho_2)] = 16k^4 \rho_2$$

or

$$\rho_2 = \frac{\alpha^2(h^2 + k^2)}{4k^2(\alpha^2 + 4k^2)} . \tag{h}$$

Since $\rho_2 = s(1 - s)$ the formula (h) is valid iff there is an $s \in [0, 1]$ for which

$$s(1 - s) = \frac{\alpha^2(h^2 + k^2)}{4k^2(\alpha^2 + 4k^2)} .$$

Now, the range of $s(1 - s)$ for $s \in [0, 1]$ is $[0, 1/4]$. Thus

$$\frac{\alpha^2(h^2+k^2)}{4k^2(\alpha^2+4k^2)} \le \frac{1}{4}$$

must hold in order that (h) be valid. Simplifying this inequality, we obtain $2k^2 \ge \alpha |h|$. If we substitute (h) in (f) we obtain a critical value of d(s) as

$$d(s) = \left(\frac{4k^2+\alpha^2}{\alpha}\right)\sqrt{p_2}$$

$$= \left(\frac{4k^2+\alpha^2}{\alpha}\right)\frac{\alpha}{2|k|}\frac{\sqrt{h^2+k^2}}{\sqrt{\alpha^2+4k^2}}$$

$$= \frac{\sqrt{h^2+k^2}}{|k|}\frac{\sqrt{\alpha^2+4k^2}}{2}$$

$$= \frac{|\lambda|}{|k|}\frac{\sqrt{\alpha^2+4k^2}}{2}. \tag{i}$$

This is the value of d(s) for which $d'(s) = 0$, $s \in [0, 1]$.

Of course, (i) is a satisfactory formula but it is interesting to interpret it geometrically. According to (37) the length of the semi-major axis of the ellipse (a1) in Fig. 2 is

$$\frac{(v^2 - 2\text{Re}(\lambda\bar{\mu}))^{1/2}}{2} = \frac{(|\lambda|^2 + |\mu|^2 + \alpha^2 - 2\text{Re}(\lambda^2))^{1/2}}{2}$$

$$= \frac{(2(h^2+k^2) + \alpha^2 - 2(h^2-k^2))^{1/2}}{2}$$

$$= \frac{\sqrt{\alpha^2+4k^2}}{2}. \tag{j}$$

Thus the critical value of d can be expressed as follows:

$$d(s) = \frac{\text{(distance from origin to focus)}}{\text{(distance from center to focus)}} \times \text{(semi-major axis)}. \tag{k}$$

For case (II) we have confirmed that one of the possibilities for the maximum of $d(s)$, $s \in [0, 1]$ is the formula (i). The other possibilities are $d(0)$, $d(1)$, and $d(1/2)$. From the formula for $d(s)$, it is clear that $d(0) = d(1) = \sqrt{h^2 + k^2} = |\lambda|$. Since λ is an interior point of the ellipse ($\alpha > 0$), this cannot be a maximum value of $d(s)$. Next, observe that $d(1/2) = |h| + \alpha/2$ which is the distance from the origin to the end of the minor axis of the ellipse. We show that the value of $d(s)$ in (i) is at least $|h| + \alpha/2$, i.e.,

$$\frac{|\lambda|}{|k|} \frac{\sqrt{\alpha^2 + 4k^2}}{2} \geq |h| + \frac{\alpha}{2}.$$

The above inequality is equivalent to each of the following:

$$\frac{\sqrt{h^2 + k^2}}{|k|} \frac{\sqrt{\alpha^2 + 4k^2}}{2} \geq |h| + \frac{\alpha}{2},$$

$$\sqrt{h^2 + k^2} \sqrt{\alpha^2 + 4k^2} \geq 2|h||k| + \alpha|k|,$$

$$(h^2 + k^2)(\alpha^2 + 4k^2) \geq \alpha^2 k^2 + 4\alpha|h|k^2 + 4h^2 k^2$$

$$h^2 \alpha^2 + 4k^4 \geq 4\alpha|h|k^2,$$

$$4k^4 - 4\alpha|h|k^2 + |h|^2 \alpha^2 \geq 0,$$

$$(2k^2 - |h|\alpha)^2 \geq 0.$$

This concludes the proof of case (II).

To proceed to the proof of case (III), we first observe, from (h) and the two formulas that follow it, that if $2k^2 < \alpha|h|$ (the situation in case III)) there is no $s \in [0, 1]$ for which the formula in the second factor on the right in the expression (b) for $d'(s)$ can be 0. Hence the only possibilities for the maximum are $d(0)$, $d(1)$, and $d(1/2)$. As above, $d(0) = d(1) = |\lambda|$ and λ is an interior point of the ellipse, and thus not a maximum for $d(s)$. Thus the only remaining alternative is $d(1/2) = |h| + \alpha/2$.

21. Show by example that both possibilities in #20 (II) and (III) can occur.

Hint: Take $\alpha = 1$, $h = 1$, $k = 1.01$. Then $|h| + \alpha/2 = 3/2$ and

$$\frac{|\lambda|}{|k|} \frac{\sqrt{\alpha^2 + 4k^2}}{2} = 1.59 > \frac{3}{2}.$$

Moreover, this second eventuality is valid since $2k^2 = 2.0402 > \alpha|h| = 1$. Note that the distance from the origin to the end of the major axis is

$$\sqrt{h^2 + \left(\frac{\sqrt{\alpha^2 + 4k^2}}{2}\right)^2} = \sqrt{1 + \frac{1 + 4(1.01)^2}{4}}$$

$$= \sqrt{1.25 + (1.01)^2}$$

$$= 1.51 < 1.59.$$

This shows that the most distant point on the ellipse from the origin is not at the ends of either axis of the ellipse.

If we next take $\alpha = 2.1$, $h = k = 1$ then the condition $2k^2 \geq \alpha|h|$ fails and $w(A) = |h| + \alpha/2 = 1 + (2.1)/2 = 2.05$.

22. With the same assumptions on A as in #20, show that $w(A)$ is the distance from the origin O to the end of one of the axes of $W(A)$ iff one of the following conditions holds

(i). λ and μ are real.

(ii). λ and μ are complex conjugates and $h\alpha = 0$.

(iii). λ and μ are complex conjugates and $2k^2 < \alpha|h|$.

Hint: If (i) holds then #20 (I) must be the case, so that $w(A)$ is the distance from the origin to an end of the major axis of $W(A)$.

If (ii) holds then obviously $2k^2 \geq \alpha |h| = 0$, and by #20 (II),

$$w(A) = \frac{|\lambda|}{|k|} \frac{\sqrt{\alpha^2 + 4k^2}}{2}. \tag{l}$$

There are two possibilities: $\alpha = 0$ or $h = 0$. Suppose $\alpha = 0$. Then the right side of (l) is $|\lambda|$. In this case $W(A)$ is a line segment joining λ and $\mu = \bar{\lambda}$ and hence $w(A) = |\lambda|$. But this is the distance to the end of an axis (degenerate). Next, suppose $h = 0$. Then the right side of (l) becomes

$$\frac{\sqrt{\alpha^2 + 4k^2}}{2}$$

which is the length of the semi-major axis (see #20, item (j)). But $W(A)$ has its major axis along the y-axis (i.e., $h = 0$) and $w(A)$ is the distance from the origin to an end of the major axis.

If (iii) holds then #20 (III) shows $w(A) = \alpha |h| + \alpha/2$, the distance from the origin to the end of the minor axis of $W(A)$.

Suppose, conversely, that $w(A)$ is the distance from the origin to the end of an axis of $W(A)$. Under this circumstance we will show that if (iii) fails to hold then (i) or (ii) must hold. If (iii) fails it means that λ and μ are real, or λ and μ are complex conjugates and $\alpha k^2 \geq \alpha |h|$. In the first alternative we have (i). If the second alternative holds then by #20 (II)

$$w(A) = \frac{|\lambda|}{|k|} \frac{\sqrt{\alpha^2 + 4k^2}}{2}.$$

Now refer to the diagram (a1) in #20, the present situation. The semi-major axis has length $\frac{\sqrt{\alpha^2 + 4k^2}}{2}$ (see item (j) in the Hint for #20). Thus the distance from the origin to either end of the major axis is

$$(h^2 + (\sqrt{\alpha^2 + 4k^2}/2)^2)^{1/2}. \tag{m}$$

9.2 Exercises

Since we are assuming that $w(A)$ is the number in (m), we have

$$\frac{|\lambda|}{|k|} \frac{\sqrt{\alpha^2 + 4k^2}}{2} = \left(h^2 + \left(\sqrt{\alpha^2 + 4k^2}/2\right)^2\right)^{1/2},$$

$$\frac{|\lambda|^2}{k^2} \frac{(\alpha^2 + 4k^2)}{4} = h^2 + \frac{(\alpha^2 + 4k^2)}{4},$$

$$(h^2 + k^2)(\alpha^2 + 4k^2) = 4k^2h^2 + k^2\alpha^2 + 4k^4,$$

$$h^2\alpha^2 + 4h^2k^2 + k^2\alpha^2 + 4k^4 = 4k^2h^2 + k^2\alpha^2 + 4k^4,$$

$$h^2\alpha^2 = 0.$$

Thus the alternative (ii) must hold.

23. Let $A \in M_2(\mathbb{C})$. Prove: $\overline{W(A)} = W(A)$ iff A is unitarily similar to a real matrix.

Hint: If A is unitarily similar to a real matrix then #16 shows that $\overline{W(A)} = W(A)$. Conversely, assume that $\overline{W(A)} = W(A)$. By Theorem 4, $W(A)$ is an ellipse, possibly degenerate. Since $\overline{W(A)} = W(A)$, either the major or the minor axis of $W(A)$ must lie along the x-axis. (If $W(A)$ is a circular disk then the foci are collapsed to the center and there is a diameter along the x-axis.) It follows that the eigenvalues λ and μ of A, foci of $W(A)$, are either both real or are complex conjugates.

Case 1. λ and μ are real. Then A is unitarily similar to

$$\begin{bmatrix} \lambda & \alpha \\ 0 & \mu \end{bmatrix}$$

and by a diagonal unitary similarity, $\alpha \geq 0$.

Case 2. $\lambda = h + ik$, $\mu = h - ik$. Then A is unitarily similar to

$$\begin{bmatrix} h + ik & \alpha \\ 0 & h - ik \end{bmatrix} = hI_2 + k\begin{bmatrix} i & \alpha/k \\ 0 & -i \end{bmatrix}$$

where h and k are real, $\alpha \geq 0$. Clearly $\overline{W(A)} = W(A)$ iff $\overline{W(B)} = W(B)$ where

$$B = \begin{bmatrix} i & \alpha \\ 0 & -i \end{bmatrix}$$

in which α/k has been replaced by α for notational simplicity. Define a real matrix

$$R = \begin{bmatrix} \dfrac{\alpha}{2} & -\dfrac{(4+\alpha^2)^{1/2}}{2} \\ \dfrac{(4+\alpha^2)^{1/2}}{2} & -\dfrac{\alpha}{2} \end{bmatrix}.$$

It is easy to check that the characteristic polynomial of R is $\lambda^2 + 1$ so that the eigenvalues of R are $\pm i$. Hence by Schur's theorem again, R is unitarily similar to

$$C = \begin{bmatrix} i & \rho \\ 0 & -i \end{bmatrix}, \quad \rho \geq 0.$$

Moreover, by Theorem 4, with R playing the role of A,

$$\rho = (v^2 - |i|^2 - |-i|^2)^{1/2}$$

$$= \left(\frac{\alpha^2}{4} + \frac{\alpha^2}{4} + \frac{4+\alpha^2}{2} - 2 \right)^{1/2}$$

$$= \alpha.$$

Thus C = B and B is unitarily similar to the real matrix R.

9.2 MatLab

1. Devise a sequence of MatLab commands to graph W(A) for

$$A = \begin{bmatrix} 0 & 1 & 0 \\ 0 & 0 & 1 \\ 1 & 0 & 0 \end{bmatrix}.$$

   Hint: First review Theorem 2.

   ```
   A = [0 1 0; 0 0 1; 1 0 0];
   ev = eig(A);
   ev = ev.';
   axis('square');
   plot([ev ev(1)]);
   ```

2. Devise a sequence of MatLab commands to compute the numerical radius w(A) of an arbitrary hermitian matrix A.

   Hint:
   ```
   ev = eig(A);
   ev = abs(ev);
   w = max(ev);
   ```

3. Write a MatLab function called up that returns a vector u consisting of those eigenvalues $\lambda_1, ..., \lambda_p$ of an n-square complex matrix A that lie in the closed upper half-plane, i.e., $Im(\lambda_t) \geq 0$, $t = 1, ..., p$. Moreover, $Re(\lambda_1) \geq \cdots \geq Re(\lambda_p)$. Save up in a library called field (i.e., field of values).

   Hint:
   ```
   function u=up(A)
   %UP accepts an n-square complex A
   %and returns the p-tuple of
   %eigenvalues of A that lie in
   %the closed upper half-plane,
   %arranged in decreasing order of
   ```

```
%real parts.
ev=eig(A);
f=find(imag(ev)>=0);        %1
lam=ev(f);                  %2
p=length(lam);
[Y, I]=sort(real(lam));     %3
u1=lam(I);                  %4
u=u1(p:-1:1);               %5
```

%1   f is a vector of those indices j for which ev(j) >= 0.
%2   lam is the vector of eigenvalues of A that lie in closed upper half-plane.
%3   I defines the indices for which real(lam) is in increasing order.
%4   u1 is the vector consisting of these eigenvalues of A in the upper half-plane arranged in ascending order of their real parts.
%5   u reverses the ordering in u1.

4. Write a MatLab function called down that returns a vector v consisting of those eigenvalues $\mu_1, \ldots, \mu_q$ of an n-square complex matrix A that lie in the open lower half-plane, i.e., $Im(\mu_t) < 0$, $t = 1, \ldots, q$. Moreover $Re(\mu_1) \leq \cdots \leq Re(\mu_q)$. Save down in field.

Hint:
```
function v=down(A)
%DOWN accepts an n-square
%complex matrix A and
%returns the q-tuple of
%eigenvalues of A that lie
%in the open lower half-plane,
%arranged in increasing order
%of real parts.
ev=eig(A);
f=find(imag(ev)<0);
mu=ev(f);
q=length(mu);
[Y, I]=sort(real(mu));
v=mu(I);
```

9.2 MatLab

5. Use up and down in #3 and #4 to write a MatLab script called unitrng that accepts an n-square complex unitary matrix A and plots W(A) in the Graph window. Save unitrng in field.

    Hint:

    %UNITRNG plots the numerical
    %range of a unitary matrix.
    A=input('Enter an n-square unitary matrix A: ');
    u=up(A);
    v=down(A);
    w=[u.' v.' u(1)];
    axis('square');
    plot(w);

6. Run unitrng for several unitary matrices generated by the function urand (see §8.1 MatLab, #6.). What property do the resulting plots have in common? Why?

    Hint: All plots are polygons with vertices on the unit circle. See Theorem 2.

7. Write a MatLab function called nrad that accepts an n-square complex normal matrix A and returns the numerical radius of A. Save nrad in field.

    Hint:

    function w=nrad(A)
    %NRAD accepts an n-square
    %complex normal matrix A
    %and returns the numerical
    %radius w(A).
    ev=eig(A);
    ev=abs(ev);
    w=max(ev);

8. Write a MatLab script called normrng that accepts an n-square complex normal matrix A and then plots the eigenvalues of A and the circle centered at the origin with radius w(A). Use the point type '*' for the eigenvalues of A. Save normrng in field.

Hint:

```
%NORMRNG accepts an n-square
%complex normal A. Then eig(A)
%is plotted together with a
%circle of radius w(A) centered
%at the origin.
A=input('Enter an n-square normal A: ');
w=nrad(A);
v=0:2*pi/100:2*pi;
C=w*exp(i*v);
axis('square');
plot(C);
hold on;
plot(eig(A),'*');
hold off;
```

Remark: If the Graph window is opened to full size the plot appears properly.

9. Run normrng in #8 for several normal matrices A generated by the function normrand (see §8.1 MatLab, #9). What do you observe about the location of the eigenvalues with respect to the circle $|z| = w(A)$?

10. Review Theorem 5. Then write a MatLab function called nrad2 that accepts an arbitrary 2-square complex matrix A and returns w(A), the numerical radius of A. Save nrad2 in field.

Hint:

```
function w=nrad2(A)
%NRAD2 accepts an arbitrary
%2-square complex matrix A
%and returns w(A) the
%numerical radius of A.
[U, T]=schur(A);
[U, T]=rsf2csf(U, T);
a=abs(T(1,2));
D=[ ];
```

```
            for s=0:.01:1
                ds=abs(s*T(1,1)+(1-s)*T(2,2))+a*sqrt(s*(1-s));
                D = [D ds];
            end
            w=max(D);
```

11. Run nrad2 on each of the following matrices A and confirm that the approximation used in #10 in computing w(A) is reasonable.

    (a) $A = \begin{bmatrix} 0 & 1 \\ 0 & 0 \end{bmatrix}$

    (b) $A = \begin{bmatrix} 1 & i \\ -i & 1 \end{bmatrix}$

    (c) A = normrand(2)

    (d) A = urand(2)

12. Review §9.2 Exercises, #20. Then write a MatLab function called nrad3 that accepts any 2-square complex matrix A which is unitarily similar to a real matrix and returns w(A). Save nrad3 in field.

    Hint:
```
            function w=nrad3(A)
            %NRAD3 accepts a 2-square
            %complex matrix A that
            %is unitarily similar to
            %a real matrix and returns
            %w(A), the numerical radius
            %of A.
            [U, T]=schur(A);
            [U, T]=rsf2csf(U,T);
            lam=T(1,1);
            mu=T(2,2);
            h=real(lam);
            k=imag(lam);
```

```
        h1=real(mu);
        k1=imag(mu);
        a=abs(T(1,2));
        if abs(k)+abs(k1) < 1e-10    %Test for lam and mu to be real.
                part = sqrt((h-h1)^2 + a^2);
                plus = abs(h + h1 + part)/2;
                minus = abs(h + h1 - part)/2;
                w = max([plus, minus]);
        elseif 2*(k^2) >= a*abs(h)
                w=(abs(lam)*sqrt(a^2+4*(k^2)))/(2*abs(k));
        else
                w=abs(h) + a/2;
        end
```

13. Write a MatLab script called elplot that

    - calls for the input of the semi-axis a, parallel to the x-axis

    - calls for the input of the semi-axis b, parallel to the y-axis

    - calls for the input of the center of the ellipse

    - plots the corresponding ellipse in the Graph window

    Save elplot in field.

    Hint: The parametric equations for an ellipse centered at the origin are

    $$x(s) = a\cos(s),$$
    $$y(s) = b\sin(s),$$

    $0 \le s \le 2\pi$. Thus if $z(s) = x(s) + iy(s)$ then the complex numbers $z(s)$ run over the ellipse for $0 \le s \le 2\pi$.

    ```
            %ELPLOT calls for the semi-axes
            %and center and plots the corresponding
            %ellipse.
            z0=input('Enter the center as a complex number z0: ');
    ```

```
a=input('Enter the horizontal semi-axis a: ');
b=input('Enter the vertical semi-axis b: ');
s=0:.2*pi/100:2*pi;
E=a*cos(s) + i*b*sin(s);
E=z0+E;
axis([1 2 3 4]);
axis;
plot(E)
```

You will have to adjust the size of the Graph window.

9.2 Glossary

closed half-plane	529
convex combination	530
convex hull	530
convex set	528
convexity	528
elliptical range theorem	539
envelope	535
field of values	526
$H(X)$	530
negative open half-plane	529
numerical radius	526
numerical range	526
positive open half-plane	529
$\sigma(A)$	528
spectrum	528
$W(A)$	526
$w(A)$	526

9.3 The Toeplitz-Hausdorff Theorem

In §9.2 the principal emphasis was on the set of values taken on by x^*Ax for $A \in M_2(\mathbb{C})$. It is interesting that the general case, $A \in M_n(\mathbb{C})$, can be dealt with in terms of 2-square matrices. A result of this development is the important Toeplitz–Hausdorff theorem concerning the convexity of $W(A)$.

Recall that if $u_1, \ldots, u_p$ are in $M_{n,1}(\mathbb{C})$ then these vectors are orthonormal (o.n.) if

$$u_s^* u_t = \delta_{st}, \quad s, t = 1, \ldots, p,$$

(see §8.2 Theorem 1 et seq.). If $u_1, \ldots, u_p$ are in $M_{n,1}(\mathbb{R})$ then the condition for orthonormality becomes

$$u_s^T u_t = \delta_{st}, \quad s, t = 1, \ldots, p.$$

If $A \in M_n(\mathbb{C})$ and x and v are an o.n. pair of vectors in $M_{n,1}(\mathbb{C})$ then the 2-square matrix

$$A_{xv} = \begin{bmatrix} x^*Ax & x^*Av \\ v^*Ax & v^*Av \end{bmatrix} \quad (1)$$

is called a 2-*dimensional orthogonal compression* of A. The definition (1) will be less obscure if it is realized that A_{xv} is simply a 2-square principal submatrix of a matrix unitarily similar to A (see §9.3 Exercises, #1). The following result is interesting because it shows that the numerical range of any matrix is the union of numerical ranges of 2×2 matrices.

Theorem 1.

Let $A \in M_n(\mathbb{C})$. Then the numerical range $W(A)$ is the union

$$\cup W(A_{xv}) \quad (2)$$

as x and v run over all pairs of o.n. vectors in $M_{n,1}(\mathbb{R})$.

9.3 The Toeplitz-Hausdorff Theorem

Proof.

It is important to note that although $W(A)$ consists of complex numbers of the form y^*Ay, $y \in M_{n,1}(\mathbb{C})$, it is nevertheless the case that only *real* x and v are required in (2).

Let $w = u + iv \in M_{n,1}(\mathbb{C})$, u and v in $M_{n,1}(\mathbb{R})$. Then if $w^*w = 1$ it follows that

$$1 = w^*w = u^T u + v^T v.$$

If $v = ru$ for some $r \in \mathbb{R}$ then $w = u + iru = (1 + ir)u$,

$$w^*Aw = (|1 + ir|u)^T A(|1 + ir|u),$$

and, obviously, $x = |1 + ir|u \in M_{n,1}(\mathbb{R})$ satisfies

$$w^*Aw = x^T Ax. \tag{3}$$

Choose $v \in M_{n,1}(\mathbb{R})$ so that x and v are o.n. (see §7.2 Theorem 1, et seq.). Then clearly (3) states that

$$w^*Aw \in W(A_{xv}). \tag{4}$$

If u is a real multiple of v the argument is identical and results in (4) again.

Thus we can assume that neither u nor v is a multiple of the other. The Gram-Schmidt process of §7.2 can then be applied to the pair of real vectors u, v to obtain the real o.n. pair

$$\gamma = u/c \in M_{n,1}(\mathbb{R}) \tag{5}$$

and

$$\tau = [v - (\gamma^T v)\gamma]/d \in M_{n,1}(\mathbb{R}) \tag{6}$$

where $c = (u^T u)^{1/2} > 0$ and $d = ([v - (\gamma^T v)\gamma]^T [v - (\gamma^T v) \gamma])^{1/2} > 0$ (see §9.3 Exercises, #2). Then (5) and (6) become

$$u = c\gamma \tag{7}$$

and

$$v = (\gamma^T v)\gamma + d\tau . \tag{8}$$

From (7) and (8) we compute that

$$w^* A w = (u + iv)^* A(u + iv)$$

$$= u^* A u + v^* A v + i u^* A v - i v^* A u$$

$$= c^2 \gamma^T A \gamma + [d\tau + (\gamma^T v)\gamma]^T A[d\tau + (\gamma^T v)\gamma]$$

$$+ i(c\gamma)^T A[d\tau + (\gamma^T v)\gamma] - i[d\tau + (\gamma^T v)\gamma]^T A(c\gamma)$$

$$= c^2 \gamma^T A \gamma + d^2 \tau^T A \tau + d(\gamma^T v)\gamma^T A \tau + d(\gamma^T v)\tau^T A \gamma + (\gamma^T v)^2 \gamma^T A \gamma$$

$$+ icd\gamma^T A \tau + ic(\gamma^T v)\gamma^T A \gamma - id c\tau^T A \gamma - ic(\gamma^T v)\gamma^T A \gamma$$

$$= [c^2 + (\gamma^T v)^2]\gamma^T A \gamma + d^2 \tau^T A \tau$$

$$+ [d(\gamma^T v) - idc]\tau^T A \gamma + [d(\gamma^T v) + icd]\gamma^T A \tau. \tag{9}$$

As we noted in (5) and (6), the pair γ and τ are o.n. so that

$$A_{\gamma\tau} = \begin{bmatrix} \gamma^T A \gamma & \gamma^T A \tau \\ \tau^T A \gamma & \tau^T A \tau \end{bmatrix} \tag{10}$$

is a real 2-dimensional orthogonal compression of A. Define

9.3 The Toeplitz-Hausdorff Theorem

$$\xi = \gamma^T v - ic \in \mathbb{C} \tag{11}$$

so that

$$|\xi|^2 = (\gamma^T v)^2 + c^2 \tag{12}$$

and let $y \in M_{2,1}(\mathbb{C})$ be the vector

$$y = [\xi \ d]^T. \tag{13}$$

Then,

$$y^*y = |\xi|^2 + d^2$$

$$= (\gamma^T v)^2 + c^2 + d^2$$

$$= (\gamma^T v)^2 + c^2 + v^T v - (\gamma^T v)^2 - (\gamma^T v)(v^T \gamma) + (\gamma^T v)^2 \gamma^T \gamma$$

$$= u^T u + v^T v \qquad \text{(from } \gamma^T \gamma = 1\text{)}$$

$$= 1. \qquad \text{(from } w^*w = 1\text{)} \tag{14}$$

Also, using (11) we compute that

$$y^* A_{\gamma \tau} y = [\bar{\xi} \ d] \begin{bmatrix} \gamma^T A \gamma & \gamma^T A \tau \\ \tau^T A \gamma & \tau^T A \tau \end{bmatrix} \begin{bmatrix} \xi \\ d \end{bmatrix}$$

$$= |\xi|^2 \gamma^T A \gamma + \bar{\xi} d \gamma^T A \tau + d \xi \tau^T A \gamma + d^2 \tau^T A \tau$$

$$= [(\gamma^T v)^2 + c^2] \gamma^T A \gamma + d^2 \tau^T A \tau$$

$$+ [d(\gamma^T v) - idc] \tau^T A \gamma + [d(\gamma^T v) + icd] \gamma^T A \tau. \tag{15}$$

But the right side of (15) is precisely the right side of (9) and hence

$$w^*Aw = y^*A_{\gamma\tau}\, y, \quad y^*y = 1$$

so that

$$w^*Aw \in W(A_{\gamma\tau}). \tag{16}$$

Thus, from (16) we conclude that $W(A)$ is certainly contained in the union of all $W(A_{xy})$ where x and y are real o.n. vectors in $M_{n,1}(\mathbb{R})$. On the other hand, by §9.3 Exercises, #1, any A_{xy} is a 2-square principal submatrix of a matrix orthogonally similar to A. We can then use §9.2 formulas (9) and (10), to complete the proof. ∎

The next result is the single most important geometric fact concerning the numerical range of a general matrix.

Theorem 2 (Toeplitz-Hausdorff Theorem).

For any $A \in M_n(\mathbb{C})$, $W(A)$ is a convex subset of $\mathbb{C}$.

Proof.

The idea of the proof is simple. Given any two vectors x and y in $M_{n,1}(\mathbb{C})$, $x^*x = y^*y = 1$, we want to prove that the line segment joining x^*Ax and y^*Ay lies entirely in $W(A)$. What we will do is show that both x^*Ax and y^*Ay lie in an appropriate $W(A_{xy})$ where A_{xy} is a two dimensional orthogonal compression of A. By §9.2 Theorem 4, the elliptical range theorem, we know that $W(A_{xy})$ is convex so that the line segment $[x^*Ax, y^*Ay]$ is a subset of $W(A_{xy})$. But by §9.3 Exercises, #1, A_{xy} is a principal submatrix of a matrix unitarily similar to A and hence by §9.2 formulas (9) and (10),

$$[x^*Ax, y^*Ay] \subset W(A). \tag{17}$$

Thus our task is to construct A_{xy}. But this is easy. If y is a multiple of x, $y^*Ay = x^*Ax$ (why?) and the line segment in (17) is the single point x^*Ax in $W(A)$. If neither x nor y is a multiple of the other, then the Gram-Schmidt process (see §7.2 Theorem 1 et seq.) can be applied to produce the pair of o.n. vectors x and

9.3 The Toeplitz-Hausdorff Theorem

$$v = (y - (x^*y)x)/d, \tag{18}$$

where

$$d = ([y - (x^*y)x]^* [y - (x^*y)x])^{1/2} > 0. \tag{19}$$

Let $B = A_{xv}$ and define $z \in M_{2,1}(\mathbb{C})$ by

$$z = \begin{bmatrix} x^*y \\ d \end{bmatrix}. \tag{20}$$

We can then compute from (19) that

$$d^2 = y^*y - (x^*y)(y^*x) - \overline{(x^*y)}(x^*y) + |(x^*y)|^2$$

$$= 1 - |(x^*y)|^2. \tag{21}$$

Hence from (20) it follows that

$$z^*z = 1. \tag{22}$$

From (18), $y = (x^*y)x + dv$ and hence

$$y^*Ay = [(x^*y)x + dv]^* A[(x^*y)x + dv]$$

$$= |(x^*y)|^2 x^*Ax + \overline{(x^*y)} \, dx^*Av + d(x^*y)v^*Ax + d^2 v^*Av. \tag{23}$$

But

$$z^*Bz = z^*A_{xv}z$$

$$= \begin{bmatrix} \overline{x^*y} & d \end{bmatrix} \begin{bmatrix} x^*Ax & x^*Av \\ v^*Ax & v^*Av \end{bmatrix} \begin{bmatrix} x^*y \\ d \end{bmatrix}$$

$$= |(x^*y)|^2 x^*Ax + \overline{(x^*y)} \, dx^*Av + d(x^*y)v^*Ax + d^2 v^*Av. \tag{24}$$

The right sides of (23) and (24) are the same and hence

$$y^*Ay = z^*A_{xv}z$$

$$\in W(A_{xv}).$$

It is obvious (why?) that $x^*Ax \in W(A_{xv})$ and the proof is complete. ∎

The following result can be used to illustrate the rich variety of plane point-sets that can occur as numerical ranges.

Theorem 3.

Let $A = B \oplus C \in M_n(\mathbb{C})$ be a direct sum of two square matrices B and C. Then

$$W(A) = H(W(B) \cup W(C)). \tag{25}$$

In other words, W(A) is the smallest convex set containing both W(B) and W(C).

Proof.

Assume that B is p-square and C is q-square where $p + q = n$. If $x \in M_{n,1}(\mathbb{C})$ let $u \in M_{p,1}(\mathbb{C})$ and $v \in M_{q,1}(\mathbb{C})$ consist respectively of the first p components of x and the remaining q components of x. By block multiplication,

$$x^*Ax = u^*Bu + v^*Cv. \tag{26}$$

If $x^*x = 1$ then $1 = x^*x = u^*u + v^*v$ and hence letting $u_0 = u/(u^*u)^{1/2}$ and $v_0 = v/(v^*v)^{1/2}$, (26) becomes

$$x^*Ax = (u^*u)u_0^*Bu_0 + (v^*v)v_0^*Cv_0. \tag{27}$$

9.3 The Toeplitz-Hausdorff Theorem

Moreover, $u_0^*Bu_0 \in W(B)$ and $v_0^*Cv_0 \in W(C)$. The computation (27) is valid if $u \neq 0$ and $v \neq 0$. Of course, if $u = 0$ then $v^*v = 1$ and $x^*Ax = v^*Cv \in W(A)$. Similarly if $v = 0$ then $u^*u = 1$ and $x^*Ax = u^*Bu \in W(B)$. If (27) holds then x^*Ax is seen to be a convex combination of $u_0^*Bu_0 \in W(B)$ and $v_0^*Cv_0 \in W(C)$ and hence in $H(W(B) \cup W(C))$. We have confirmed that

$$W(A) \subset H(W(B) \cup W(C)). \tag{28}$$

On the other hand, §9.2 formulas (9), (10) imply that $W(B) \subset W(A)$ and $W(C) \subset W(A)$ so that

$$W(B) \cup W(C) \subset W(A).$$

Since $W(A)$ is convex by Theorem 2, the inclusion

$$H(W(B) \cup W(C)) \subset W(A) \tag{29}$$

also holds. Combining (28) and (29) produces (25). ∎

As an example, consider the matrix

$$A = \begin{bmatrix} 0 & 1 & 0 & 0 & 0 \\ 0 & 0 & 1 & 0 & 0 \\ 1 & 0 & 0 & 0 & 0 \\ 0 & 0 & 0 & 0 & 1 \\ 0 & 0 & 0 & 0 & 0 \end{bmatrix}. \tag{30}$$

The matrix A in (30) is the direct sum of

$$B = \begin{bmatrix} 0 & 1 & 0 \\ 0 & 0 & 1 \\ 1 & 0 & 0 \end{bmatrix}$$

and

$$C = \begin{bmatrix} 0 & 1 \\ 0 & 0 \end{bmatrix}. \tag{31}$$

From the narrative in §9.2, Fig. 2, W(B) is an equilateral triangle with vertices at 1, $e^{i2\pi/3}$, $e^{i4\pi/3}$. Also, W(C) is a disk of radius 1/2 centered at the origin. It is not difficult to check that W(C) ⊂ W(B) and hence, by Theorem 3, W(A) is the triangle W(B). Notice that A is not a normal matrix. Nevertheless, in conformity with §9.2 Theorem 2, W(A) is the convex hull of the spectrum σ(A). It follows from this example that W(A) = H(σ(A)) is not a sufficient condition to conclude the normality of A. Notice however that each vertex of W(A) is an eigenvalue of A. There is an interesting result about the vertices of W(A) proved by W. F. Donoghue in 1957. In order to state Donoghue's result it is necessary to understand what is meant by the *boundary* of a convex set S in the plane. A point P ∈ S is a boundary point of S if it is not in the interior of any line segment joining two points that are strictly inside of S. We remark that this is not the most general definition possible for a boundary point, but it is certainly correct for all situations encountered in studying the numerical range. The totality of boundary points of S is denoted traditionally by ∂S.

Notice for the matrix A in the narrative for §9.2, Fig. 2 that 1, $e^{i2\pi/3}$, $e^{i4\pi/3}$ are points in ∂W(A) at which W(A) does not have a tangent line. In fact, every other point of ∂W(A) has a tangent line. On the other hand, the set ∂W(C) for the matrix in (31) consists of the entire circle of radius 1/2 centered at the origin, and a tangent line obviously exists for each point on this circle.

Theorem 4 (Donoghue's Theorem).

Let $A \in M_n(\mathbb{C})$. If z ∈ ∂W(A) and ∂W(A) does not have a tangent line at z then z is an eigenvalue of A.

Proof.

The assumption that z ∈ ∂W(A) and ∂W(A) does not have a tangent line at z means that no non-degenerate ellipse entirely contained in W(A) can contain or pass through z (see Fig. 1)

9.3 The Toeplitz-Hausdorff Theorem

Fig. 1

Now let $z = x^*Ax$, $x^*x = 1$, let x be the first column of a unitary matrix U, and note that

$$z = (U^*AU)_{11}.$$

Since $W(U^*AU) = W(A)$ and $\sigma(U^*AU) = \sigma(A)$, we shall assume that z is the 1,1 entry of A itself. Consider B, a 2-square principal submatrix of A with z as its 1,1 entry. For simplicity we take $B = A[1, 2 \mid 1, 2]$ in the following argument with no loss of generality. Thus

$$B = \begin{bmatrix} z & a_{12} \\ a_{21} & a_{22} \end{bmatrix}. \tag{32}$$

Clearly $z \in W(B) \subset W(A)$. If z were not also on $\partial W(B)$ it would be in the interior of a line segment joining two points in $W(B)$ and hence in the interior of $W(A)$. It follows that $z \in \partial W(B)$. Moreover, $W(B)$ cannot be a non-degenerate elliptical region as noted above in Fig. 1. Thus $W(B)$ is a line segment, possibly a point, and $z \in \partial W(B)$, i.e., it is an endpoint of the line segment $W(B)$. On the other hand, we know that the foci of $W(B)$ are the eigenvalues of B. Thus z must be an eigenvalue of B. But we also know from §9.2 Theorem 4 that B is unitarily similar to an upper triangular matrix whose 1, 2 entry is the length of the minor axis of $W(B)$. Since $W(B)$ is a line segment (or point), this 1, 2 entry must be 0. Thus B is unitarily similar to a diagonal matrix and must therefore be a normal matrix. To summarize so far:

$$B = \begin{bmatrix} z & a_{12} \\ a_{21} & a_{22} \end{bmatrix} \tag{33}$$

is normal, and $z \in \sigma(B) = \{z, \mu\}$. We know that

$$\mathrm{tr}(B) = z + \mu$$

so that $a_{22} = \mu$, i.e., the eigenvalues of the normal matrix B are z and a_{22}. Since B is unitarily similar to $\mathrm{diag}(z, a_{22})$

$$|z|^2 + |a_{22}|^2 = \mathrm{tr}(B^*B)$$
$$= |z|^2 + |a_{22}|^2 + |a_{12}|^2 + |a_{21}|^2$$

and hence

$$a_{12} = a_{21} = 0.$$

Precisely the same argument shows that

$$a_{1j} = a_{j1} = 0, \quad j = 2, \ldots, n,$$

by replacing $A[1, 2 \mid 1, 2]$ with $A[1, j \mid 1, j]$. Thus

$$A = z \oplus A(1 \mid 1)$$

and hence z is an eigenvalue of A. ■

A point of $\partial W(A)$ at which there is no tangent line is called an *exceptional point* of $\partial W(A)$. If S is any convex set of points in the plane, then a line L is called a *support line* of S if L contains at least one point $p \in \partial S$ and S itself is contained in one of the closed half-planes defined by L. The closed half-plane containing S is called a *supporting half-plane*. Thus a support line of S is a tangent line to ∂S if such exists.

9.3 The Toeplitz-Hausdorff Theorem

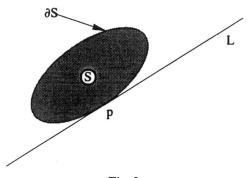

Fig. 2

If S is a convex set in $\mathbb{C}$ which is closed, i.e., contains all its limit points, and which is bounded, then S is the intersection of all its supporting half-planes (see Fig. 3).

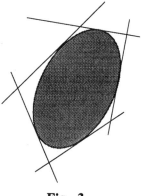

Fig. 3

This fact is geometrically evident and it is not difficult to prove. We shall not do so here. However, the geometry of this situation is very useful in developing an algorithm for constructing W(A). Thus, let W(A) be the numerical range of an arbitrary $A \in M_n(\mathbb{C})$.

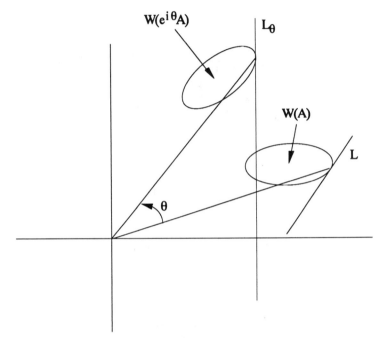

Fig. 4

The idea is simple: we want to construct a relatively dense set of support lines for W(A). Then W(A) will be accurately depicted as the intersection of the corresponding support half-planes. In fact, simply drawing a sufficiently dense set of such support lines will define W(A) with accuracy.

Of course, the problem is to devise a computationally reasonable method of determining the support lines L. We will make our method depend on computing the smallest and largest eigenvalues of a sequence of appropriate hermitian matrices. Thus let L be a fixed but arbitrary support line for W(A). Then perform a counterclockwise rotation around the origin in the plane through an angle θ chosen so that the rotated image of L, call it L_θ, is perpendicular to the x-axis (see Fig. 4). Clearly, each such support line L determines a unique L_θ and, conversely, each such vertical line L_θ can be rotated clockwise into the support line L. Now, L_θ is a support line for $W(e^{i\theta}A) = e^{i\theta}W(A)$. Write $A(\theta) = e^{i\theta}A$ and let $H(\theta)$ and $K(\theta)$ be the hermitian parts of $A(\theta)$:

$$A(\theta) = H(\theta) + iK(\theta). \qquad (34)$$

9.3 The Toeplitz-Hausdorff Theorem

Then if $u^*u = 1$,

$$u^*A(\theta)u = u^*H(\theta)u + iu^*K(\theta)u$$

and

$$\text{Re}(u^*A(\theta)u) = u^*H(\theta)u .$$

Hence

$$\max_{u^*u=1} \text{Re}(u^*A(\theta)u) = \max_{u^*u=1} u^*H(\theta)u . \qquad (35)$$

But $H(\theta)$ is hermitian and §8.3 Theorem 1 can be applied to conclude that

$$\max_{u^*u=1} u^*H(\theta)u = h_M(\theta) \qquad (36)$$

where h_M is the largest eigenvalue of $H(\theta)$. In fact, the maximizing u in (36) must be an eigenvector of $H(\theta)$ corresponding to $h_M(\theta)$. Assume that it is feasible to compute $h_M(\theta)$. Then the equation of L_θ in Fig. 4 is obviously

$$x = h_M(\theta) . \qquad (37)$$

By an almost exact repetition of the preceding argument, we can very easily construct a rectangle whose sides are parallel to the coordinate axes and that snugly fits around $W(e^{i\theta}A) = W(A(\theta))$ as depicted in Fig. 5.

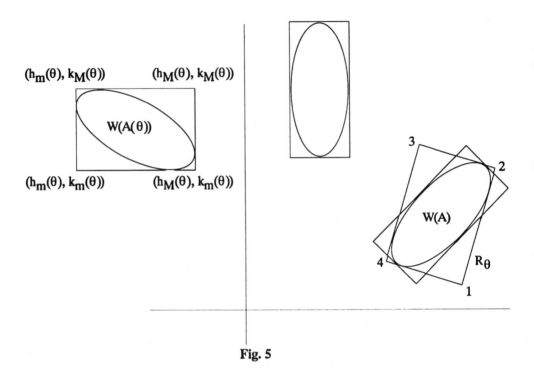

Fig. 5

In Fig. 5, $h_M(\theta)$ and $h_m(\theta)$ are the largest and smallest eigenvalues of $H(\theta)$; $k_M(\theta)$ and $k_m(\theta)$ are the largest and smallest eigenvalues of $K(\theta)$, respectively. If we imagine the rectangle surrounding $W(A(\theta))$ in Fig. 5 rotated clockwise through the angle θ, the resulting rectangle, labelled R_θ, will, of course, snugly fit around $W(A)$. The vertices of R_θ, designated 1, 2, 3, and 4 in Fig. 5, have coordinates (in that order):

$$(x_1(\theta), y_1(\theta)) = (\cos(\theta)h_M(\theta) + \sin(\theta)k_M(\theta), -\sin(\theta)h_M(\theta) + \cos(\theta)k_M(\theta)), \quad (38)$$

$$(x_2(\theta), y_2(\theta)) = (\cos(\theta)h_m(\theta) + \sin(\theta)k_M(\theta), -\sin(\theta)h_m(\theta) + \cos(\theta)k_M(\theta)), \quad (39)$$

$$(x_3(\theta), y_3(\theta)) = (\cos(\theta)h_m(\theta) + \sin(\theta)k_m(\theta), -\sin(\theta)h_m(\theta) + \cos(\theta)k_m(\theta)), \quad (40)$$

$$(x_4(\theta), y_4(\theta)) = (\cos(\theta)h_M(\theta) + \sin(\theta)k_m(\theta), -\sin(\theta)h_M(\theta) + \cos(\theta)k_m(\theta)). \quad (41)$$

Suppose then that we specify a sequence of N values of θ,

9.3 The Toeplitz-Hausdorff Theorem

$$\theta_k = k\frac{2\pi}{N}, \quad k = 0, \ldots, N-1. \tag{42}$$

With each such θ_k we can then construct the rectangle containing $W(A)$ whose vertices are the points (38) - (41). Then $W(A)$ can be depicted as the intersection of these N rectangular regions. If N is large, then we will obtain a reasonably accurate representation of $W(A)$. We will pursue this line of thought in §9.3 MatLab.

The method for depicting $W(A)$ just described can also be used to determine the numerical radius $w(A)$. In fact, we can easily verify that

$$w(A) = \max_{\theta \in [0, 2\pi]} h_M(\theta). \tag{43}$$

To prove (43), first let $u_0 \in M_{n,1}(\mathbb{C})$ be chosen so that $u_0^* u_0 = 1$ and

$$|u_0^* A u_0| = w(A). \tag{44}$$

Then

$$w(A) = e^{i\theta_0} u_0^* A u_0 \tag{45}$$

for some $\theta_0 \in [0, 2\pi]$. We have

$$\begin{aligned}
h_M(\theta_0) &= \max_{u^*u=1} u^* H(\theta_0) u \\
&= \max_{u^*u=1} \text{Re}(u^* A(\theta_0) u) \\
&\geq \text{Re}(u_0^* A(\theta_0) u_0) \\
&= \text{Re}(u_0^* (e^{i\theta_0} A) u_0) \\
&= \text{Re}(e^{i\theta_0} u_0^* A u_0) \\
&= \text{Re}(w(A)) \qquad \text{(from (45))} \\
&= w(A).
\end{aligned}$$

Hence $h_M(\theta_0) \geq w(A)$ so that

$$\max_{\theta \in [0, 2\pi]} h_M(\theta) \geq h_M(\theta_0)$$

$$\geq w(A). \qquad (46)$$

On the other hand,

$$h_M(\theta) = \max_{u^*u=1} u^* H(\theta) u$$

$$= \max_{u^*u=1} \text{Re}(u^* A(\theta) u) \qquad \text{(from (35))}$$

$$= \max_{u^*u=1} \text{Re}(u^* (e^{i\theta} A) u)$$

$$= \max_{u^*u=1} \text{Re}(e^{i\theta} u^* A u)$$

$$\leq \max_{u^*u=1} | e^{i\theta} u^* A u |$$

$$= \max_{u^*u=1} | u^* A u |$$

$$= w(A),$$

and hence

$$\max_{\theta \in [0, 2\pi]} h_M(\theta) \leq w(A). \qquad (47)$$

Combining (46) and (47) produces (43).

In Theorem 1 we saw that the numerical range of an arbitrary $A \in M_n(\mathbb{C})$ is the union of the numerical ranges of all 2-square principal submatrices of matrices orthogonally similar to A. The fact that real orthogonal matrices can be used has important simplifying implications in designing software to exhibit W(A) graphically and compute w(A). However, we will not pursue this approach here.

9.3 The Toeplitz-Hausdorff Theorem

Our final result in this section will state that any real normal matrix can be reduced to a simple "nearly" diagonal form by an orthogonal similarity. This fact, combined with Theorem 3, allows us to graphically depict a rich variety of sets that are numerical ranges of real normal matrices.

Let A be an n-square real normal matrix with nonreal eigenvalues $a_t \pm ib_t$, ($b_t \neq 0$), $t = 1, \ldots, q$, and real eigenvalues $\alpha_1, \ldots, \alpha_{n-2q}$. The *real normal form* of A is the matrix

$$\text{diag}(\alpha_1, \ldots, \alpha_{n-2q}) \oplus \sum_{t=1}^{q} \oplus \begin{bmatrix} a_t & -b_t \\ b_t & a_t \end{bmatrix}. \tag{48}$$

The main result concerning real normal matrices is the following.

Theorem 5. Let A and B be commuting real normal matrices. Then there exists a real orthogonal matrix Q such that $Q^{-1}AQ$ is in real normal form and $Q^{-1}BQ$ can be obtained from the real normal form of B by a reordering on the main diagonal of the 2 × 2 blocks and the real eigenvalues.

We will omit the proof of Theorem 5. The proof is not particularly difficult but it is rather long and somewhat specialized. However, we remark that the 2 × 2 and 1 × 1 blocks in the normal form for A can be made to appear in any specified order. In general, however, we cannot be assured that the blocks in B also occur in any particular order. However, it is not difficult to show that if

$$\begin{bmatrix} a & -b \\ b & a \end{bmatrix}$$

occurs in a particular block position on the main diagonal of a normal form for A (i.e., a direct sum of 2 × 2 and 1 × 1 blocks), then in the corresponding block position in a normal form for B we will have either a real block of the form

$$\begin{bmatrix} \beta & 0 \\ 0 & \beta \end{bmatrix}$$

or a block of the form

$$\begin{bmatrix} c & -d \\ d & c \end{bmatrix},$$

$d \neq 0$.

9.3 Exercises

1. Prove that the matrix in §9.3 formula (1) is a 2-square principal submatrix of a matrix unitarily similar to A.

   Hint: We know that a pair of o.n. vectors x and v can be embedded as the first two columns in a unitary matrix U. In fact, if x and v are real then U can be chosen as a real orthogonal matrix (because the Gram-Schmidt process does not leave $\mathbb{R}$ once it starts in $\mathbb{R}$). Now

   $$(U^*AU)_{11} = x^*Ax,$$

   $$(U^*AU)_{12} = x^*Av,$$

   $$(U^*AU)_{21} = v^*Ax,$$

   $$(U^*AU)_{22} = v^*Av.$$

   Hence

   $$A_{xv} = (U^*AU)[1, 2 \mid 1, 2].$$

2. Let u and v be a pair of vectors in $M_{n,1}(\mathbb{C})$. Assume that neither u nor v is a multiple of the other. Define in succession

$$c = (u^*u)^{1/2} > 0,$$

$$\gamma = u/c,$$

$$w = v - (\gamma^*v)\gamma,$$

$$d = (w^*w)^{1/2} > 0,$$

$$\tau = w/d.$$

Prove that γ and τ are o.n.

3. Let $A \in M_n(\mathbb{C})$ and let $\sigma(A) = \{\lambda_1, ..., \lambda_n\}$, the spectrum of A. Prove: if the λ_t are distinct and each λ_t is on $\partial W(A)$ then A is normal.

   Hint: Let $U^*AU = T$ where T is upper triangular with $\lambda_1, ..., \lambda_n$ appearing down the main diagonal of T. Let $i < j$ and define $B = T[i, j | i, j]$:

   $$B = \begin{bmatrix} \lambda_i & t_{ij} \\ 0 & \lambda_j \end{bmatrix}.$$

   Since $\lambda_i \in \partial W(A)$, it follows that no non-degenerate elliptical subset of W(A) can contain λ_i in its interior. Hence W(B) must be degenerate. It follows that $t_{ij} = 0$ (why? see §9.2 formula (54) et seq.). Thus all the off-diagonal entries of T are 0, i.e., T is diagonal, and it follows that A is normal. Note that only n - 1 of the λ_t need be points of $\partial W(A)$ and precisely the same result holds.

4. Prove: if $A = \sum_{k=1}^{p} \oplus A_k$ then $W(A) = H(\bigcup_{k=1}^{p} W(A_k))$.

   Hint: Use Theorem 3 and induction on p.

5. Exhibit a 3-square real matrix whose numerical range is

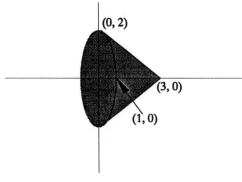

Fig. 1

Hint: From Theorem 3 it suffices to find a 2-square real matrix A whose numerical range is the indicated elliptical region. For then A ⊕ [3] is the required matrix. The foci of the ellipse are easily computed to be $\pm \sqrt{3}\, i$. The matrix

$$B = \begin{bmatrix} \sqrt{3}\, i & 2 \\ 0 & -\sqrt{3}\, i \end{bmatrix}$$

has as its numerical range the required elliptical region, i.e., $\alpha = 2$ is the length of the minor axis in §9.2 Theorem 4 and $\alpha/2 = 1$ is the semi-minor axis. let

$$C = \frac{1}{\sqrt{3}} B$$

$$= \begin{bmatrix} i & \frac{2}{\sqrt{3}} \\ 0 & -i \end{bmatrix}.$$

By §9.2 Exercises, #23 the real matrix

$$R = \begin{bmatrix} \frac{1}{\sqrt{3}} & -\frac{(4 + (2/\sqrt{3})^2)^{1/2}}{2} \\ \frac{(4 + (2/\sqrt{3})^2)^{1/2}}{2} & -\frac{1}{\sqrt{3}} \end{bmatrix}$$

9.3 Exercises

$$= \begin{bmatrix} \frac{1}{\sqrt{3}} & -\frac{2}{\sqrt{3}} \\ \frac{2}{\sqrt{3}} & -\frac{1}{\sqrt{3}} \end{bmatrix}$$

$$= \frac{1}{\sqrt{3}} \begin{bmatrix} 1 & -2 \\ 2 & -1 \end{bmatrix}$$

is unitarily similar to C. But then

$$W(R) = W(C) = W\left(\frac{1}{\sqrt{3}} B\right) = \frac{1}{\sqrt{3}} W(B)$$

or

$$W(B) = \sqrt{3}\, W(R) = W(\sqrt{3}\, R).$$

The required matrix is

$$A = \sqrt{3}\, R$$

$$= \begin{bmatrix} 1 & -2 \\ 2 & -1 \end{bmatrix}$$

and hence the matrix

$$\begin{bmatrix} 1 & -2 & 0 \\ 2 & -1 & 0 \\ 0 & 0 & 3 \end{bmatrix}$$

has the depicted numerical range.

6. Let X and Y be two convex sets in $\mathbb{C}$. Show that any point $z \in H(X \cup Y)$ is of the form $z = (1 - s)x + sy$ for appropriate $x \in X$ and $y \in Y$, and $0 \leq s \leq 1$.

 Hint: Any point $z \in H(X \cup Y)$ is by definition a convex combination of points $x_1, \ldots, x_p$ in X and $y_1, \ldots, y_q$ in Y, say

$$z = \sum_{t=1}^{p} \sigma_t x_t + \sum_{t=1}^{q} \tau_t y_t .$$

Let

$$s = \sum_{t=1}^{q} \tau_t$$

so that

$$1 - s = \sum_{t=1}^{p} \sigma_t .$$

If either s or $1 - s$ is 0 then

$$z = \sum_{t=1}^{p} \sigma_t x_t = x \in X$$

or

$$z = \sum_{t=1}^{p} \tau_t y_t = y \in Y .$$

Otherwise, it is easy to check that

$$z = (1 - s) \sum_{t=1}^{p} \frac{\sigma_t}{1 - s} x_t + s \sum_{t=1}^{q} \frac{\tau_t}{s} y_t = (1 - s)x + sy .$$

The convexity of X and Y ensures that $x \in X$ and $y \in Y$.

7. Let $A \in M_n(\mathbb{C})$ and assume that $A = B \oplus C$. Using the results of this section prove that:

$$w(A) = \max\{w(B), w(C)\} .$$

Hint: Let $z \in W(A)$ be a point for which $|z| = w(A)$. By Theorem 3 and #6 there exist points $p \in W(B)$, $q \in W(C)$ and a number s, $0 \leq s \leq 1$, such that

$z = (1 - s)p + sq$. Then $w(A) = |z| \leq (1 - s)|p| + s|q| \leq (1 - s)w(B) + sw(C) \leq \max\{w(B), w(C)\}$. On the other hand, $W(B) \subset W(A)$ and $W(C) \subset W(A)$, so that $w(B) \leq w(A)$, $w(C) \leq w(A)$ and $\max\{w(B), w(C)\} \leq w(A)$.

8. Find formulas for $H(\theta)$ and $K(\theta)$ in (34).

   Hint: We compute that

   $$A(\theta) = e^{i\theta}(H + iK) = (\cos \theta + i \sin \theta)(H + iK)$$

   $$= ((\cos \theta)H - (\sin \theta)K) + i ((\sin \theta)H + (\cos \theta)K).$$

   Then by the uniqueness of the decomposition into hermitian parts,

   $$H(\theta) = (\cos \theta)H - (\sin \theta)K, \ K(\theta) = (\sin \theta)H + (\cos \theta)K.$$

9. Let A and B be matrices in $M_n(\mathbb{C})$. Prove that there exists a nonzero $v \in M_{n, 1}(\mathbb{C})$ such that Av is a multiple of Bv or Bv is a multiple of Av.

   Hint: If A is singular choose v such that $Av = 0$. If A is not singular, consider the equation

   $$\det(\lambda A - B) = 0.$$

   This equation is equivalent to

   $$\det(\lambda I_n - A^{-1}B) = 0.$$

   Let λ_0 be an eigenvalue of $A^{-1}B$ with corresponding eigenvector v. Then

   $$A^{-1}Bv = \lambda_0 v,$$

   $$Bv = \lambda_0 Av.$$

10. Let A and B be matrices in $M_n(\mathbb{C})$. Show that there exist unitary matrices Q and W in $M_n(\mathbb{C})$ such that QAW and QBW are both upper triangular.

Hint: By #9, let $v_1 \neq 0$ be such that one of Av_1 and Bv_1 is a multiple of the other and assume $v_1^* v_1 = 1$. Choose a unitary matrix U such that the first column of U is a multiple of both Av_1 and Bv_1. Also, choose a unitary V whose first column is v_1. Then

$$(U^*AV)_{k,1} = U^{(k)*}(Av_1) = 0, \quad k \geq 2,$$

and

$$(U^*BV)_{k,1} = U^{(k)*}(Bv_1) = 0, \quad k \geq 2.$$

In other words,

$$U^*AV = \begin{bmatrix} \alpha & x \\ 0 & \\ \vdots & A_1 \\ 0 & \end{bmatrix}$$

and

$$U^*BV = \begin{bmatrix} \beta & y \\ 0 & \\ \vdots & B_1 \\ 0 & \end{bmatrix}$$

for appropriate numbers α and β and rows x and y. The remainder of the proof uses induction on n and block multiplication.

9.3 MatLab

1. Note that if z = (a, b) is a point in the cartesian plane, then the point w = (a cos θ + b sin θ, - a sin θ + b cos θ) is obtained from z by rotating z around the origin in a clockwise direction through an angle θ. Use this fact to write a MatLab function named rotr that

    - accepts two real n-tuples xth and yth and an angle th (in radians)

    - produces the pair of n-tuples x and y for which (x(t), y(t)) is obtained from (xth(t), yth(t)), t = 1, ..., n, by a clockwise rotation around the origin through an angle th.

    Save rotr in the folder field.

    Hint:
    ```
    function [x, y] = rotr(xth, yth, th)
    %ROTR accepts two n-tuples xth and yth
    %and an angle th and returns the n-tuples
    %x and y for which (x(t), y(t)) is obtained from
    %(xth(t), yth(t)) by a clockwise rotation thru th.
    x = cos(th)*xth + sin(th)*yth;
    y = -sin(th)*xth + cos(th)*yth;
    ```

2. Review the Help entries for axis, sign, max, min, and plot(X, Y). Then write a function named axel that

    - accepts two m × n real matrices X and Y

    - computes m (M) the least (largest) entry in X or Y

    - produces a 1 × 4 matrix ax = [L R L R] where L = (1−.05*sign(m))*m, R = (1+.05*sign(M))*M.

    The purpose of this function is to provide an axis scaling to accommodate graphing X against Y.

Hint:

```
function ax = axel(X, Y)
%AXEL(X, Y) delivers a 1 x 4 matrix to be used in the
%axis command, i.e., axis(ax), in graphing X against Y.
m = min(min([X  Y]));
M = max(max([X  Y]));
L = (1 - .05*sign(m))*m;
R = (1 + .05*sign(M))*M;
ax = [L  R  L  R];
```

3. Review the Help entries for plot and axis. Then write a MatLab script named swing that

- calls for the input of 4 points z1, z2, z3, z4 in the Cartesian plane and an angle th (in radians)

- draws the quadrilateral Q determined by z1, z2, z3, z4 (in that order)

- draws in the same Graph window the quadrilateral determined by rotating Q clockwise around the origin through the angle th.

Write swing so that the plot command is used only once in the script. Also use the function axel in #2 to scale the axes in the graph so that the quadrilaterals appear in proper scale. The following is a typical dialog and output.

Enter a 1 x 4 complex matrix: [1+5*i   4+5*i   4+2*i   1+3*i]

Enter an angle th in radians: pi

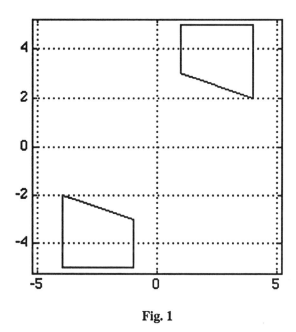

Fig. 1

Save swing in field.

Hint:

%SWING calls for the input of 4 points as a 1 x 4 complex
%matrix as well as an angle th. It then graphs the quadrilateral
%Q determined by the four points together with the quadrilateral
%obtained by rotating Q clockwise around the origin through th.
Z = input('Enter a 1 x 4 complex matrix: ');
th = input('Enter an angle th in radians: ');
Q = [Z Z(1)];
W = exp(-i*th)*Q;
A = [Q.' W.'];
ax = axel(real(A), imag(A));
axis('square');
axis(ax);
plot(A);

4. What should the inputs to swing be in order that the graphical output have the form (a) or (b)?

(a)

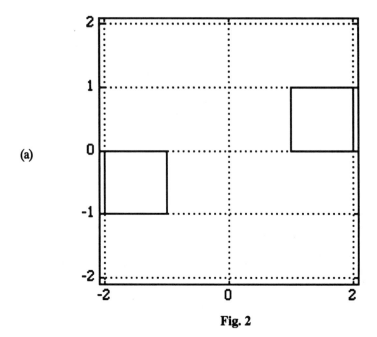

Fig. 2

(b)

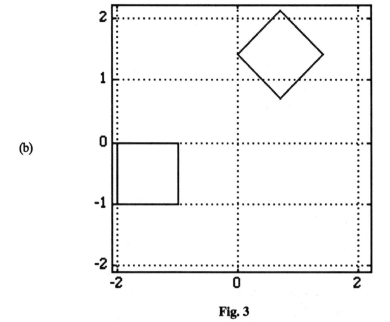

Fig. 3

Hint: (a) Z = [-1 -2 -2 - i -1 - i], th = π.

(b) Z = [-1 -2 -2 - i -1 - i], th = 3π/4.

5. Review the Help entries for eig, min, and max. Then write a MatLab function named rect that

- accepts an n-square complex matrix A

- computes the hermitian parts of A, A = H + iK

- computes hM, hm, kM, km, the maximum and minimum eigenvalues of H and K, respectively

- returns two 5-tuples

$$x = [hM\ hm\ hm\ hM\ hM]$$
and
$$y = [kM\ kM\ km\ km\ kM].$$

The purpose of this function is to provide the vertices of the rectangle depicted in Fig. 5 in the narrative part of §9.3. Save rect in field.

Hint:
```
function [x, y]=rect(A)
%RECT returns two 5-tuples used to determine
%the enclosing rectangle for W(A(theta)). The vertices are
%(hM, kM), (hm, kM), (hm, km) and (hM, km).
H = (A + A')/2;
K = (A - A')/(2*i);
evh = real(eig(H));
evk = real(eig(K));
hM = max (evh);
hm = min(evh);
kM = max(evk);
km = min(evk);
```

x = [hM hm hm hM hM]';
y = [kM kM km km kM]';

6. In this problem we again refer to Fig. 5 in the narrative part of §9.3. Write a MatLab script called numrange that

- calls for the input of an n-square complex matrix A

- calls for the input of an integer N

- constructs each rectangle R_θ as θ varies from 0 to 2π in increments of $2\pi/N$ and holds it in the Graph window.

Use rotr in #1 and rect in #5. Also, obtain the time elapsed in executing the program. Save numrng in field.

Hint:
```
%NUMRANGE calls for the input of an n-square
%complex matrix A and a positive integer N.
%The program then plots the numerical range W(A)
%as the intersection of a family of N enclosing
%rectangles.
A = input('Enter a complex matrix A: ');
N = input('Enter the number of rectangles N: ');
t0 = clock;
X = [ ]; Y = [ ];
for th = 0:2*pi/N:2*pi
        [xth, yth] = rect(exp(i*th)*A);
        [x y] = rotr(xth, yth, th);
        X = [X x]; Y = [Y y];
end
axis('square');
plot(X, Y);
etime(clock, t0)
```

In each of the problems #7 - 11 the input and output of numrange for various matrices A is depicted. Explain the geometric properties of each figure.

7.

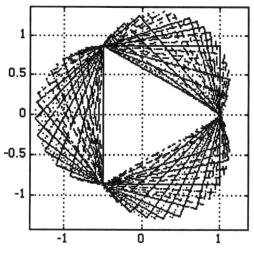

Fig. 4

$$A = \begin{bmatrix} 0 & 1 & 0 \\ 0 & 0 & 1 \\ 1 & 0 & 0 \end{bmatrix}, \quad N = 50$$

8.

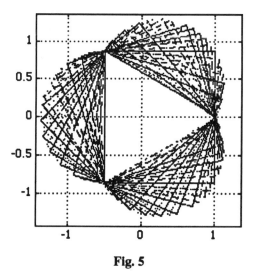

Fig. 5

$$A = \begin{bmatrix} 0 & 1 & 0 & 0 & 0 \\ 0 & 0 & 1 & 0 & 0 \\ 1 & 0 & 0 & 0 & 0 \\ 0 & 0 & 0 & 0 & 1/2 \\ 0 & 0 & 0 & 0 & 0 \end{bmatrix}, \quad N = 50$$

9.

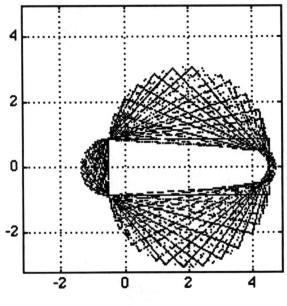

Fig. 6

$$A = \begin{bmatrix} 0 & 1 & 0 & 0 & 0 \\ 0 & 0 & 1 & 0 & 0 \\ 1 & 0 & 0 & 0 & 0 \\ 0 & 0 & 0 & 4 & 1 \\ 0 & 0 & 0 & 0 & 4 \end{bmatrix}, \quad N = 50$$

10.

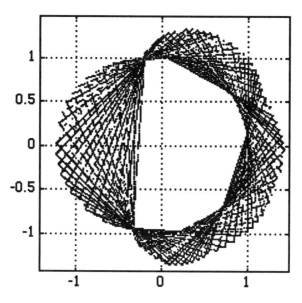

Fig. 7

A = urand(7),   N = 100

11.

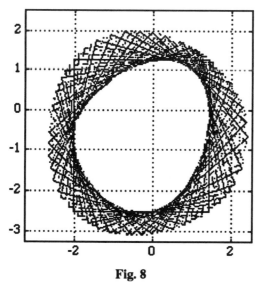

Fig. 8

A = arand(5, 5),   N = 50

12. Write a MatLab function named nrad4 that

   - accepts an n-square matrix A and a positive integer N

   - returns an approximation to the numerical radius w(A) based on formula (43) in the narrative of §9.3; the interval $[0, 2\pi]$ is divided into N equal subintervals.

   Save nrad4 in field.

   Hint:
   ```
   function w = nrad4(A, N)
   %NRAD4 accepts an n-square complex matrix
   %A and a positive integer N.  It returns an
   %approximation to w(A).
   hM = [ ];
   for k = 0:(N- 1)
        th = 2*k*pi/N;
        Ath = exp(i*th)*A;
        Hth = (Ath + Ath')/2;
        hM = [hM max(real(eig(Hth)))];
   end
   w = max(hM);
   ```

13. Use nrad4 to compute w(A) for the matrices in #7 - 10 and explain why the output is correct. Use N = 50 in each case.

9.3 Glossary

2-dimensional orthogonal compression	570
boundary	578
Donoghue's theorem	578
∂S	578
exceptional point	580
real normal form	587
support line	580
supporting half-plane	580
Toeplitz-Hausdorff theorem	574

Chapter 10

Matrix Polynomials and Similarity

Topics
- *Smith normal form*
- *elementary divisors*
- *minimal polynomial*
- *Frobenius normal form*
- *reducibility*
- *Jordan normal form*

10.1 Equivalence

Let $A_0, A_1, \ldots, A_p$ be matrices in $M_n(R)$ where R is a *field* of numbers. For our purposes R will be $\mathbb{Q}$ (the rational numbers), $\mathbb{R}$ (the real numbers) or $\mathbb{C}$ (the complex numbers). In general, the results we obtain will be valid for any one of these three possibilities for R. Let λ be a variable and set

$$f(\lambda) = \lambda^p A_p + \lambda^{p-1} A_{p-1} + \lambda^{p-2} A_{p-2} + \cdots + \lambda A_1 + A_0. \tag{1}$$

Then $f(\lambda)$ is a polynomial with matrix coefficients, or a *matrix polynomial*. In substituting a matrix $A \in M_n(R)$ for λ in (1) it is important to realize that A may or may not commute with every coefficient A_i. Thus two *specializations* of $f(\lambda)$ to a matrix A are distinguished: the *right-hand value*,

$$f_r(A) = A_p A^p + A_{p-1} A^{p-1} + \cdots A_1 A + A_0 \qquad (2)$$

and the *left-hand value*

$$f_L(A) = A^p A_p + A^{p-1} A_{p-1} + \cdots A A_1 + A_0 . \qquad (3)$$

We remark that the powers of λ that appear in the polynomial (1) can appear on either side of the matrix coefficients. It is the specialization that depends on commutativity. If $A_t = a_t I_n$, $a_t \in R$, $t = 1, \ldots, n$, then of course $f_r(A) = f_L(A) = f(A)$ and we have

$$f(A) = a_p A^p + a_{p-1} A^{p-1} + \cdots + a_1 A + a_0 I_n. \qquad (4)$$

We adhere to the following standard notation: $R[\lambda]$ denotes the totality of polynomials in the variable λ whose coefficients come from R. The totality of fractions $\dfrac{f(\lambda)}{g(\lambda)}$, $g(\lambda) \neq 0$, in which $f(\lambda)$ and $g(\lambda)$ are in $R[\lambda]$ is called the *rational function field* determined by $R[\lambda]$ and is denoted by $R(\lambda)$. We assume that the reader is acquainted with the usual rules for adding and multiplying polynomial fractions.

The expression (1) can also be regarded as an $n \times n$ matrix whose entries are polynomials (a *polynomial matrix*), i.e., $f(\lambda) \in M_n(R[\lambda])$. For example, suppose

$$f(\lambda) = \lambda^2 \begin{bmatrix} 1 & 2 \\ 3 & -2 \end{bmatrix} + \lambda \begin{bmatrix} 2 & 1 \\ 0 & 0 \end{bmatrix} + \begin{bmatrix} 1 & 2 \\ 1 & 1 \end{bmatrix}$$

$$= \begin{bmatrix} \lambda^2 + 2\lambda + 1 & 2\lambda^2 + \lambda + 2 \\ 3\lambda^2 + 1 & -2\lambda^2 + 1 \end{bmatrix} .$$

Clearly we can go back and forth in this way between polynomials in λ with matrix coefficients and matrices with entries that are scalar polynomials in λ.

If $P(\lambda)$ is a matrix in $M_n(R[\lambda])$ then $P(\lambda)$ is a *unit matrix* if $P(\lambda)^{-1}$ exists and has entries in $R[\lambda]$, i.e., $P(\lambda)^{-1} \in M_n(R[\lambda])$. For example, if

10.1 Equivalence

$$P(\lambda) = \begin{bmatrix} \lambda + 1 & \lambda \\ \lambda & \lambda - 1 \end{bmatrix}$$

then

$$P(\lambda)^{-1} = \begin{bmatrix} 1 - \lambda & \lambda \\ \lambda & -(1 + \lambda) \end{bmatrix}.$$

Note that $\det(P(\lambda)) = -1$. In general, we have:

Theorem 1.

If $P(\lambda) \in M_n(R[\lambda])$ then $P(\lambda)$ is a unit matrix iff $\det(P(\lambda))$ is a nonzero constant in R.

Proof.

If $\det(P(\lambda)) = c \neq 0$, $c \in R$, then the entries of

$$c^{-1} \text{adj}(P(\lambda)) \tag{5}$$

are clearly polynomials in λ. But (5) is simply the formula for $P(\lambda)^{-1}$ (see §6.2, Theorem 4). Conversely, if $P(\lambda)^{-1}$ has polynomial entries then $\det(P(\lambda)^{-1})$ is a polynomial and we have

$$1 = \det(I_n) = \det(P(\lambda)^{-1}) \det(P(\lambda)).$$

But if the product of two polynomials is 1, both must be constants. ∎

The three types of elementary row operations on matrices with polynomial entries are virtually the same as those for constant matrices:

- Type I: interchange two rows: $I_{(i),(j)}$

- Type II: add to a row a polynomial $p(\lambda)$ times another row: $II_{(i)+p(\lambda)(j)}$

- Type III: multiply a row by a nonzero constant $r \in R$: $III_{r(i)}$

The notations for the corresponding elementary matrices in $M_n(R[\lambda])$ are:

$$E_{(i),(j)}, \quad E_{(i)+p(\lambda)(j)}, \quad E_{r(i)}. \tag{6}$$

Note that

$$\det(E_{(i),(j)}) = -1, \quad \det(E_{(i)+p(\lambda)(j)}) = 1, \quad \det(E_{r(i)}) = r.$$

Thus by Theorem 1 any product of elementary matrices is a unit matrix. The elementary column operations are defined similarly and can be achieved by right multiplication by elementary matrices.

Two matrices $A(\lambda)$ and $B(\lambda)$ in $M_{m,n}(R[\lambda])$ are said to be *equivalent*, written $A(\lambda) \stackrel{E}{=} B(\lambda)$, if there exist unit matrices $P(\lambda) \in M_m(R[\lambda])$ and $Q(\lambda) \in M_n(R[\lambda])$, such that

$$P(\lambda) A(\lambda) Q(\lambda) = B(\lambda). \tag{7}$$

One of the goals of the present chapter is to determine a canonical form for polynomial matrices under equivalence.

Let $A(\lambda) \in M_{m,n}(R[\lambda])$ and let $1 \le k \le \min\{m, n\}$. Define

$$d_k(A(\lambda)) = \gcd\{\det(A(\lambda)[\alpha \mid \beta]) \mid \alpha \in Q_{k,m}, \beta \in Q_{k,n}\} \tag{8}$$

when $\rho(A(\lambda)) \ge k$ and set $d_k(A(\lambda)) = 0$ if $\rho(A(\lambda)) < k$. In other words, $d_k(A(\lambda))$ is the monic gcd of all k-square subdeterminants of $A(\lambda)$. The polynomial $d_k(A(\lambda))$ is called the kth *determinantal divisor* of $A(\lambda)$. For $k = 0$ we make the convention that $d_0(A(\lambda)) = 1$. Recall that a monic polynomial $d(\lambda)$ is the gcd of a set of polynomials $M \ne \{0\}$ if:

- $d(\lambda)$ divides every $f(\lambda) \in M$, (notationally, $d(\lambda) \mid f(\lambda)$),

and

- if $\delta(\lambda)$ divides every $f(\lambda) \in M$ then $\delta(\lambda) \mid d(\lambda)$.

10.1 Equivalence

We remark that $\rho(A(\lambda))$ is simply the size of the largest nonzero subdeterminant of $A(\lambda)$ (see §6.2, Theorem 6).

Theorem 2.

Let $A(\lambda) \in M_{m,n}(R[\lambda])$. Then

(i) if $d_k(A(\lambda)) \neq 0$ then $d_j(A(\lambda)) \neq 0$, $j = 1, \ldots, k$;

(ii) if $r = \rho(A(\lambda))$ then $d_r(A(\lambda)) \neq 0$ and $d_k(A(\lambda)) = 0$, $k > r$;

(iii) if $d_k(A(\lambda)) \neq 0$ then $d_k(A(\lambda)) \mid d_{k+1}(A(\lambda))$;

(iv) if $B(\lambda) \in M_{m,n}(R[\lambda])$ and $A(\lambda) \stackrel{E}{\equiv} B(\lambda)$ then $d_k(A(\lambda)) = d_k(B(\lambda))$, $k = 1, \ldots, \min\{m, n\}$.

Proof.

(i) By the Laplace expansion theorem, $d_k(A(\lambda))$ is a sum of polynomial multiples of $(k - 1)^{st}$ order subdeterminants of $A(\lambda)$. Hence if $d_k(A(\lambda)) \neq 0$, not all $(k - 1)^{st}$ order subdeterminants can be 0, so that $d_{k-1}(A(\lambda)) \neq 0$.

(ii) If $r = \rho(A(\lambda))$, then the size of the largest nonzero subdeterminant of $A(\lambda)$ is r and every subdeterminant of order $r + t$, $t \geq 1$, is 0.

(iii) As we know from the elementary algebra of polynomials (i.e., the Euclidean algorithm), the gcd of any finite set of polynomials M is a sum of polynomial multiples of the polynomials in M. Thus $d_{k+1}(A(\lambda))$ is a sum of polynomial multiples of the $(k + 1)^{st}$ order subdeterminants of $A(\lambda)$. But each of these is a sum of polynomial multiples of kth order subdeterminants (Laplace expansion) and it follows that

$$d_k(A(\lambda)) \mid d_{k+1}(A(\lambda)).$$

(iv) Let $A(\lambda) = P(\lambda) B(\lambda) Q(\lambda)$. By §6.3 Theorem 1, for any k satisfying

$$1 \leq k \leq \rho(A(\lambda)) = \rho(B(\lambda)),$$

we have

$$C_k(A(\lambda)) = C_k(P(\lambda)) \, C_k(B(\lambda)) \, C_k(Q(\lambda)).$$

Hence any common divisor of the entries of $C_k(B(\lambda))$ must divide the entries of $C_k(A(\lambda))$. Thus $d_k(B(\lambda)) \mid d_k(A(\lambda))$. Similarly, since $B(\lambda) = P(\lambda)^{-1} A(\lambda) Q(\lambda)^{-1}$ and $P(\lambda)$ and $Q(\lambda)$ are unit matrices, it follows that $d_k(A(\lambda)) \mid d_k(B(\lambda))$. Since both $d_k(A(\lambda))$ and $d_k(B(\lambda))$ are monic, they must be equal. If $k > \rho(A(\lambda)) = \rho(B(\lambda))$ then obviously $d_k(A(\lambda)) = 0$ and $d_k(B(\lambda)) = 0$, by definition. ∎

Notation: frequently we write P instead of $P(\lambda)$ for a matrix with polynomial entries (a polynomial matrix) if no confusion can arise.

10.1 Exercises

1. Write each of the following as a polynomial with matrix coefficients:

   (a) $\begin{bmatrix} \lambda^2 & \lambda \\ \lambda - 1 & \lambda + 1 \end{bmatrix}$

   (b) $\begin{bmatrix} \lambda & 0 & 0 \\ 0 & \lambda & 0 \\ 0 & 1 & \lambda \end{bmatrix}$

   (c) $\begin{bmatrix} \lambda^3 + 2\lambda + 1 & 2\lambda \\ \lambda - 1 & 0 \end{bmatrix}$

2. Write each of the following as a matrix with polynomial entries:

   (a) $\lambda^2 \begin{bmatrix} 1 & 0 \\ -1 & 2 \end{bmatrix} + \lambda \begin{bmatrix} 2 & 1 \\ 3 & 5 \end{bmatrix} + \begin{bmatrix} 1 & 0 \\ 0 & 0 \end{bmatrix}$

   (b) $\left(\lambda^2 \begin{bmatrix} 2 & 3 \\ -1 & 0 \end{bmatrix} + \lambda \begin{bmatrix} 1 & 2 \\ 1 & 1 \end{bmatrix} \right) \left(\lambda \begin{bmatrix} 1 & 0 \\ 0 & -1 \end{bmatrix} + \begin{bmatrix} 1 & 0 \\ 0 & 0 \end{bmatrix} \right).$

3. In general, the *product of two polynomials* with matrix coefficients is defined by

$$\sum_{k=0}^{m} A_k \lambda^k \sum_{k=0}^{n} B_k \lambda^k = A_m B_n \lambda^{m+n} + (A_m B_{n-1} + A_{m-1} B_n) \lambda^{m+n-1} + \cdots + A_0 B_0.$$

If $A_m \neq 0$ then m is called the *degree* of $A(\lambda) = \sum_{k=0}^{m} A_k \lambda^k$, written $\deg(A(\lambda))$.

Prove: $\deg(A(\lambda) B(\lambda)) \leq \deg(A(\lambda)) + \deg(B(\lambda))$. Find an example in which the inequality is strict.

4. Prove: if $A(\lambda)$ is n × n, $\deg(A(\lambda)) = k$, then $\deg(\det(A(\lambda))) \leq nk$.

5. Prove: if $A(\lambda)$ is n × n and $A(\lambda) = \lambda^k A_k + \lambda^{k-1} A_{k-1} + \cdots + A_0$ then the coefficient of λ^{nk} in $\det(A(\lambda))$ is $\det(A_k)$.

6. Let $A(\lambda) = \lambda^k I_n + \lambda^{k-1} A_{k-1} + \lambda^{k-2} A_{k-2} + \cdots + \lambda A_1 + A_0$, an n × n matrix. Show that if

$$C = C(A(\lambda)) = \begin{bmatrix} 0 & 0 & \cdots & 0 & -A_0 \\ I_n & 0 & \cdots & 0 & -A_1 \\ 0 & I_n & 0 & \cdots & 0 & -A_2 \\ \vdots & & \ddots & & \vdots \\ 0 & \cdots & & 0 & I_n & -A_{k-1} \end{bmatrix}, \text{ kn × kn,}$$

then

$$\det(\lambda I_{kn} - C) = \det(A(\lambda))$$

(if k = 1 then $C = -A_0$).

Hint: The case k = 4 exhibits the general idea. Thus assume that

$$A(\lambda) = \lambda^4 I_n + \lambda^3 A_3 + \lambda^2 A_2 + \lambda A_1 + A_0.$$

Then write

$$X = \lambda I_{4n} - C = \begin{bmatrix} \lambda I_n & 0 & 0 & A_0 \\ -I_n & \lambda I_n & 0 & A_1 \\ 0 & -I_n & \lambda I_n & A_2 \\ 0 & 0 & -I_n & \lambda I_n + A_3 \end{bmatrix}.$$

To block row 1 add λ times block row 2 and expand the determinant of the resulting matrix by block column 1, using the Laplace expansion theorem (see §6.2, formula (16)). We then obtain

$$\det(X) = (-1)^{s(\alpha)+s(\beta)} \det(-I_n) \det\left(\begin{bmatrix} \lambda^2 I_n & 0 & \lambda A_1 + A_0 \\ -I_n & \lambda I_n & A_2 \\ 0 & -I_n & \lambda I_n + A_3 \end{bmatrix}\right)$$

where $\alpha = (1, \cdots, n)$, $\beta = (n+1, \cdots, 2n)$. Next, to block row 1 add λ^2 times block row 2 of the preceding matrix and expand by block column 1 to obtain

$$\det(X) = (-1)^{s(\alpha)+s(\beta)}(-1)^n(-1)^{s(\alpha)+s(\beta)} \det(-I_n) \det\left(\begin{bmatrix} \lambda^3 I_n & \lambda^2 A_2 + \lambda A_1 + A_0 \\ -I_n & \lambda I_n + A_3 \end{bmatrix}\right)$$

$$= \det\left(\begin{bmatrix} \lambda^3 I_n & \lambda^2 A_2 + \lambda A_1 + A_0 \\ -I_n & \lambda I_n + A_3 \end{bmatrix}\right).$$

In this last matrix, to block row 1 add λ^3 times block row 2 to obtain

$$\det(X) = \det\left(\begin{bmatrix} 0 & A(\lambda) \\ -I_n & \lambda I_n + A_3 \end{bmatrix}\right) = (-1)^{s(\alpha)+s(\beta)} \det(-I_n) \det(A(\lambda))$$

$$= (-1)^p \det(A(\lambda)) = \det(A(\lambda))$$

(where it is easily checked that $p = n(n+1)/2 + n^2 + n(n+1)/2 + n$ is even).

7. Let

$$A(\lambda) = \lambda^p A_p + \lambda^{p-1} A_{p-1} + \cdots + A_0$$

and

$$B(\lambda) = \lambda^m B_m + \lambda^{m-1} B_{m-1} + \cdots + B_0,$$

and assume that $\det(B_m) \neq 0$ and $A_p \neq 0$. Show that there exist unique matrix polynomials $Q(\lambda)$ and $R(\lambda)$ such that

$$A(\lambda) = Q(\lambda) B(\lambda) + R(\lambda)$$

where either $R(\lambda) = 0$ or $\deg(R(\lambda)) < \deg(B(\lambda)) = m$. This result is called the *division algorithm* for matrix polynomials. Show that a similar result holds for "left" division:

$$A(\lambda) = B(\lambda) Q_1(\lambda) + R_1(\lambda),$$

where either $R_1(\lambda) = 0$ or $\deg(R_1(\lambda)) < \deg(B(\lambda))$.

Hint: If $m \leq p$, write

$$A_1(\lambda) = A(\lambda) - A_p B_m^{-1} \lambda^{p-m} B(\lambda)$$

and proceed by induction on $\deg(A(\lambda))$. If $m > p$, set $Q(\lambda) = 0$, $R(\lambda) = A(\lambda)$. To prove uniqueness, suppose

$$Q(\lambda)B(\lambda) + R(\lambda) = Q_2(\lambda)B(\lambda) + R_2(\lambda).$$

Then $[Q(\lambda) - Q_2(\lambda)] B(\lambda) = R_2(\lambda) - R(\lambda)$. Since B_m is nonsingular, if the left side is not zero, it has degree $m + \deg(Q(\lambda) - Q_2(\lambda))$, while the right side has degree less than m. Thus $Q(\lambda) - Q_2(\lambda)$ must be 0, and uniqueness follows.

8. Let

$$A(\lambda) = \begin{bmatrix} 2 & 1 \\ 0 & 0 \end{bmatrix} \lambda^2 + \begin{bmatrix} 0 & 0 \\ 1 & 0 \end{bmatrix} \lambda + \begin{bmatrix} 1 & 0 \\ 0 & 1 \end{bmatrix}$$

and

$$B(\lambda) = \begin{bmatrix} 0 & 1 \\ -1 & 0 \end{bmatrix} \lambda + \begin{bmatrix} 1 & 1 \\ 1 & 1 \end{bmatrix}.$$

Find $Q(\lambda)$ and $R(\lambda)$ as in #7.

9. Let $A(\lambda)$, $B(\lambda)$, and $E(\lambda)$ be in $M_n(R[\lambda])$. Prove:

$$\begin{bmatrix} A(\lambda) & E(\lambda) \\ 0 & B(\lambda) \end{bmatrix} \overset{E}{\equiv} \begin{bmatrix} A(\lambda) & 0 \\ 0 & B(\lambda) \end{bmatrix}$$

if there exist $X(\lambda)$ and $Y(\lambda)$ in $M_n(R[\lambda])$ for which $A(\lambda)X(\lambda) + Y(\lambda)B(\lambda) = E(\lambda)$.

Hint: Consider

$$\begin{bmatrix} I_n & -Y \\ 0 & I_n \end{bmatrix} \begin{bmatrix} A & E \\ 0 & B \end{bmatrix} \begin{bmatrix} I_n & -X \\ 0 & I_n \end{bmatrix} = \begin{bmatrix} I_n & -Y \\ 0 & I_n \end{bmatrix} \begin{bmatrix} A & -AX + E \\ 0 & B \end{bmatrix}$$

$$= \begin{bmatrix} A & -AX + E - YB \\ 0 & B \end{bmatrix}$$

$$= \begin{bmatrix} A & 0 \\ 0 & B \end{bmatrix}$$

and hence

$$\begin{bmatrix} A & E \\ 0 & B \end{bmatrix} \overset{E}{\equiv} \begin{bmatrix} A & 0 \\ 0 & B \end{bmatrix}.$$

10. Let $A(\lambda)$ be an $n \times n$ matrix polynomial and let C be a constant matrix in $M_n(R)$. Prove that there exist unique matrix polynomials $Q(\lambda)$ and $K(\lambda)$ such that

$$A(\lambda) = Q(\lambda)(\lambda I_n - C) + A_r(C)$$

and

$$A(\lambda) = (\lambda I_n - C)K(\lambda) + A_L(C).$$

10.1 Exercises

This result is called the *Remainder Theorem* for matrix polynomials.

Hint: According to #7, there is a $D \in M_n(R)$ such that

$$A(\lambda) = Q(\lambda)(\lambda I_n - C) + D.$$

Write $Q(\lambda) = Q_m \lambda^m + Q_{m-1} \lambda^{m-1} + \cdots + Q_0$. Then

$$Q(\lambda)(\lambda I_n - C) = (Q_m \lambda^m + Q_{m-1} \lambda^{m-1} + \cdots + Q_0)(\lambda I_n - C)$$

$$= Q_m(\lambda^{m+1} I_n - C\lambda^m) + Q_{m-1}(\lambda^m I_n - C\lambda^{m-1}) + \cdots + Q_0(\lambda I_n - C).$$

Thus

$$[Q(\lambda)(\lambda I_n - C)]_r(C) = Q_m(C^{m+1} - CC^m) + Q_{m-1}(C^m - CC^{m-1}) + \cdots + Q_0(C - C)$$

$$= 0.$$

Thus $A_r(C) = [Q(\lambda)(\lambda I_n - C)]_r(C) + D = D.$

11. Let $Q(\lambda)$ be an $n \times n$ matrix polynomial and let $C \in M_n(R)$. Prove:

$$[Q(\lambda)(\lambda I_n - C)]_r(C) = 0.$$

Hint: See #10.

12. Calculate the determinantal divisors of each of the following matrices:

(a) $\begin{bmatrix} \lambda & \lambda \\ \lambda & \lambda \end{bmatrix}$

(b) $\lambda I_3 - \begin{bmatrix} 0 & 0 & 1 \\ 1 & 0 & 2 \\ 0 & 1 & 3 \end{bmatrix}$

(c) $\begin{bmatrix} \lambda - 1 & 0 & 0 & 0 \\ 0 & \lambda - 1 & 0 & 0 \\ 0 & 0 & \lambda + 1 & \lambda \\ 0 & 0 & \lambda & \lambda \end{bmatrix}$

13. Let $C(f(\lambda))$ be the *companion matrix* of the scalar polynomial

$$f(\lambda) = \lambda^m + c_{m-1}\lambda^{m-1} + \cdots + c_1\lambda + c_0$$

(see §7.1, Exercises #17):

$$C(f(\lambda)) = \begin{bmatrix} 0 & 0 & \cdots & 0 & -c_0 \\ 1 & 0 & \cdots & 0 & -c_1 \\ 0 & 1 & \cdots & 0 & -c_2 \\ 0 & 0 & & & \vdots \\ \vdots & \vdots & \ddots & 0 & \vdots \\ 0 & 0 & \cdots & 0 & 1 & -c_{m-1} \end{bmatrix}.$$

Show that the determinantal divisors of $\lambda I_m - C(f(\lambda))$ are

$$d_0 = d_1 = \cdots = d_{m-1} = 1, \quad d_m(\lambda) = f(\lambda).$$

Hint: If we take the case $n = 1$ (i.e., $f(\lambda) \in R[\lambda]$) and apply the result in #6 we conclude that $\det(\lambda I_m - C(f(\lambda))) = f(\lambda)$. Thus $d_m(\lambda) = f(\lambda)$. On the other hand, for $k = 1, \ldots, m-1$, it is obvious that $\lambda I_m - C(f(\lambda))$ has a subdeterminant with the value of 1. Hence $d_k = 1$, $k = 1, \ldots, m-1$, and $d_0 = 1$ by definition.

14. Prove: every matrix $A \in M_n(R)$ satisfies its characteristic equation. This result is known as the *Cayley-Hamilton Theorem*. We will encounter it again in this chapter.

Hint: Let $f(\lambda) = \lambda^n + c_{n-1}\lambda^{n-1} + \cdots + c_1\lambda + c_0$ be the characteristic polynomial of A. Then, by §6.2, Theorem 4,

$$(\lambda I_n - A) \operatorname{adj}(\lambda I_n - A) = (\det(\lambda I_n - A))I_n = f(\lambda)I_n. \tag{1}$$

Now, $\operatorname{adj}(\lambda I_n - A)$ is a matrix whose entries are determinants (multiplied by ± 1) of $(n-1)$-square submatrices of $\lambda I_n - A$. Hence, $\operatorname{adj}(\lambda I_n - A)$ is a matrix whose entries are polynomials of degree not exceeding $n - 1$. Thus

$$\text{adj}(\lambda I_n - A) = B_{n-1}\lambda^{n-1} + \cdots + B_1\lambda + B_0$$

where the B_j are matrices with scalar entries (i.e., entries in R). Hence (1) can be written as

$$(\lambda I_n - A)(B_{n-1}\lambda^{n-1} + \cdots + B_1\lambda + B_0)$$
$$= (\lambda^n + c_{n-1}\lambda^{n-1} + \cdots + c_1\lambda + c_0)I_n. \qquad (2)$$

Comparing the matrix coefficients of $\lambda^n, \ldots, \lambda, \lambda^0$ on both sides of (2) we obtain

$$B_{n-1} = I_n$$

$$B_{n-2} - AB_{n-1} = c_{n-1}I_n$$

$$B_{n-3} - AB_{n-2} = c_{n-2}I_n$$

$$\vdots$$

$$B_0 - AB_1 = c_1 I_n$$

$$-AB_0 = c_0 I_n.$$

Multiply the first of these equalities on the left by A^n, the second on the left by A^{n-1}, ..., the next to the last on the left by A, and the last by $I_n = A^0$. Then add the resulting equalities to obtain

$$0 = A^n + c_{n-1}A^{n-1} + \cdots + c_1 A + c_0 I_n = f(A).$$

10.1 MatLab

1. Recall *Horner's method* for computing the value of the polynomial

$$f(x) = c_n x^n + \cdots + c_1 x + c_0. \qquad (1)$$

The idea is to initialize f to be c_n and then execute the loop

>   for k = 1:n
>   f = f*x + c_{n-k}
>   end

The polynomial f is represented by the matrix

$$f = [c_n \ c_{n-1} \ \cdots \ c_1 \ c_0]. \qquad (2)$$

Write a MatLab function named polyv that accepts a scalar x and a polynomial f and returns the value f(x), based on Horner's method. Save polyv in a library called pmatrix.

Hint:
```
function fx = polyv(f,x)
%POLYV accepts the scalar polynomial f
%and the scalar x and returns f(x),
%using Horner's method.
n=length(f) - 1;
if n==0
    fx=f;
else
    val=f(1);
    for k=1:n
        val=val*x+f(k+1);
    end
    fx=val;
end
```

2. Compare polyv(f, x) with the built-in function polyval(f, x) (see §7.2 MatLab, #8) for several random f and x.

3. The sequence of m-square matrix coefficients $C_n, C_{n-1}, \ldots, C_0$ in the matrix polynomial

10.1 MatLab

$$F(\lambda) = C_n\lambda^n + C_{n-1}\lambda^{n-1} + \cdots + C_1\lambda + C_0 \qquad (3)$$

can be represented in MatLab as an $m \times m(n+1)$ matrix

$$F = [C_n \ C_{n-1} \ \cdots \ C_1 \ C_0]. \qquad (4)$$

Write a MatLab function called rpolyv that accepts a matrix polynomial F in the form (4) and an m-square matrix X and returns the right-hand value of F at X, $F_r(X)$. Save rpolyv in pmatrix.

Hint:
```
function FX=rpolyv(F,X)
%RPOLYV accepts the matrix
%polynomial F and the conformal
%matrix X. The function then
%returns the right-hand value
%of F at X.
[m, M]=size(F);    %F is mxm(n+1) so M=m(n+1)
n=round(M/m)-1;
if n==0
    FX=F;
else
    val=F(:,1:m);    %val = Cn
    for k=1:n
        val=val*X+F(:,(k*m+1):((k+1)*m));
    end
    FX=val;
end
```

4. Write a MatLab function called lpolyv that returns $F_L(X)$ where the notation is the same as in #3. Save lpolyv in pmatrix.

Hint:
```
function FX=lpolyv(F,X)
%LPOLYV accepts the matrix
%polynomial F and the conformal
```

```
%matrix X. The function then
%returns the left-hand value
%of F at X.
[m, M]=size(F);   %F is mxm(n+1) so M=m(n+1)
n=round(M/m)-1;
if n==0
        FX=F;
else
        val=F(:,1:m);   %val = Cn
        for k=1:n
                val=X*val+F(:,(k*m+1):((k+1)*m));
        end
        FX=val;
end
```

5. Let

$$F(\lambda) = \begin{bmatrix} 1 & -1 \\ 1 & 1 \end{bmatrix}\lambda^3 + \begin{bmatrix} 2 & 0 \\ 1 & 0 \end{bmatrix}\lambda^2 + \begin{bmatrix} -5 & 3 \\ 2 & 1 \end{bmatrix}\lambda + \begin{bmatrix} 2 & 1 \\ 1 & -1 \end{bmatrix}.$$

Use rpolyv and lpolyv to compute $F_r(X)$ and $F_L(X)$ for each of the following X:

(a) $X = \begin{bmatrix} 0 & 0 \\ 0 & 0 \end{bmatrix}$

(b) $X = \begin{bmatrix} 1 & 1 \\ 1 & 1 \end{bmatrix}$

(c) $X = \begin{bmatrix} 1 & -1 \\ 1 & -1 \end{bmatrix}$

6. Let $F(\lambda)$ be the matrix polynomial in (3). Write a MatLab function named subm that

- accepts an m × m(n + 1) matrix polynomial $F(\lambda)$ and a pair of sequences a and b in $Q_{r, m}$, $1 \le r \le m$

- returns the r × r matrix polynomial that lies in rows a and columns b of $F(\lambda)$.

Save subm in pmatrix.
Hint:
```
function S=subm(F,a,b)
%SUBM accepts an mxm(n+1)
%matrix polynomial F and
%a pair of sequences a and
%b in Qr,m, 1≤ r ≤ m. The
%function then returns the
%rxr matrix polynomial that
%lies in rows a and columns
%b of F.
[m, M]=size(F);
n=round(M/m)-1;
sub=[];
for k=0:n
        C=F(:,(k*m+1):((k+1)*m));
        sub=[sub C(a,b)];
end
S=sub;
```

7. Review §1.2 MatLab, #8. Then write a MatLab function named rndpoly that

- accepts positive integers m, n, and N

- returns an m × m(n+1) matrix polynomial F, as in (3), whose coefficients are integral matrices with random integer entries in the range [-N, N].

Save rndpoly in pmatrix.

Hint:
```
function F=rndpoly(m,n,N)
%RNDPOLY accepts positive integers
%m, n, N and returns an mxm(n+1)
%matrix polynomial whose coefficients
```

```
%are integral matrices with random
%entries in the range [-N, N].
F=floor(-N+(2*N+1)*rand(m,m*(n+1)));
```

8. Review the Help entry for conv. Then write a MatLab function called pdprd that

- accepts an m × m(n+1) matrix polynomial $F(\lambda)$ (see (3)) and a permutation $\sigma \in S_m$

- returns the diagonal product

$$\prod_{t=1}^{m} F(\lambda)_{t,\,\sigma(t)} \,. \tag{5}$$

Note that each factor in (5) is a scalar polynomial represented as a 1 × (n + 1) row matrix. Thus, the product (5) is a polynomial represented as a 1 × m(n + 1) row matrix of scalars. Save pdprd in pmatrix.

Hint:
```
function d=pdprd(F, s)
%PDPRD accepts an mxm(n+1)
%matrix polynomial F and a
%permutation s in Sm. The
%function then returns the
%diagonal product
%F1,s(1)*...*Fm,s(m) as a 1xm(n+1)
%row matrix of scalars.
[m, N]=size(F);
dp=subm(F, 1, s(1));
for k=2:m
    dp=conv(dp,subm(F,k,s(k)));
end
d=dp;
```

9. Review §5.2 MatLab, #2 and §5.3 MatLab, #1. Then use pdprd in #8 and subm in #6 to write a MatLab function called psubd (polynomial subdeterminant) that

- accepts an m × m(n + 1) matrix polynomial F(λ) and a pair of sequences a and b in $Q_{r,m}$, $1 \leq r \leq m$

- returns the r × r subdeterminant of F(λ), det(F(λ)[a | b])

Note that an r × r subdeterminant of F(λ) is a scalar polynomial represented as a 1 × (rn + 1) row matrix of scalars. Save psubd in pmatrix.

Hint:
```
function d=psubd(F,a,b)
%PSUBD accepts an mxm(n+1)
%matrix polynomial F and a
%pair of sequences a and b
%in Qr,m. The function then
%returns the subdeterminant
%of F lying in rows a and
%columns b.
[m, N]=size(F);
r=length(a);
S=subm(F,a,b);
sm=0;                          %cumulative sum of products
x=1:r;                         %initalize x to identity perm
dp=pdprd(S,x);                 %diag prod in S for x
sm=sm+dp;                      %first term in det(S)
p=Lt(x);                       %this generates Sr
while p>1                      %         *
    q=Lr(x,p);                 %         *
    x=swap(x,p-1,q);           %         *
    x=rev(x,p);                %as in Allperms
    dp=psn(x)*pdprd(S,x);      %next signed diag prod
    sm=sm+dp;                  %accumulate in sm
    p=Lt(x);                   %when Lt(x) drops below 2
end                            %all of Sr has been generated
d=sm;
```

10. (a) Let $f_1(\lambda)$, ..., $f_p(\lambda)$ be scalar polynomials in λ. If d(λ) is a monic

polynomial such that $d \mid f_t$, $t = 1, \ldots, p$, and if $\delta \mid f_t$, $t = 1, \ldots, p$, implies that $\delta \mid d$, then d is called the *greatest common divisor* of $f_1, \ldots, f_p$, $d = \gcd(f_1, \ldots, f_p)$. Show that

$$d = \gcd(\gcd(f_1, \ldots, f_{p-1}), f_p).$$

(b) Review the Help entries for any and find. Then write a MatLab function called top that accepts

- an m × n matrix A

- returns zero if A = zeros(m, n); otherwise top returns a p × n matrix B whose rows are the nonzero rows of A in the same relative order in which they appear in A.

Save top in pmatrix.

(c) Let A be an m × (N + 1) matrix each of whose rows represents a scalar polynomial. That is, the kth row of A,

$$A(k, :) = [A(k, 1) \; A(k, 2) \; \ldots \; A(k, N+1)],$$

represents the polynomial

$$A(k, 1)\lambda^N + A(k, 2)\lambda^{N-1} + \cdots + A(k, N)\lambda + A(k, N+1).$$

Review the functions strip and gcdp in §4.1 MatLab, #22(b). Then use the function top in part(b) of this problem to write a MatLab function called gcdm that accepts an m × (N + 1) matrix A and returns 0 if A = zeros(m, N + 1) and otherwise returns the greatest common divisor of the polynomials represented by the nonzero rows of A. Save gcdm in pmatrix.

Hint: (a) Let $\delta = \gcd(f_1, \ldots, f_{p-1})$ and $\Delta = \gcd(\delta, f_p)$. Then

$$\Delta \mid \delta \mid f_t, \; t = 1, \ldots, p - 1$$

10.1 MatLab

and

$$\Delta \mid f_p,$$

so $\Delta \mid d$. Similarly, $d \mid f_t$, $t = 1, ..., p$, so $d \mid \Delta$. Since both Δ and d are monic, $d = \Delta$.

(b)   function B=top(A)
          %TOP accepts an mxn matrix
          %A and returns 0 if
          %A = zeros(m,n).
          %Otherwise the function
          %returns a pxn matrix B
          %consisting of the non-zero
          %rows of A.
          Ind=find(any(abs(A')));
          if length(Ind)==0
              B=0;
          else
              B=A(Ind,:);
          end

(c)   function d=gcdm(A)
          %GCDM accepts an mx(N+1)
          %matrix A, returns zero if
          %A is zero, otherwise returns
          %the gcd of the polynomials
          %represented by the non-zero rows of A.
          [m,L]=size(A);   %This section of code zeros
          Ind=find(all(abs(A')<1e-6));   %out rows of A close to zero,
          LL=length(Ind);   %deletes them, and puts the
          A(Ind,:)=zeros(LL,L);   %remaining rows into B.
          B=top(A);

```
            if B==0                              %if B is 0, d is 0
                    div=0;
            else
                    [p,L]=size(B);
                    if L==1
                            div=1;               %otherwise if L=1, d=1
                    elseif p==1
                            div=B;               %otherwise if p=1, div=B
                    else
                            div=B(1,:);          %This section of code
                            for k=2:p            %computes the gcd of
                                    div=gcdp(div,B(k,:));  %the polynomials represented
                            end                  %by the rows of B. The gcdp
                    end                          %function takes care of the
            end                                  %situation in which B(k,:)
            d=div;                               %has a leading coeff of 0.
```

11. Recall that if $A(\lambda)$ is an m-square matrix with polynomials entries, then the kth determinantal divisor of $A(\lambda)$, $d_k(A(\lambda))$, is the gcd of all $k \times k$ subdeterminants of $A(\lambda)$, if any of these are nonzero. Otherwise, $d_k(A(\lambda))$ is 0 (see (8) in the narrative of §10.1). Write a MatLab function called ddiv that

    - accepts an $m \times m(n+1)$-square scalar matrix F representing $A(\lambda)$ (see formula (3) above) and a nonnegative integer k, $0 \le k \le m$

    - returns the kth determinantal divisor of $A(\lambda)$.

    We remark that ddiv assumes that $A(\lambda)$ is m-square, the case of prinicipal importance in the similarity theory developed in §10.2. However, only a slight modification in the code is necessary to deal with a rectangular $A(\lambda)$. We leave this to the reader. Save ddiv in pmatrix.

    Hint:
```
        function d = ddiv(F,k)
        %DDIV accepts an mxm(n+1) matrix F
        %representing an mxm matrix A with
```

```
%polynomial entries, with n the largest degree
%of any polynomial in A. The function also
%accepts a non-negative integer k, 0 ≤ k ≤ m,
%and returns the gcd of all the k-square
%subdeterminants in A.
[m,N]=size(F);
if k==0
    dk=1;
else
    cm=cmb(k,m);
    [M, L]=size(cm);
    D=[ ];
    for p=1:M
        for q=1:M
            a=cm(p,:); b=cm(q,:);
            subdet=psubd(F,a,b);
            D=[D;subdet];
        end
    end
    dk=gcdm(D);
end
dk=strip(dk,1e-6);
if dk==0
    d=0;
else
    d=dk/dk(1);
end
```

12. Use ddiv to compute each of the following determinantal divisors:

(a) $A(\lambda) = \begin{bmatrix} \lambda^3 - 3\lambda - 2 & 2\lambda^3 + 3\lambda^2 - 1 \\ \lambda^3 - \lambda^2 - 5\lambda - 3 & \lambda^4 + 2\lambda^3 + \lambda^2 \end{bmatrix}$, $d_1(A(\lambda))$

(b) $\lambda I_3 - A$ and $\lambda I_3 - B$ where

$$A = \begin{bmatrix} 1 & -1 & 4 \\ 3 & 2 & -1 \\ 2 & 1 & -1 \end{bmatrix}, \quad B = \begin{bmatrix} 3 & 0 & 0 \\ 0 & -2 & 0 \\ 0 & 0 & 1 \end{bmatrix}, \quad d_1, d_2, d_3$$

(c) $\lambda I_2 - A$ and $\lambda I_2 - B$ where

$$A = \begin{bmatrix} -4 & 15 \\ -2 & 7 \end{bmatrix}, \quad B = \begin{bmatrix} 1 & 0 \\ 0 & 2 \end{bmatrix}, \quad d_1, d_2$$

(d) $\lambda I_2 - A$ and $\lambda I_2 - B$ where

$$A = \begin{bmatrix} 3 & 5 \\ 3 & 1 \end{bmatrix}, \quad B = \begin{bmatrix} 2 & 4 \\ 4 & 2 \end{bmatrix}, \quad d_1, d_2$$

(e) $\lambda I_3 - A$ and $\lambda I_3 - B$ where

$$A = \begin{bmatrix} 3 & 2 & 1 \\ 0 & 2 & 0 \\ 1 & 2 & 3 \end{bmatrix}, \quad B = \begin{bmatrix} 2 & 0 & 0 \\ 0 & 2 & 0 \\ 0 & 0 & 4 \end{bmatrix}, \quad d_1, d_2, d_3$$

(f) $\lambda I_2 - A$ and $\lambda I_2 - B$ where

$$A = \begin{bmatrix} 2 & -3 \\ 1 & -2 \end{bmatrix}, \quad B = \begin{bmatrix} 1 & 0 \\ 0 & -1 \end{bmatrix}, \quad d_1, d_2$$

(g) $\lambda I_2 - A$ and $\lambda I_2 - B$ where

$$A = \begin{bmatrix} 2 & -5 \\ 1 & -2 \end{bmatrix}, \quad B = \begin{bmatrix} i & 0 \\ 0 & -i \end{bmatrix}, \quad d_1, d_2$$

(h) $\lambda I_p - J_p, d_1, \ldots, d_p$ where $J_p = \text{ones}(p)$, $p = 2, 3$ and 4.

10.1 Glossary

$\underline{E}$ 608
$C(A(\lambda))$ 611
$C(f(\lambda))$ 616

10.2 Similarity

Cayley-Hamilton theorem 616
companion matrix 616
degree 611
determinantal divisor 608
division algorithm 613
$d_k(A(\lambda))$ 608
equivalent 608
field 605
greatest common divisor 624
Horner's method 617
left-hand value 606
matrix polynomial 605
polynomial matrix 606
polynomial product 610
rational function field 606
Remainder Theorem 615
right-hand value 605
specialization 605
Type I 607
Type II 607
Type III 607
unit matrix 606

10.2 Similarity

The main result concerning matrices in $M_{m,n}(R[\lambda])$ is the following:

Theorem 1 (Smith Normal Form).

Let $A(\lambda)$ be a nonzero m × n matrix of polynomials in $R[\lambda]$, and assume that $\rho(A(\lambda)) = r$. Then

$$A(\lambda) \stackrel{E}{\equiv} \begin{bmatrix} q_1(\lambda) & & 0 & & 0 \\ & \ddots & & & 0 \\ 0 & & q_r(\lambda) & & \\ \hline & 0 & & & 0 \end{bmatrix} \quad (1)$$

where $q_i(\lambda) \mid q_{i+1}(\lambda)$, $i = 1, \ldots, r-1$, and $q_i(\lambda) \neq 0$, $i = 1, \ldots, r$, are monic polynomials in $R[\lambda]$. The polynomials $q_i(\lambda)$ are uniquely determined by $A(\lambda)$.

Proof.

By type I operations bring a nonzero entry to the (1, 1) position. Scan the entries in the first column for the one of least degree and by a type I operation bring it to the (1, 1) position; call this element $a_1(\lambda)$. By dividing each of the entries of the first column by $a_1(\lambda)$ we obtain a sequence of type II operations that reduce the entries in the first column below the (1, 1) entry to polynomials of degree less that $\deg(a_1(\lambda))$. Again scan column 1 for the entry of least degree, bring it to the (1, 1) position, and repeat the process. Ultimately the entries in the first column below the (1, 1) entry must become 0 because their degrees are decreasing after each such "sweep." Thus we can assume that we are now dealing with a matrix that has the form

$$\begin{bmatrix} b_1 & b_2 & \cdots & b_n \\ 0 & * & & * \\ \vdots & & & \\ 0 & * & & * \end{bmatrix}. \quad (2)$$

Replace b_1 by the entry of least degree in row 1 and perform analogous elementary column operations to ultimately reduce to 0 all the elements in row 1 except the (1, 1) entry. Of course, the entries in column 1 may no longer be 0, but the (1, 1) entry has no larger degree than $\deg(b_1)$. The sweep through column 1 can then be repeated. Finally, a (1, 1) entry is obtained that divides all the other entries in row 1 and column 1, and then these can be made 0 by type II operations. The matrix now on hand has the form

10.2 Similarity

$$\begin{bmatrix} \beta & 0 & \cdots & 0 \\ 0 & & & \\ \vdots & & C & \\ 0 & & & \end{bmatrix}. \tag{3}$$

If there is an entry c in C not divisible by β, say it is in row i of the matrix (3), then add row i to row 1 and repeat the entire process that led from (2) to (3). Since β did not divide c, the entry that now appears in the (1, 1) position has degree less than $\deg(\beta)$. After a finite number of such reductions to the form (3), the entry appearing in the (1, 1) position must divide every other entry in the matrix. This is because the degrees of the (1, 1) entries strictly decrease at the end of each reduction. Thus, we may assume in (3) that β divides every entry in C. Now, by induction on the size, the matrix C can be brought to a form analogous to (1) by elementary row and column operations on the matrix (3) that do not effect either the first row or the first column. These operations also do not alter the fact that β divides every entry in the matrix. Thus $A(\lambda)$ can be reduced to the form (1) by elementary row and column operations, and by type III operations we can make each of the non-zero entries monic. Denote by $S(A)$ the matrix on the right in (1). Observe next that since

$$q_1(\lambda) \mid q_2(\lambda) \mid \cdots \mid q_r(\lambda)$$

it follows that

$$d_k(S(A)) = q_1(\lambda) \cdots q_k(\lambda), \quad 1 \le k \le r,$$

$$d_k(S(A)) = 0, \quad k > r,$$

$$d_0(S(A)) = 1.$$

Thus

$$q_k(\lambda) = \frac{d_k(S(A))}{d_{k-1}(S(A))}, \quad k = 1, \ldots, r. \tag{4}$$

However, by §10.1, Theorem 2 (iv), $d_k(S(A)) = d_k(A(\lambda))$ and hence (4) implies that the $q_k(\lambda)$ are uniquely determined by $A(\lambda)$. ∎

The matrix $S(A)$ on the right in (1) is called the *Smith normal form* of A; the nonzero polynomials $q_1(\lambda), \ldots, q_r(\lambda)$ are called the *invariant factors* of A. If we factor each $q_k(\lambda)$,

$$q_k(\lambda) = p_1(\lambda)^{e_{k,1}} \cdots p_m(\lambda)^{e_{k,m}}, \quad k = 1, \ldots, r, \tag{5}$$

where $p_1(\lambda), \ldots, p_m(\lambda)$ are distinct irreducible (i.e., have no proper factors in $R[\lambda]$) polynomials and the $e_{k,j}$ are nonnegative integers, then any prime power

$$p_j^{e_{k,j}}, \quad e_{k,j} > 0, \tag{6}$$

appearing in a factorization (5) is called an *elementary divisor* of A. Note that since $q_k \mid q_{k+1}$ it follows that

$$e_{k,j} \leq e_{k+1,j}. \tag{7}$$

The *list of elementary divisors* of $A(\lambda)$ is the totality of elementary divisors, each counted the number of times it appears among the factorizations (5).

For example, consider the matrix

$$C = \begin{bmatrix} 0 & 0 & 0 & 4 \\ 1 & 0 & 0 & -4 \\ 0 & 1 & 0 & -3 \\ 0 & 0 & 1 & 4 \end{bmatrix}$$

and let $A(\lambda)$ be the *characteristic matrix* of C: $A(\lambda) = \lambda I_4 - C$. Then C is the companion matrix (see §10.1 Exercises, #13) of the polynomial

$$f(\lambda) = \lambda^4 - 4\lambda^3 + 3\lambda^2 + 4\lambda - 4 = (\lambda - 1)(\lambda + 1)(\lambda - 2)^2.$$

From §10.1 Exercises, #13 the determinantal divisors of $A(\lambda)$ are $d_0 = d_1 = d_2 = d_3 = 1$ and $d_4 = f(\lambda)$. Hence the invariant factors are $q_1 = d_1/d_0 = 1$, $q_2 = d_2/d_1 = 1$, $q_3 = d_3/d_2 = 1$, $q_4 = d_4/d_3 = f(\lambda)$. The list of elementary divisors is

10.2 Similarity

$$\lambda - 1,\ \lambda + 1,\ (\lambda - 2)^2.$$

Observe that, in general, if the rank and the list of elementary divisors of a matrix of polynomials $A(\lambda) \in M_{m,n}(R[\lambda])$ are given, then $S(A)$ can be reconstructed. For example, suppose the list of elementary divisors of a 5×7 matrix of polynomials $A(\lambda)$ is

$$(\lambda - 2)^3, (\lambda - 2)^4, (\lambda - 2)^4, \lambda - 3, (\lambda - 3)^3, (\lambda - 3)^3, (\lambda - 5)^2, (\lambda - 5)^5, \lambda - 7, (\lambda - 7)^2 \quad (8)$$

and moreover, $\rho(A(\lambda)) = 4$. Since the rank is 4, there are 4 invariant factors of $A(\lambda)$. Moreover, $q_4(\lambda)$ must be the product of all the highest powers of the distinct primes that appear in the list (8):

$$q_4(\lambda) = (\lambda - 2)^4 (\lambda - 3)^3 (\lambda - 5)^5 (\lambda - 7)^2.$$

Then $q_3(\lambda)$ must be the product of all the highest powers of the distinct primes left in the list after those used to construct $q_4(\lambda)$ have been deleted:

$$q_3(\lambda) = (\lambda - 2)^4 (\lambda - 3)^3 (\lambda - 5)^2 (\lambda - 7).$$

Similarly,

$$q_2(\lambda) = (\lambda - 2)^3 (\lambda - 3).$$

The list has now been exhausted so that $q_1(\lambda) = 1$. Note also that $d_1(A) = q_1$, $d_2(A) = q_1 q_2$, ..., $d_r(A) = q_1 \cdots q_r$ so that the determinantal divisors can also be obtained from the list of elementary divisors. Summarizing these remarks we have:

Theorem 2.

The rank, dimensions, and list of elementary divisors (or invariant factors, or determinantal divisors) comprise a complete set of invariants with respect to equivalence. That is, two matrices in $M_{m,n}(R[\lambda])$ are equivalent iff they have the same rank and the same list of elementary divisors (or invariant factors, or determinantal divisors).

The following result is important in determining $S(A)$ in the event $A(\lambda)$ has nonzero entries only in positions (t, t), $t = 1, \ldots, k$.

Theorem 3.

Let $D(\lambda) \in M_{m,n}(R[\lambda])$ and assume D has nonzero entries $h_1(\lambda), \ldots, h_k(\lambda)$ in positions $(1, 1), \ldots, (k, k)$, respectively, and 0 elsewhere. Then the list of elementary divisors of D is the list of all prime power factors of any of the $h_1(\lambda), \ldots, h_k(\lambda)$, each appearing in the list the number of times it occurs as a factor.

Proof.

Let $p_1(\lambda), \ldots, p_m(\lambda)$ be all of the distinct prime (i.e., irreducible) divisors of any of the $h_i(\lambda)$, $i = 1, \ldots, k$. Then write

$$h_t(\lambda) = p_1(\lambda)^{e_{t,1}} p_2(\lambda)^{e_{t,2}} \cdots p_m(\lambda)^{e_{t,m}}, \quad t = 1, \ldots, k, \tag{9}$$

where the $e_{t,j}$ are nonnegative integers. Let s be a fixed integer, $1 \leq s \leq m$, and by reordering the $h_t(\lambda)$ we may assume

$$e_{1,s} \leq e_{2,s} \leq \cdots \leq e_{k,s}, \tag{10}$$

i.e., we can assume $h_1(\lambda), \ldots, h_k(\lambda)$ are arranged in order of increasing powers of $p_s(\lambda)$. Then

$$h_t(\lambda) = p_s(\lambda)^{e_{t,s}} a_t(\lambda) \tag{11}$$

where $p_s(\lambda)$ (being a prime) and $a_t(\lambda)$ are *relatively prime polynomials* (i.e., have no proper common factors), $t = 1, \ldots, k$. If $1 \leq r \leq k$, $\omega \in Q_{r,k}$, then

$$\det(D[\,\omega\,|\,\omega\,]) = h_{\omega(1)}(\lambda) \cdots h_{\omega(r)}(\lambda)$$

$$= p_s(\lambda)^{m(\omega,s)} a_{(\omega)}(\lambda) \qquad \text{(from (11))}$$

where

$$m(\omega,s) = e_{\omega(1),s} + \cdots + e_{\omega(r),s}$$

10.2 Similarity

and

$$a_\omega(\lambda) = a_{\omega(1)}(\lambda) \cdots a_{\omega(r)}(\lambda).$$

Clearly $a_\omega(\lambda)$ and $p_s(\lambda)$ are relatively prime, and moreover the power of $p_s(\lambda)$ occurring in

$$d_r(D(\lambda)) = \gcd \{ \det(D[\,\alpha\,|\,\beta\,]) \,|\, \alpha, \beta \in Q_{r,k} \}$$

is (from (10))

$$\min_{\omega \in Q_{r,k}} m(\omega, s) = e_{1,s} + \cdots + e_{r,s}. \tag{12}$$

Since the rth invariant factor of $D(\lambda)$ is

$$q_r(\lambda) = \frac{d_r(D(\lambda))}{d_{r-1}(D(\lambda))}, \tag{13}$$

it follows from (12) and (13) that the power of $p_s(\lambda)$ appearing in $q_r(\lambda)$ is

$$(e_{1,s} + \cdots + e_{r,s}) - (e_{1,s} + \cdots + e_{r-1,s}) = e_{r,s}.$$

We have proved that in the complete factorization of $q_r(\lambda)$, the polynomial $p_s(\lambda)$ appears with power $e_{r,s}$. Hence the list of elementary divisors of $D(\lambda)$ consists of all $p_s(\lambda)^{e_{r,s}}$, $e_{r,s} > 0$. ∎

As an example of Theorem 3, suppose

$$D(\lambda) = \text{diag}((\lambda - 3)^2,\ (\lambda - 7)^3,\ (\lambda - 2)^3,\ (\lambda - 3)).$$

According to Theorem 3, the list of elementary divisors of $D(\lambda)$ is

$$\lambda - 3,\ (\lambda - 3)^2,\ (\lambda - 7)^3,\ (\lambda - 2)^3.$$

Since $\rho(D(\lambda)) = 4$ there are 4 invariant factors:

$$q_4(\lambda) = (\lambda - 3)^2 (\lambda - 2)^3 (\lambda - 7)^3,$$
$$q_3(\lambda) = \lambda - 3,$$
$$q_1(\lambda) = q_2(\lambda) = 1.$$

The determinantal divisors of $D(\lambda)$ are

$$d_1(\lambda) = d_2(\lambda) = 1,$$
$$d_3(\lambda) = q_1(\lambda)\, q_2(\lambda) q_3(\lambda) = \lambda - 3,$$
$$d_4(\lambda) = q_1(\lambda)\, q_2(\lambda)\, q_3(\lambda) q_4(\lambda) = (\lambda - 3)^3 (\lambda - 2)^3 (\lambda - 7)^3 .$$

As an easy corollary to Theorem 3 we have:

Theorem 4.

Suppose $A(\lambda) \in M_{m,n}(R[\lambda])$ has the following form:

$$A(\lambda) = \left[\begin{array}{c|c} B(\lambda) & 0 \\ \hline 0 & C(\lambda) \end{array}\right],$$

where $B(\lambda)$ is $p \times q$ and $C(\lambda)$ is $r \times s$. Then the list of elementary divisors of $A(\lambda)$ is the combined list of elementary divisors of $B(\lambda)$ and $C(\lambda)$.

Proof.

Let $q_1(\lambda), \ldots, q_k(\lambda)$ and $f_1(\lambda), \ldots, f_m(\lambda)$ be the two sets of invariant factors of $B(\lambda)$ and $C(\lambda)$ respectively. Then obviously $A(\lambda)$ is equivalent to a matrix whose only nonzero entries are $q_1(\lambda), \ldots, q_k(\lambda), f_1(\lambda), \ldots, f_m(\lambda)$ appearing in positions (t, t), $t = 1, \ldots, k + m$. According to Theorem 3, the list of elementary divisors of $A(\lambda)$ is the list of all prime power factors of any of the preceding $k + m$ polynomials. But this is just the combined list of elementary divisors of $B(\lambda)$ and $C(\lambda)$. ∎

Clearly, the result in Theorem 4 can be extended by a trivial induction to a direct sum of any number of polynomial matrices.

10.2 Similarity

In the narrative of §7.1, matrices A and B in $M_n(R)$ were defined as being *similar* over R, written

$$A \stackrel{S}{\equiv} B, \tag{14}$$

if there exists a nonsingular $S \in M_n(R)$ such that

$$B = S^{-1}AS. \tag{15}$$

The key notion we will investigate now is the connection between similarity of A and B over the field R and equivalence of $\lambda I_n - A$ and $\lambda I_n - B$ as matrices in $M_n(R[\lambda])$.

If $A \in M_n(R)$ then the characteristic matrix of A is the matrix

$$\lambda I_n - A \in M_n(R[\lambda]), \tag{16}$$

and

$$\det(\lambda I_n - A) \in R[\lambda] \tag{17}$$

is the *characteristic polynomial* of A (see §7.1 formula (3)). Also recall that the roots $\lambda_1, \ldots, \lambda_n$ of the characteristic polynomial are called the *characteristic roots* or *eigenvalues* of A. Thus

$$\det(\lambda_j I_n - A) = 0, \quad j = 1, \ldots, n. \tag{18}$$

In general, the numbers $\lambda_1, \ldots, \lambda_n$ may not be in R even though $A \in M_n(R)$; e.g.,

$$A = \begin{bmatrix} 0 & 1 \\ -1 & 0 \end{bmatrix} \in M_2(\mathbb{R}),$$

$\det(\lambda I_2 - A) = \lambda^2 + 1$, but the roots of the characteristic polynomial are $\pm i$.

The next result tells the important story relating equivalence and similarity.

Theorem 5.

Let A and B be in $M_n(R)$. Then $A \stackrel{S}{\equiv} B$ iff $\lambda I_n - A \stackrel{E}{\equiv} \lambda I_n - B$, where $\stackrel{E}{\equiv}$ is equivalence in $M_n(R[\lambda])$.

Proof.

Assume that $\lambda I_n - A \overset{E}{\equiv} \lambda I_n - B$ and let $P = P(\lambda)$ and $Q = Q(\lambda)$ be unit matrices in $M_n(R[\lambda])$ for which

$$\lambda I_n - A = P(\lambda I_n - B)Q.$$

Recall that P is a unit matrix iff $0 \neq \det(P) \in R$. Let $L = Q^{-1} \in M_n(R[\lambda])$ and write $L = L(\lambda)$ as a polynomial in λ with coefficients in $M_n(R)$:

$$L = \sum_{t=0}^{m} L_{m-t}\lambda^{m-t}, \quad L_{m-t} \in M_n(R), \; t = 0, \ldots, m. \tag{19}$$

Then

$$I_n = Q^{-1}Q$$

$$= \sum_{t=0}^{m} L_{m-t}\lambda^{m-t}Q$$

$$= \sum_{t=0}^{m} L_{m-t}Q\lambda^{m-t}. \tag{20}$$

Now write $Q = Q(\lambda)$ as a polynomial in λ with coefficients in $M_n(R)$ and let

$$W = Q_r(A), \tag{21}$$

i.e., W is the right-hand value of $Q(\lambda)$ at A. Then, since (see §10.2 Exercises, #24)

$$(Q(\lambda)\lambda^{m-t})_r(A) = Q_r(A) A^{m-t}$$

$$= WA^{m-t},$$

we can evaluate the right-hand value of (20) to obtain

10.2 Similarity

$$I_n = \sum_{t=0}^{m} L_{m-t}(Q(\lambda)\lambda^{m-t})_r (A)$$

$$= \sum_{t=0}^{m} L_{m-t} W A^{m-t}. \tag{22}$$

Now

$$P^{-1}(\lambda I_n - A) = (\lambda I_n - B) Q(\lambda)$$

$$= \lambda Q(\lambda) - B Q(\lambda)$$

$$= Q(\lambda)\lambda - B Q(\lambda). \tag{23}$$

We have seen (§10.1 Exercises, #11) that the right-hand value of $P^{-1}(\lambda I_n - A)$ at A is 0 and thus from (23)

$$0 = (Q(\lambda)\lambda)_r (A) - (BQ(\lambda))_r (A)$$

$$= Q_r(A)A - BQ_r (A)$$

$$= WA - BW,$$

or

$$WA = BW. \tag{24}$$

Thus, from (24) we have (see §10.2 Exercises, #24),

$$WA^{m-t} = B^{m-t}W, \quad t = 0, \ldots, m, \tag{25}$$

and substituting (25) into (22) produces

$$I_n = \left(\sum_{t=0}^{m} L_{m-t} B^{m-t}\right) W.$$

We conclude that $W = Q_r(A) \in M_n(R)$ is nonsingular. From (24) we have

$$A = W^{-1}BW \qquad (26)$$

and hence $A \stackrel{S}{\equiv} B$.

Conversely, suppose that $A \stackrel{S}{\equiv} B$, so that $A = S^{-1}BS$, $S \in M_n(R)$. Then

$$\lambda I_n - A = \lambda I_n - S^{-1}BS$$

$$= S^{-1}(\lambda I_n - B)S \qquad (27)$$

and since S is nonsingular in $M_n(R)$, it is a unit matrix in $M_n(R[\lambda])$, i.e., $0 \neq \det(S) \in R$. But then (27) implies that

$$\lambda I_n - A \stackrel{E}{\equiv} \lambda I_n - B. \qquad \blacksquare$$

It should be observed that the proof of Theorem 5 provides a constructive procedure for obtaining a nonsingular matrix W in $M_n(R)$ for which

$$A = W^{-1}BW,$$

if it is the case that $\lambda I_n - A \stackrel{E}{\equiv} \lambda I_n - B$:

- obtain $Q = Q(\lambda)$ and $P(\lambda)$, products of elementary matrices in $M_n(R[\lambda])$, for which

$$\lambda I_n - A = P(\lambda I_n - B) Q ;$$

- write $Q(\lambda)$ as a polynomial in λ with coefficients in $M_n(R)$;

- evaluate $W = Q(\lambda)_r(A)$.

As an example, we show that

10.2 Similarity

$$A = \begin{bmatrix} 1 & 1 \\ 1 & 1 \end{bmatrix}$$

and

$$B = \begin{bmatrix} 2 & 0 \\ 0 & 0 \end{bmatrix}$$

are similar in $M_n(\mathbb{Q})$. We have

$$\lambda I_2 - A = \begin{bmatrix} \lambda - 1 & -1 \\ -1 & \lambda - 1 \end{bmatrix}$$

and

$$\lambda I_2 - B = \begin{bmatrix} \lambda - 2 & 0 \\ 0 & \lambda \end{bmatrix}.$$

There is no difficulty in confirming that

$$P_1(\lambda)\,(\lambda I_2 - A)\,Q_1(\lambda) = \begin{bmatrix} 1 & 0 \\ 0 & \lambda^2 - 2\lambda \end{bmatrix},$$

where $P_1(\lambda) = E_{(2)+(1-\lambda)(1)}\,E_{1(1)}\,F_{(1),(2)}$, and $Q_1(\lambda) = E_{(1)+(\lambda-1)(2)}$. Also

$$P_2\,(\lambda I_2 - B)\,Q_2(\lambda) = \begin{bmatrix} 1 & 0 \\ 0 & \lambda^2 - 2\lambda \end{bmatrix}$$

where

$$P_2(\lambda) = E_{2(2)}\,E_{(2)+(2-\lambda)(1)}\,E_{-\frac{1}{2}(1)}\,E_{(1)+\,1(2)}\,E_{(1),(2)}$$

and

$$Q_2(\lambda) = E_{(1),(2)}E_{(1)+-1(2)}\,E_{(1)+(\lambda-1)(2)}\,E_{(1),(2)}\,E_{(1)\,+\,\frac{\lambda}{2}(2)}.$$

Hence

$$P_1(\lambda)\,(\lambda I_2 - A)\,Q_1(\lambda) = P_2(\lambda)\,(\lambda I_2 - B)\,Q_2(\lambda),$$

and

$$\lambda I_2 - A = P_1(\lambda)^{-1}\,P_2(\lambda)\,(\lambda I_2 - B)\,Q_2(\lambda)\,Q_1(\lambda)^{-1}.$$

Thus,

$$Q(\lambda) = Q_2(\lambda)\,Q_1(\lambda)^{-1}$$
$$= E_{(1),(2)} E_{(1)+-1(2)}\, E_{(1),(2)}\, E_{(1)+\frac{1}{2}(2)}\, E_{(1)+(1-\lambda)(2)}$$

$$= \begin{bmatrix} 1 & 1-\frac{\lambda}{2} \\ -1 & \frac{\lambda}{2} \end{bmatrix}$$

$$= \begin{bmatrix} 1 & 1 \\ -1 & 0 \end{bmatrix} + \begin{bmatrix} 0 & -\frac{1}{2} \\ 0 & -\frac{1}{2} \end{bmatrix}\lambda.$$

Finally,

$$W = Q_r(\lambda)(A)$$

$$= \begin{bmatrix} 1 & 1 \\ -1 & 0 \end{bmatrix} + \begin{bmatrix} 0 & -1/2 \\ 0 & 1/2 \end{bmatrix}\begin{bmatrix} 1 & 1 \\ 1 & 1 \end{bmatrix}$$

$$= \frac{1}{2}\begin{bmatrix} 1 & 1 \\ -1 & 1 \end{bmatrix}.$$

We check that

$$WA = \begin{bmatrix} 1 & 1 \\ 0 & 0 \end{bmatrix}$$

$$= BW$$

and since W is nonsingular, $A = W^{-1}BW$. We remark that it is not necessary to find W explicitly in order to conclude that A and B are similar over $R = \mathbb{Q}$. For, $\lambda I_2 - A$ and $\lambda I_2 - B$ obviously have the same determinantal divisors and hence are equivalent over $\mathbb{Q}[\lambda]$. It follows from Theorem 5 that $A \stackrel{s}{=} B$ over $\mathbb{Q}$.

Let $f(\lambda) \in R[\lambda]$ be a monic polynomial,

$$f(\lambda) = \lambda^m + c_{m-1}\lambda^{m-1} + c_{m-2}\lambda^{m-2} + \cdots + c_1\lambda + c_0. \tag{28}$$

Recall (§7.1 Exercises, #17) that the *companion matrix* of $f(\lambda)$, denoted by $C(f(\lambda)) \in M_n(R)$, is defined by

$$C(f(\lambda)) = \begin{bmatrix} 0 & 0 & \cdots & 0 & -c_0 \\ 1 & 0 & & & -c_1 \\ 0 & 1 & & & -c_2 \\ 0 & 0 & \ddots & & \vdots \\ \vdots & \vdots & & 0 & -c_{m-2} \\ 0 & 0 & \cdots & 0 & 1 & -c_{m-1} \end{bmatrix} \tag{29}$$

(if $m = 1$ so that $f(\lambda) = \lambda + c_0$ then $C(f(\lambda)) = [-c_0]$).

Theorem 6.

Let

$$f(\lambda) = \lambda^m + c_{m-1}\lambda^{m-1} + \cdots + c_1\lambda + c_0.$$

Then the determinantal divisors of $\lambda I_m - C(f(\lambda))$ are $d_0 = \cdots = d_{m-1} = 1$ and $d_m = f(\lambda)$. Thus $\lambda I_m - C(f(\lambda))$ has only one nonunit invariant factor, namely $f(\lambda)$.

Proof.

In §7.1 Exercises, #17 we saw that $\det(\lambda I_m - C(f(\lambda))) = f(\lambda)$. On the other hand, it is obvious from (29) that $\lambda I_m - C(f(\lambda))$ has a k-square subdeterminant with value 1, $k = 1, \ldots, m - 1$. ∎

We are now able to develop various normal forms for matrices with respect to similarity. In the next result the field R is arbitrary. Theorem 7 is a famous classical result in this subject. It is a *constructive* theorem in that the elementary divisors of the characteristic matrix can be rationally computed. Remember that two matrices A and B in $M_n(R)$ are *similar over* R iff $S^{-1}AS = B$ for some nonsingular matrix S in $M_n(R)$.

Theorem 7 (Frobenius Normal Form).

If $A \in M_n(R)$ then A is similar over R to the direct sum of the companion matrices of the elementary divisors of $\lambda I_n - A$.

Proof.

Let $h_1(\lambda), \ldots, h_p(\lambda)$ be the complete list of elementary divisors of $\lambda I_n - A$, each $h_t(\lambda)$ a power of a prime (i.e., *irreducible*) polynomial in $R[\lambda]$. Let $m_t = \deg(h_t(\lambda))$. Since the product of all the invariant factors of a matrix is its determinant, it follows that the product $h_1(\lambda) \cdots h_p(\lambda)$ is the characteristic polynomial of A and hence

$$m_1 + \cdots + m_p = n.$$

By Theorem 6 the determinantal divisors of $\lambda I_{m_t} - C(h_t(\lambda))$ are 1 ($m_t - 1$ times) and $h_t(\lambda)$. Since $h_t(\lambda)$ is a power of a prime it is obvious that $h_t(\lambda)$ is the single elementary divisor of $\lambda I_{m_t} - C(h_t(\lambda))$. Consider the direct sum matrix

$$\lambda I_n - \sum_{t=1}^{p} \oplus\, C(h_t(\lambda)) = \sum_{t=1}^{p} \oplus\, (\lambda I_{m_t} - C(h_t(\lambda))). \tag{30}$$

The list of elementary divisors of the right side of (30) is precisely

$$h_1(\lambda), \ldots, h_p(\lambda)$$

by Theorem 4. Thus the characteristic matrices, $\lambda I_n - A$ and $\lambda I_n - \sum_{t=1}^{p} \oplus C(h_t(\lambda))$, have the same elementary divisors, are both n-square, and are both of rank n in $M_n(R[\lambda])$ (i.e., they both have determinants equal to the characteristic polynomial). Thus, from Theorem 2,

$$\lambda I_n - A \stackrel{E}{\equiv} \lambda I_n - \sum_{t=1}^{p} \oplus C(h_t(\lambda)).$$

But then Theorem 5 implies that

$$A \stackrel{S}{\equiv} \sum_{t=1}^{p} \oplus C(h_t(\lambda))$$

over R. ∎

The direct sum of the companion matrices of the elementary divisors of $\lambda I_n - A$ in Theorem 7 is called the *Frobenius Normal Form* of A.

If $A \in M_n(R)$ then the *minimal polynomial* of A is the monic polynomial $m(\lambda)$ of least degree such that $m(A) = 0$. The *Cayley-Hamilton Theorem*, (see §10.1 Exercises, #14) shows that the characteristic polynomial of A, $f(\lambda)$, is an *annihilating polynomial* for A, i.e., $f(A) = 0$. It follows that $m(\lambda) | f(\lambda)$. For, write

$$f(\lambda) = q(\lambda) m(\lambda) + r(\lambda)$$

where $\deg(r(\lambda)) < \deg(m(\lambda))$ or $r(\lambda) = 0$. But then

$$0 = f(A)$$

$$= q(A) m(A) + r(A)$$

$$= r(A) \quad (m(A) = 0).$$

Hence if $r(\lambda) \neq 0$ it is a polynomial of degree less than the degree of $m(\lambda)$, and $r(A) = 0$. This contradicts the definition of $m(\lambda)$ as the polynomial of least degree annihilating A.

We see then that under any circumstances, if $f(\lambda)$ is the characteristic polynomial of A, then

$$m(\lambda) \mid f(\lambda) \tag{31}$$

and

$$\deg(m(\lambda)) \le n. \tag{32}$$

Theorem 8.

If $A \in M_n(R)$, then A is similar over R to the direct sum of the companion matrices of the nonunit invariant factors of $\lambda I_n - A$. Moreover, the minimal polynomial of A is $q_n(\lambda)$, the nth invariant factor of $\lambda I_n - A$.

Proof.

The blocks in a direct sum can be arranged in any desired order by a permutation matrix similarity (see §10.2 Exercises, #33). Thus we can rearrange the blocks in the Frobenius normal form so that the companion matrices of all the highest degree elementary divisors of $\lambda I_n - A$ (i.e., highest powers of distinct primes) come first, then the companion matrices of all the remaining highest degree elementary divisors come next, etc. Let A_1 denote the direct sum of this first chain of companion matrices, A_2 the direct sum of the second chain, etc., so that

$$A \stackrel{s}{=} A_1 \oplus A_2 \oplus \cdots \oplus A_k. \tag{33}$$

By Theorem 4 the elementary divisors of $\lambda I_p - A_1$ (I_p is an appropriate size identity matrix) are all the highest degree elementary divisors of $\lambda I_n - A$. On the other hand, if $q_n(\lambda)$ is the highest degree invariant factor of $\lambda I_n - A$ then $\deg(q_n(\lambda)) = p$ because $q_n(\lambda)$ is the product of the highest degree elementary divisors of $\lambda I_n - A$. Also, $\lambda I_p - C(q_n(\lambda))$ has the single nonunit invariant factor $q_n(\lambda)$ and hence its elementary divisors are precisely the highest degree elementary divisors of $\lambda I_n - A$. Thus $\lambda I_p - C(q_n(\lambda))$ and $\lambda I_p - A_1$ have the same elementary divisors and consequently are equivalent. Hence $A_1 \stackrel{s}{=} C(q_n(\lambda))$. By using precisely the same argument, $A_2 \stackrel{s}{=} C(q_{n-1}(\lambda))$, ..., $A_k \stackrel{s}{=} C(q_{n-k+1}(\lambda))$, and

10.2 Similarity

$q_n(\lambda), \ldots, q_{n-k+1}(\lambda)$ account for all the nonunit invariant factors of $\lambda I_n - A$ (i.e., they are obtained from the list of elementary divisors of $\lambda I_n - A$). It follows that

$$A_1 \oplus \cdots \oplus A_k \stackrel{s}{=} C(q_n(\lambda)) \oplus \cdots \oplus C(q_{n-k+1}(\lambda)) \tag{34}$$

and (33) and (34) yield the first statement in the theorem.

Write

$$q_n(\lambda) = \lambda^p + c_{p-1}\lambda^{p-1} + c_{p-2}\lambda^{p-2} + \cdots + c_1\lambda + c_0 \tag{35}$$

and let B be the n-square matrix

$$B = C(q_n(\lambda)) \oplus O_{n-p}. \tag{36}$$

If we let $e_1, \ldots, e_n$ be the usual unit vectors (i.e., the columns of I_n) then from the form of the companion matrix (see (29)) it should be clear that

$$Be_1 = e_2$$

$$\vdots$$

$$Be_{p-2} = e_{p-1}$$

$$Be_{p-1} = e_p$$

$$Be_p = -c_0 e_1 - c_1 e_2 - \cdots - c_{p-1} e_p. \tag{37}$$

Now $e_2 = Be_1$, $e_3 = Be_2 = B^2 e_1$, ..., $e_p = B^{p-1}e_1$, so that the last equation in (37) becomes

$$B^p e_1 = -c_0 e_1 - c_1 Be_1 - \cdots - c_{p-1} B^{p-1} e_1,$$

or

$$q_n(B)e_1 = 0.$$

Also,

$$q_n(B)e_k = q_n(B)B^{k-1}e_1$$
$$= B^{k-1}q_n(B)e_1$$
$$= 0, \quad k = 1, \ldots, p.$$

Thus $q_n(B) = 0$. Suppose now that there is a polynomial $g(\lambda) \in R[\lambda]$, $\deg(g(\lambda)) = r < p$, and $g(B) = 0$. Then

$$g(B)e_1 = 0$$

so that writing $g(\lambda) = d_r\lambda^r + d_{r-1}\lambda^{r-1} + \cdots + d_0$ we have

$$d_r B^r e_1 + d_{r-1} B^{r-1} e_1 + \cdots + d_0 e_1 = 0,$$

or

$$d_r e_{r+1} + d_{r-1} e_r + \cdots + d_0 e_1 = 0, \quad r+1 \leq p.$$

Thus $d_0 = \cdots = d_r = 0$ and it follows that $g(\lambda)$ must be the zero polynomial. We conclude that $q_n(\lambda)$ is the polynomial of least degree in $R[\lambda]$ for which $q_n(B) = 0$. If $h(\lambda)$ is any polynomial for which $h(B) = 0$ then we can divide $h(\lambda)$ by $q_n(\lambda)$ to obtain $h(\lambda) = k(\lambda) q_n(\lambda) + g(\lambda)$ where $g(\lambda) = 0$ or $\deg(g(\lambda)) < p$. But then

$$0 = h(B) = k(B) q_n(B) + g(B) = g(B)$$

and we just saw that since $\deg(g(\lambda)) < p$, we can conclude that $g(\lambda) = 0$. We have proved: $q_n(B) = 0$, and if $g(B) = 0$, where $g(\lambda) \in R[\lambda]$, then $q_n(\lambda) \mid g(\lambda)$. If we next write

$$q_{n-1}(\lambda) = \lambda^m + a_{m-1}\lambda^{m-1} + \cdots + a_1\lambda + a_0,$$

and define

$$C = 0_p \oplus C(q_{n-1}(\lambda)) \oplus 0_{n-(m+p)}$$

10.2 Similarity

we can argue as above to conclude that $q_{n-1}(C) = 0$ and if $g(C) = 0$, where $g(\lambda) \in R[\lambda]$, then $q_{n-1}(\lambda) \mid g(\lambda)$. We repeat the argument for each block on the right in (34). Now if D is the matrix on the right in (34) and $f(\lambda)$ is a polynomial for which $f(D) = 0$, then clearly $f(B) = 0$, $f(C) = 0$ etc., and hence $f(\lambda)$ is divisible by $q_n(\lambda), q_{n-1}(\lambda), \ldots, q_{n-k+1}(\lambda)$.

On the other hand, $q_n(\lambda)$ itself is divisible by $q_{n-1}(\lambda), \ldots, q_{n-k+1}(\lambda)$, so $q_n(D) = 0$. Since D is similar to A, $q_n(A) = 0$. Thus $q_n(\lambda)$ annihilates A and divides any other polynomial that annihilates A, i.e., it is the minimal polynomial of A. ∎

Note that the Cayley-Hamilton Theorem also follows from Theorem 8, i.e., if $A \in M_n(R)$ and $f(\lambda) = \det(\lambda I_n - A)$, then $f(A) = 0$. For, the characteristic polynomial $f(\lambda)$ is the product of the nonunit invariant factors of $\lambda I_n - A$. Hence $q_n(\lambda) \mid f(\lambda)$, and it follows from Theorem 8 that $f(A) = 0$. The Cayley-Hamilton Theorem is often expressed by saying: every matrix satisfies its own characteristic equation.

We say $A \in M_n(R)$ is *reducible* over R if it is similar over R to a direct sum of smaller square matrices, and *irreducible* over R otherwise. It should not be assumed that a companion matrix of a monic polynomial is necessarily irreducible. For example, consider the 6 × 6 matrix over the rational number field $\mathbb{Q}$, $A = C(f(\lambda))$, where

$$f(\lambda) = (\lambda - 1)^2 (\lambda^2 + 1)^2.$$

According to Theorem 6, the single nonunit invariant factor of $\lambda I_6 - A$ is $f(\lambda)$. Thus the list of elementary divisors in $\mathbb{Q}[\lambda]$ of $\lambda I_6 - A$ is

$$h_1(\lambda) = (\lambda - 1)^2$$
$$= \lambda^2 - 2\lambda + 1,$$
$$h_2(\lambda) = (\lambda^2 + 1)^2$$
$$= \lambda^4 + 2\lambda^2 + 1.$$

Hence from Theorem 7, $A \stackrel{s}{\equiv} C(h_1(\lambda)) \oplus C(h_2(\lambda))$.

The preceding example suggests the following result.

Theorem 9.

A matrix $A \in M_n(R)$ is irreducible iff $\lambda I_n - A$ has precisely one elementary divisor in $R[\lambda]$.

Proof.

Suppose A is reducible so that $A \overset{S}{\equiv} B \oplus C$. Then by Theorem 5

$$\lambda I_n - A \overset{E}{\equiv} \lambda I_n - (B \oplus C) = (\lambda I_p - B) \oplus (\lambda I_q - C)$$

where B is $p \times p$ and C is $q \times q$. Then by Theorem 4, the elementary divisors of $\lambda I_n - A$ are those of $\lambda I_p - B$ and $\lambda I_q - C$ taken together. Hence $\lambda I_n - A$ has at least two elementary divisors. Conversely, if $\lambda I_n - A$ has at least two elementary divisors then the Frobenius normal form of A (Theorem 7) is a direct sum of at least two matrices. ∎

Theorem 10.

Any matrix $A \in M_n(R)$ is similar over R to a direct sum of irreducible matrices:

$$A \overset{S}{\equiv} A_1 \oplus \cdots \oplus A_p.$$

If

$$A \overset{S}{\equiv} B_1 \oplus \cdots \oplus B_q$$

is another such representation, then $p = q$ and there exists a permutation $\sigma \in S_p$ such that $A_i \overset{S}{\equiv} B_{\sigma(i)}$, $i = 1, ..., p$.

Proof.

The Frobenius normal form provides one such representation as a direct sum. For, each direct summand in the Frobenius normal form is the companion matrix of a power of an irreducible polynomial and hence, by Theorem 6, its characteristic matrix has a single elementary divisor - apply Theorem 9. Now suppose the Frobenius normal form of A is given by

$$A \stackrel{s}{=} A_1 \oplus \cdots \oplus A_p \tag{38}$$

where $h_1(\lambda), \ldots, h_p(\lambda)$ are the elementary divisors of $\lambda I_n - A$ and $A_i = C(h_i(\lambda))$, $i = 1, \ldots, p$. Suppose also that

$$A \stackrel{s}{=} B_1 \oplus \cdots \oplus B_q \tag{39}$$

where each B_t is irreducible and B_t is $m_t \times m_t$, $t = 1, \ldots, q$. Then by Theorem 4, the elementary divisors of $\lambda I_n - A$ are the elementary divisors of all the $\lambda I_{m_t} - B_t$ taken together. By Theorem 9, each $\lambda I_{m_t} - B_t$ has precisely one elementary divisor, say $k_t(\lambda)$. It follows that the two lists, $h_1(\lambda), \ldots, h_p(\lambda)$ and $k_1(\lambda), \ldots, k_q(\lambda)$ are identical. Thus $p = q$ and there exists a permutation $\sigma \in S_p$ such that

$$h_i(\lambda) = k_{\sigma(i)}(\lambda), \ i = 1, \ldots, p.$$

But $B_i \stackrel{s}{=} C(k_i(\lambda))$ because the characteristic matrix of B_i has $k_i(\lambda)$ as its only elementary divisor (i.e., $C(k_i(\lambda))$ is the Frobenius normal form of B_i). Thus

$$B_{\sigma(i)} \stackrel{s}{=} C(k_{\sigma(i)}(\lambda))$$
$$= C(h_i(\lambda))$$
$$\stackrel{s}{=} A_i, \ i = 1, \ldots, p. \qquad \blacksquare$$

It is customary to list the elementary divisors of $\lambda I_n - A$, $A \in M_n(R)$, in a standard way. Let $p_1(\lambda), \ldots, p_r(\lambda)$ be the list of distinct monic prime polynomials that occur in the factorizations of all the invariant factors of $\lambda I_n - A$. Then we can write out the complete list of elementary divisors of $\lambda I_n - A$:

$$p_1(\lambda)^{e_{1,1}}, p_1(\lambda)^{e_{1,2}}, \ldots, p_1(\lambda)^{e_{1,n_1}}, \quad 0 < e_{1,1} \leq e_{1,2} \leq \cdots \leq e_{1,n_1};$$

$$p_2(\lambda)^{e_{2,1}}, p_2(\lambda)^{e_{2,2}}, \ldots, p_2(\lambda)^{e_{2,n_2}}, \quad 0 < e_{2,1} \leq \cdots \leq e_{2,n_2}; \quad (40)$$

$$\vdots$$

$$p_r(\lambda)^{e_{r,1}}, p_r(\lambda)^{e_{r,2}}, \ldots, p_r(\lambda)^{e_{r,n_r}}, \quad 0 < e_{r,1} \leq \cdots \leq e_{r,n_r}.$$

Note that if the characteristic polynomial of A factors into a product of linear factors in $R[\lambda]$ then each $p_i(\lambda)$ is a monic binomial of the form $\lambda - \alpha_i$.

Theorem 11.

If $A \in M_n(R)$ then A is similar over R to a diagonal matrix D iff the elementary divisors of $\lambda I_n - A$ are all linear.

Proof.

If the elementary divisors of $\lambda I_n - A$ are all linear then the Frobenius normal form is a diagonal matrix similar to A. Conversely if $A \stackrel{S}{\equiv} \mathrm{diag}(\alpha_1, \ldots, \alpha_n)$ then

$$\lambda I_n - A \stackrel{E}{\equiv} \mathrm{diag}(\lambda - \alpha_1, \ldots, \lambda - \alpha_n)$$

and the elementary divisors of $\lambda I_n - A$ are $\lambda - \alpha_1, \ldots, \lambda - \alpha_n$. ∎

Theorem 12.

If $A \in M_n(R)$ then A is similar over R to a diagonal matrix D iff the minimal polynomial of A, $m(\lambda)$, factors into distinct linear factors in $R[\lambda]$.

Proof.

We know from Theorem 8 that the minimal polynomial is the nth invariant factor of $\lambda I_n - A$. Thus, in the notation (40),

$$m(\lambda) = p_1(\lambda)^{e_{1,n_1}} p_2(\lambda)^{e_{2,n_2}} \cdots p_r(\lambda)^{e_{r,n_r}}.$$

10.2 Similarity

Suppose $m(\lambda)$ is a product of distinct linear factors in $R[\lambda]$. Then we must have $e_{1,n_1} = \cdots = e_{r,n_r} = 1$ and $p_t(\lambda) = \lambda - \alpha_t$, $\alpha_i \neq \alpha_j$ for $i \neq j$. But then all the elementary divisors in the list (40) are linear and Theorem 11 implies that A is similar to a diagonal matrix.

Conversely, if A is similar to a diagonal matrix, then all the elementary divisors of $\lambda I_n - A$ are linear, again from Theorem 11. The last invariant factor, i.e., $m(\lambda)$, is the product of all the distinct highest degree elementary divisors of $\lambda I_n - A$ and hence $m(\lambda)$ is a product of distinct linear factors. ∎

It should be noted that the elements $\alpha_1, \ldots, \alpha_n$ appearing along the main diagonal in the proofs of Theorems 11 and 12 are precisely the eigenvalues of A.

In order to discuss another of the standard canonical forms for matrices over a field we recall from §7.1 Exercises, #16 that the m-square *auxiliary unit matrix* U_m has 1's on the first superdiagonal, i.e., in positions $(1, 2), (2, 3), \ldots, (m - 1, m)$, and 0's elsewhere. (If $m = 1$, $U_m = [0]$.) If $g(\lambda) = (\lambda - \alpha)^m$, $\alpha \in R$, also recall that the *hypercompanion matrix* of $g(\lambda)$ is defined as

$$H(g(\lambda)) = \alpha I_m + U_m.$$

Theorem 13.

If $g(\lambda) = (\lambda - \alpha)^m$ then

$$H(g(\lambda)) \stackrel{s}{\equiv} C(g(\lambda)).$$

Proof.

If $m = 1$ then $H(g(\lambda)) = [\alpha] = C(g(\lambda))$. If $m > 1$ then clearly

$$\det((\lambda I_m - H(g(\lambda)))(m \mid 1)) = (-1)^{m-1}.$$

Thus $d_0 = \cdots = d_{m-1} = 1$ where the d_i are the determinantal divisors of $\lambda I_m - H(g(\lambda))$. Also, it is obvious that

$$\det(\lambda I_m - H(g(\lambda))) = (\lambda - \alpha)^m = g(\lambda)$$

and hence the invariant factors of $\lambda I_m - H(g(\lambda))$ are 1 ($m - 1$ times) and $g(\lambda)$. But according to Theorem 6 these are precisely the invariant factors of $\lambda I_m - C(g(\lambda))$. Thus by Theorem 5, $H(g(\lambda)) \stackrel{s}{=} C(g(\lambda))$. ∎

The following theorem is one of the most famous and widely used in mathematics.

Theorem 14 (Jordan Normal Form).

Let $A \in M_n(R)$ and assume that R contains all the characteristic roots of A. Then the polynomials $p_1(\lambda), \ldots, p_r(\lambda)$ in the list (40) are of the form

$$p_t(\lambda) = \lambda - \alpha_t, \quad t = 1, \ldots, r, \tag{41}$$

and the matrix A is similar over R to the direct sum of the hypercompanion matrices of the polynomials (40).

Proof.

The product of the elementary divisors of $\lambda I_n - A$ is the characteristic polynomial of A, and by hypothesis this factors into linear factors in $R[\lambda]$. The factorization of a monic polynomial into prime monic factors is unique and since each of the elementary divisors is a power of a prime polynomial in $R[\lambda]$, it follows that each $p_t(\lambda)$ is linear. The Frobenius normal form of A is

$$\sum_{i=1}^{r} \oplus \sum_{j=1}^{n_i} \oplus C((\lambda - \alpha_i)^{e_{ij}}),$$

which by Theorem 13 is similar over R to

$$\sum_{i=1}^{r} \oplus \sum_{j=1}^{n_i} \oplus H((\lambda - \alpha_i)^{e_{ij}}). \quad \blacksquare \tag{42}$$

The matrix (42) is called the *Jordan normal form* of A.

We end this section by considering several typical examples:

10.2 Similarity

(a) Suppose $R = \mathbb{R}$, the real number field. Then the irreducible monic polynomials in $\mathbb{R}[\lambda]$ are of the form $\lambda - \alpha$ and $\lambda^2 + c_1\lambda + c_0$, $c_1^2 - 4c_0 < 0$. Thus the possibilities for the elementary divisors of the characteristic matrix of a matrix in $M_2(\mathbb{R})$ are:

$$\lambda - \alpha_1, \quad \lambda - \alpha_2, \quad \alpha_i \in \mathbb{R};$$

$$(\lambda - \alpha)^2, \quad \alpha \in \mathbb{R};$$

$$\lambda^2 + c_1\lambda + c_0, \quad c_1^2 - 4c_0 < 0.$$

Thus (why?) any $A \in M_2(\mathbb{R})$ is similar over $\mathbb{R}$ to one of the following three types of matrices:

$$\begin{bmatrix} \alpha_1 & 0 \\ 0 & \alpha_2 \end{bmatrix}, \begin{bmatrix} \alpha & 1 \\ 0 & \alpha \end{bmatrix}, \begin{bmatrix} 0 & -c_0 \\ 1 & -c_1 \end{bmatrix}.$$

(b) We compute the Jordan normal form over $\mathbb{C}$ of the matrix $A = C(f(\lambda))$, $f(\lambda) = (\lambda - 1)^2 (\lambda^2 + 1)$. The single nonunit determinantal divisor of $\lambda I_4 - A$ is $f(\lambda)$ and hence the only nonunit invariant factor is $f(\lambda)$. Factoring $f(\lambda)$ in $\mathbb{C}[\lambda]$ we have

$$f(\lambda) = (\lambda - 1)^2 (\lambda + i)(\lambda - i)$$

and hence the list (40) of elementary divisors of $\lambda I_4 - A$ is

$$(\lambda - 1)^2,$$

$$\lambda + i,$$

$$\lambda - i.$$

We have

$$H((\lambda-1)^2) = \begin{bmatrix} 1 & 1 \\ 0 & 1 \end{bmatrix},$$

$$H(\lambda+i) = [\,-i\,],$$

$$H(\lambda-i) = [\,i\,].$$

Thus the Jordan normal form of A is

$$\begin{bmatrix} 1 & 1 \\ 0 & 1 \end{bmatrix} \oplus [\,-i\,] \oplus [\,i\,].$$

(c) We compute the Jordan normal form over $\mathbb{R}$ of

$$A = \begin{bmatrix} 1 & 3 & -1 \\ 0 & 1 & 2 \\ 0 & 0 & 1 \end{bmatrix}.$$

The characteristic matrix is

$$\lambda I_3 - A = \begin{bmatrix} \lambda-1 & -3 & 1 \\ 0 & \lambda-1 & -2 \\ 0 & 0 & \lambda-1 \end{bmatrix}.$$

Thus $d_0 = 1$, $d_1 = 1$ and two of the 2 × 2 subdeterminants are $(\lambda-1)^2$ and $\lambda - 7$ (to within constant multiples). It follows that $d_2 = 1$, and clearly $d_3 = (\lambda-1)^3$. The invariant factors are $q_1 = 1$, $q_2 = 1$, $q_3 = (\lambda-1)^3$. The single elementary divisor is $(\lambda-1)^3$. Hence

$$A \stackrel{s}{=} H((\lambda-1)^3)$$

$$= I_3 + U_3$$

$$= \begin{bmatrix} 1 & 1 & 0 \\ 0 & 1 & 1 \\ 0 & 0 & 1 \end{bmatrix}.$$

10.2 Exercises

Exercises #1 - 5 use a common notation.

1. Recall that a relation or pairing "~" of elements from a set X is called an *equivalence relation* if:

    (i)   x ~ x, all x ∈ X   (reflexive)

    (ii)   x ~ y implies y ~ x   (symmetric)

    (iii)   x ~ y and y ~ z implies x ~ z   (transitive)

    Prove that if $X = M_{m,n}(R[\lambda])$ then " $\stackrel{E}{=}$ " is an equivalence relation.

2. An *equivalence class* in X to which x belongs is the totality of y ∈ X such that y ~ x. Prove: X is the union of all the equivalence classes and any two distinct equivalence classes are disjoint.

3. Describe the equivalence classes in $M_{m,n}(R[\lambda])$ with respect to the equivalence relation " $\stackrel{E}{=}$ ".

    Hint: Two matrices are in the same equivalence class iff they have the same rank and determinantal divisors.

4. Prove: if $S(\lambda)$ and $T(\lambda)$ in $M_{m,n}(R[\lambda])$ are two matrices in Smith normal form, and if they are equivalent, then $S(\lambda) = T(\lambda)$.

    Hint: $S(\lambda) \stackrel{E}{=} T(\lambda)$ implies that the invariant factors are equal.

5. A *system of distinct representatives* (SDR) or *transversal* for "~" in X is a set consisting of precisely one element from each equivalence class. Prove that the totality of matrices in Smith normal form in $M_{m,n}(R[\lambda])$ is an SDR for "$\stackrel{E}{\equiv}$".

Exercises #6 - 22 use a common notation.

6. Let $P \in M_m(\mathbb{Z})$, i.e., P is a matrix with integer entries. Prove: P^{-1} exists and is in $M_n(\mathbb{Z})$ iff $\det(P) = \pm 1$. The matrix P is called a *unit* or *unimodular* matrix.

7. In $M_{m,n}(\mathbb{Z})$ define elementary row operations as follows: Type I - interchange two rows; Type II - add to a row an integral multiple of another row; Type III - multiply a row by ± 1. Prove: any elementary row operation on a matrix $A \in M_{m,n}(\mathbb{Z})$ can be effected with left multiplication by a corresponding elementary matrix in $M_m(\mathbb{Z})$.

8. Prove: any elementary matrix in $M_m(\mathbb{Z})$ is a unit matrix.

9. State and prove a result analogous to #7 for elementary column operations.

10. Define equivalence of matrices in $M_{m,n}(\mathbb{Z})$ in analogy to §10.1, formula (7).

11. Define determinantal divisors for a matrix A in $M_{m,n}(\mathbb{Z})$ precisely as in §10.1, formula (8), i.e., $d_k(A)$ is the gcd in $\mathbb{Z}$ of all of the kth order subdeterminants of A. State and prove a result analogous to §10.1 Theorem 2 for $M_{m,n}(\mathbb{Z})$.

12. State and prove a result analogous to §10.2 Theorem 1 for $M_{m,n}(\mathbb{Z})$.

13. Define invariant factors and elementary divisors for matrices in $M_{m,n}(\mathbb{Z})$.

14. State and prove a result analogous to §10.2 Theorem 2 for $M_{m,n}(\mathbb{Z})$.

15. State and prove a result analogous to §10.2 Theorem 3 for $M_{m,n}(\mathbb{Z})$.

10.2 Exercises

16. State and prove a result analogous to §10.2 Theorem 4 for $M_{m,n}(\mathbb{Z})$.

17. Find unit matrices $P \in M_2(\mathbb{Z})$ and $Q \in M_3(\mathbb{Z})$ such that

$$A = \begin{bmatrix} 7 & 3 & 15 \\ 1 & 5 & 3 \end{bmatrix}$$

is in Smith normal form $S(A)$ in $M_{2,3}(\mathbb{Z})$. What is $S(A)$?

Hint: Verify that $PAQ = S(A)$ where

$$P = \begin{bmatrix} 0 & 1 \\ -1 & 7 \end{bmatrix},$$

$$Q = \begin{bmatrix} 1 & 10 & -33 \\ 0 & 1 & -3 \\ 0 & -5 & 16 \end{bmatrix},$$

$$S(A) = \begin{bmatrix} 1 & 0 & 0 \\ 0 & 2 & 0 \end{bmatrix}.$$

18. Let $A \in M_{m,n}(\mathbb{Z})$, $b \in M_{m,1}(\mathbb{Z})$, $x \in M_{n,1}(\mathbb{Z})$. A system of *linear diophantine equations* is a system of equations $Ax = b$ for the determination of the integer vector x. Devise a constructive method of solving $Ax = b$ using the Smith normal form of A.

Hint: Suppose that $PAQ = S(A)$ where P and Q are units. Set $x = Qy$ so that $x \in M_{n,1}(\mathbb{Z})$ iff $y \in M_{n,1}(\mathbb{Z})$ (why?). Then $Ax = b$ becomes $AQy = b$ or $PAQy = Pb$, $S(A)y = c$, $c = Pb$. Let $q_1, \ldots, q_r$ be the invariant factors (in $\mathbb{Z}$) of A. Then

$$q_1 y_1 = c_1, \ldots, q_r y_r = c_r, 0 = c_{r+1}, \ldots, 0 = c_m,$$

where $r = \rho(A)$. Thus $Ax = b$ has a solution iff $q_t \mid c_t$, $t = 1, \ldots, r$, and

$$c_{r+1} = \cdots = c_m = 0.$$

The totality of solutions x is then obtained by letting $y_{r+1}, \ldots y_n$ be arbitrary integers in the vector

$$x = Q[\, c_1/q_1, c_2/q_2, \ldots, c_r/q_r, y_{r+1}, \ldots, y_n \,]^T.$$

19. Find all solutions of the linear diophantine system

$$7x_1 + 3x_2 + 15x_3 = 100$$

$$x_1 + 5x_2 + 3x_3 = 120.$$

Hint: Using the results of #17, and the method described in #18, the system becomes

$$y_1 = 120, \; 2y_2 = 740.$$

Thus $y_1 = 120$, $y_2 = 370$ and the totality of solutions x has the form

$$x = Q[\, 120 \;\; 370 \;\; y_3 \,]^T = [\, 3820 - 33y_3 \;\; 370 - 3y_3 \;\; -1850 + 16y_3 \,]^T.$$

If we set $y_3 = 116 - d$ where d runs through $\mathbb{Z}$ we have:

$$x_1 = -8 + 33d, \; x_2 = 22 + 3d, \; x_3 = 6 - 16d.$$

20. Let $P \in M_m(\mathbb{Z})$ be a unit matrix. Prove: adj(P) is a unit matrix.

Hint: We have $(adj(P))P = (det(P)I_m) = \pm I_m$.

21. If $2^3, 2^4, 2^4\, 3, 3^3, 3^3, 5^2, 5^5, 7, 7^2$ is the list of elementary divisors of $A \in M_{5,7}(\mathbb{Z})$ and $\rho(A) = 4$, find $S(A)$.

Hint: We have $q_4 = 2^4\, 3^3\, 5^5\, 7^2$, $q_3 = 2^4\, 3^3\, 5^2\, 7$, $q_2 = 2^3\, 3$, $q_1 = 1$.

10.2 Exercises

22. Find S(A) for

$$A = \begin{bmatrix} 7 & 4 & 1 \\ 8 & 5 & 2 \\ 9 & 6 & 3 \end{bmatrix} \in M_3(\mathbb{Z}).$$

Find unit matrices $P, Q \in M_3(\mathbb{Z})$ such that $PAQ = S(A)$.

Hint: We have $S(A) = \text{diag}(1, 3, 0)$,

$$P = \begin{bmatrix} 1 & 0 & 0 \\ 2 & -1 & 0 \\ 1 & -2 & 1 \end{bmatrix},$$

$$Q = \begin{bmatrix} 0 & 0 & 1 \\ 0 & 1 & -2 \\ 1 & -4 & 1 \end{bmatrix}.$$

23. Verify that similarity of matrices in $M_n(R)$ is an equivalence relation. The equivalence classes are called *similarity classes*.

24. Let $Q(\lambda)$ be a matrix polynomial. Prove: $(Q(\lambda)\lambda^k)_r(A) = Q_r(A)A^k$. Also, show that if W, A, B are in $M_n(R)$ and $WA = BW$ then $WA^k = B^kW$, $k = 0, 1, 2, \ldots$.

Hint: Write $Q(x) = A_0 + A_1\lambda + \cdots + A_m\lambda^m$. Then

$$(Q(\lambda)\lambda^k)_r(A) = (A_0\lambda^k + A_1\lambda^{k+1} + \cdots + A_m\lambda^{k+m})_r(A) = A_0A^k + \cdots + A_mA^{k+m}$$

$$= (A_0 + A_1A + \cdots + A_mA^m)A^k = Q_r(A)A^k.$$

Also: $WA^2 = WAA = BWA = BBW = B^2W$, etc.

25. Using Theorem 5 (or otherwise) decide which of the following pairs of matrices are similar.

(a) $\begin{bmatrix} 1 & 1 \\ 0 & 1 \end{bmatrix}, \begin{bmatrix} 1 & 0 \\ 0 & 1 \end{bmatrix}$ (b) $\begin{bmatrix} 1 & 1 \\ 1 & 1 \end{bmatrix}, \begin{bmatrix} 2 & 0 \\ 0 & 0 \end{bmatrix}$

(c) $\begin{bmatrix} 1 & 1 \\ 0 & 2 \end{bmatrix}, \begin{bmatrix} 1 & 0 \\ 0 & 2 \end{bmatrix}$ (d) $\begin{bmatrix} 1 & 1 & 0 \\ 0 & 1 & 1 \\ 0 & 0 & 1 \end{bmatrix}, \begin{bmatrix} 1 & 1 & 0 \\ 0 & 1 & 0 \\ 0 & 0 & 1 \end{bmatrix}$

(e) $\begin{bmatrix} 1 & 1 & 0 \\ 0 & 2 & 1 \\ 0 & 0 & 3 \end{bmatrix}, \begin{bmatrix} 1 & 0 & 0 \\ 0 & 2 & 0 \\ 0 & 0 & 3 \end{bmatrix}$.

26. Prove: if $A, B \in M_n(R)$ and A is nonsingular then, as a matrix in $M_n(R[\lambda])$, $\lambda A + B$ satisfies $\rho(\lambda A + B) = n$.

 Hint: We have

$$\det(\lambda A + B) = \det(A(\lambda I_n + A^{-1}B)) = \det(A)\det(\lambda I_n + A^{-1}B)$$
$$= (\det(A))(\lambda^n + \cdots) \neq 0.$$

27. Let $f(\lambda), g(\lambda) \in R[\lambda]$ and $A \in M_n(R)$. If $h(\lambda) = f(\lambda) + g(\lambda)$ and $k(\lambda) = f(\lambda)g(\lambda)$, show that $h(A) = f(A) + g(A)$ and $k(A) = f(A)g(A)$.

28. Compute the elementary divisors of the characteristic matrices of each of the following matrices:

 (a) $\begin{bmatrix} 0 & -2 \\ 1 & 3 \end{bmatrix}$ (b) $\begin{bmatrix} 0 & -1 \\ 1 & 0 \end{bmatrix}$

 (c) $\begin{bmatrix} 0 & 0 & 2 \\ 1 & 0 & -5 \\ 0 & 1 & 4 \end{bmatrix}$.

29. Let $A \in M_n(\mathbb{C})$ be a normal matrix. Prove: the elementary divisors of $\lambda I_n - A$ are linear polynomials.

 Hint: Apply §8.2 Theorem 1, and §10.2 Theorem 11.

10.2 Exercises

30. Let $f(\lambda) = \det(\lambda I_n - A)$. Find a new proof that $f(A) = 0 \in M_n(R)$. Recall that the result is universally named the Cayley-Hamilton Theorem.

    Hint: Write $f(\lambda)I_n = \det(\lambda I_n - A)I_n = (\text{adj}(\lambda I_n - A))(\lambda I_n - A)$. Thus division of $f(\lambda)I_n$ by $\lambda I_n - A$ leaves a 0 remainder. However, by the Remainder Theorem (§10.1 Exercises, #10), $0 = f_r(A)I_n = f(A)$.

31. Let A, B, C, D be in $M_n(R)$ and assume A is nonsingular. Prove: if $\lambda A + B \stackrel{E}{=} \lambda C + D$ then C is non-singular.

    Hint: Since $\lambda A + B \stackrel{E}{=} \lambda C + D$, $\det(\lambda A + B) = a \det(\lambda C + D)$, $0 \neq a \in R$. But the coefficient of λ^n in $\det(\lambda A + B)$ is $\det(A)$ and the coefficient of λ^n in $\det(\lambda C + D)$ is $\det(C)$. Thus $\det(A) \neq 0$ implies $\det(C) \neq 0$.

32. Let A be an n-square matrix partitioned as follows:

    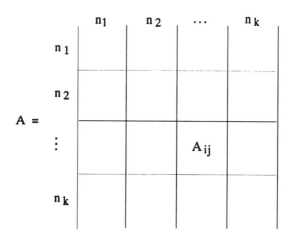

    Let P be an n-square permutation matrix with the partitioning

$$P = \begin{array}{c} {}_{n\,\sigma(1)} \\ {}_{n\,\sigma(2)} \\ \vdots \\ {}_{n\,\sigma(k)} \end{array} \begin{array}{cccc} {}^{n_1} & {}^{n_2} & \cdots & {}^{n_k} \\ \hline & & & \\ \hline & & & \\ \hline & & & \\ \hline & & & \\ \hline \end{array}$$

in which $\sigma \in S_k$ and in block row s every block is 0 except for the block column $t = \sigma(s)$, and there the identity matrix $I_{n_t} = I_{n_{\sigma(s)}}$ is found. Similarly, for $\mu \in S_k$ define an n-square permutation matrix Q by the partitioning

$$Q = \begin{array}{c} {}_{n_1} \\ {}_{n_2} \\ \vdots \\ {}_{n_k} \end{array} \begin{array}{cccc} {}^{n\,\mu(1)} & {}^{n\,\mu(2)} & \cdots & {}^{n\,\mu(k)} \\ \hline & & & \\ \hline & & & \\ \hline & & & \\ \hline & & & \\ \hline \end{array}$$

where block column t has 0 blocks in it, except in block row $s = \mu(t)$, and there the identity matrix $I_{n_s} = I_{n_{\mu(t)}}$ is found. Show that the (s, t) block in the partitioned matrix PAQ is $A_{\sigma(s), \mu(t)}$, s, t, = 1, ..., k.

10.2 Exercises

33. Let A_{ii} be an n_i-square matrix, $i = 1, \ldots, k$, and let $\sigma \in S_k$. Prove:

$$\sum_{i=1}^{k} \oplus A_{\sigma(i), \sigma(i)} \stackrel{s}{=} \sum_{i=1}^{k} \oplus A_{ii}.$$

Hint: Construct the matrix P in #32 corresponding to σ and also construct the matrix Q corresponding to σ. Then, by #32,

$$P(\sum_{i=1}^{k} \oplus A_{ii}) Q = \sum_{i=1}^{k} \oplus A_{\sigma(i), \sigma(i)}.$$

Note that, according to #32, the blocks in block row s of P are all 0 except $I_{n_{\sigma(s)}}$ appearing in block column $\sigma(s)$. Similarly block column t of Q has 0 blocks in it except in block row $\sigma(t)$ and there $I_{n_{\sigma(t)}}$ is found. Now P and Q are partitioned conformally for multiplication. By block multiplication, the (s, t) block in PQ is 0 unless $\sigma(s) = \sigma(t)$, i.e., $s = t$, and then the (s, s) block is $I_{n_{\sigma(s)}}$. Hence $PQ = I_n$.

34. If A_{11} is 2×2, A_{22} is 3×3, describe the similarity that makes $A_{11} \oplus A_{22}$ similar to $A_{22} \oplus A_{11}$.

Hint: Here $\sigma = (1\ 2)$, $n_1 = 2$, $n_2 = 3$ and

$$P = \begin{bmatrix} 0 & 0 & 1 & 0 & 0 \\ 0 & 0 & 0 & 1 & 0 \\ 0 & 0 & 0 & 0 & 1 \\ 1 & 0 & 0 & 0 & 0 \\ 0 & 1 & 0 & 0 & 0 \end{bmatrix}, \quad Q = P^{-1} = \begin{bmatrix} 0 & 0 & 0 & 1 & 0 \\ 0 & 0 & 0 & 0 & 1 \\ 1 & 0 & 0 & 0 & 0 \\ 0 & 1 & 0 & 0 & 0 \\ 0 & 0 & 1 & 0 & 0 \end{bmatrix}.$$

35. Let $A = \text{diag}(\lambda_1, \lambda_2, \lambda_3)$. Describe the similarity that makes A similar to $\text{diag}(\lambda_2, \lambda_3, \lambda_1)$. Perform the multiplication.

Hint: Here $\sigma = (1\ 2\ 3)$ and

$$P = \begin{bmatrix} 0 & 1 & 0 \\ 0 & 0 & 1 \\ 1 & 0 & 0 \end{bmatrix}, \quad Q = P^T.$$

Then

$$PAP^T = \begin{bmatrix} 0 & 1 & 0 \\ 0 & 0 & 1 \\ 1 & 0 & 0 \end{bmatrix} \begin{bmatrix} \lambda_1 & 0 & 0 \\ 0 & \lambda_2 & 0 \\ 0 & 0 & \lambda_3 \end{bmatrix} \begin{bmatrix} 0 & 0 & 1 \\ 1 & 0 & 0 \\ 0 & 1 & 0 \end{bmatrix} = \begin{bmatrix} 0 & \lambda_2 & 0 \\ 0 & 0 & \lambda_3 \\ \lambda_1 & 0 & 0 \end{bmatrix} \begin{bmatrix} 0 & 0 & 1 \\ 1 & 0 & 0 \\ 0 & 1 & 0 \end{bmatrix}$$

$$= \begin{bmatrix} \lambda_2 & 0 & 0 \\ 0 & \lambda_3 & 0 \\ 0 & 0 & \lambda_1 \end{bmatrix}.$$

36. Prove:

$$\begin{bmatrix} 0 & 1 & 0 \\ 0 & 0 & 1 \\ 1 & 0 & 0 \end{bmatrix} \stackrel{s}{\equiv} \begin{bmatrix} 1 & 0 & 0 \\ 0 & 0 & -1 \\ 0 & 1 & -1 \end{bmatrix}$$

as matrices in $M_3(\mathbb{R})$.

37. Prove: if A is an n × n permutation matrix then A is similar over $\mathbb{R}$ to a direct sum of matrices.

   Hint: Observe that if $e = [\frac{1}{\sqrt{n}} \ \frac{1}{\sqrt{n}} \ \cdots \ \frac{1}{\sqrt{n}}]^T$ then $Ae = e$. Let U be an orthogonal matrix with e as first column and compute that $U^TAU = [1] \oplus B$.

38. Prove: if $A \stackrel{s}{\equiv} C \oplus D$ and A is nonsingular, then C and D are nonsingular and $A^{-1} \stackrel{s}{\equiv} C^{-1} \oplus D^{-1}$.

   Hint: If $P^{-1}AP = C \oplus D$ then $P^{-1}A^{-1}P = C^{-1} \oplus D^{-1}$.

39. Show that any two similar matrices have the same minimal polynomial.

   Hint: This follows easily from the fact that $p(SAS^{-1}) = Sp(A)S^{-1}$ for any polynomial $p(\lambda) \in R[\lambda]$.

10.2 Exercises

40. Let $p(\lambda) \in R[\lambda]$ and let $m(\lambda)$ be the minimal polynomial of $A \in M_n(R)$. Prove that if $r = \deg m(\lambda)$, then there exists $d(\lambda) \in R[\lambda]$ with $\deg d(\lambda) \le r - 1$ such that $p(A) = d(A)$.

    Hint: Write $p(\lambda) = m(\lambda)q(\lambda) + d(\lambda)$, $\deg d(\lambda) \le r - 1$. Then $p(A) = d(A)$.

41. Let $P \in M_n(\mathbb{C})$, $P = Q + iM$ where $P, Q \in M_n(\mathbb{R})$. Show that $Q = \dfrac{P + \bar{P}}{2}$, $M = \dfrac{P - \bar{P}}{2i}$.

42. With the same notation as #41 show that if P is nonsingular then $Q + rM$ is singular for at most n values of $r \in \mathbb{R}$.

    Hint: Let $f(\lambda) = \det(Q + \lambda M) \in \mathbb{R}[\lambda]$. Then $f(i) = \det(P) \ne 0$ so $f(\lambda)$ is not the zero polynomial and clearly $\deg f(\lambda) \le n$. Hence $f(\lambda)$ has at most n roots in $\mathbb{R}$.

43. Let $A, B \in M_n(\mathbb{R})$ and suppose $PAP^{-1} = B$ where $P \in M_n(\mathbb{C})$. Prove that there exists a nonsingular $S \in M_n(\mathbb{R})$ such that $SAS^{-1} = B$.

    Hint: Write P as in #41 so that $PA = BP$, $(Q + iM)A = B(Q + iM)$, $QA = BQ$, $MA = BM$. Hence for any real r, $(Q + rM)A = B(Q + rM)$. By #42 choose r such that $Q + rM$ is nonsingular.

44. Let $g(\lambda) = (\lambda - \alpha)^m$ and let $A = H(g(\lambda))$ be the hypercompanion matrix of $g(\lambda)$. Prove that $\lambda I_m - A$ has the single elementary divisor $g(\lambda)$.

45. With A as in #44 prove that $A \stackrel{s}{=} A^T$.

46. Evaluate characteristic and minimal polynomials for each of the following matrices:

    (a) $0_{n,n}$ (b) I_n

(c)   diag($\lambda_1, \ldots, \lambda_n$), where $\lambda_i \neq \lambda_j$, $i \neq j$ (d)   A as in #44

(e) A^T, where A is the matrix in #44

(f) J_n, the n × n matrix each of whose entries is 1

(g) $\begin{bmatrix} -2 & 1 & 1 \\ 1 & -2 & 1 \\ 1 & 1 & -2 \end{bmatrix}$.

47.   Let $A \in M_n(\mathbb{R})$ satisfy $A^2 = -I_n$. Prove: n is even.

     Hint: Let $f(\lambda) = \lambda^2 + 1$. Then $f(A) = 0$, so that $f(\lambda)$ must be a multiple of the minimal polynomial. Since $f(\lambda)$ is irreducible, i.e., cannot be factored in $\mathbb{R}[\lambda]$, it follows that $f(\lambda)$ is the minimal polynomial of A. By Theorem 8, $f(\lambda) = q_n(\lambda)$ is the nth invariant factor of $\lambda I_n - A$. By the divisibility properties of the invariant factors, every nonunit invariant factor is a divisor of $\lambda^2 + 1$ and hence must equal $\lambda^2 + 1$. But then the companion matrix of each nonunit invariant factor is $\begin{bmatrix} 0 & -1 \\ 1 & 0 \end{bmatrix}$, so that by Theorem 7, n is even.

48.   Let A be as in #47 and n = 2p. Prove that A is similar over $\mathbb{R}$ to a direct sum of p 2 × 2 matrices $\begin{bmatrix} 0 & -1 \\ 1 & 0 \end{bmatrix}$.

     Hint: Use #47.

49.   Let $A \in M_p(R)$, $B \in M_q(R)$, $C = A \oplus B \in M_{p+q}(R)$. Prove: the minimal polynomial of C is lcm($m_A(\lambda)$, $m_B(\lambda)$) (i.e., *least common multiple* of $m_A(\lambda)$, $m_B(\lambda)$), where $m_A(\lambda)$ and $m_B(\lambda)$ are the minimal polynomials of A and B, respectively. Remember that the lcm of two polynomials is a multiple of each polynomial and any polynomial divisible by the two polynomials is divisible by the lcm.

     Hint: Let n = p + q. The elementary divisors of $\lambda I_n - C$ are those of $\lambda I_p - A$ and $\lambda I_q - B$ taken together. It follows that the highest degree invariant factor $q_n(\lambda)$

10.2 Exercises

of λI_n - C is formed by multiplying together all the different highest degree elementary divisors that appear in the combined lists for λI_p - A and λI_q - B. Now, by Theorem 8, $m_A(\lambda)$ (the pth invariant factor of λI_p - A) is the product of all the highest degree polynomials in the elementary divisor list for λI_p - A, and a similar statement is true for $m_B(\lambda)$. It follows, again by Theorem 8, that $q_n(\lambda)$ is the minimal polynomial for C and by construction $q_n(\lambda) = \text{lcm}(m_A(\lambda), m_B(\lambda))$.

50. Let A_1, ..., A_k be square matrices over R. Show that the minimal polynomial of $C = A_1 \oplus \cdots \oplus A_k$ is lcm $(m_{A_1}(\lambda), ..., m_{A_k}(\lambda))$, where $m_{A_i}(\lambda)$ is the minimal polynomial of A_i, i = 1, ..., k.

    Hint: Use #49 and induction.

51. Prove: if $f(\lambda)$ and $g(\lambda)$ are two relatively prime monic polynomials in $R[\lambda]$ and $A = C(f(\lambda))$, $B = C(g(\lambda))$, then the minimal polynomial of $A \oplus B$ is $f(\lambda)g(\lambda)$.

    Hint: Use #49.

52. Prove: if $\alpha, \beta \in R$, $\alpha \neq \beta$, then the minimal polynomial of

    $$H((\lambda - \alpha)^p) \oplus H((\lambda - \beta)^q)$$

    is $(\lambda - \alpha)^p (\lambda - \beta)^q$.

    Hint: Use #49.

53. Find the minimal polynomial of each of the following matrices:

    (a) $\begin{bmatrix} 0 & 0 & -1 \\ 1 & 0 & 2 \\ 0 & 1 & 0 \end{bmatrix}$

    (b) $\begin{bmatrix} 0 & 0 & -1 & 0 \\ 1 & 0 & 2 & 0 \\ 0 & 1 & 0 & 0 \\ 0 & 0 & 0 & 1 \end{bmatrix}$

(c) $\begin{bmatrix} 2 & 1 & 0 & 0 & 0 \\ 0 & 2 & 1 & 0 & 0 \\ 0 & 0 & 2 & 0 & 0 \\ 0 & 0 & 0 & 2 & 1 \\ 0 & 0 & 0 & 0 & 2 \end{bmatrix}$

54. Let
$$A = \begin{bmatrix} 2 & 1 & 0 \\ 0 & 2 & 1 \\ 0 & 0 & 2 \end{bmatrix}.$$

Express A^{-1} in the form $p(A)$ where $p(\lambda)$ is a polynomial in $R[\lambda]$ of degree 3.

Hint: In this problem, $A = H((\lambda - 2)^3)$ so that the minimal polynomial of A is $(\lambda - 2)^3 = \lambda^3 - 6\lambda^2 + 12\lambda - 8$. Hence $A^3 - 6A^2 + 12A - 8I_4 = 0$. Multiplying by A^{-1} we have $8A^{-1} = A^2 - 6A + 12I_4$ or $A^{-1} = p(A)$ where $p(\lambda) = \frac{1}{8}\lambda^2 - \frac{3}{4}\lambda + \frac{3}{2}$.

55. Let $A = \begin{bmatrix} 0 & -1 \\ 1 & 2 \end{bmatrix}$. Find A^5.

Hint: Divide λ^5 by $\lambda^2 - 2\lambda + 1$, the minimal polynomial of A. The remainder is $5\lambda - 4$. Thus $A^5 = 5A - 4I_2 =$

$$\begin{bmatrix} 0 & -5 \\ 5 & 10 \end{bmatrix} - \begin{bmatrix} 4 & 0 \\ 0 & 4 \end{bmatrix} = \begin{bmatrix} -4 & -5 \\ 5 & 6 \end{bmatrix}.$$

56. Find the minimal polynomial of
$$\begin{bmatrix} 0 & 1 & 0 & 0 \\ 0 & 0 & 1 & 0 \\ -1 & 2 & 0 & 0 \\ 0 & 0 & 0 & 1 \end{bmatrix}.$$

Hint: This is the transpose of the matrix in #53(b).

57. Prove: if $A \in M_n(R)$ then $A \stackrel{S}{\equiv} A^T$.

   Hint: Consider the two characteristic matrices $\lambda I_n - A$ and $\lambda I_n - A^T = (\lambda I_n - A)^T$. It is obvious that the k-square subdeterminants of $\lambda I_n - A$ and $(\lambda I_n - A)^T$ are the same (i.e., $C_k(X^T) = C_k(X)^T$, see §6.3 Theorem 2(a)). Hence the determinantal divisors, invariant factors, and elementary divisors of $\lambda I_n - A$ and $\lambda I_n - A^T$ are identical. Hence $\lambda I_n - A \stackrel{E}{\equiv} \lambda I_n - A^T$. Apply Theorem 5.

58. Find the Jordan canonical form of $A = C(f(\lambda))$ where $f(\lambda) = (\lambda + 3)^3 (\lambda + 5)(\lambda - 1)^2$.

   Hint: The single invariant factor of $\lambda I_6 - A$ is $f(\lambda)$. Thus the elementary divisors are $(\lambda + 3)^3, (\lambda + 5), (\lambda - 1)^2$. Hence the Jordan canonical form of A is

$$H((\lambda + 3)^3) \oplus H((\lambda - 1)^2) \oplus H(\lambda + 5) = \begin{bmatrix} -3 & 1 & 0 & 0 & 0 & 0 \\ 0 & -3 & 1 & 0 & 0 & 0 \\ 0 & 0 & -3 & 0 & 0 & 0 \\ 0 & 0 & 0 & 1 & 1 & 0 \\ 0 & 0 & 0 & 0 & 1 & 0 \\ 0 & 0 & 0 & 0 & 0 & -5 \end{bmatrix}.$$

59. Prove: if $A \in M_n(R)$ has n distinct eigenvalues in R then A is similar over R to a diagonal matrix.

   Hint: Apply Theorem 12.

60. Show that there is no matrix $A \in M_2(R)$ such that

$$A^2 = \begin{bmatrix} 0 & 1 \\ 0 & 0 \end{bmatrix}.$$

   Hint: Both eigenvalues of A must be 0 (i.e., any matrix is similar to an upper triangular matrix and hence the eigenvalues of A are λ_i^2 where λ_i, $i = 1, 2$ are the eigenvalues of A). Now the Jordan form of A must be

$$J_A = \begin{bmatrix} 0 & 1 \\ 0 & 0 \end{bmatrix}.$$

But then $J_A^2 = 0$ so $A^2 = 0$ (why?).

61. Show that there is no matrix $A \in M_3(\mathbb{R})$ such that

$$A^3 = \begin{bmatrix} 0 & 1 & 0 \\ 0 & 0 & 1 \\ 0 & 0 & 0 \end{bmatrix}.$$

Hint: As in #60, the eigenvalues of A are all 0 so that the Jordan normal form J_A must be one of

$$\begin{bmatrix} 0 & 1 & 0 \\ 0 & 0 & 1 \\ 0 & 0 & 0 \end{bmatrix}, \begin{bmatrix} 0 & 1 & 0 \\ 0 & 0 & 0 \\ 0 & 0 & 0 \end{bmatrix}.$$

But $J_A^3 = 0$ for both of these matrices.

62. Find the elementary divisors of the characteristic matrix of

$$A = \begin{bmatrix} 2 & -1 & 1 \\ 2 & 2 & -1 \\ 1 & 2 & -1 \end{bmatrix}$$

and compute both the Frobenius and Jordan normal forms of A.

63. Find the Jordan normal form over $\mathbb{C}$ of $A = C(f(\lambda))$ where $f(\lambda) = (\lambda - 1)^2 (\lambda^2 + 1)^2$.

Hint: The elementary divisors of $\lambda I_4 - A$ are $(\lambda - 1)^2$, $(\lambda + i)^2$, $(\lambda - i)^2$. Thus the Jordan normal form of A is

10.2 Exercises

$$\begin{bmatrix} 1 & 1 \\ 0 & 1 \end{bmatrix} \oplus \begin{bmatrix} -i & 1 \\ 0 & -i \end{bmatrix} \oplus \begin{bmatrix} i & 1 \\ 0 & i \end{bmatrix}.$$

64. Find the Frobenius normal form of

$$A = \begin{bmatrix} 0 & 0 & a_1 \\ 0 & a_2 & 0 \\ a_3 & 0 & 0 \end{bmatrix}.$$

Discuss cases.

Hint: If $a_1 = a_3 = 0$ there is no problem. By performing a similarity if necessary, $P^{-1}AP$,

$$P = \begin{bmatrix} 0 & 0 & 1 \\ 0 & 1 & 0 \\ 1 & 0 & 0 \end{bmatrix},$$

we can assume $a_3 \neq 0$. Then

$$\lambda I_3 - A = \begin{bmatrix} \lambda & 0 & -a_1 \\ 0 & \lambda - a_2 & 0 \\ -a_3 & 0 & \lambda \end{bmatrix}.$$

65. Prove that if r is an eigenvalue of $A \in M_n(R)$ and $m(\lambda)$ is the minimal polynomial of A then $m(r) = 0$.

Hint: By Theorem 8, $m(\lambda)$ is the nth invariant factor $q_n(\lambda)$ of $\lambda I_n - A$. The facts that $\det(\lambda I_n - A) = q_n(\lambda) \cdots q_1(\lambda)$ and $q_1(\lambda) | q_2(\lambda) | \cdots | q_n(\lambda)$ completes the proof.

66. Prove: if $A \in M_n(R)$, $a \in R$, then $p(\lambda)$ is an elementary divisor of $\lambda I_n - A$ iff $p(\lambda + a)$ is an elementary divisor of $\lambda I_n - (A - aI_n)$.

Hint: $\lambda I_n - (A - aI_n) = (\lambda + a)I_n - A$. Let $\mu = \lambda + a$ so that $p(\mu)$ is an elementary divisor of $\mu I_n - A$. But $p(\lambda + a) = p(\mu)$.

67. As in #66, if $a \neq 0$ then $p(\lambda)$ is an elementary divisor of A iff $p(\lambda/a)$ is an elementary divisor of $\lambda I_n - aA$.

Hint: $\lambda I_n - aA \stackrel{E}{\equiv} \frac{\lambda}{a} I_n - A$. Let $\mu = \lambda/a$ and argue as in the Hint in #66.

68. If $A \in M_n(R)$ and $A^2 = A$, discuss the possibilities for the Frobenius normal form of A.

Hint: Since $A^2 - A = 0$ it follows that the minimal polynomial of A, $m(\lambda)$, is a divisor of $\lambda^2 - \lambda$. Thus the possibilities for $m(\lambda)$ are:

(i) $m(\lambda) = \lambda$;
(ii) $m(\lambda) = \lambda - 1$;
(iii) $m(\lambda) = \lambda(\lambda - 1)$.

In case (i), $A = 0$; in (ii), $A - I_n = 0$, i.e., $A = I_n$; in (iii) the highest degree elementary divisors are λ and $\lambda - 1$, i.e., the factors of $m(\lambda)$. Thus the Frobenius normal form of A is a direct sum of companion matrices of λ and $\lambda - 1$. Thus A is similar over R to a direct sum of 1-square matrices of the form [0] or [1]. But then

$$A \stackrel{S}{\equiv} I_r \oplus 0_{n-r}$$

where $r = \rho(A)$, I_r is the r-square identity matrix and 0_{n-r} is the (n - r)-square zero matrix.

69. Let $B \in M_2(R)$ be the companion matrix of the irreducible binomial

$$f(\lambda) = \lambda^2 + c_1\lambda + c_0 \in R[\lambda],$$

$$B = \begin{bmatrix} 0 & -c_0 \\ 1 & -c_1 \end{bmatrix}.$$

Define $A \in M_{2p}(R)$ to be the matrix

$$\begin{bmatrix} B & \begin{smallmatrix}0&0\\1&0\end{smallmatrix} & & & & \\ & B & \begin{smallmatrix}0&0\\1&0\end{smallmatrix} & & & \\ & & B & \begin{smallmatrix}0&0\\1&0\end{smallmatrix} & & \\ & & & B & \ddots & \\ & & & & \ddots & \begin{smallmatrix}0&0\\1&0\end{smallmatrix} \\ & & & & & B \end{bmatrix}$$

in which there are p occurrences of B down the main diagonal of A, 1's occur in positions (2, 3), (4, 5), ..., (2p - 2, 2p - 1), and all other entries of A are 0. Prove

$$A \overset{s}{\equiv} C(f(\lambda)^p).$$

Hint: It is easy to see that

$$\det((\lambda I_{2p} - A)(2p \mid 1)) = \pm c_0^p$$

and, since $f(\lambda)$ is irreducible, $c_0 \neq 0$. Hence $d_{2p-1} = 1$ and the determinantal divisors of $\lambda I_{2p} - A$ are $d_0 = \cdots = d_{2p-1} = 1$, $d_{2p} = \det(\lambda I_{2p} - A) = (\det(\lambda I_2 - B))^p = f(\lambda)^p$. Thus $\lambda I_{2p} - A$ has the single elementary divisor $f(\lambda)^p$, as does the characteristic matrix of $C(f(\lambda)^p)$.

70. Let A and B be matrices in $M_n(R)$. Let F be an *extension field* of R, i.e., F is a field containing R and the operations in F are the same as those in R when applied to elements of R. For example: $\mathbb{C}$ is an extension field of $\mathbb{R}$; $\mathbb{R}$ is an extension field of $\mathbb{Q}$. Suppose that there exists a nonsingular matrix $S \in M_n(F)$ such that $S^{-1}AS = B$. Then show that there exists a nonsingular matrix $P \in M_n(R)$ such that $P^{-1}AP = B$.

Hint: The characteristic matrices $\lambda I_n - A$ and $\lambda I_n - B$ are equivalent as matrices over $F[\lambda]$ and hence have the same determinantal divisors. However, these determinantal divisors are gcd's of polynomials in $R[\lambda]$, and hence are in $R[\lambda]$ (why?). Thus, as matrices over $R[\lambda]$, $\lambda I_n - A$ and $\lambda I_n - B$ have the same determinantal divisors. Thus they are equivalent over $R[\lambda]$ so that by Theorem 5, A and B are similar over R.

71. Suppose that A and B have the same irreducible characteristic polynomial. Prove that A and B are similar.

Hint: If a matrix has an irreducible characteristic polynomial, then that polynomial is the only nontrivial invariant factor of the characteristic matrix. Thus, $\lambda I_n - A$ and $\lambda I_n - B$ have the same invariant factors, i.e., A and B are similar.

72. Let R be a field, and let $A(\lambda) \in M_m(R[\lambda])$, $B(\lambda) \in M_n(R[\lambda])$. Let $p = p(\lambda)$ in $R[\lambda]$ be an irreducible polynomial and suppose $p^{e_1}, \ldots, p^{e_r}$ are all the elementary divisors of $A(\lambda)$ involving $p(\lambda)$ while $p^{f_1}, \ldots, p^{f_s}$ are all the elementary divisors of $B(\lambda)$ involving p. Prove that the elementary divisors of the Kronecker product $A(\lambda) \otimes B(\lambda)$ involving p are $p^{e_i+f_j}$, $i = 1, \ldots, r$, $j = 1, \ldots, s$.

Hint: First note that $A(\lambda) \otimes B(\lambda) \stackrel{E}{=} S(A) \otimes S(B)$ where $S(A)$ and $S(B)$ are the Smith normal forms of $A(\lambda)$ and $B(\lambda)$ respectively. For, suppose $PAQ = S(A)$ and $DBG = S(B)$. Then, $(P \otimes D)(A \otimes B)(Q \otimes G) = S(A) \otimes S(B)$. But $S(A) \otimes S(B)$ is diagonal, so its elementary divisors are just the prime power factors of the main diagonal elements. Hence the elementary divisors of $A(\lambda) \otimes B(\lambda)$ involving $p = p(\lambda)$ are $p^{e_i+f_j}$, $i = 1, \ldots, r$, $j = 1, \ldots, s$. (We allow the exponents e_i, f_j to be zero, if necessary.)

73. Let A be an n-square upper triangular matrix and let a be a characteristic root of A of algebraic multiplicity r. Assume that

$$A[\,i, i+1 \mid i, i+1\,] = \begin{bmatrix} a & x \\ 0 & a \end{bmatrix}, x \neq 0.$$

Prove that $\lambda I_n - A$ has an elementary divisor of the form $(\lambda - a)^e$, $e \geq 2$.

Hint: The matrix $\lambda I_n - A$ is upper triangular, so that

10.2 MatLab

$$\det(\lambda I_n - A) = (\lambda - a)^r p(\lambda),$$

where gcd $(\lambda - a, p(\lambda)) = 1$ and $r \geq 2$. Observe that $(\lambda I_n - A)(i + 1 \mid i)$ is upper triangular, and that the $(n-1)^{st}$ determinantal divisor of $\lambda I_n - A$, $d_{n-1}(\lambda)$, satisfies

$$d_{n-1}(\lambda) \mid \det((\lambda I_n - A)(i + 1 \mid i)) = x(\lambda - a)^{r-2} p(\lambda).$$

Hence

$$(\lambda - 2)^2 \mid q_n(\lambda) = \frac{d_n(\lambda)}{d_{n-1}(\lambda)}.$$

74. Let $f(\lambda)$ and $g(\lambda)$ be monic and assume that gcd $(f(\lambda), g(\lambda)) = 1$ in $R[\lambda]$. Prove that

$$C(f(\lambda)g(\lambda)) \overset{S}{\equiv} C(f(\lambda)) \oplus C(g(\lambda)).$$

Hint: The list of elementary divisors of the characteristic matrix of $C(f(\lambda)g(\lambda))$ is obtained by factoring its single nontrivial invariant factor, $f(\lambda)g(\lambda)$. The list of elementary divisors of the characteristic matrix of $C(f(\lambda)) \oplus C(g(\lambda))$ is obtained by putting together the lists of elementary divisors of the characteristic matrices of $C(f(\lambda))$ and $C(g(\lambda))$, i.e., by listing the factors of $f(\lambda)$ and those of $g(\lambda)$. Since gcd$(f(\lambda), g(\lambda)) = 1$, these lists are the same.

10.2 MatLab

1. We know from §10.2 Theorem 5 that if A and B are two n-square matrices in $M_n(R)$, then $A \overset{S}{\equiv} B$ iff $\lambda I_n - A \overset{E}{\equiv} \lambda I_n - B$. But then §10.2 Theorem 2 shows that $A \overset{S}{\equiv} B$ iff the determinantal divisors of $\lambda I_n - A$ and $\lambda I_n - B$ are identical. Use the function ddiv in §10.1 MatLab, #11 to write a MatLab function named sym that accepts two n-square matrices A and B and returns the value 1 if they are similar over R, 0 otherwise. Save sym in pmatrix.

Hint:
```
function s=sym(A, B);
%SYM accepts two n-square matrices
%and returns 1 if they are similar,
%0 if they are not similar.
n=length(A);
F=[eye(n) -A];
G=[eye(n) -B];
dkF=[ ];
dkG=[ ];
for k=1:n
    dkF=[dkF ddiv(F,k)];
    dkG=[dkG ddiv(G,k)];
end
p=length(dkF);
q=length(dkG);
if (p~=q)
    s=0;
elseif (dkF~=dkG)
    s=0;
else
    s=1;
end
```

2. Use sym to check which of the pairs of matrices A, B appearing in §10.1 MatLab #12(b) - (g) are similar.

3. Use sym to show that $J_p \stackrel{s}{=} \mathrm{diag}(p, 0, \ldots, 0)$ for $p = 2, 3, 4$. Notation: J_p = ones(p). State and prove a generalization of these results.

4. Review the functions hrand (see §8.1 MatLab, #4) and ddiv (§10.1, MatLab, #11). Explain the following dialog in the Command window.

```
>>A = hrand(2);               %1
>>F = [eye(2) -A];
>>ddiv(F, 2)                  %2
```

ans =

        1.0000   0.5886   -1.0727   %3

```
>>d = eig(A);                        %4
>>D = diag(d);                       %5
>>G = [eye(2) -D];                   %6
>>ddiv(G, 2)                         %7
```

ans =

    1.0000   0.5886 - 0.0000i   -1.0727 - 0.0000i   %8

Hint: First, a random 2-square hermitian A is generated (%1); then, (%2) $d_2(\lambda I_2 - A)$ is computed. The result (%3) is the polynomial

$$p(\lambda) = \lambda^2 - .5886\lambda - 1.0727.$$

In (%4) the eigenvalues of A are computed, say λ_1, and λ_2, and then in (%5) and (%6) $\lambda I_2 - \text{diag}(\lambda_1, \lambda_2)$ is obtained. In (%7) $d_2(\lambda I_2 - D)$ is computed yielding the polynomial $p(\lambda)$. The reason for this is that A and D are similar.

5. Use MatLab to compute the minimal polynomials of each of the following matrices.

(a) J_3 (b) J_4

(c) $A = \begin{bmatrix} 1 & 1 \\ 0 & 1 \end{bmatrix}$ (d) $A = \begin{bmatrix} 0 & 1 & 0 \\ 0 & 0 & 1 \\ 0 & 0 & 0 \end{bmatrix}$

(e) $A = \begin{bmatrix} 0 & 0 & 0 \\ 1 & 0 & 0 \\ 0 & 1 & 0 \end{bmatrix}$ (f) $A = \begin{bmatrix} 0 & 0 & 1 \\ 0 & 0 & 0 \\ 0 & 0 & 0 \end{bmatrix}$

(g) $A = \begin{bmatrix} 3 & 0 & 0 & 0 \\ 2 & 3 & 0 & 0 \\ 0 & 0 & 3 & 0 \\ 1 & 0 & -1 & 3 \end{bmatrix}$

Hint:

(a)   >>A = ones(3);
           >>F = [eye(3) -A];
           >>d3 = ddiv(F, 3);
           >>d2 = ddiv(F, 2);
           >>p = deconv(d3, d2)

           p =

             1   -3   0

Thus $\lambda(\lambda - 3)$ is the minimal polynomial of J_3. The result can be confirmed by evaluating polyvalm(p, A).

(g)   >>A = [3 0 0 0; 2 3 0 0; 0 0 3 0; 1 0 -1 3];
           >>F = [eye(4) -A];
           >>d4 = ddiv(F, 4);
           >>d3 = ddiv(F, 3);
           >>p = deconv(d4, d3)

           p =

             1   -6   9

           >>polyvalm(p, A)

           ans =

$$\begin{matrix} 0 & 0 & 0 & 0 \\ 0 & 0 & 0 & 0 \\ 0 & 0 & 0 & 0 \\ 0 & 0 & 0 & 0 \end{matrix}$$

Thus the minimal polynomial of A is $\lambda^2 - 6\lambda + 9$.

6. Use MatLab to obtain the Frobenius normal form of A in #5(g).

Hint: The determinantal divisors and invariant factors of $\lambda I_4 - A$ can be computed with the following script entered in the Edit window (see §10.2 formula (4)).

```
A = [3 0 0 0; 2 3 0 0; 0 0 3 0; 1 0 -1 3];
F = [eye(4) -A];
d4 = ddiv(F, 4);
d3 = ddiv(F, 3);
d2 = ddiv(F, 2);
d1 = ddiv(F, 1);
q4 = deconv(d4, d3)
q3 = deconv(d3, d2)
q2 = deconv(d2, d1)
q1 = d1
d4
```

The resulting output indicates that q4 = [1 -6 9] = q3, q2 = q1 = 1, and d4 = [1 –12 54 -108 81] in which d4 is the characteristic polynomial of A. Thus the two nonunit invariant factors of $\lambda I_4 - A$ are $\lambda^2 - 6\lambda + 9$, twice. The companion matrix of this polynomial is

$$C = \begin{bmatrix} 0 & -9 \\ 1 & 6 \end{bmatrix}$$

and thus the Frobenius normal form of A is $C \oplus C$.

7. What are the elementary divisors of $\lambda I_4 - A$ for the matrix A in #5(g)? What is the Jordan normal form?

   Hint: Since the two nonunit invariant factors are both
   $$\lambda^2 - 6\lambda + 9 = (\lambda - 3)^2,$$
   it follows that the list of elementary divisors is just
   $$(\lambda - 3)^2, \ (\lambda - 3)^2.$$
   The Jordan form is then $C \oplus C$, where
   $$C = \begin{bmatrix} 3 & 1 \\ 0 & 3 \end{bmatrix}.$$

8. Use the function sym to determine if the matrices
   $$A = I_4 + (C \oplus C) \text{ and } B = I_4 + U$$
   are similar. Here
   $$C = \begin{bmatrix} 0 & 1 \\ 0 & 0 \end{bmatrix}, \quad U = \begin{bmatrix} 0 & 1 & 0 & 0 \\ 0 & 0 & 1 & 0 \\ 0 & 0 & 0 & 1 \\ 0 & 0 & 0 & 0 \end{bmatrix}.$$
   Explain the output.

   Hint: The output is 0, indicating that A and B are not similar. Obviously A and B can be similar iff $C \oplus C$ and U are. But the elementary divisors of $\lambda I_4 - (C \oplus C)$ are λ^2, twice, whereas the single elementary divisor of $\lambda I_4 - U$ is λ^4.

9. Use the Command window to construct several random complex square matrices A whose maximum modulus eigenvalue is strictly less than 1. Use eig to confirm

that the matrices satisfy $|\lambda_1(A)| < 1$. Compute the Frobenius norm of A^n for several increasing values of n. What seems to be the case, based on the output?

Hint: The following is a typical dialog.

>>A = arand(4, 4);
>>L = max(abs(eig(A)));

L =

   0.7921

>>norm(A^2, 'fro')

ans =

   0.9623

>>norm(A^100, 'fro')

ans =

   9.1728e-11

It appears that if $|\lambda_1(A)| < 1$ then A^n is approaching the zero matrix as n increases. In general, a *convergent matrix* is one that satisfies $\lim_{n \to \infty} A^n = 0$. The preceding limit means that every entry of A^n approaches 0 as $n \to \infty$.

10. Prove: if A is a square complex matrix then A is convergent (see #9) iff $|\lambda_1(A)| < 1$.

Hint: According to Theorem 14, A is similar over $\mathbb{C}$ to a direct sum of matrices of the form

$$H((\lambda - \alpha)^m) = \alpha I_m + U_m \quad (1)$$

where α is an eigenvalue of A and U_m is the m-square auxiliary unit matrix. Thus it is clear that A is convergent iff each of the matrices (1) is convergent (why?). Since I_m commutes with any matrix, it is simple to see that

$$(\alpha I_m + U_m)^p = \sum_{k=0}^{p} \binom{p}{p-k} \alpha^{p-k} U_m^k . \qquad (2)$$

Although the number of summands on the right side of (2) appears to increase with p, this is not actually the case because $U_m^m = 0$. Since α^p appears down the main diagonal of (2), it is clear that $|\alpha| < 1$ is necessary for (1) to be convergent. Conversely, suppose $|\alpha| < 1$. Consider

$$\left| \binom{p}{p-k} \alpha^{p-k} \right| = \left| \frac{p(p-1)\cdots(p-k+1)}{k! \, \alpha^k} \alpha^p \right|$$

$$\leq \frac{p^k |\alpha|^p}{k! \, \alpha^k} .$$

Thus we need only show that $p^k |\alpha|^p \to 0$ as $p \to \infty$. But

$$\log (p^k |\alpha|^p) = k \log p + p \log |\alpha|$$

$$= p(k \frac{\log p}{p} + \log |\alpha|) . \qquad (3)$$

From calculus we know that $(\log p)/p \to 0$ as $p \to \infty$ so that the right side of (3) goes to $-\infty$ as $p \to \infty$ (i.e., $\log |\alpha| < 0$). Hence $p^k |\alpha|^p \to 0$ as $p \to \infty$.

11. Let A be an n-square complex matrix with no eigenvalues lying on the imaginary axis in the complex plane. Suppose that

$$A = S J_A S^{-1} \qquad (4)$$

where J_A is the Jordan normal form of A (see Theorem 14). Each Jordan block down the main diagonal of J_A has the form (1) (if $m = 0$, $U_m = 0$). Replace each Jordan block (1) in J_A by $\varepsilon(\alpha) I_m$ where $\varepsilon(\alpha) = 1$ if $\text{Re}(\alpha) > 0$ and $\varepsilon(\alpha) = -1$ if $\text{Re}(\alpha) < 0$. Denote the resulting matrix by $\varepsilon(J_A)$. The matrix

10.2 MatLab

$$\varepsilon(A) = S\varepsilon(J_A)S^{-1} \qquad (5)$$

is called the *sign of A* and $\varepsilon(A)$ is called the matrix sign function. It is frequently denoted as sgn(A).

(a) Write a single command that returns sgn(J_A), given A as described above. Read the Help entry for sign.

Hint:

$$\text{diag(sign(sort(real(eig(A)))))} \,. \qquad (6)$$

(b) Prove: $A + \varepsilon(A)$ is nonsingular.

Hint: From (4) and (5),

$$A + \varepsilon(A) = SJ_AS^{-1} + S\varepsilon(J_A)S^{-1}$$
$$= S(J_A + \varepsilon(J_A))S^{-1} \,. \qquad (7)$$

The eigenvalues of $J_A + \varepsilon(J_A)$ are of the form $\alpha + \varepsilon(\alpha)$ which always has a non-zero real part.

(c) Prove:

$$B = (A - \varepsilon(A))(A + \varepsilon(A))^{-1} \qquad (8)$$

has eigenvalues of the form

$$\beta = \frac{\alpha - \varepsilon(\alpha)}{\alpha + \varepsilon(\alpha)} \,. \qquad (9)$$

Hint: Observe that (4) and (5) imply that

$$S^{-1}BS = (S^{-1}AS - S^{-1}\varepsilon(A)S)(S^{-1}AS + S^{-1}\varepsilon(A)S)^{-1}$$
$$= (J_A + \varepsilon(J_A))(J_A + \varepsilon(J_A))^{-1}. \tag{10}$$

The matrix $J_A + \varepsilon(J_A)$ is upper triangular with numbers of the form $\alpha + \varepsilon(\alpha)$ on the main diagonal. Similarly for $(J_A + \varepsilon(J_A))^{-1}$. Hence the eigenvalues of B are of the form (9).

(d)   Prove: the eigenvalues β of B satisfy $|\beta| < 1$ and hence B is convergent.

Hint: Write $\alpha = a + ib$ so that

$$\alpha - \varepsilon(\alpha) = a + ib - \frac{a}{|a|},$$

$$\alpha + \varepsilon(\alpha) = a + ib + \frac{a}{|a|}.$$

Then

$$|\alpha - \varepsilon(\alpha)|^2 = a^2(1 - \frac{1}{|a|})^2 + b^2,$$

$$|\alpha + \varepsilon(\alpha)|^2 = a^2(1 + \frac{1}{|a|})^2 + b^2.$$

Thus $|\beta| < 1$ iff

$$(1 - \frac{1}{|a|})^2 < (1 + \frac{1}{|a|})^2$$

which is simple to square out and confirm.

(e)   Define two sequences of matrices as follows: $A_1 = A$,

$$A_{k+1} = (A_k + A_k^{-1})/2, \quad k = 1, 2, \ldots \tag{11}$$

and

$$B_k = (A_k - \varepsilon(A_k))(A_k + \varepsilon(A_k))^{-1}, \quad k = 1, 2, \ldots. \tag{12}$$

Prove: the eigenvalues of A_k have nonzero real parts. Also show that $\varepsilon(A_{k+1}) = \varepsilon(A_k)$ and hence $\varepsilon(A_k) = \varepsilon(A)$ for $k = 1, 2, \ldots$.

Hint: Use induction on k. The assumption about $A_1 = A$ is that every eigenvalue α has a nonzero real part. We need only show that consequently $\alpha + \frac{1}{\alpha}$ has a non-zero real part. But

$$\begin{aligned} \mathrm{Re}(\alpha + \frac{1}{\alpha}) &= \mathrm{Re}(a + ib + \frac{a - ib}{|\alpha|^2}) \\ &= a(1 + \frac{1}{|\alpha|^2}) \neq 0. \end{aligned} \tag{13}$$

However (13) also shows that the sign of $\mathrm{Re}(\alpha + \frac{1}{\alpha})$ is the same as the sign of $\mathrm{Re}(\alpha)$. Thus

$$\varepsilon(J_{A_2}) = \varepsilon(J_{A_1}),$$

$$\varepsilon(J_{A_3}) = \varepsilon(J_{A_2}), \text{ etc.}$$

But by the definition (5),

$$\begin{aligned} \varepsilon(A_k) &= S\varepsilon(J_{A_k})S^{-1} \\ &= S\varepsilon(J_{A_{k-1}})S^{-1} \\ &\vdots \\ &= S\varepsilon(J_{A_1})S^{-1} \\ &= \varepsilon(A). \end{aligned}$$

This simplifies the definition of the sequence B_k to

$$B_k = (A_k - \varepsilon(A))(A_k + \varepsilon(A))^{-1}. \qquad (14)$$

Also note that since $\varepsilon(A)$ commutes with A, it also commutes with every A_k (an easy induction argument).

(f)   Prove: $B_{k+1} = B_k^2$.

Hint: First, it is obvious that $\varepsilon(A)^2 = I_n$. Then, from (14) and the fact that $\varepsilon(A)$ commutes with every A_k, we have

$$B_k^2 = (A_k - \varepsilon(A))^2 (A_k + \varepsilon(A))^{-2}$$

$$= (A_k^2 - 2\varepsilon(A)A_k + I_n)(A_k^2 + 2\varepsilon(A)A_k + I_n)^{-1}$$

$$= (\frac{A_k + A_k^{-1}}{2} - \varepsilon(A))(2A_k)((\frac{A_k + A_k^{-1}}{2} + \varepsilon(A))(2A_k))^{-1}$$

$$= (\frac{A_k + A_k^{-1}}{2} - \varepsilon(A))(\frac{A_k + A_k^{-1}}{2} + \varepsilon(A))^{-1}$$

$$= (A_{k+1} - \varepsilon(A))(A_{k+1} + \varepsilon(A))^{-1}$$

$$= B_{k+1}.$$

(g)   Prove: $\lim_{k \to \infty} B_k = 0$.

Hint: We have (from (f))

$$B_{k+1} = B_k^2 = (B_{k-1}^2)^2 = B_{k-1}^4 = \cdots = B^{2^k},$$

and by (d), $B^{2^k} \to 0$ (rapidly!)

(h)   Prove: $\lim_{k \to \infty} A_k = \varepsilon(A)$.

Hint: From (14), a little algebra shows that

$$(I_n - B_k)A_k = (B_k + I_n)\varepsilon(A).\tag{15}$$

Since $B_k \to 0$ as $k \to \infty$ it follows immediately from (15) that

$$A_k \to \varepsilon(A)$$

as $k \to \infty$.

(i) Write a MatLab program called sgn that accepts an n-square complex matrix A with no eigenvalues on the imaginary axis, and returns the matrix sgn(A). Use (h) and the recursion (11). Save sgn in pmatrix.

Hint:
```
function [E,m]=sgn(A)
%SGN accepts a complex
%matrix A,with no eigenvalues
%on the imaginary axis,and
%and returns sgn(A).
X=A;
Y=(X+inv(X))/2;
k=1;
while norm(Y-X,'fro')>=1e-10
    X=Y;
    Y=(X+inv(X))/2;
    k=k+1;
end
m=k;
E=Y;
```

(j) Prove: if A has no eigenvalues on the imaginary axis then pos = $\rho(I_n + \varepsilon(A))$ is the number of eigenvalues of A with positive real part.

Hint: Since $I_n + \varepsilon(A) = I_n + S\varepsilon(J_A)S^{-1} = S(I_n + \varepsilon(J_A))S^{-1}$, it follows that $\rho(I_n + \varepsilon(A)) = \rho(I_n + \varepsilon(J_A))$. The result should now be clear.

(k) Use the result in (j) to compute the number of eigenvalues of A with positive real parts for various random A. Confirm your answer by evaluating sign(real(eig(A))).

Hint: The following script and output is typical.

```
u = [ ]; v = [ ];
for k = 1:10
    A = arand(10, 10);
    u = [u  rank(eye(10) + sgn(A))];
    v = [v  length(find(real(eig(A)) > 0))];
end
u
v
```

```
>>
u =

    7 4 6 4 6 5 5 7 5 5


v =

    7 4 6 4 6 5 5 7 5 5
```

10.2 Glossary

annihilating polynomial	645
auxiliary unit matrix	653
Cayley-Hamilton Theorem	645
characteristic matrix	632
characteristic polynomial	637
characteristic roots	637
companion matrix	643
constructive	644
convergent matrix	683
$\varepsilon(A)$	685
eigenvalues	637
elementary divisor	631
equivalence class	657
equivalence relation	657

10.2 Glossary

extension field 675
Frobenius normal form 645
hypercompanion matrix 653
invariant factor 632
irreducible 644
irreducible matrix 649
Jordan normal form 654
lcm 668
least common multiple 668
linear diophantine equations 659
list of elementary divisors 632
matrix sign function 685
minimal polynomial 645
reducible matrix 649
SDR 658
sgn(A) 685
sign 685
similar 644
similar over R 644
similarity classes 661
Smith normal form 632
system of distinct representatives 658
transversal 658
unimodular matrix 658
unit matrix 658

REFERENCES

Thomas F. Coleman, Charles F. Van Loan, *Handbook for Matrix Computations*, Society for Industrial and Applied Mathematics, Philadelphia, 1988.

Philip E. Gill, Walter Murray, Margaret H. Wright, *Numerical Linear Algebra and Optimization, Vol. 1,* Addison-Wesley Publishing Company, Reading, Massachusetts, 1991.

Jack L. Goldberg, *Matrix Theory with Applications*, McGraw-Hill, New York, 1991.

Gene H. Golub, Charles F. Van Loan, *Matrix Computations*, 2nd ed., The Johns Hopkins University Press, Baltimore, 1989.

David R. Hill, *Experiments in Computational Matrix Algebra,* Random House: Birkhauser Mathematics Series, New York, 1988.

Roger A. Horn, Charles R. Johnson, *Matrix Analysis*, Cambridge University Press, New York, 1985.

Roger A Horn, Charles R. Johnson, *Topics in Matrix Analysis*, Cambridge University Press, New York, 1991.

Steven J. Leon, *Linear Algebra with Applications,* 3rd ed., Macmillan Publishing Company, New York, 1990.

Marvin Marcus, Henryk Minc, *A Survey of Matrix Theory and Matrix Inequalities,* Dover Publications, New York, 1992.

Marvin Marcus, Henryk Minc, *Introduction to Linear Algebra,* Dover Publications, New York, 1988.

Ben Noble, James Daniel, *Applied Linear Algebra,* 3rd ed., Prentice Hall, Englewood Cliffs, New Jersey, 1988.

James M. Ortega, *Numerical Analysis, A Second Course*, Society for Industrial and Applied Mathematics, Philadelphia, 1990.

Louis L. Scharf, Richard T. Behrens, *A First Course in Electrical and Computer Engineering with MATLAB™ Programs and Experiments*, Addison-Wesley Publishing Company, Reading, Massachusetts, 1990.

The MathWorks, Inc., *MatLab™ for Macintosh Computers, User's Guide*, The MathWorks, Inc., Natick, Massachusetts, 1991.

John Todd, *Basic Numerical Mathematics, Vol. 2, Numerical Algebra*, Academic Press, New York, 1978.

David S. Watkins, *Fundamentals of Matrix Computations*, John Wiley & Sons, New York, 1991.

SYMBOL INDEX

&, 108, 144

', 13

$\sum$, 23

$\sum \oplus$, 29

$\sum\cdot$, 30

$\oplus$, 29

$\dot{+}$, 30

$\otimes$, 28

*, 10, 37

+, 10, 37, 108

^, 10, 37, 65

~, 132, 144, 180

-, 10, 37

., 14, 37, 65

/, 10, 168, 175

:, 13, 197

;, 9

<, 144

<=, 144

= =, 44, 108, 144

=, 44, 108

~=, 144

>, 144

>=, 144

>>, 7

[ ], 42

\, 175

|, 132, 144, 197

∅, 234

α', 121

A^*, 93

A^T, 4

$\|A\|_F$, 226

A^+, 208

A^{-1}, 172

$A^{[k]}$, 157

$A(\alpha, \beta)$, 121

$A(\alpha \mid \beta)$, 121

$A(\alpha \mid \beta]$, 121

$A[\alpha \mid \beta)$, 121

$A[\alpha \mid \beta]$, 121

$A \stackrel{c}{\equiv} B$, 506

$A \stackrel{S}{\equiv} B$, 637

$A(\lambda) \stackrel{E}{\equiv} B(\lambda)$, 608

$A \cdot B$, 30

$A \otimes B$, 28

$A * B$, 48

$A.* B$, 48

$A.\char`\^ r$, 48

$A\char`\^ p$, 48

$A > 0$, 316

$A > B$, 438

$A \geq 0$, 316

$A \geq B$, 438

$A(:)$, 127, 165

$A(:, j)$, 127

$\overline{A}$, 93

$A(i, :)$, 127

$A_{(p)}$, 3

$A^{(q)}$, 4

$A \backslash b$, 176

$c(A)$, 127
$c(X)$, 149
$C_k(A)$, 335
$C(f(\lambda))$, 368, 616, 643
$<CR>$, 11
$\mathbb{C}$, 53, 172
$\mathbb{C}(x)$, 173
$\mathbb{C}[x]$, 172
δ_{st}, 124, 133
$\partial(A)$, 508
∂S, 578
$\partial W(A)$, 578
$d_k(\lambda)$, 608
d_χ, 278
$\varepsilon(A)$, 685
$\varepsilon(\sigma)$, 247
$E^{(i),(j)}$, 196
$E_{(i),(j)}$, 194
$E_{(i)+c(j)}$, 194
$E^{(i)r}$, 196
$E_{c(i)}$, 194
$E_k(r)$, 185, 364
e_j^n, 153
$E_k(\lambda_1, \ldots, \lambda_n)$, 364
$f_L(A)$, 606
$F_L(X)$, 619
$f_r(A)$, 606
$F_r(X)$, 619
$\Gamma(n_1, \ldots, n_p)$, 120
$\Gamma_{r,n}$, 120

$G_{r,n}$, 120
$H(g(\lambda))$, 368, 653
$H(X)$, 530
i, 54
Im, 56
I_n, 5
$In(A)$, 508
$I^{(i),(j)}$, 196
$I_{(i),(j)}$, 193
$II_{(i)+p(\lambda)(j)}$, 607
$III_{r(i)}$, 193
J_A, 684
$M_{m,n}(R)$, 5
$M_n(R)$, 5
$\binom{n}{r}$, 124
$\nu(A)$, 508
$\nu(\sigma)$, 247
$0_{m,n}$, 5
0_n, 5
π, 11
$\pi(A)$, 508
$Q_{r,n}$, 120
$\mathbb{Q}$, 172
$\mathbb{Q}(x)$, 173
$\mathbb{Q}[x]$, 172
$\rho(A)$, 206
$R(\lambda)$, 606
$R[\lambda]$, 606
$\mathbb{R}$, 51, 172
$\mathbb{R}(x)$, 173
$\mathbb{R}[x]$, 172

Symbol Index

Re, 56
$\sigma(A)$, 528
$\sigma[i]$, 234
$\sigma \sim \phi$, 252
$S(A)$, 631
s_λ, 253
S_n, 230
U_m, 368, 653
$\|u\|$, 96, 223
$W(A)$, 113, 526
$w(A)$, 461, 526
$[x, y]$, 529
$\bar{z}$, 56
$|z|$, 55
$\mathbb{Z}$, 172
$\mathbb{Z}(x)$, 173
$\mathbb{Z}[x]$, 172

INDEX

About MacII-MatLab, 8
abs, 64, 85
absolute value, 55
acos, 85
acosh, 85
addition, 10
additive identity, 55
adj, 308
adjacent interchange, 271
adjoin, 120
adjugate, 308
algebraic multiplicity, 377
all, 142
angle, 52, 65
annihilating polynomial, 645
ans, 9
any, 108, 142
Apple menu, 8
arg, 66, 90, 104
Argand diagram, 53
argument, 56
array, 1
asin, 85
asinh, 85
associativity of addition, 55
assignment operator, 44
assignment, 9
associative law for addition, 35
associative law for
    multiplication, 35, 55, 119

atan, 69, 85
atan2(y, x), 93
atan2, 85
atanh, 85
auxiliary unit matrix, 368, 653
axis('normal'), 19
axis('square'), 18
axis, 66, 106

Bellman, Richard, vii
Behrens, Richard T., 694
bilinear form, 496
binomial coefficient, 124, 185
binomial theorem, 161
boundary, 578
Built-in commands, 9

canonical form, 201
casesen, 12
Cauchy index, 247
Cauchy-Binet theorem, 300, 337
Cauchy-Schwarz inequality, 390
Cayley parameterization, 437
Cayley-Hamilton theorem, 616, 645
ceil, 166, 177
characteristic matrix, 632
characteristic polynomial, 357, 359, 637
characteristic root, 357, 637
characteristic vector, 377
chol, 329

Cholesky factorization theorem, 316
clc, 12
clock, 47, 75
closed half-plane, 529
Coleman, Thomas F., 693
colon notation, 13, 39
column form, 127, 149
column vector, 3
column, 2
columnwise sum, 25
Command window, 7
commutative law of addition, 55
commutative law of
    multiplication, 55
commutative, 27
companion matrix, 132, 368, 616, 643
complementary sequence, 121
complex conjugate, 56
complex number addition, 52
complex number multiplication, 52
complex number, 53
components, 2
compound matrix, 335
conformally partitioned matrices, 136
congruent matrices, 498
conj, 65
conjugacy, 251
conjugate permutation, 251
conjugate transpose, 93
conjugate-bilinear property, 96
conjugate-symmetric property, 96

conjunctive, 504
constructive, 644
continuation, 11
conv, 189, 622
convergent matrix, 683
convex combination, 261, 530
convex hull, 530
convex set, 116, 528
convexity, 528
Corgain, Gregory A., iv
cos, 85
cosh, 85
Cramer's rule, 310
cumprod, 39
cumsum, 39
cycle notation, 233
cycle structure, 253

d.s., 262
Daniel, James, 693
De Moivre's theorem, 59, 82
decimal numbers, 51
deconv, 189
deg, 611
degree, 611
det, 278, 291, 326
determinant, 278
determinantal divisor, 608
diag, 5, 37, 292
diagonal matrix, 5

Index

diagonal product, 277
diff, 39
dimension, 2
direct product, 28
direct sum, 29
discriminant of a
    quadratic form, 497
disjoint cycles, 234
disp, 21, 43, 65
distributive law, 55
division algorithm
    for integers, 177
division algorithm for matrix
    polynomials, 613
division, 10
Donoghue's theorem, 578
Donoghue, W. F., 535
double commute, 471
double sum, 25
doubly stochastic, 262

Echo Scripts, 16
Edit1:Unititled, 14
eig, 374, 406
eigenvalue, 289, 357, 637
eigenvector, 289, 377
elementary column matrices, 196
elementary column operations, 196
elementary divisors, 631
elementary row matrices, 194
elementary row operations, 193

elementary symmetric
    function, 185, 364
elements, 2
elliptical range theorem, 539
empty matrix, 42, 128
empty set, 234
entries, 2
envelope, 535
eps, 47, 108
equality relation, 44
equivalence class, 657
equivalence relation, 251, 657
equivalent, 608
etime, 47, 75
Euclidean algorithm, 177, 178
eval, 492
even permutation, 247
exceptional point, 580
exp, 14, 85
expansion by column, 307
expansion by row, 307
expm, 492
exponentiation, 10
extension field, 675
eye, 37, 175

field of numbers, 605
field of values, 113, 461, 526
File menu, 12
find, 142, 404, 517,

fix, 142, 177
floor, 45, 139, 165, 177
flops, 261
for..end, 141
form, 495
Format menu, 10
format long E, 11
format long, 11
format short E, 11
format short, 11
four-quadrant arctangent function, 93
fractal, 73
Franklin, Susan L., iv
Frobenius norm, 226, 487
Frobenius Normal Form, 645
full cycle permutation matrix, 256
function, 69, 71
fundamental theorem of algebra, 358
funm, 492

gamma, 265, 356
Gaussian elimination, 200
gcd, 83, 624
generalized inverse, 208
generalized matrix function, 278
geometric multiplicity, 377
Gill, Philip E., 693
Goldberg, Jack L., 693
Golub, Gene H., 693
Gram-Schmidt process, 357, 378

Graph window, 7
greatest common divisor, 83, 179, 624
grid, 329

Hadamard product, 30, 48
hadamard determinant inequality, 320
hankel, 145
Hausdorff, F., 116
Help, 8
Hermite interpolating polynomial, 186
Hermite normal form, 199
hermitian congruence, 504
hermitian form, 503
hermitian matrix, 94
hilb, 291, 489
Hilbert matrix, 131
hilbert norm, 458
Hill, David R., 693
HNF, 199
hold, 69, 106
homogeneous, 495
homogeneous equation, 219, 221
homogeneous of degree k, 335
Horn, Roger A., 693
Horner's method, 130, 617
Householder matrix, 130, 197
Householder, Alston S., 197
hyperbolic functions, 88
hypercompanion matrix, 368, 653

identity matrix, 5

Index

identity, 171
if, 108, 197
iff, 206
Im, 56
imag, 108, 144
imaginary part, 56
imaginary unit, 54
In, 508
incidence matrix, 238
inertia, 508
inf, 47
input, 41, 44, 106
integer quotient, 177
integer remainder, 177
interchange, 234
inv, 175, 326
invariant factor, 632
inverse, 171, 172, 232
inversion, 270
invertible matrix, 172
irreducible matrix, 644
irreducible polynomial, 634, 644
iteration, 20

Jacobi determinant
    theorem, 344
Johnson, Charles, R., 693
Jordan normal form, 654
Julia set, 73
Julia, Gaston, 73

k-cycle, 233
Klein four group, 274
Koseluk, William G., iv
kron, 44, 405
Kronecker delta, 124, 133, 149
Kronecker product, 28
kronecker power, 157
kronecker product associativity, 153
k^{th} compound, 335

Labels, 20
Lagrange interpolation polynomial, 441
Laplace expansion theorem, 303
lcm, 668
least common multiple, 668
least squares, 224, 445
left division, 613
left inverse, 205
left-hand value, 606
length of orbit, 234
length, 130, 139
Leon, Steven J., 693
lexicographic order, 120
line, 4
linear diophantine equations, 659
list of elementary divisors, 632
ln, 22
log, 14, 22, 85
log10, 14, 85
logical operator, 144

logm, 492
Long, 9
lower triangular, 163, 287
LR factorization, 316
LU factorization, 316
lu, 326

M-File, 16, 69
main diagonal, 5
Marcus, Marvin, 693
MatLab Programs
    Allperms, 243, 245
    anyf, 493
    arand, 485
    Arithmetic, 181
    axel, 595
    canf, 222
    cbin, 324
    chg, 275
    chl, 329
    clsunit, 488
    cmb, 265
    cmpl, 325, 354
    cmpnd, 298, 351
    cmpr, 240
    cnj, 263
    coef, 189
    commondiv, 183
    comp, 404
    comph, 352

compinv, 352
compn, 353
compu, 352
congru, 518
ctyp1, 199
ctyp2, 199
ctyp3, 199
cube, 70
cycle, 245, 246
cyst, 260
ddiag, 293
ddiv, 626
determ, 293
dherm, 187
direc, 295
disj, 245, 246, 260
down, 564
dprd, 292, 293
eigcube, 372
eigen, 371
eigsq, 371
Elemops, 197
elplot, 568
esf, 185, 372
euclid, 178
factr, 489, 491
field, 563
First, 15
formdet, 524
fr, 245

gcd, 179
gcdd, 180
gcdm, 624, 625
gcdp, 190, 191, 626
gcds, 183
hadamard, 328, 329
hankint, 147
herm, 187
Hermite Polynomials, 187
Hermite, 187
hrand, 428
Incidence, 242
iner, 518, 519
inerh, 519
inerttest, 520
invs, 291, 292
isom, 404, 405
jacobi, 355
julia, 73
kronpwr, 167
kronsv, 405
lapl, 325
lpolyv, 619
Lr, 243, 244
lsqrs, 223
Lt, 243, 244
Lxc, 264, 265
matroot, 168
movieunit, 450, 451
normrand, 430
normrng, 565, 566

Norms, 223
notneg, 108
nrad, 565
nrad2, 566
nrad3, 567
nrad4, 604
nrand, 353, 565
nrange, 111
nrdg, 430, 432
nroot, 445
nroottest1, 446
nroottest2, 447
nrunit, 486, 487
numrange, 600
orand, 428, 429
pdprd, 622
perdir, 295, 296
perm, 261, 293
permpos, 294, 295
Permute, 240
pmat, 240
pmatrix, 618
pmtly, 240
pnv, 241
polar1, 448
polar2, 448, 449
polyherm, 449, 450
polymat, 405
Polynomial, 189
polynorm, 446, 447

polyv, 618
posidet, 326
powr, 241
ppwr, 240
prand, 242
prime, 183
primelist, 184
psdet, 297, 372
psinv, 486
psn, 275
psper, 297
psubd, 623
ptrans, 226
quadr, 71, 72
rankola, 222
rankr, 491
ranktest, 223
ratadd, 180
ratdiff, 181
ratdiv, 182
ratmult, 182
rcmb, 265, 266
rcmbtest, 266
rdsn, 262
rect, 599
rev, 243, 244
rk, 222
rndpoly, 621
rotr, 595
rpolyv, 619
rtyp1, 197

rtyp2, 198
rtyp3, 198
scmpinv, 354, 355
scmpnd, 354
sgn, 689
shur, 428
sinis, 146
skhrand, 430
skrand, 429
smax, 223
sn, 263
srand, 428
sroot, 167
sschur, 433
strip, 190
subm, 620, 621
sumdet, 298
swap, 243, 244
swing, 596, 597
sylfrank, 356
sym, 677, 678
top, 624, 625
unitrng, 565
up, 563
urand, 429
vdw, 262
MatLab_Toolbox, 14
matrix addition, 26
matrix conjugate, 93
matrix diagonal, 277

matrix equality, 2
matrix equation, 152
matrix inverse, 172
matrix of a form, 496
matrix polynomial, 605
matrix product, 27
matrix sign function, 685
matrix sum, 26
matrix, 1
max, 143
mesh, 329
meshdom, 329, 331
min, 143
Minc, Henryk, 693
minimal polynomial, 645
modulus, 55
Moler, Cleve, vii
monic, 190, 359
Moore-Penrose inverse, 210, 458
multiplication, 10
multiplicative Identity, 55
Murray, Walter, 693

n-square, 4
n-tuples, 3
n.n.h., 316
nan, 47
natural logarithm, 22
negative open half-plane, 529
New, 12
Noble, Ben, 693

nonnegative hermitian, 316
nonsingular, 172
norm, 96, 223, 351, 458
norm(A, 'fro'), 226
norm(X, 'fro'), 352
normal form, 256
normal matrix, 95, 127
numerical radius, 461, 526
numerical range, 113, 526

o.n., 435
odd permutation, 247
ones(t), 19
ones, 20, 37, 175
Open, 16
orbit, 73, 234
Ortega, James M., 694
orth, 352
orthogonal matrix, 95, 126
orthonormal, 435

p.d.h., 316
partial isometry, 405
partitioned matrix, 122
per, 261, 278
period notation, 13
permanent, 261, 278
permutation, 132, 229
permutation identity, 231
permutation inverse, 232

permutation matrix, 133, 238
permutation product, 133, 230
pi, 11, 47
pinv, 224
plane rotation, 196
plot, 18, 66, 106, 595
polar, 68
polar factorization, 440
polar form, 56
poly, 184, 369
polyfit, 444
polynomial matrix, 606
polynomial specialization, 605
polyval, 188, 406, 445, 618
polyvalm, 189, 406
positive definite hermitian, 316
positive definite property, 96
positive definite, 438
positive open half-plane, 529
positive semi-definite hermitian, 316
positive semi-definite, 438
prime polynomial, 634
principal subdeterminant, 285
principal submatrix, 121
principal value of log, 88
principal value of power, 90
principal value, 66
prod, 39, 261, 292
product of polynomials, 610

QR factorization, 387
qr, 402
quadratic form, 497
quadratic formula, 77
quadratic Plücker relations, 286
quantic, 495

rand, 45, 108, 165
random permutation, 241
rank, 206, 222
rank of a form, 496
rank of a quadratic form, 497
rational function, 173
rational function field, 606
rational power of a complex number, 82
Re, 56
real, 134
real normal form, 587
real normal matrix, 127
real numbers, 51
real part, 56
real quaternions, 36, 171
Redmond, Kathleen A., iv
reduced row echelon form, 199
reducible matrix, 649
reflexive, 252, 657
regular matrix, 172
relational operator, 144
relatively prime polynomials, 634
rem, 142, 166, 177

Index

remainder theorem, 615
right division, 613
right inverse, 205
right-hand value, 605
root multiplicity, 358
roots, 185, 369
roots of a complex number, 78
roots of unity, 83
roots1, 369
rot90, 329
round, 177, 222
row vector, 3
row, 1
rref, 199, 404
rsf2csf, 426
Run Script, 16
Run, 16

Save and Go, 15
scalar product, 26
scalar, 26
Scharf, Louis L., 694
Schlieman, Scott J., iv
schur, 426
Schur, I., 409
Schur product, 30
Schur triangularization theorem, 410
script, 16, 37, 69
SDR, 658
sequence sets, 120

sgn of A, 685
Short, 10
sign, 517, 595, 685
sign of a matrix, 685
sign of a permutation, 247
similar, 359, 644
similar over R, 359, 644
similarity classes, 661
sin, 85
singular matrix, 172
singular value decomposition, 456
singular values, 393, 452
sinh, 85
sinister diagonal, 145
sinister diagonal matrix, 145
size, 2, 39, 44, 167, 197
skew-hermitian matrix, 95
skew-symmetric matrix, 5, 95
Smith normal form, 629, 632
sort, 405, 433
specialization, 605
spectrum, 528
sqrt, 10, 14, 39, 85
sqrtm, 492
square matrix, 4
standard inner product, 95
subdeterminant, 285
submatrix, 121
subscripts, 2
subtraction, 10
sum, 39, 130, 187

superdiagonal, 368
supplementary compound matrix, 340
support line, 580
supporting half-plane, 580
SVD, 456
svd, 484
Sylvester's law of inertia, 509
Sylvester-Franke theorem, 349
symmetric, 5, 252, 657
symmetric group of degree n, 230
symmetric matrix, 5, 94
system of distinct representatives, 658

tan, 85
tanh, 85
Taylor series, 86
tensor product, 28
text, 329
The MathWorks, Inc., 694
title, 329
Todd, John, 694
Toeplitz, O., 116
toeplitz, 140
Toeplitz-Hausdorff theorem, 574
Topics, 9
Topics button, 9
tr, 363
trace, 363
transitive, 252, 657
transpose, 4

transposition, 234
transversal, 658
triangle inequality, 58
tril, 46, 144
triu, 46, 144
two-dimensional orthogonal
 compression, 570
Type I, 193, 607
Type II, 193, 607
Type III, 193, 607

unimodular matrix, 658
unit, 171
unit matrix, 172, 606, 658
unitary completion, 384
unitary dilation, 459
unitary matrix, 95, 126
upper triangular, 163, 287

Van Loan, Charles F., 693
van der Waerden conjecture, 262
vector, 3

Watkins, David S., 694
while..end, 21, 39, 41
Wielandt deflation, 399
Witt cancellation theorem, 516
Wright, Margaret H., 693

zero matrix, 5
zeros, 37